THE INTERNATIONAL SERIES OF MONOGRAPHS ON CHEMISTRY

THE INTERNATIONAL SERIES OF MONOGRAPHS ON CHEMISTRY

1 J.D. Lambert: *Vibrational and rotational relaxation in gases*
2 N.G. Parsonage and L.A.K. Staveley: *Disorder in crystals*
3 G.C. Maitland, M. Rigby, E.B. Smith, and W.A. Wakeham: *Intermolecular forces: their origin and determination*
4 W.G. Richards, H.P. Trivedi, and D.L. Cooper: *Spin-orbit coupling in molecules*
5 C.F. Cullis and M.M. Hirschler: *The combustion of organic polymers*
6 R.T. Bailey, A.M. North, and R.A. Pethrick: *Molecular motion in high polymers*
7 Atta-ur-Rahman and A. Basha: *Biosynthesis of indole alkaloids*
8 J.S. Rowlinson and B. Widom: *Molecular theory of capillarity*
9 C.G. Gray and K.E. Gubbins: *Theory of molecular fluids, volume 1: Fundamentals*
10 C.G. Gray and K.E. Gubbins: *Theory of molecular fluids, volume 2: Applications* (in preparation)
11 S. Wilson: *Electron correlations in molecules*
12 E. Haslam: *Metabolites and metabolism: a commentary on secondary metabolism*
13 G.R. Fleming: *Chemical applications of ultrafast spectroscopy*
14 R.R. Ernst, G. Bodenhausen, and A. Wokaun: *Principles of nuclear magnetic resonance in one and two dimensions*
15 M. Goldman: *Quantum description of high-resolution NMR in liquids*
16 R.G. Parr and W. Yang: *Density-functional theory of atoms and molecules*
17 J.C. Vickerman, A. Brown, and N.M. Reed (editors): *Secondary ion mass spectrometry: principles and applications*
18 F.R. McCourt, J. Beenakker, W.E. Köhler, and I. Kuščer: *Nonequilibrium phenomena in polyatomic gases, volume 1: Dilute gases*
19 F.R. McCourt, J. Beenakker, W.E. Köhler, and I. Kuščer: *Nonequilibrium phenomena in polyatomic gases, volume 2: Cross-sections, scattering, and rarefied gases*
20 T. Mukaiyama: *Challenges in synthetic organic chemistry*
21 P. Gray and S.K. Scott: *Chemical oscillations and instabilities: non-linear chemical kinetics*
22 R.F.W. Bader: *Atoms in molecules: a quantum theory*
23 J.H. Jones: *The chemical synthesis of peptides*
24 S.K. Scott: *Chemical chaos*
25 M.S. Child: *Semiclassical mechanics with molecular applications*
26 D.T. Sawyer: *Oxygen chemistry*
27 P.A. Cox: *Transition metal oxides: an introduction to their electronic structure and properties*
28 B.R. Brown: *The organic chemistry of aliphatic nitrogen compounds*
29 Y. Yamaguchi, Y. Osamura, J.D. Goddard, and H.F. Schaeffer: *A new dimension to quantum chemistry: analytic derivative methods in* ab initio *molecular electronic structure theory*
30 P.W. Fowler and D.E. Manolopoulos: *An atlas of fullerenes*
31 T. Baer and W.L. Hase: *Unimolecular reaction dynamics: theory and experiments*

UNIMOLECULAR REACTION DYNAMICS

Theory and Experiments

Tomas Baer
and
William L. Hase

New York Oxford
OXFORD UNIVERSITY PRESS
1996

Oxford University Press

Oxford New York
Athens Auckland Bangkok Bombay
Calcutta Cape Town Dar es Salaam Delhi
Florence Hong Kong Istanbul Karachi
Kuala Lumpur Madras Madrid Melbourne
Mexico City Nairobi Paris Singapore
Taipei Tokyo Toronto

and associated companies in
Berlin Ibadan

Published by Oxford University Press, Inc.,
198 Madison Avenue, New York, New York 10016

Baer, Tomas.
Unimolecular reaction dynamics : theory and experiments /
by Tomas Baer and William L. Hase.
p. cm. — (The international series of monographs on chemistry)
Includes bibliographical references and index.
ISBN 0-19-507494-7
1. Unimolecular reactions. I. Hase, William L. II. Title.
III. Series.
QD501.B1618 1996 95-10954
541.3′93—dc20

Printing: 9 8 7 6 5 4 3 2 1
Printed in the United States of America
on acid-free paper

Preface and Acknowledgments

The field of unimolecular reactions has been revolutionized during the past 20 years by a series of new developments in both experimental techniques and theoretical advances. The precise information about the dissociation of large and small molecules and ions that is now being routinely collected has permitted the testing of the most detailed aspects of the theory. These advances have come about through a wonderful interplay between experimentalists and theoreticians. We hope that this book transmits not only information, but also the excitement in the field that has kept so many of us absorbed.

In many ways, this book is a community effort. The field of unimolecular reactions has progressed to this point through the combined efforts of many scientists located in all parts of the globe. In the early stages of this work, we received numerous and excellent suggestions about the contents and structure of the book. Finally, many individuals read sections of the manuscript while we were preparing them. Among these people are Elliot Bernstein, Terry Gough, Stephen Gray, William Green, Paul Houston, Steve Klippenstein, Cornelius Klots, Kevin Lehmann, Chava Lifshitz, Robert Marquardt, Roger Miller, Will Polik, Hanna Reissler, Ned Sibert, Kihyung Song, and Howard Taylor. Special thanks go to some who have made major contributions to this book; among these are Jon Booze, Martin Hunter, Gilles Peslherbe, and Ling Zhu, whose insightful comments and suggestions have profoundly influenced the content of chapters 7 and 9. As with any work of this sort which attempts to review a large field of research, we have attempted to refer to as many studies as seems reasonable. However, we are fully aware that much excellent work has not been cited. To the extent possible, we have referred to studies which have been addressed by both experiment and theory.

TB is grateful to the UNC departments of chemistry and romance languages, which permitted him to spend all of 1992 with the UNC Year in Montpellier, France exchange program where the first drafts of several chapters were written. Reading and digesting the many excellent papers between croissant breakfasts and beers at sidewalk cafes is a memory not readily forgotten. This volume was written by the authors working in separate locations, a mode made possible by the marvels of electronic mail and FAX. However, a week spent together in the Virginia mountains was an essential and memorable element for progressing past various bottlenecks and completing the manuscript.

WLH wishes to acknowledge the inspiration provided by his research advisors John W. Simons and Don L. Bunker, who acquainted him with unimolecular reactions and pointed out the beautiful detail and nuances of their dynamics. We both acknowledge our debt to the many graduate students and postdocs who have passed through our laboratories. The stimulation provided by students is something that all of us in the academic life share and treasure. The writing of this book has been a major preoccupation for us during the past three years. We are especially grateful to Carol and Cris for their generous understanding, their encouragement, and above all their faith that we would some day finish this book.

Chapel Hill, NC
August 1995

T.B.

Detroit, MI
August 1995

W.L.H.

Contents

Unimolecular Reaction Dynamics

1

Introduction

The field of unimolecular reactions has witnessed impressive advances in both experimental and theoretical techniques during the past 20 years. These developments have resulted in experimental measurements that finally permit critical tests of the major assumptions made more than 60 years ago when Rice and Ramsperger (1927, 1928) and Kassel (1928) first proposed their statistical RRK theory of unimolecular decay. At the heart of these advances is our ability to prepare molecules in narrow ranges of internal energy, even in single quantum states, at energies below and above the dissociation limit. This has led to detailed spectroscopic studies of intramolecular vibrational energy redistribution (IVR), a process that is intimately related to the assumption of random energy flow in the statistical theory of unimolecular decay.

This book is devoted exclusively to the study of state- or energy-selected systems. However, in order to place these studies in the context of the much larger field of unimolecular reactions in general, we provide a brief background of the field up to about 1970. The experimental studies of unimolecular reactions developed in three stages. The early studies involved strictly thermal systems in which molecules were energized by heating the sample either in a bulb (Chambers and Kistiakowsky, 1934; Schlag and Rabinovitch, 1960; Flowers and Frey, 1962; Schneider and Rabinovitch, 1962), or by more sophisticated methods such as shock tubes which were applied to unimolecular reactions by Tsang (1965, 1972, 1978, 1981) and others (Astholz et al., 1979; Brouwer et al.,1983). The drawback of these studies is that molecules were prepared in a very broad (albeit well characterized) distribution of internal energy states. A major advance was the use of chemical activation in the early 1960s in which a species such as CH_2 reacted with a molecule, thereby forming an energized species which could either isomerize or be stabilized by collisions (Rabinovitch and Flowers, 1964; Rabinovitch and Setser, 1964; Kirk et al., 1968; Hassler and Setser, 1966; Simons and Taylor, 1969). This approach permitted the reacting species to be prepared in a narrow range of internal energies. However, the number of reactions that could be investigated was small, the energy selection was not very precise, and the range of internal energies was very limited. In addition, in both the thermal and chemical activation studies, the time base for measuring the dissociation rate was the rather imprecise collisional relaxation rate. Many of the early studies dealing with thermal systems and chemical activation have been the subject of two excellent books on unimolecular reactions (Forst, 1973; Robinson and Holbrook, 1972). More recently a book which focuses primarily on thermal unimolecular and the related three-body recombination reactions, as well as collisional energy transfer processes, has appeared (Gilbert and Smith, 1990).

The third stage in the study of unimolecular reactions (the subject of this book) began in the early 1970s with the development of more general and precise methods of energy and state preparation as well as the development of absolute time bases for measuring reaction rates. Molecules and ions can be prepared above their dissociation limits by laser excitation with wavenumber resolution. With the use of two or more pulsed lasers, the reaction can be monitored as a function of time from milliseconds to femtoseconds. Less dramatic advances in photoelectron photoion coincidence (PEPICO) allow ionic dissociation rates to be studied with an energy resolution of about 100 cm^{-1} over a time range from 10^{-4} to 10^{-7} sec.

Our theoretical understanding of unimolecular reactions has kept pace with the advances in experimental techniques. As a result of powerful computers, the geometry of the reacting molecule as well as its vibrational frequencies can now be calculated as a function of the reaction coordinate from the beginning to the end of the reaction. The results of these calculations provide sufficient information to treat the reaction with much greater sophistication than was previously possible. In addition, accurate potential energy surfaces have permitted reactions to be investigated by classical trajectory and quantum dynamics methods, approaches that are not inherently limited by the assumptions of the statistical theory.

1.1 WHAT IS A UNIMOLECULAR REACTION?

A unimolecular reaction is defined as any system that evolves in time as a result of some prior stimulus or excitation step. Thus, both dissociation and isomerization are examples of unimolecular processes. Phenomenologically, a unimolecular reaction, $A \rightarrow$ products, is written as

$$-\frac{d[A]}{dt} = k[A], \tag{1.1}$$

which, when integrated, gives rise to the time dependence of the concentration of $A(t)$:

$$[A] = [A_0]e^{-kt}, \tag{1.2}$$

where k is the unimolecular rate constant with units of reciprocal time, and $[A_0]$ is the concentration of species A at time, $t = 0$. The rate constant k depends on the internal energy of A, or in the case of an equilibrium ensemble of A, its temperature. For a given state of excitation, the exponential decay is a result of the assumption that the rate is a function only of the concentration of A.

The important questions in the study of unimolecular reactions are (a) what is the initial state produced in the excitation step, (b) how fast does the system evolve toward products, (c) what are the reaction products, and (d) what are the product energy states? Up until about 1975, the first and last questions could not be addressed experimentally. Most experiments were carried out with the reacting system specified in terms of a temperature with its attendant distribution of initial states. From the very beginning, it was recognized that a dissociation rate depends on the internal energy of the molecule (Hinshelwood, 1926). Thus, all detailed statistical theories of unimolecular reactions begin with the calculation of $k(E)$, the rate constant as a function of the internal energy, E.

The connection between $k(T)$, often called the canonical rate constant, and $k(E)$, the microcanonical rate constant, involves averaging $k(E)$ over the energy distribution

$$k(T) = \int_{E_0}^{\infty} P(E,T)k(E)dE, \tag{1.3}$$

where E_0 is the activation energy and $P(E,T)$ is the distribution of internal energies at a given temperature, T. In many applications the quantity of interest is $k(T)$ because most natural systems can be described adequately by a thermal distribution characterized by a given temperature. However, there are many important exceptions such as flames, discharges, and explosions. For such systems it is important to know $k(E)$ as well as $P(E,T)$. Perhaps a more important reason for studying $k(E)$ rather than $k(T)$ is that the theory can be tested adequately only by comparing the measured and calculated $k(E)$. Once this has been accomplished, the rate constant of any system with a known distribution of internal states can be calculated.

1.2 THERMALLY ACTIVATED EXPERIMENTS AND THE LINDEMANN MECHANISM

Our basic understanding of unimolecular reactions can be traced to Lindemann (1922), who proposed that reacting molecules in a thermal system are energized by collisions and that the mechanism for the reaction can be expressed as

$$A + M \xrightarrow{k_1} A^* + M, \tag{1.4}$$

$$A^* + M \xrightarrow{k_{-1}} A + M, \tag{1.5}$$

$$A^* \xrightarrow{k_2} \text{Products}, \tag{1.6}$$

where A^* represents a molecule that is sufficiently excited to react with no further input of energy. M is any molecule in the vessel. Assuming a steady-state production of A^*, we can readily show that the overall rate of product formation is given by

$$\text{Rate} = k_2[A^*] = \frac{k_1k_2[A][M]}{k_{-1}[M] + k_2} = k_{\text{uni}}[A] \tag{1.7}$$

The interesting feature of this rate law is that the phenomenological unimolecular rate constant, k_{uni}, changes from second order (in A and M) at low pressures where $k_{-1}[M] << k_2$, to first order in $[A]$ at high pressures. The high-pressure rate constant is then simply $k_\infty = k_1k_2/k_{-1}$. When data are plotted as k_{uni}/k_∞ versus the pressure $[M]$, the famous "fall-off" plots are obtained. An example of such a plot is shown in figure 1.1 for the case of cyclopropane isomerization (Prichard et al., 1953). The change from second order (the straight-line portion at low pressures) to unimolecular (independent of the pressure) is shown very convincingly in these results. This reaction, and variations thereof, were studied more than any other during the 1950s and 1960s (Butler and Kistiakowsky, 1960; Slater, 1953; Prichard et al., 1953; Schlag and Rabinovitch, 1960).

According to Eq. (1.7), the fall-off plot depends not only on the value of the unimolecular rate constant, k_2, but also on the collisional activation, k_1, and deactivation, k_{-1}, rate constants. The bimolecular reactions both complicate and aid the analysis of the rate data. On the one hand, three rate constants must be determined. On the other hand, with appropriate assumptions, the bimolecular collisions serve as a time

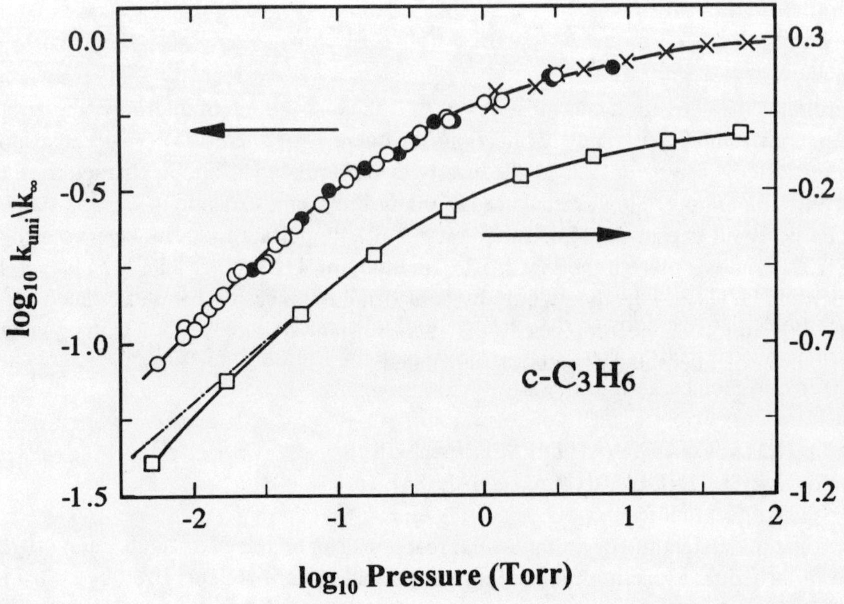

Figure 1.1 Pressure dependence of the unimolecular rate constant for c-C_3H_6 → H_2═CH—CH_3. The open and closed circles are data from Prichard et al. (1953) while the x's are data of Chambers and Kistiakowsky (1934). The lower curves are displaced down by 0.3 log units. The solid lines are the experimental results, the open squares are calculated by Slater (1953) assuming 13 active oscillators (in place of the full 21), and the dashed curve is a Kassel or RRK calculation with 13 oscillators by Prichard et al. (1953). Taken in modified form, and with permission, from Prichard et al. (1953).

base so that k_2 can be determined relative to the collision frequency. However, to do so requires assumptions about the collisional deactivation efficiency, $k_{-1}[M]$. The crudest assumption is the strong-collision assumption in which it is supposed that a single collision will deactivate A^*. With this assumption, $k_{-1}[M]$ is just given by the collision frequency, ω. In light of recent collisional energy transfer studies (Gilbert et al., 1983; Hippler et al., 1981, 1983; Jalenak et al., 1988; Weston and Flynn, 1992; Gilbert and Smith, 1990), the deactivation step can now be treated with considerably greater sophistication.

At this point it is useful to consider again Eq. (1.2), which states that the time evolution of an ensemble of molecules is given by an exponential function. This has been amply verified by numerous experiments in thermal systems. Yet we note that a canonical distribution, which contains molecules in a large distribution of internal energies, is not characterized by a single rate constant and therefore cannot evolve by a single exponential decay. Rather, the time behavior of such a system should be described by a weighted sum of exponential decays given by

$$[A(t)] = [A_0]\int P(E,T)k(E)e^{-k(E)t}dE. \tag{1.8}$$

The explanation for this paradox is that a canonical or thermal system maintains the equilibrium distribution of internal energy states through collisions. Thus, Eq. (1.8) does not describe the time behavior of a canonical ensemble. What it does describe is a hypothetical system in which thermal equilibrium has been established at some time, but which from then on evolves with no collisions. In such a system, the energy-rich molecules will decay first with short lifetimes, while the lower-energy molecules decay with longer lifetimes.

1.3 THE RRK AND RRKM/QET THEORIES

The Lindemann mechanism consists of three reaction steps. Reactions (1.4) and (1.5) are bimolecular reactions so that the true unimolecular step is reaction (1.6). Because the system described by Eqs. (1.4)–(1.6) is at some equilibrium temperature, the high-pressure unimolecular rate constant is the canonical $k(T)$. This can be derived by transition state theory in terms of partition functions. However, in order to illustrate the connection between microcanonical and canonical systems, we consider here the case of $k(E)$ and use Eq.(1.3) to convert to $k(T)$.

The conceptual framework for understanding the unimolecular reaction (1.6) was developed by Rice and Ramsperger (1927, 1928) and Kassel (1928). The dissociating system is treated as an assembly of s identical harmonic oscillators, one of which is truncated at an energy E_o, the activation energy for dissociation. If the dissociative, or critical oscillator happens to have an energy, ϵ, in excess of the activation energy, the molecule dissociates. A fundamental assumption of the theory is that energy flows statistically among all of the oscillators and that the chance of finding the system with a particular arrangement of its internal energy is equivalent to any other. The dissociation rate is then proportional to the probability that for a given total energy, $E = nh\nu$, the critical oscillator contains an energy equal to, or greater than $E_o = mh\nu$. The problem reduces to a combinatorial determination of fitting n quanta into the various oscillators. The number of ways of distributing n quanta (the total energy) among the s oscillators is $(n + s - 1)!/n!(s - 1)!$. This is the degeneracy of a state with n quanta of s equally spaced oscillators. The number of ways that the n quanta can be placed in the molecule such that at least m quanta are in the critical oscillator is given by $(n - m + s - 1)!/(n - m)!(s - 1)!$. The probability of the molecule being in a dissociative state is then just the ratio of these probabilities, or

$$\text{Probability} = \frac{(n - m + s - 1)!\, n!}{(n - m)!(n + s - 1)!}. \tag{1.9}$$

This quantum expression can be simplified by assuming that the spacing is very small so that n and m are very large. With this assumption we can apply Sterling's approximation for the factorial functions, which is $h! = h^h/e^h$. If in addition the number of quanta is much larger than the number of oscillators, that is $(n - m) >> s$, then $(n - m + s - 1)$ can be replaced by $(n - m)$ (but not in the exponent!). The probability then reduces to

$$\text{Probability} = \left(\frac{n - m}{n}\right)^{s-1}. \tag{1.10}$$

This expression is converted to a rate constant by multiplying the probability by a rate of passage to products. In the spirit of this model, the rate is just the vibrational frequency ν. Thus the unimolecular rate constant is given by

$$k(E) = \nu\left(\frac{n - m}{n}\right)^{s-1} = \nu\left(\frac{E - E_o}{E}\right)^{s-1} \tag{1.11}$$

The expression in terms of the energies are obtained by multiplying top and bottom by $(h\nu)^{s-1}$.

This RRK expression for the rate contains two important features. First, it predicts that the rates are a strong function of the number of vibrational oscillators s. Second, it predicts that the rate increases rapidly with excess energy, $(E - E_o)$. Although both of these conclusions are correct, Eq. (1.11) unfortunately is incapable of giving the correct rate, even within an order of magnitude. There are several reasons. First, the assumption of classical oscillators with $(n - m) >> s$ is inappropriate for most chemical systems. For instance, a molecule such as benzene, with its 12 atoms, has $s = 30$. At an energy of 20,000 cm^{-1} above the dissociation limit, $n - m \approx 20$ (assuming an average vibrational frequency of 1000 cm^{-1}). In most chemically interesting situations, $s \approx n - m$. This has forced workers to use s values that are artificially lower than the number of oscillators. Reasonable agreement with experiment is generally obtained when s is equal to about one half of the total oscillators [see fig. 1.1 and also Weston (1986)]. A second shortcoming of the classical RRK theory is the neglect of the zero point energy. However, the theory can be fixed up to account for the zero point energy, a process that led Whitten and Rabinovitch (1963, 1964) to propose a very useful and interesting method for determining accurate densities of states.

The problems associated with the classical RRK expression in Eq. (1.11) were eliminated by Marcus and Rice (1951) and by Rosenstock, Wallenstein, Wahrhaftig, and Eyring (1952). These quantum theories, which treat the vibrational (and rotational) degrees of freedom in detail, became known as the RRKM and the quasi-equilibrium theory (QET), respectively. The RRKM/QET expression, which will be discussed in detail in chapter 6, is given by

$$k(E) = \frac{\sigma N^{\ddagger}(E - E_o)}{h\rho(E)}, \tag{1.12}$$

where $\rho(E)$ is the density of vibrational states at the energy E, $N^{\ddagger}(E - E_o)$ is the sum of the vibrational states from 0 to $E - E_o$ in the transition state, h is Planck's constant, and σ is the reaction symmetry factor. For simplicity, the role of rotations is not included in Eq. (1.12). A major difference between RRKM and RRK theory is that the former depends strongly on the values of the vibrational frequencies, while the latter does not. It is this RRKM/QET formulation which is the starting point for modern statistical theories.

For a decade, a rival theory due to Slater (1955, 1959) provided considerable motivation for more detailed experimental as well as theoretical investigations. This very interesting and elegant theory, which is discussed in more detail by Robinson and Holbrook (1972) and Nikitin (1974), as well as in chapter 8, is more akin to a dynamical than a statistical theory. Because the Slater theory treats the vibrations classically, it also requires the use of fewer oscillators to fit the experiment (see fig. 1.1). Its flawed fundamental hypothesis that the molecule's modes were strictly harmonic, thereby preventing energy flow among them, and its failure to account quantitatively for the experimentally measured rates led to its being quickly overshadowed by the successes of the RRKM/QET theory.

1.4. RRKM THEORY IN THE FALL-OFF REGION

Until about 1960, fall-off plots were the major source of quantitative information about unimolecular reactions. In order to adapt the microcanonical RRKM theory to these thermal data, it was necessary to take into account the internal energy of the reactants by rewriting Eqs. (1.4)–(1.6) as follows:

$$A + M \xrightarrow{dk_1} A^*(E,E + dE) + M \tag{1.13}$$

$$A^*(E,E + dE) + M \xrightarrow{k_{-1}} A + M \tag{1.14}$$

$$A^*(E, E + dE) \xrightarrow{k(E)} \text{Products.} \tag{1.15}$$

In this mechanism, dk_1 is the differential rate constant for producing the excited molecule in a range of energies from E to $E + dE$, and the energy dependence on the unimolecular decay is explicitly stated in $k(E)$. Imposing steady state on A^* leads to

$$dk_{\text{uni}} = \frac{dk_1 k(E)[M]}{k_{-1}[M] + k(E)} = \frac{k(E)(dk_1/k_{-1})}{1 + k(E)/(k_{-1}[M])}. \tag{1.16}$$

The total thermal unimolecular rate constant is obtained by integrating Eq. (1.16) over the energy range from E_o to ∞. As pointed out by Steinfeld et al. (1989), dk_1/k_{-1} is the equilibrium probability that the reactant molecule A^* has an energy in the range from E to $E + dE$. Thus, $dk_1/k_{-1} = P(E,T)dE$, where $P(E,T)$ is the thermal energy distribution of the molecule. The total unimolecular rate constant thus becomes

$$k_{\text{uni}}(T) = \int_{E_o}^{\infty} \frac{k(E)P(E,T)dE}{1 + k(E)/\omega}, \tag{1.17}$$

where ω is the collision frequency, $k_{-1}[M]$. This assumes that each collision between an excited A^* and a bath molecule results in a deactivation, in other words, the strong collision assumption.

Equation (1.17) can be evaluated using either the RRK or RRKM theory expressions. Proceeding with the latter, we note that

$$P(E,T) = \frac{\rho(E)e^{-E/k_BT}}{\int_0^{\infty} \rho(E)e^{-E/k_BT}dE} = \frac{1}{Q(T)}\rho(E)e^{-E/k_BT}, \tag{1.18}$$

where $Q(T)$ is the vibrational partition function. When the RRKM expression in Eq. (1.12) is substituted for $k(E)$ and Eq. (1.18) is used for $P(E,T)$, the unimolecular rate constant in Eq. (1.17) becomes

$$k_{\text{uni}}(T) = \frac{\sigma}{hQ(T)} \int_{E_o}^{\infty} \frac{N^{\ddagger}(E - E_o)e^{-E/k_BT}}{1 + k(E)/\omega} dE \tag{1.19}$$

For simplicity, the role of the rotations is ignored in Eq. (1.19). A similar expression is derived by Steinfeld et al. (1989) which includes the rotational energy explicitly.

The total unimolecular decay rate constant depends upon the gas pressure through the collision frequency ω. At high pressures, where it is independent of ω, it becomes the canonical transition state theory rate constant (see chapter 6). However, in the interesting fall-off region Eq. (1.19) must evaluated numerically. Equation (1.19) is capable of fitting experimental fall-off curves with the use of reasonable frequencies for the molecule and the transition state. An example for the $CH_3NC \rightarrow CH_3CN$ isomerization reaction is shown in figure 1.2 (Schneider and Rabinovitch 1962). Experiments of this type, carried out with various isotopic substituents, provided excellent agreement between theory and experiments without an arbitrary reduction in the

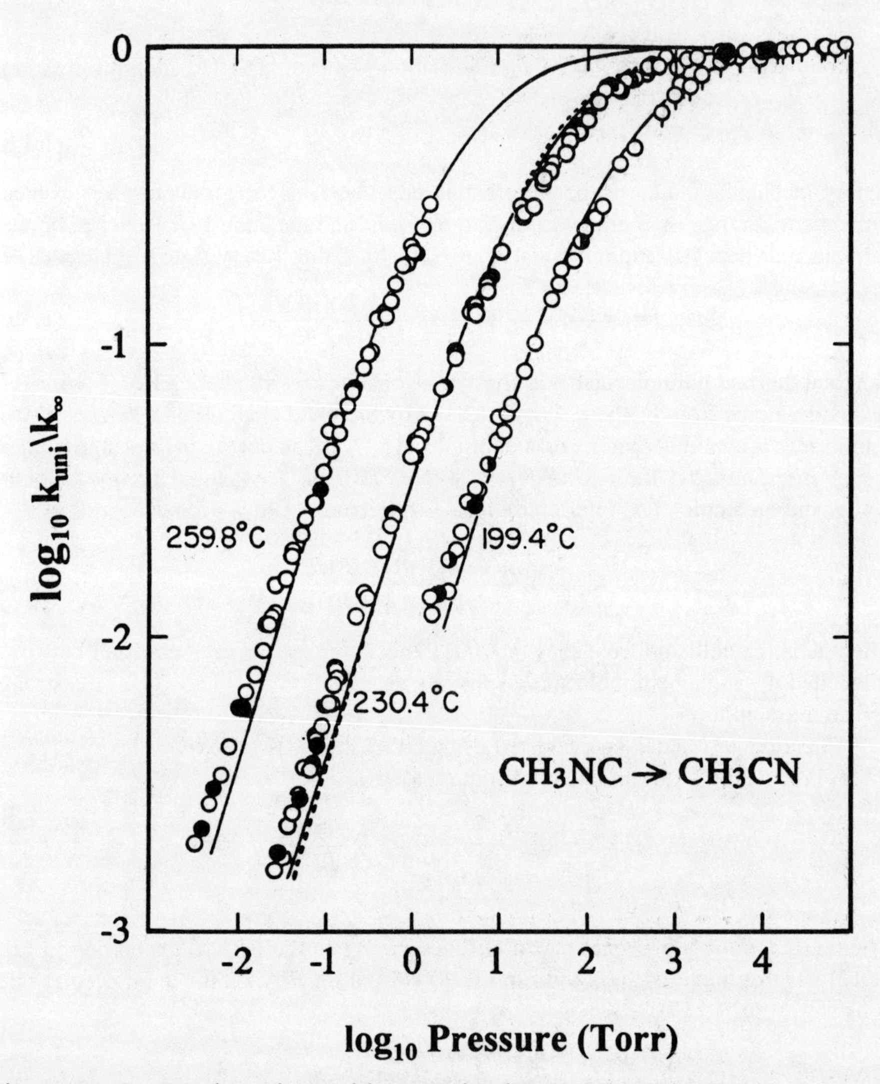

Figure 1.2 Pressure dependence of the unimolecular rate constant for $CH_3NC \rightarrow CH_3CN$ isomerization. For clarity, the 260° and 200° curves are displaced by one log unit to the left and right, respectively. (All three curves are nearly identical.) The solid lines through the data are RRKM calculated rates with the strong collision assumption. Taken with permission from Schneider and Rabinovitch (1962).

number of oscillators. However, the thermal nature of the experiments as well as the required integration over the energies reduced considerably the precision with which the theory could be tested.

1.5 CHEMICAL ACTIVATION EXPERIMENTS: A FIRST ATTEMPT AT $k(E)$

One of the major problems in the interpretation of the thermal data was that instead of depending on one rate constant, the overall rate was a function of three constants, none of which was easy to calculate. As a result, much effort was placed on developing simple models for the activation and deactivation steps so that interesting information about the unimolecular rate constant, $k(E)$ could be extracted. One of the big questions for many years was the number of collisions required to deexcite the activated molecule, A^*. According to the strong-collision model, a single collision sufficed, while other models attempted to treat the problem more delicately, but with no significantly greater success. Although the distinction between A and A^* is reasonably evident from the point of view of reactions 1.13–1.15, it is very crude when dealing with activation and deactivation steps. Thus, it was recognized that a different means for activating molecules must be found in order to isolate the unimolecular rate constant, $k(E)$.

A successful approach to energy selection was achieved by a variety of groups using chemical activation (Butler and Kistiakowsky, 1960; Rabinovitch and Flowers, 1964; Rabinovitch and Setser, 1964; Kirk et al., 1968; Hassler and Setser, 1966; Rabinovitch et al., 1963; Simons and Taylor, 1969). In these experiments such high energy species as H, CH_2, or C_2H_5 radicals were added to molecules, thereby producing an activated molecule with a somewhat well defined energy content. In one of the more famous experiments, a methylene group (CH_2) was added to ethylene, thereby forming an excited cyclopropane molecule. The internal energy is sufficient to cause the molecule to isomerize to propene. However, collisional deactivation could also stabilize the hot cyclopropane. This experiment not only removed the excitation step k_1, it also replaced $k(T)$ by $k(E)$. By using methylene from the photolysis of either ketene or diazomethane, it was possible to adjust somewhat the internal energy of the CH_2 and thereby the total energy in the nascent cyclopropane. However, uncertainties in the state (singlet or triplet) of the CH_2 prevented clear interpretations of these energy-dependent studies. The more interesting parameter was the inert gas pressure. By adjusting the pressure, it was possible to vary the ratio of stabilized to isomerized product. The rate of stabilization increased with the pressure while the isomerization reaction (being unimolecular) was independent of the pressure. In this fashion, it was determined that A^* decays exponentially in time.

One of the major assumptions in the statistical theory of unimolecular decay is that energy flows or equilibrates rapidly among all of the oscillators prior to dissociation or isomerization. Several chemical activation experiments addressed this question directly and the results mostly confirmed the hypothesis. In one such study, Butler and Kistiakowsky (1960) chemically activated methylcyclopropane in two different ways:

$$\begin{array}{l} CH_2 + CH_2CH{=}CH_2 \searrow \\ \qquad\qquad\qquad\qquad c\text{-}C_3H_6\text{—}CH_3^* \begin{array}{l}\nearrow c\text{-}C_3H_6\text{—}CH_3 \\ \searrow \text{butenes}\end{array} \\ CH_2 + c\text{-}C_3H_6 \nearrow \end{array} \tag{1.20}$$

In the first reaction, the methylene is added across a double bond, while in the second reaction, it is inserted in the C—H single bond. The other difference in the two methods for preparing the excited methylcyclopropane molecule is in the total energy. The heat of formation of propene is 8 kcal/mol less than that of cyclopropane so that second preparation of $C_4H_8^*$ gives an excited species with 8 kcal/mol more excitation. However, in view of the total energy deposited, this small difference was deemed unimportant. Initially, the energy is certainly deposited in different parts of the molecule. Yet, the fraction of butenes to methylcyclopropane was nearly the same at all pressures investigated. From the upper limit of the experiment, it was concluded that the energy is distributed randomly in less than 10^{-11} sec. Numerous such experiments lead to similar conclusions. In a limited number of cases, it was possible to vary the molecule's internal energy by preparing $C_2H_5F^*$ with a variety of reactions such as $CH_2 + CH_3F \rightarrow C_2H_5F^*$ and $C_2H_5 + F_2 \rightarrow C_2H_5F^* + F$ (Kirk et al., 1968). A similar approach by Rabinovitch et al. (1963) provided a microcanonical rate constant for the butyl radical decomposition over a 4 kcal/mol energy range. Agreement with RRKM theory in both cases was noted.

The search for nonrandom energy flow finally succeeded when Rynbrandt and Rabinovitch (1970, 1971a,b) reacted singlet methylene with hexafluorovinylcyclopropane to produce a bi-cyclic excited molecule.

$$^1CD_2 + CF_2\text{—}CF\text{—}CF\text{=}CF_2 \;(\text{ring: } CF_2, CF, CH_2) \rightarrow [CF_2\text{—}CF\text{—}CF\text{—}CF_2]^* \;(\text{rings: } CH_2, CD_2) \tag{1.21}$$

The excited molecule could decompose by ring opening, or it could be stabilized. It was noted that at pressures below about 100 Torr, the two rings opened with nearly equal probability which indicates that the energy initially deposited in the nascent ring has been equilibrated. However, as shown in figure 1.3, above a pressure of about 100 Torr the ring opening was more dominant in the c-$C_3F_3D_2$ ring. The collisional deactivation rate apparently becomes greater than the energy relaxation rate at pressures in excess of 100 Torr. This indicates that in this molecule energy flow between the two cyclopropyl rings takes place with a rate of only 3×10^9 sec^{-1}. This result is not only of historical significance, it also remains today as one of the very few examples of incomplete energy randomization in the dissociation of molecules. In a related chemical activation study involving F plus tetra-allyl tin, Rogers et al. (1982) found that apparently the heavy tin atom located at the center of the molecule prevented rapid energy flow between the allyl units thereby increasing the dissociation rates by a factor of 1000. The most numerous examples of hindered energy flows are found not in normal molecules or ions, but in loosely bound dimers which require typically less than 1000 cm^{-1} to dissociate. The reason for these non RRKM dissociations are discussed fully in chapters 6 and 10.

1.6 TRAJECTORY STUDIES OF UNIMOLECULAR DECOMPOSITION

While the statistical RRKM/QET theory was being used to fit fall-off curves and the results of chemical activation studies, its fundamental assumptions were also being tested by classical trajectory calculations, an approach pioneered by Bunker (1962,

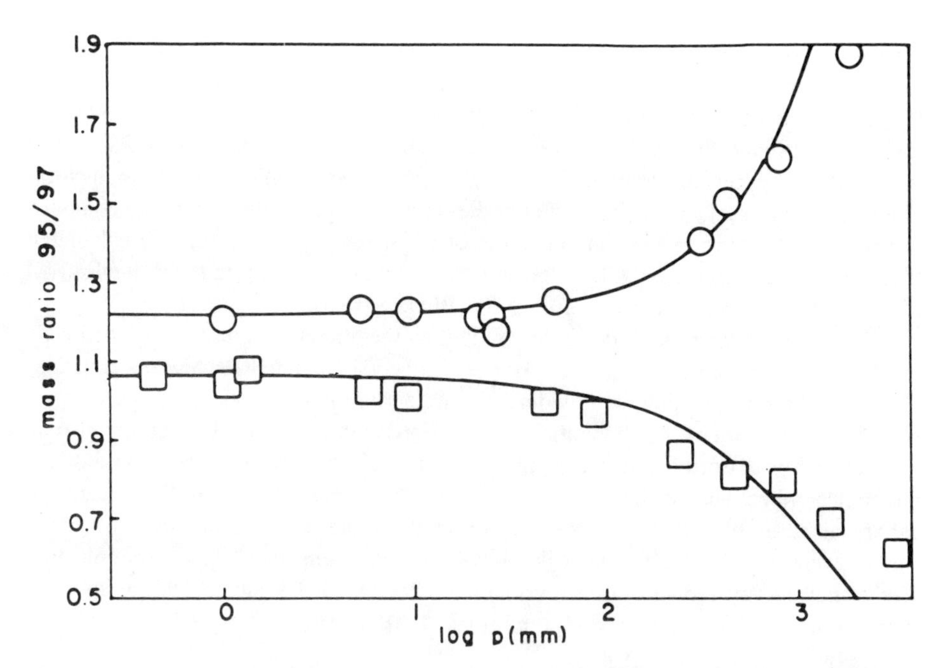

Figure 1.3 Ratio of products from the chemical activation of hexafluorovinylcyclopropane. ○ are for CPO + CD_2, while □ are for CPO-d_2 + CH_2. In the mass spectral analysis, the partially fluorinated cyclopropyl rings with *m/z* 95 ($C_3F_3H_2$) and *m/z* 97 ($C_3F_3D_2$) were used to identify the ring position after the reaction. The deviation of the *m/z* 95/97 ratio above 100 Torr pressure indicates that the unactivated ring tended to remain intact. Taken with permission from Rynbrandt and Rabinovitch (1971a).

1964, 1966, 1970) and co-workers (Bunker and Pattengill, 1968; Bunker and Hase, 1973; Harris and Bunker, 1971). If a molecule is excited randomly at constant total energy, so that a microcanonical ensemble of molecular states is prepared, RRKM theory predicts the ensemble will decay exponentially:

$$N(t) = N(0)e^{-k(E)t}, \tag{1.22}$$

where $N(t)$ is the concentration at time t, and $k(E)$ is the RRKM rate constant. The lifetime distribution $P(t)$, which is the probability that decomposition occurs at time t, is given by

$$P(t) = \frac{-1}{N(0)} \frac{dN(t)}{dt}. \tag{1.23}$$

Hence, according to RRKM theory, $P(t)$ is given by

$$P(t) = k(E)e^{-k(E)t}. \tag{1.24}$$

Thus if a molecule's unimolecular decomposition is in accord with RRKM theory, the $t = 0$ intercept of its $P(t)$ will equal $k(E)$, and its $P(t)$ will be exponential with a decay constant equal to $k(E)$ (Bunker, 1966).

Bunker and co-workers (Bunker, 1962, 1964; Bunker and Pattengill, 1968; Harris and Bunker, 1971; Bunker and Hase, 1973) used classical trajectory simulations to study the nature of $P(t)$ for a variety of model molecules. Since the initial states for the classical trajectories are chosen from a microcanonical ensemble, the $t = 0$ intercept of the trajectory $P(t)$ is the classical RRKM unimolecular rate constant. For this equivalency to be strictly correct, there can be no recrossing of the transition state by the trajectories (Steinfeld et al., 1989) which was shown to be the case for the classical trajectories (Bunker and Pattengill, 1968). The full anharmonicity of the molecule's potential energy surface can be treated in the trajectory simulations, so that a comparison of $k(E)$ determined from the intercept of the trajectory $P(t)$ with $k(E)$ calculated from classical harmonic RRKM theory gives the anharmonic correction to the RRKM rate constant (Bunker and Pattengill, 1968; Bunker, 1970).

For many of the model molecules studied by the trajectory simulations, the decay of $P(t)$ was exponential with a decay constant equal to the RRKM rate constant. However, for some models with widely disparate vibrational frequencies and/or masses, decay was either nonexponential or exponential with a decay constant larger than $k(E)$ determined from the intercept of $P(t)$. This behavior occurs when some of the molecule's vibrational states are inaccessible or only weakly coupled. Thus, a microcanonical ensemble is not maintained during the molecule's decomposition. These studies were a harbinger for what is known now regarding inefficient intramolecular vibrational energy redistribution (IVR) in weakly coupled systems such as van der Waals molecules and mode-specific unimolecular dynamics.

1.7 CONCLUSION

Prior to about 1970, classical experiments, in which the samples were in a thermal heat bath, resulted in an uncertain energy content of the molecule, as well as uncertain collisional deactivation rates. Although RRKM theory provided an excellent framework for a qualitative, as well as semiquantitative, understanding of the unimolecular reactions, the experiments failed to provide firm evidence for the fundamental correctness of the statistical assumptions.

Since 1970, direct photolysis of molecules or ions in low-pressure, collisionless environments, has permitted molecules to be excited to well-defined energy levels, while the use of pulsed lasers or coincidence techniques has provided an accurate external time base with which to measure the dissociation rate constants over many orders of magnitude. It is often the case that more precise experimental results lead to fundamental changes in the theoretical models which describe the phenomena. This has not happened in the case of unimolecular reactions. The statistical theory has remained surprisingly robust. Most molecular systems that dissociate on a bound potential energy surface do so in a statistical fashion. What has changed in the past 25 years is our ability to apply the statistical theory. It is now possible to calculate unimolecular rate constants with essentially no adjustable parameters and which are in quantitative agreement with experiments.

REFERENCES

Astholz, D.C., Troe, J., and Wieters, W. (1979). *J. Chem. Phys.* **70**, 5107.
Brouwer, L., Muller-Markgraf, W., and Troe, J. (1983). *Ber. Bunsenges. Phys. Chem.* **87**, 1031.

Bunker, D.L. (1962). *J. Chem. Phys.* **37**, 393.
Bunker, D.L. (1964). *J. Chem. Phys.* **40**, 1946.
Bunker, D.L. (1966). *Theory of Elementary Gas Reaction Rates*. Pergamon Press, New York.
Bunker, D.L. (1970). In Schlier, C., Ed., Proceedings of the International School of Physics Enrico Fermi Course XLIV: Molecular Beam and Reaction Kinetics, pp. 315–319. Academic Press, New York.
Bunker, D.L., and Hase, W.L. (1973). *J. Chem. Phys.* **54**, 4621.
Bunker, D.L., and Pattengill, M. (1968). *J. Chem. Phys.* **48**, 772.
Butler, J.N., and Kistiakowsky, G.B. (1960). *J. Am. Chem. Soc.* **82**, 759.
Chambers, T.S., and Kistiakowsky, G.B. (1934). *J. Am. Chem. Soc.* **56**, 399.
Flowers, M.C., and Frey, H.M. (1962). *J. Chem. Soc.* 1157.
Forst, W. (1973). *Theory of Unimolecular Reactions*. Academic Press, New York.
Gilbert, R.G., Luther, K., and Troe, J. (1983). *Ber. Bunsenges. Phys. Chem.* **87**, 169.
Gilbert, R.G., and Smith, S.C. (1990). *Theory of Unimolecular and Recombination Reactions*. Blackwell Scientific, Oxford.
Harris, H.H., and Bunker, D.L. (1971). *Chem. Phys. Lett.* **11**, 433.
Hassler, J.C., and Setser, D.W. (1966). *J. Chem. Phys.* **45**, 3237.
Hinshelwood, C.N. (1926). *Proc. R. Soc. Lond. A* **113**, 230.
Hippler, H., Troe, J., and Wendelken, H.J. (1981). *Chem. Phys. Lett.* **84**, 257.
Hippler, H., Troe, J., and Wendelken, H.J. (1983). *J. Chem. Phys.* **78**, 6709.
Jalenak, W., Weston, R.E., Sears, T.J., and Flynn, G.W. (1988). *J. Chem. Phys.* **89**, 2015.
Kassel, L.S. (1928). *J. Phys. Chem.* **32**, 225.
Kirk, A.W., Trotman-Dickenson, A.F., and Trus, B.L. (1968). *J. Chem. Soc.* 3058.
Lindemann, F.A. (1922). *Trans. Faraday Soc.* **17**, 598.
Marcus, R.A., and Rice, O.K. (1951). *J. Phys. Colloid Chem.* **55**, 894.
Nikitin, E.E. (1974). *Theory of Elementary Atomic and Molecular Processes in Gases*. Clarendon Press, Oxford.
Prichard, H.O., Sowden, R.G., and Trotman-Dickenson, A.F. (1953). *Proc. R. Soc. Lond. A* **217**, 563.
Rabinovitch, B.S., and Flowers, M.C. (1964). *Quart. Rev.* **18**, 122.
Rabinovitch, B.S., and Setser, D.W. (1964). *Adv. Photochem.* **3**, 1.
Rabinovitch, B.S., Kubin, R.F., and Harrington, R.E. (1963). *J. Chem. Phys.* **38**, 405.
Rice, O.K., and Ramsperger, H.C. (1927). *J. Am. Chem. Soc.* **49**, 1617.
Rice, O.K., and Ramsperger, H.C. (1928). *J. Am. Chem. Soc.* **50**, 617.
Robinson, P.J., and Holbrook, K.A. (1972) *Unimolecular Reactions*. Wiley-Interscience, London.
Rogers, P.J., Montague, D.C., Frank, J.P., Tyler, S.C., and Rowland, F.S. (1982). *Chem. Phys. Lett.* **89**, 9.
Rosenstock, H.M., Wallenstein, M.B., Wahrhaftig, A.L., and Eyring, H. (1952). *Proc. Nat. Acad. Sci.* **38**, 667.
Rynbrandt, J.D., and Rabinovitch, B.S. (1970). *J. Phys. Chem.* **74**, 4175.
Rynbrandt, J.D., and Rabinovitch, B.S. (1971a). *J. Phys. Chem.* **75**, 2164.
Rynbrandt, J.D., and Rabinovitch, B.S. (1971b). *J. Chem. Phys.* **54**, 2275.
Schlag, E.W., and Rabinovitch, B.S. (1960). *J. Am. Chem. Soc.* **82**, 5996.
Schneider, F.W., and Rabinovitch, B.S. (1962). *J. Am. Chem. Soc.* **84**, 4215.
Simons, J.W., and Taylor, G.W. (1969). *J. Phys. Chem.* **73**, 1274.
Slater, N.B. (1953). *Proc. R. Soc. Lond. A* **218**, 224.
Slater, N.B. (1955). *Proc. Roy. Soc. Edinburgh A* **64**, 161.
Slater, N.B. (1959). *Theory of Unimolecular Reactions*. Cornell University Press, Ithaca.
Steinfeld, J.I., Francisco, J.S., and Hase, W.L. (1989). *Chemical Kinetics and Dynamics*. Prentice-Hall, Englewood Cliffs, N.J.
Tsang, W. (1965). *J. Chem. Phys.* **43**, 352.

Tsang, W. (1972). *J. Phys. Chem.* **76**, 143.
Tsang, W. (1978). *Int. J. Chem. Kin.* **10**, 41.
Tsang, W. (1981). In A. Lifshitz, Ed., *Shock Tubes in Chemistry.* Marcel Decker, New York, pp. 59–129.
Weston, R.E. (1986). *Int. J. Chem. Kin.* **18**, 1259.
Weston, R.E., and Flynn, G.W. (1992). *Ann. Rev. Phys. Chem.* **43**, 559.
Whitten, G.Z., and Rabinovitch, B.S. (1963). *J. Chem. Phys.* **38**, 2466.
Whitten, G.Z., and Rabinovitch, B.S. (1964). *J. Chem. Phys.* **41**, 1883.

2

Vibrational/Rotational Energy Levels

The first step in a unimolecular reaction is the excitation of the reactant molecule's energy levels. Thus, a complete description of the unimolecular reaction requires an understanding of such levels. In this chapter molecular vibrational/rotational levels are considered. The chapter begins with a discussion of the Born-Oppenheimer principle (Eyring, Walter, and Kimball, 1944), which separates electronic motion from vibrational/rotational motion. This is followed by a discussion of classical molecular Hamiltonians, Hamilton's equations of motion, and coordinate systems. Hamiltonians for vibrational, rotational, and vibrational/rotational motion are then discussed. The chapter ends with analyses of energy levels for vibrational/rotational motion.

2.1 BORN-OPPENHEIMER PRINCIPLE

The Born-Oppenheimer principle assumes separation of nuclear and electronic motions in a molecule. The justification in this approximation is that motion of the light electrons is much faster than that of the heavier nuclei, so that electronic and nuclear motions are separable. A formal definition of the Born-Oppenheimer principle can be made by considering the time-independent Schrödinger equation of a molecule, which is of the form

$$\hat{H}\Psi = (\hat{T}_{\mathrm{N}} + \hat{T}_{\mathrm{e}} + V_{\mathrm{NN}} + V_{\mathrm{ee}} + V_{\mathrm{Ne}})\Psi = E\Psi \tag{2.1}$$

where $\hat{T}_{\mathrm{N}}$ is the kinetic energy operator of the nuclei, $\hat{T}_{\mathrm{e}}$ is the kinetic energy operator of the electrons, V_{NN} is the repulsive electrostatic potential energy of the nuclei, V_{ee} is the repulsive electrostatic potential energy of the electrons, and V_{Ne} is the attractive electrostatic potential energy between the nuclei and electrons. The Born-Oppenheimer principle allows one to write the total molecular wave function Ψ as the product of an electronic wave function Ψ_{e} and a nuclear wave function Ψ_{N}:

$$\Psi = \Psi_{\mathrm{e}}(R,r)\ \Psi_{\mathrm{N}}(R), \tag{2.2}$$

where Ψ_{e} depends on both the nuclear coordinates R and electronic coordinates r, while Ψ_{N} only depends on the R coordinates. The electronic wave function Ψ_{e} is a solution to the electronic Schrödinger equation

$$(\hat{T}_{\mathrm{e}} + \hat{V}_{\mathrm{ee}} + \hat{V}_{\mathrm{Ne}})\ \Psi_{\mathrm{e}}(R,r) = E_{\mathrm{e}}(R)\ \Psi_{\mathrm{e}}(R,r). \tag{2.3}$$

This equation describes the motion of the electrons for a fixed nuclear configuration R and, when solved, gives the electronic energy $E_{\mathrm{e}}(R)$ for that configuration. By varying R, the electronic energy can be determined as a function of the nuclear coordinates. This type of calculation is considered in chapter 3.

The wave function $\Psi_N(R)$ is the solution to the nuclear Schrödinger equation

$$[\hat{T}_N + V_{NN}(R) + E_e(R)]\, \Psi_N(R) = E\, \Psi_N(R), \tag{2.4}$$

where $E_e(R)$, the electronic energy versus R, is found from Eq. (2.3). The sum $V_{NN}(R) + E_e(R)$ represents the potential energy $V(R)$ of the nuclei as a function of their coordinates, which is known as the *potential energy surface*. For a nonlinear molecule consisting of N atoms, the potential energy surface depends on $3N - 6$ independent coordinates. An analytic function which represents a potential energy surface is called a *potential energy function*. Substituting $V(R)$ into Eq. (2.4) gives

$$[\hat{T}_N + V(R)]\, \Psi_N(R) = E\, \Psi_N(R), \tag{2.5}$$

where the term $[\hat{T}_N + V(R)]$ is the quantum mechanical Hamiltonian operator for the translational, vibrational, and rotational motion of the molecule. Thus, if the center of mass translation motion is removed from the nuclear kinetic energy $\hat{T}_N$, solving Eq. (2.5) yields the vibrational/rotational energy levels, E, and wave functions, $\Psi_N(R)$, for the molecule (Hirst, 1985). However, before considering such a solution to Eq. (2.5), it is useful to review classical Hamiltonians and coordinate systems for the nuclear motion of a molecule.

2.2 CLASSICAL HAMILTONIANS, HAMILTON'S EQUATIONS OF MOTION, AND COORDINATE SYSTEMS

The classical mechanical energy for the nuclear motion of a molecule is given by the Hamiltonian H (Goldstein, 1950) which depends on the nuclear coordinates $\mathbf{q}$ and their conjugate momenta $\mathbf{p}$ and is a sum of kinetic and potential energies:

$$H = T + V \tag{2.6}$$

This equation is the classical analog of the quantum mechanical operator $\hat{T}_N + V(R)$ in Eq. (2.5). A generalized momentum is defined by

$$p_i = \partial L / \partial \dot{q}_i, \tag{2.7}$$

where L, the Lagrangian, is defined as $L = T - V$.

In laboratory-based Cartesian coordinates T is only a function of the Cartesian momenta and is written as

$$T = \sum_{i=1}^{N} (p_{x_i}^2 + p_{y_i}^2 + p_{z_i}^2)/2m_i \tag{2.8}$$

for a molecule with N atoms. On the other hand, the potential energy V is a function of the Cartesian coordinates. The nuclear motion for the molecule can be determined from H by simultaneously solving Hamilton's equations of motion for each coordinate q_i and its conjugate momentum p_i:

$$\partial q_i / \partial t = \partial H / \partial p_i, \qquad \partial p_i / \partial t = -\partial H / \partial q_i \tag{2.9}$$

For laboratory-based Cartesian coordinates the index i extends from 1 to $3N$.

Though the above Hamiltonian in Cartesian momenta and coordinates is certainly correct, it is not very useful for representing the energy of a molecule since properties of the potential energy are usually not apparent in Cartesian coordinates and the Cartesian kinetic energy expression does not distinguish rotational, vibrational, and coriolis kinetic energies. To represent these chemically more interesting molecular energies, it is necessary to transform from Cartesian coordinates to molecular-type coordinates. In the following, molecular-type coordinates are considered for both diatomic and polyatomic molecules.

2.2.1 Diatomic Molecule

For a diatomic molecule the Cartesian Hamiltonian can be transformed to a Hamiltonian dependent on relative (or internal) and center-of-mass coordinates and momenta. The relative coordinates x, y, z and their conjugate momenta are defined by

$$\begin{aligned} x &\equiv x_2 - x_1, \quad & y &\equiv y_2 - y_1, \quad & z &\equiv z_2 - z_1 \\ p_x &\equiv \mu \mathrm{v}_x, & p_y &\equiv \mu \mathrm{v}_y, & p_z &\equiv \mu \mathrm{v}_z, \end{aligned} \tag{2.10}$$

where $\mathrm{v}_x = dx/dt$ and the reduced mass is defined by

$$\mu \equiv \frac{m_1 m_2}{m_1 + m_2} \tag{2.11}$$

The center-of-mass coordinates and momenta are

$$\begin{aligned} X &\equiv \frac{m_1 x_1 + m_2 x_2}{M}, \quad & Y &\equiv \frac{m_1 y_1 + m_2 y_2}{M}, \quad & Z &\equiv \frac{m_1 z_1 + m_2 z_2}{M}, \\ P_X &\equiv M\mathrm{v}_X, & P_Y &\equiv M\mathrm{v}_Y, & P_Z &\equiv M\mathrm{v}_Z, \end{aligned} \tag{2.12}$$

where $M = m_1 + m_2$. If the diatomic Hamiltonian is expressed in terms of relative and center-of-mass coordinates, it turns out that

$$H = \left[\frac{1}{2\mu}(p_x^2 + p_y^2 + p_z^2) + V(x, y, z) \right] + \left[\frac{1}{2M}(P_x^2 + P_y^2 + P_z^2) \right] \tag{2.13}$$

The first term in brackets represents the vibrational/rotational motion of the molecule, while the second term represents the molecule's translational motion; i.e., $H = H_{\mathrm{vr}} + H_{\mathrm{t}}$. Since there are no couplings between the two terms, they can be treated separately.

The vibrational/rotational motion of a diatomic molecule can be confined to one plane (e.g., the x,y-plane) and the relative coordinates and momenta in Eq. (2.10) can be transformed to polar coordinates and momenta by the relations

$$\begin{aligned} r &= (x^2 + y^2)^{1/2}, \quad \theta = \tan^{-1}(y/x), \\ p_r &= p_x \cos\theta + p_y \sin\theta, \\ p_\theta &= -p_x\, r \sin\theta + p_y\, r \cos\theta. \end{aligned} \tag{2.14}$$

Here, p_r and p_θ are the radial and angular momentum. In polar coordinates the diatomic Hamiltonian becomes

$$H = p_r^2/2\mu + p_\theta^2/2\mu r^2 + V(r). \tag{2.15}$$

According to Hamilton's equations of motion, Eq. (2.9), $\partial H/\partial\theta = -\partial p_\theta/\partial t$. Since H is independent of θ, $\partial H/\partial\theta = 0$ and p_θ, the angular momentum, is a constant of the motion. This angular momentum p_θ represents rotation about a Cartesian axis, and it is common practice to use the symbol j for p_θ. The second term in Eq. (2.15), which depends on the angular momentum and the bond length r, is considered the rotational kinetic energy at small r and the centrifugal potential at large r. The concept of a centrifugal barrier is discussed in chapter 7 when considering angular momentum for dissociating molecules.

In a more complete treatment than that given in the preceding paragraph, the vibrational/rotational Hamiltonian in x, y, and z coordinates is transformed to the following vibrational/rotational Hamiltonian in r, θ, and ϕ spherical polar coordinates (Pauling and Wilson, 1935; Davidson, 1962):

$$H = p_r^2/2\mu + p_\theta^2/2\mu r^2 + p_\phi^2/2\mu r^2\sin^2\theta + V(r). \tag{2.16}$$

The relations between p_r, p_θ, p_ϕ and p_x, p_y, p_z are given in Davidson (1962).

2.2.2 Polyatomic Molecules

To write a Hamiltonian that distinguishes vibrational, rotational, and coriolis kinetic energies for a polyatomic molecule it is necessary to consider figure 2.1, which illustrates the relationship between fixed laboratory-based and rotating center-of-mass-based coordinate systems for a molecule with N atoms (Wilson et al., 1955; Califano, 1976). The instantaneous position of the ith atom in the rotating system is given by a vector $\mathbf{r}_i$ with components x_i, y_i, z_i and the equilibrium position by a vector $\mathbf{r}_i^0$ with components x_i^0, y_i^0, z_i^0 (these latter components are the equilibrium coordinates with

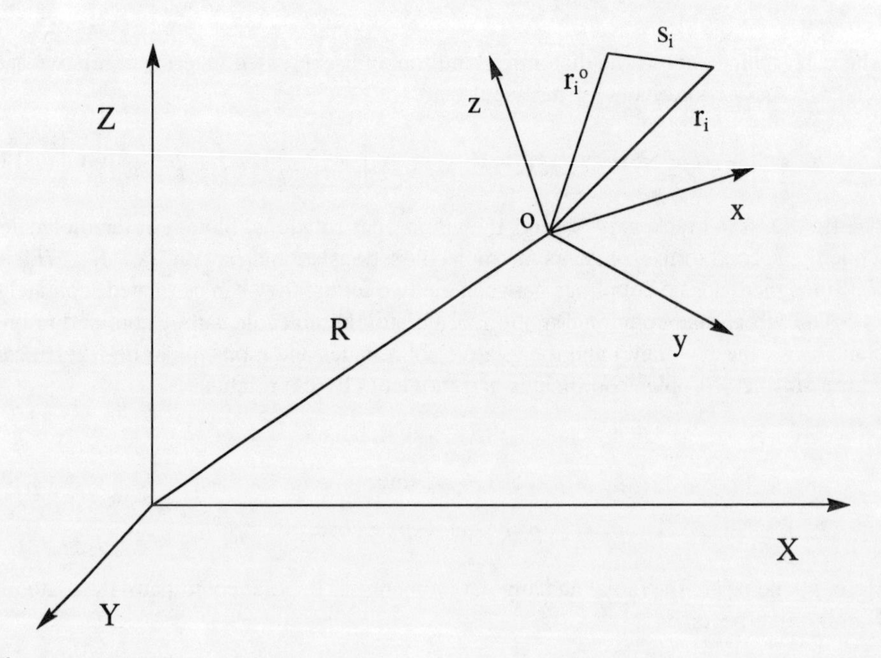

Figure 2.1 Fixed (X, Y, Z) and rotating (x, y, z) axes systems. Taken from Califano (1976).

respect to the rotating system). The instantaneous displacement of the ith atom from its equilibrium position in the rotating system is then given by

$$\mathbf{s}_i = \mathbf{r}_i - \mathbf{r}_i^{\circ}. \tag{2.17}$$

The velocity of the ith atom in a molecule with respect to a fixed-laboratory framework can be written as

$$\mathbf{v}_i = \dot{\mathbf{R}} + \boldsymbol{\omega} \times \mathbf{r}_i + \dot{\mathbf{s}}_i, \tag{2.18}$$

where $\dot{\mathbf{R}}$ is the velocity vector of the center of mass and $\boldsymbol{\omega}$ is the angular velocity of the rotating coordinate system attached to the molecule with respect to the fixed-laboratory frame.

The kinetic energy of the molecule is found by summing the kinetic energies of the individual atoms, that is,

$$2T = \Sigma m_i \mathbf{v}_i \cdot \mathbf{v}_i = \dot{\mathbf{R}}^2 \Sigma m_i + \Sigma m_i (\boldsymbol{\omega} \times \mathbf{r}_i) \cdot (\boldsymbol{\omega} \times \mathbf{r}_i) + \Sigma m_i \dot{s}_i^2$$
$$+ 2\dot{\mathbf{R}} \cdot \boldsymbol{\omega} \times \Sigma m_i \mathbf{r}_i + 2\dot{\mathbf{R}} \cdot m_i \dot{\mathbf{s}}_i + 2\Sigma m_i \boldsymbol{\omega} \times \mathbf{r}_i \cdot \dot{\mathbf{s}}_i. \tag{2.19}$$

The first three terms in Eq. (2.19) represent the pure translational, rotational, and vibrational kinetic energies while the remaining three represent the corresponding interaction energies. From the definition of the center of mass in the rotating frame,

$$\sum_i m_i \mathbf{r}_i = 0. \tag{2.20}$$

By differentiating this equation with respect to time, it follows that

$$\sum_i m_i \dot{\mathbf{r}}_i = \sum_i m_i (\boldsymbol{\omega} \times \mathbf{r}_i + \dot{\mathbf{s}}_i) = \boldsymbol{\omega} \times \sum_i m_i \mathbf{r}_i + \sum_i m_i \dot{\mathbf{s}}_i = \sum_i m_i \dot{\mathbf{s}}_i. \tag{2.21}$$

Equation (2.20) is called the first Sayvetz (or Eckart) condition (Wilson et al., 1955; Califano, 1976) and specifies that during a molecular vibration the center of mass of the molecule remains constant. Introducing Eqs. (2.20) and (2.21) into Eq. (2.19) causes the first two interaction terms to vanish.

The second Sayvetz (or Eckart) condition is chosen so "that whenever the atomic displacements in a molecular vibration tend to produce a rotation of the molecule, the rotating system reorients in order to eliminate this component of the motion" (Califano, 1976). This condition, which is given by

$$\Sigma m_i \mathbf{r}_i^{\circ} \times \dot{\mathbf{s}}_i = 0, \tag{2.22}$$

means that during a molecular vibration there must be no zero-order vibrational angular momentum. Replacing $\mathbf{r}_i$ by $\mathbf{r}_i^{\circ} + \mathbf{s}_i$ in the last term in Eq. (2.19) and introducing the condition in Eq. (2.22) into this term gives

$$2T = \dot{\mathbf{R}}^2 \Sigma m_i + \Sigma m_i (\boldsymbol{\omega} \times \mathbf{r}_i) \cdot (\boldsymbol{\omega} \times \mathbf{r}_i) + \Sigma m_i \dot{s}_i^2 + 2\boldsymbol{\omega} \cdot \Sigma (m_i \mathbf{s}_i \times \dot{\mathbf{s}}_i). \tag{2.23}$$

The first term in this equation represents the molecule's center of mass translational motion and can be ignored since it is separable from the other terms. The second and third terms represent the molecule's rotational and vibrational kinetic energies, respectively, and the last term represents the vibrational/rotational coupling often called the coriolis energy.

2.2.2.1 Vibrational motion

2.2.2.1.a Internal coordinates. Equation (2.23) is usually the starting point for formulating a molecular vibrational Hamiltonian. If the angular velocity is zero and the center of mass motion ignored, T in Eq. (2.23) becomes $\Sigma m_i \dot{s}_i^2/2$, the vibrational kinetic energy in molecular-based Cartesian coordinates. It is more meaningful to represent the vibrational energy in curvilinear internal coordinates (such as bond lengths and valence angle bends) than in Cartesian coordinates, and $\Sigma m_i \dot{\mathbf{s}}_i^2/2$ can be transformed without approximation to the following curvilinear coordinate expression for a molecule's vibrational kinetic energy (Wilson et al., 1955):

$$T = \sum_{i,j}^{n} g_{ij} p_i p_j / 2 . \tag{2.24}$$

In this equation the g-elements are effective reduced mass terms, the p_i are the internal coordinate momenta, and n is the number of internal coordinates. The specific value for a g-element depends on the coordinates to which p_i and p_j are conjugate. There are general formulas for the g-elements. A triatomic molecule has two bond lengths and a valence angle bend, and the g-elements for these types of coordinates are listed in table 2.1. As shown in table 2.1, it is common for a g-element to depend on the internal coordinates. Thus, in internal coordinates the kinetic energy depends on the internal coordinates as well as the momenta. Standard trigonometric and vector expressions are used to relate Cartesian and curvilinear internal coordinates, for example, a bond length is $r = [(x_1 - x_2)^2 + (y_1 - y_2)^2 + (z_1 + z_2)^2]^{1/2}$ and a valence angle bend is $\cos\phi = \mathbf{r}_1 \cdot \mathbf{r}_2 / |r_1|\,|r_2|$.

An internal coordinate vibrational Hamiltonian is constructed by combining Eq.

Table 2.1. G-elements for a Triatomic Molecule[a]

	r_1	r_2	ϕ
r_1	$1/\mu_1$[b]	$\cos\phi/M$	$-\sin\phi/Mr_2$
r_2	$\cos\phi/M$	$1/\mu_2$	$-\sin\phi/Mr_1$
ϕ	$-\sin\phi/Mr_2$	$-\sin\phi/Mr_1$	$\dfrac{1}{m_1 r_1^2} + \dfrac{1}{m_2 r_2^2} + \dfrac{1}{M}\left(\dfrac{1}{r_1^2} + \dfrac{1}{r_2^2} - \dfrac{2\cos\phi}{r_1 r_2}\right)$

[a]The masses of the molecule are identified as m_2-M-m_2. The bond lengths are r_1 and r_2 and ϕ is the valence angle bend.

[b]$\mu_1 = m_1 M/(m_1 + M)$.

(2.24) with the potential energy in internal coordinates. Consider a triatomic molecule with coordinates r_1, r_2, and ϕ and conjugate momenta p_1, p_2, and p_ϕ, for which the g-elements are listed in table 2.1. The internal coordinate vibrational Hamiltonian for this molecule is

$$H = p_1^2/2\mu_1 + p_2^2/2\mu_2 + p_1 p_2 \cos\phi/M + A\, p_\phi^2/2m_1 m_2 M r_1^2 r_2^2$$
$$- p_1 p_\phi \sin\phi/M r_2 - p_2 p_\phi \sin\phi/M r_1 + V(r_1, r_2, \phi), \qquad (2.25)$$

where

$$A = m_1(m_2 + M) r_1^2 + m_2(m_1 + M) r_2^2 - 2 m_1 m_2 r_1 r_2 \cos\phi, \qquad (2.26)$$

$\mu_1 = m_1 M/(m_1 + M)$, $\mu_2 = m_2 M/(m_2 + M)$, the masses m_1 and M are separated by r_1, the masses m_2 and M are separated by r_2, and ϕ is the m_1–M–m_2 bond angle. If one wishes to consider only the two stretching coordinates, $\sin\phi$ is set to its fixed value in the Hamiltonian (e.g., zero for a linear molecule) and p_ϕ is set to zero. One of the strengths of representing vibrational motion with an internal coordinate Hamiltonian is that it is straightforward to include all the coupling in the potential energy. However, as illustrated by Eqs. (2.24) and (2.25) there is also kinetic (i.e., momenta) coupling in the Hamiltonian. A more general approach for constructing a Hamiltonian with constrained coordinates is described by Hadder and Frederick (1992).

2.2.2.1.b Normal mode coordinates. Either the Cartesian coordinate or internal coordinate vibrational Hamiltonian can be transformed to a normal mode Hamiltonian by assuming infinitesimal displacements for the Cartesian (or internal) coordinates from the equilibrium geometry. Here, following Califano (1976), the transformation between Cartesian and normal mode coordinates is illustrated. Instead of using molecular-based Cartesian displacement coordinates s_i, Eq. (2.17), it is convenient to use mass-weighted Cartesian displacement coordinates which are defined by

$$q_i = \sqrt{m_i} s_i \,. \qquad (2.27)$$

The kinetic energy can then be expressed in the simpler form

$$2T = \sum_{i=1}^{3N} \dot{q}_i^2 \qquad (2.28)$$

For small displacements from the equilibrium geometry, the potential energy can be represented by the power series

$$V = V_0 + \sum_i \left(\frac{\partial V}{\partial q_i}\right)_0 q_i + 1/2 \sum_{ij} \left(\frac{\partial^2 V}{\partial q_i \partial q_j}\right)_0 q_i q_j + \cdots \qquad (2.29)$$

and without loss of generality, the potential energy can be shifted so that the minimum of the potential V_o is zero. By definition, a potential energy minimum requires

$$\left(\frac{\partial V}{\partial q_i}\right)_0 = 0 \qquad (2.30)$$

for each of the coordinates. For sufficiently small displacements of the coordinates it is only necessary to retain the quadratic terms in Eq. (2.29), so that the potential becomes

$$2V = \sum_{i,j} f_{ij} q_i q_j \,, \tag{2.31}$$

where the force constants f_{ij} are given by

$$f_{ij} = \left(\frac{\partial^2 V}{\partial q_i \partial q_j} \right)_0 . \tag{2.32}$$

Note, that since q is in mass-weighted coordinates, the units of the force constants f_{ij} are simply sec^{-2}, rather than the usual kg-sec^{-2}. A potential which only includes quadratic terms is called a *harmonic potential*. *Anharmonicity* arises from the higher-order terms.

The expressions for T, Eq. (2.28), and V, Eq. (2.31), can be written in a simpler form using matrix notation. Using the column vector $\mathbf{q}$, whose components are the $3N$ mass-weighted Cartesian displacement coordinates,

$$2T = \tilde{\dot{\mathbf{q}}}\dot{\mathbf{q}} \tag{2.33}$$

and

$$2V = \tilde{\mathbf{q}}\mathbf{F}\mathbf{q}, \tag{2.34}$$

where the symbol $\sim$ denotes the transpose vector and $\mathbf{F}$ is a $3N \times 3N$ real symmetric matrix of the force constants f_{ij}, Eq. (2.32); that is, $\mathbf{F}$ is a hermitian matrix. The F matrix is not diagonal since it contains all possible nondiagonal cross terms. However, it is interesting to consider the effect of a diagonal $\mathbf{F}$ matrix. For such a situation the vibrational energy could be expressed as the sum $E = \Sigma E_i$, where E_i would equal $(\dot{q}_i^2 + f_{ii} q_i^2)/2$. Though such a solution is not obtained for Cartesian coordinates, it is worthwhile to ask whether the energy becomes separable at small displacements for another coordinate system.

Equation (2.24) shows that, because of the kinetic energy coupling, in internal coordinates the energy is not separable at small displacements. Thus, it is necessary to search for $3N$ new coordinates Q_k called *normal mode coordinates*, which define a new column vector $\mathbf{Q}$. To begin it will be assumed that the mass-weighted Cartesian coordinates and normal mode coordinates are related by a linear transformation of the form

$$Q_k = \sum_i l'_{ki} q_i \,. \tag{2.35}$$

In matrix notation this transformation is written as

$$\mathbf{Q} = \mathbf{L}^{-1}\,\mathbf{q}, \tag{2.36}$$

where the l'_{ki} are the components of $\mathbf{L}^{-1}$. The reverse transformation is written as

$$\mathbf{q} = \mathbf{L}\,\mathbf{Q} \tag{2.37}$$

where $\mathbf{L}$ is the inverse of $\mathbf{L}^{-1}$ and $\mathbf{L}\,\mathbf{L}^{-1}$ equals the unit matrix $\mathbf{E}$. In the following it is shown that this linear transformation gives the desired solution. To solve the normal mode problem in internal coordinates, such a linear transformation is made between internal and normal mode coordinates. However, in general these internal coordinates are not the curvilinear internal coordinates described in the previous section, but are rectilinear internal coordinates (Califano, 1976). These two types of internal coordinates are only the same for infinitesimal displacements.

For the energy to be separable in normal mode coordinates requires both the kinetic and potential energies to be diagonal; that is,

$$2T = \tilde{\dot{\mathbf{Q}}}\,\dot{\mathbf{Q}} \tag{2.38}$$

and

$$2V = \tilde{\mathbf{Q}}\,\mathbf{\Lambda}\,\mathbf{Q}, \tag{2.39}$$

where $\mathbf{\Lambda}$ is a diagonal matrix with elements λ_k. Requiring both Eqs. (2.38) and (2.39) to be satisfied simultaneously allows one to determine both $\mathbf{L}$ and $\mathbf{\Lambda}$. Inserting Eq. (2.37) into Eq. (2.33) gives

$$2T = \tilde{\dot{\mathbf{Q}}}\,\tilde{\mathbf{L}}\,\mathbf{L}\,\dot{\mathbf{Q}} \tag{2.40}$$

and comparison with Eq. (2.38) shows that

$$\tilde{\mathbf{L}}\,\mathbf{L} = \mathbf{E}. \tag{2.41}$$

Thus, an important property of $\mathbf{L}$ is that it is orthogonal since

$$\tilde{\mathbf{L}} = \mathbf{L}^{-1}. \tag{2.42}$$

Similarly, inserting Eq. (2.37) into Eq. (2.34) gives

$$2V = \tilde{\mathbf{Q}}\,\tilde{\mathbf{L}}\,\mathbf{F}\,\mathbf{L}\,\mathbf{Q}, \tag{2.43}$$

and a comparison with Eq. (2.39) shows that

$$\tilde{\mathbf{L}}\,\mathbf{F}\,\mathbf{L} = \mathbf{\Lambda}. \tag{2.44}$$

Thus, another important property of $\mathbf{L}$ is that it diagonalizes the force constant matrix. Also, since $\mathbf{L}$ is an orthogonal matrix, Eq. (2.44) can be written in the form of a general eigenvalue equation, that is,

$$(\mathbf{F} - \mathbf{\Lambda})\mathbf{L} = 0. \tag{2.45}$$

Solving Eq. (2.45) is a standard problem in linear algebra [an example solution is outlined in Steinfeld et al. (1989)]. The solution gives $\mathbf{\Lambda}$, which is a diagonal matrix of the $3N$ eigenvalues λ_k and the eigenvector matrix $\mathbf{L}$ with components l_{ik}, which define the transformation between normal mode coordinates Q_k and the mass-weighted Cartesian displacement coordinates q_i, that is,

$$q_i = \sum_k l_{ik} Q_k. \tag{2.46}$$

Thus, the l_{ik} in Eq. (2.46) are one of the $3N$ rows of $\mathbf{L}$ while the l'_{ik} in Eq. (2.35) are one of the $3N$ columns. For a nonlinear molecule there are six zero eigenvalues in $\mathbf{\Lambda}$, which correspond to translation and external rotation motions. The remaining $3N - 6$ nonzero eigenvalues equal $4\pi^2\nu_k^2$, where the ν_k's are the normal mode vibrational frequencies.

A numerical solution to Eq. (2.45) for the H_2O molecule is given in table 2.2. The calculation was performed for a potential which has an OH quadratic stretching force constant of 7.60 mdyn/Å and a HOH quadratic bending force constant of 0.644 mdyn-Å/rad^2. These internal force constants are transformed to Cartesian force constants by writing the internal coordinates as functions of Cartesian coordinates as described above, following Eq. (2.24). This is a standard procedure and is used in the general dynamics computer program VENUS (Hase et al., 1996). The first six eigenvalues in table 2.2, which are for translation and rotation are only approximately zero, since the solution is obtained numerically. If the solution were obtained analytically, these six eigenvalues would be exactly zero. The remaining three eigenvalues, with frequencies of 1595, 3662, and 3716 cm^{-1} are for the three normal modes of vibration. Sketching the eigenvectors for these normal modes shows they are a symmetric bend, a symmetric stretch and an asymmetric stretch, respectively.

The normal mode Hamiltonian is the sum of Eqs. (2.38) and (2.39), and can be written as

$$H = \sum_k (P_k{}^2 + \lambda_k Q_k{}^2)/2 , \tag{2.47}$$

where the definition of the generalized momentum, Eq. (2.7), has been used to replace $\dot{Q}_k$ with P_k. Solving Hamilton's equations of motion, Eq. (2.9), for the normal mode Hamiltonian confirms that $\lambda_k = 4\pi^2\nu_k^2$. Equation (2.47) is strictly valid only for small coordinate displacements, since only the quadratic (i.e., harmonic) potential is retained in its derivation. Because the inclusion of anharmonicity in the potential energy does not affect the kinetic energy in the normal mode Hamiltonian, the anharmonic vibrational energy in normal mode coordinates is written as

$$H = \sum_i (P_i^2 + \lambda_i Q_i^2)/2 + V_{\text{anh}}(Q_1, Q_2, \ldots), \tag{2.48}$$

where V_{anh} is the anharmonic contribution to the potential expanded in higher order terms (cubic, quartic, etc.). However, in treating highly vibrationally excited molecules, one often expresses the Hamiltonian in internal coordinates instead of normal mode coordinates, since it is usually much more difficult to represent an anharmonic potential in normal mode coordinates than in curvilinear internal coordinates. Equations (2.47) and (2.48) are incomplete in that they do not include the term for vibrational angular momentum (Wilson and Howard, 1936; Darling and Dennison, 1940; Carney et al., 1978; Romanowski et al., 1985). As shown in the following section, this term is a component of the normal mode Hamiltonian even if the total rotational angular momentum of the molecule is zero.

A normal mode analysis can also be performed in curvilinear internal coordinates (Wilson et al., 1955). The approach is the same as that described above for Cartesian coordinates with one major modification. In an internal coordinate normal mode analysis, the internal coordinate cannot be simply scaled by the masses as is done for the Cartesian coordinates, Eq. (2.27), so that the masses become an explicit part of the eigenvalue problem. Thus, in an internal coordinate normal mode analysis, one does not solve Eq. (2.45), but instead solves the eigenvalue equation

$$(\mathbf{G}\,\mathbf{F} - \boldsymbol{\Lambda})\,\mathbf{L} = 0, \tag{2.49}$$

Table 2.2. Normal Mode Frequencies and Eigenvectors for H_2O.[a]

	$1.450i$[b]	$0.1168i$	0.0016	0.0350	2.671	2.947	1595	3662	3716
x_O	0.0436[c]	0.0000	0.0000	0.2367	0.0000	0.0000	0.0000	0.0000	−0.0676
y_O	0.0000	0.0033	0.2356	0.0000	0.0000	0.0000	−0.0639	−0.0540	0.0000
z_O	0.0000	0.2321	−0.0032	0.0000	0.0000	−0.0929	0.0000	0.0000	0.0000
x_{H1}	−0.3968	0.0000	0.0000	0.2261	0.0000	0.0000	−0.4544	0.5381	0.5362
y_{H1}	0.5688	0.0033	0.2356	0.0137	0.0000	0.0000	0.5071	0.4282	0.4152
z_{H1}	0.0000	0.2617	−0.0036	0.0000	0.7043	0.6539	0.0000	0.0000	0.0000
x_{H2}	−0.3968	0.0000	0.0000	0.2261	0.0000	0.0000	0.4544	−0.5381	0.5362
y_{H2}	−0.5688	0.0033	0.2356	−0.0137	0.0000	0.0000	0.5071	0.4282	−0.4152
z_{H2}	0.0000	0.2617	−0.0036	0.0000	−0.7043	0.6539	0.0000	0.0000	0.0000

[a]Eq. (2.45) is solved numerically to find the frequencies and eigenvectors. H_2O is in the x,y-plane with the O-atom at the origin. The x,y-coordinates for the H-atoms are (−0.75669, −0.58589) for H1 and (0.75669, −0.58589) for H2.

[b]The first eigenvalue, where $\nu_i = \sqrt{\lambda_i}/2\pi$. ν_i is in cm^{-1}.

[c]The eigenvector for the first eigenvalue.

where **F** is a matrix of the internal coordinate force constants and **G** is a matrix of the internal coordinate g-elements, Eq. (2.24) and table 2.1, which represent the masses. The **L**-matrix gives the transformation between normal mode and internal coordinates, as in Eqs. (2.36) and (2.37) where the q are now internal coordinates. In contrast to force constants for Cartesian coordinates, internal coordinate force constants are of chemical significance since they pertain to particular motions of a molecule. Nondiagonal internal coordinate force constants are often called "interaction terms," since they describe how the displacement of a particular internal coordinate affects the potential of the remaining coordinates. Internal coordinate quadratic force constants for H_2O and CH_4 are listed in table 2.3. The internal coordinates are depicted in figure 2.2. The nondiagonal force constants are $f_{r\bar{r}}$, $f_{r\alpha}$, and $f_{r\alpha'}$. For the latter force constant, the bond length r is not for one of the bonds which define the bend angle α'.

2.2.2.2 Rotation and vibrational/rotational motion

The simplest approach for treating the rotational motion of a molecule is to make the rigid-rotor approximation so that the third and fourth terms in Eq. (2.23) are zero. In Cartesian coordinates the second term becomes

$$T_r = \sum_{i=x,y,z} j_i^2/2I_i , \tag{2.50}$$

where the I_i are the principal moments of inertia and the j_i are the angular momenta about the Cartesian axes. The rotational Hamiltonian can also be expressed using Eulerian angles (Wilson et al., 1955), but the resulting Hamiltonian depends on whether the molecule is a spherical, symmetric, or asymmetric top. For a symmetric top the Hamiltonian is

$$T_r = \frac{p_\theta^2}{2I_A} + \frac{(p_\phi - p_\psi \cos\theta)^2}{2I_A \sin^2\theta} + \frac{p_\psi^2}{2I_C}, \tag{2.51}$$

where I_A, I_A and I_C are the three principal moments of inertia, and θ, ϕ, and ψ are the three Euler angles, with the ranges $0 \le \theta \le \pi$, $0 \le \phi \le 2\pi$, $0 \le \psi \le 2\pi$.

If the total angular momentum of the molecule is not zero, all of the terms in Eq.

Table 2.3. Internal Coordinate Force Constants for H_2O and CH_4.[a]

	H_2O[b]	CH_4[c]
rr	8.454	5.422
$r\bar{r}$	−0.101	0.0038
$\alpha\alpha$	0.761	0.5848
$r\alpha$	0.228	0.183
$r\alpha'$	——	−0.186

[a] rr and $r\bar{r}$ force constnts are in mdyn/Å, $r\alpha$, and $r\alpha'$ in mdyn/rad, and $\alpha\alpha$ in mdyn-Å/rad².

[b] B.J. Rosenberg et al. (1976).

[c] R.J. Duchovic et al. (1984).

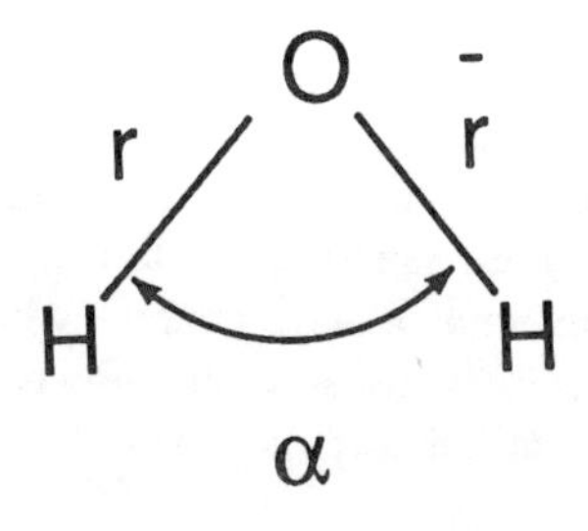

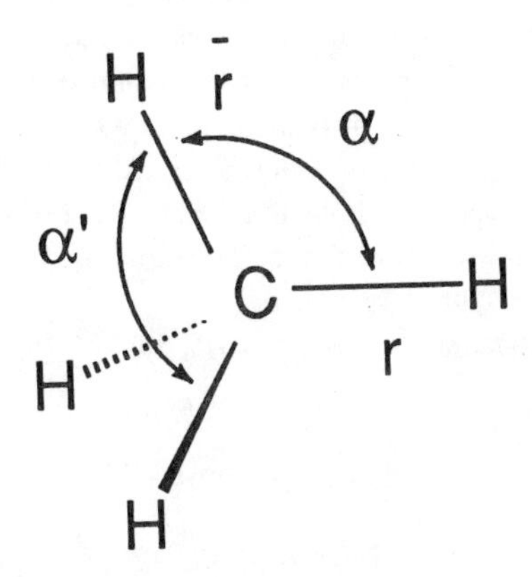

Figure 2.2 Internal coordinates for H_2O and CH_4.

(2.23) must be included in an exact calculation of simultaneous vibrational/rotational motion. Transforming the Hamiltonian in Eq. (2.23) which is for a Cartesian rotating coordinate system to a Hamiltonian of rotating internal or normal mode coordinates is a considerable task. The vibrational/rotational Hamiltonian in terms of normal mode coordinates and a rotating body-fixed axes system, chosen in accord with the Eckart (Sayvetz) conditions described above, is often called the Watson Hamiltonian (Wilson and Howard, 1936; Watson, 1968b, 1970) and is written as

$$H = \sum_k (P_k^2 + \lambda_k Q_k^2)/2 + V_{\text{anh}}(Q_1, Q_2, \ldots) + \sum_{\alpha,\beta} (j_\alpha - \pi_\alpha)\mu_{\alpha\beta}(j_\beta - \pi_\beta)/2, \tag{2.52}$$

where $\alpha(\beta) = x$, y, or z-axis, $\mu_{\alpha,\beta}$ is a component of the inverse effective moment of inertia tensor, j_α is the component of the total rotational angular momentum along the body-fixed α-axis, and π_α is the vibrational angular momentum along the α-axis which gives rise to coriolis coupling. There are two types of vibration/rotation couplings in Eq. (2.52) (Herzberg, 1945). Centrifugal coupling occurs through the coordinate dependence of the $\mu_{\alpha\beta}$ tensor. Even if the coordinate dependence of $\mu_{\alpha\beta}$ is ignored, coriolis coupling arises through the cross term $-\sum_{\alpha,\beta} \mu_{\alpha\beta} j_\alpha \pi_\beta$. If these coupling terms are removed by neglecting π_α and setting the $\mu_{\alpha\beta}$ to their values at the equilibrium geometry, the third term in Eq. (2.52) becomes equal to Eq. (2.50). This becomes apparent if the moment of inertia tensor is diagonalized, so that the nondiagonal $\mu_{\alpha\beta}$

are zero and the diagonal terms, for example, $\mu_{\alpha\alpha}$, are simply inverses of the principal moments of inertia. It is noteworthy that if the total angular momentum j equals zero, the vibrational angular momentum still contributes to the Hamiltonain through the $\pi_\alpha\mu_{\alpha\beta}\pi_\beta/2$ terms. The utility of the Watson Hamiltonian is that it has a simple expression for vibrational and rotational kinetic energy, with vibrational/rotational coupling terms which are of physical significance. Its drawback is the often exceedingly complicated expression for the anharmonic potential energy in normal mode coordinates.

The anharmonic potential energy is usually easier to represent in internal coordinates than in normal mode coordinates. However, what restricts the use of internal coordinates is the complicated expression for the vibrational/rotational kinetic energy in these coordinates (Pickett, 1972). It is difficult to write a general expression for the vibrational/rotational kinetic energy in internal coordinates and, instead, one usually considers Hamiltonians for specific molecules. For a bent triatomic molecule confined to rotate in a plane, the internal coordinate Hamiltonian is (Blais and Bunker, 1962):

$$\begin{aligned} H_{\text{vr}} &= T_{\text{v}} - p_1 p_\theta \sin\phi/2Mr_2 + p_2 p_\theta \sin\phi/2Mr_1 \\ &+ [m_1(m_2 + M)r_1^2 - m_2(m_1 + M)r_2^2]p_\phi p_\theta/2m_1 m_2 M r_1^2 r_2^2 \\ &+ [A + 4m_1 m_2 r_1 r_2 \cos\phi]p_\theta^2/8m_1 m_2 M r_1^2 r_2^2 + V(r_1, r_2, \phi) \end{aligned} \tag{2.53}$$

where T_{v} is the vibrational kinetic energy from Eq. (2.25), A is given in Eq. (2.26), and p_θ is the angular momentum. Equation (2.53) shows that p_θ couples with both the internal coordinates and momenta. The second, third, and fourth terms couple the internal coordinate momenta p_1, p_2, and p_ϕ with the rotational angular momentum. If the molecule were allowed to rotate about all three of its axes, many additional coupling terms would be added to Eq. (2.53). Internal coordinate Hamiltonians have been presented for three and four atoms with complete three-dimensional vibrational/rotational coupling (Bramley et al., 1991; Bramley and Handy, 1993). In comparing Eqs. (2.52) and (2.53), the physical picture of vibration/rotation coupling is different for the internal coordinate and normal mode Hamiltonians, and cannot be described in terms of coriolis and centrifugal couplings when using the internal coordinate Hamiltonian. Finally, since the coordinate θ does not appear explicitly in Eq. (2.53), p_θ is a constant of the motion. (The angular variable θ is measured from a fixed reference line in the plane in which the molecule is rotating.)

2.3 ENERGY LEVELS

Vibrational, rotational, and vibrational/rotational energy levels are found by first transforming the classical Hamiltonians described in the previous section to the appropriate quantum mechanical operator $\hat{H}$. The eigenvalue equation

$$\hat{H}\psi = E\psi \tag{2.54}$$

is then solved to obtain the energy levels and corresponding wave functions. In the following, solutions to Eq. (2.54) are described for both diatomic and polyatomic molecules

2.3.1 Diatomic Molecule

If p_θ in Eq. (2.15) is set equal to zero, the Hamiltonian represents just the vibrational energy for the diatomic molecule. The quantum mechanical Hamiltonian operator for this vibrational motion is

$$\hat{H} = \frac{-\hbar^2}{2\mu}\frac{\partial^2}{\partial r^2} + V(r) \tag{2.55}$$

and the eigenvalue equation, Eq. (2.54), must be solved to find the vibrational energy levels. In general analytic solutions are not obtained. However, both the harmonic oscillator potential

$$V(r) = f\,\Delta r^2/2 \tag{2.56}$$

and anharmonic Morse potential

$$V(r) = D[1 - e^{-\beta\Delta r}]^2 \tag{2.57}$$

give analytic expressions for the energy levels. The force constant f, bond dissociation energy D and exponential term β are related by $f = 2\,D\,\beta^2$. For the harmonic oscillator the energy levels are given by

$$E_n = (n + \tfrac{1}{2})\,h\nu_e, \tag{2.58}$$

where $n = 0,1,2, \ldots$ and ν_e, the harmonic vibrational frequency, is $(f/\mu)^{1/2}/2\pi$. The energy levels for the Morse potential are

$$E_n = (n + 1/2)\,h\nu_e - (n + \tfrac{1}{2})^2\,h\nu_e\chi_e, \tag{2.59}$$

where $\nu_e = \beta(2D/\mu)^{1/2}/2\pi$ and $\chi_e = h\nu_e/4D$.

Rigid-rotor rotational energy levels are found from the Hamiltonian in Eq. (2.15) by fixing r at the equilibrium bond length $r = r_e$ so that H becomes $p_\theta^2/2\mu r_e^2$. Transforming H to the Hamiltonian operator for rotational motion and solving Eq. (2.54) gives

$$E_J = J(J + 1)/\hbar^2/2I_e \tag{2.60}$$

for the rotational energy levels, where the quantum number $J = 0,1,2, \ldots,$ the moment of inertia I_e equals μr_e^2 and each level has a $2J + 1$ degeneracy. Combining either Eq. (2.58) or (2.59) with Eq. (2.60) gives approximate harmonic and anharmonic vibrational/rotational energy levels for the diatomic molecule, respectively. Since the actual moment of inertia changes as the molecule vibrates (i.e., r changes), a more accurate expression includes *vibrational/rotational coupling*. Also, the rotational motion of the molecule causes the bond to stretch, a *centrifugal distortion* effect. Using second-order perturbation theory to include such terms, yields the following expression for the anharmonic vibrational/rotational energy levels:

$$E_{nj} = (n + \tfrac{1}{2})\,h\nu_e - (n + \tfrac{1}{2})^2 h\nu_e x_e + J(J + 1)\hbar^2/2I_e$$
$$- h\alpha_e(n + \tfrac{1}{2})J(J + 1) - h\,C_e J^2(J + 1)^2. \tag{2.61}$$

The fourth and fifth terms in this expression represent the vibrational/rotational coupling and the centrifugal distortion, respectively.

2.3.2 Polyatomic Molecules

2.3.2.1 Vibrational energy levels

The Hamiltonian operator for the normal mode Hamiltonian in Eq. (2.47) is obtained by replacing P_k with the operator $-i\hbar\ \partial/\partial Q_k$. Hence

$$\hat{H}_v = \sum_k \left(\frac{-\hbar^2}{2} \frac{\partial^2}{\partial Q_k^2} + \frac{\lambda_k}{2} Q_k^2 \right). \tag{2.62}$$

Solving the eigenvalue equation, Eq. (2.54), for this operator gives the normal mode energy levels

$$E_v = \sum_i (n_i + 1/2) h\nu_i \tag{2.63}$$

and wave function

$$\psi = \prod_i \chi_i(n_i) \tag{2.64}$$

which is a product of harmonic oscillator wave functions for the individual normal modes. Equations (2.63) and (2.64) are valid at low total energies, so that the vibrational Hamiltonian is nearly separable in normal mode coordinates and the vibrational wave functions are well approximated by the product of normal mode wave functions. Thus, one can associate specific structural attributes to an energy level and assign the level names such as stretching, bending, rocking, etc. Each ψ has a different probability amplitude (given by ψ^2) on the potential energy surface. The states have easily identifiable spectral patterns and progressions which are evident to any trained spectroscopist.

In table 2.4, a comparison is made between harmonic and anharmonic frequencies determined for C_2H_4 from experiment. The anharmonic frequencies, which are given by the $n = 0 \rightarrow 1$ transitions for the vibrational modes, are 2–4% smaller than the harmonic frequencies. This small error is typical for molecules at low vibrational energies. However, for more highly excited vibrational levels the accuracy of the normal mode model decreases.

At higher levels of excitation anharmonicity has to be included to obtain accurate energy levels. Perturbation theory has been used to derive the following expression, often called a Dunham expansion (Hirst, 1985), for polyatomic anharmonic vibrational energy levels, which is similar to the Morse energy level expression Eq. (2.59), for a diatomic molecule:

$$E(\text{cm}^{-1}) = \sum_i \omega_i (n_i + \tfrac{1}{2}) + \sum_i \sum_{k \geq i} x_{ik} (n_i + \tfrac{1}{2})(n_k + \tfrac{1}{2}) \tag{2.65}$$

This equation assumes there are individual vibrational modes with identifiable quantum numbers n_i, vibrational frequencies ω_i, and anharmonicities x_{ik} for an excited molecule, in the same way quantum numbers and vibrational frequencies can be

Table 2.4. Experimental Harmonic and Anharmonic Vibrational Frequencies for C_2H_4.[a]

	Harmonic	Anharmonic
ν_1	3152	3026
ν_2	1655	1630
ν_3	1370	1342
ν_4	1044	1023
ν_5	3232	3103
ν_6	1245	1220
ν_7	969	949
ν_8	959	940
ν_9	3234	3105
ν_{10}	843	826
ν_{11}	3147	3021
ν_{12}	1473	1444

[a]The frequencies are in cm^{-1} and are taken from Duncan et al. (1973).

assigned to modes in the normal mode model; that is, Eqs. (2.63) and (2.64). However, at higher energies this equation may not fit the vibrational energy levels (Reisner et al., 1984; Abramson et al., 1985), and another approach is needed for calculating the energy levels.

The variational method (Carney et al., 1978) is a general approach for calculating vibrational energy levels. The wave function for a vibrational energy level Ψ_n is written as a linear combination of basis function ψ_i,

$$\Psi_n = \sum_i c_{in}\psi_i \tag{2.66}$$

The energies and wave functions for the energy levels are calculated by requiring the energy,

$$\int \Psi_n^* \hat{H}_v \Psi_n d\tau / \int \Psi_n^* \Psi_n d\tau , \tag{2.67}$$

to be stationary with respect to the variational parameters c_{in}, which gives an upper limit to the energy. This gives rise to a set of homogeneous simultaneous equations, secular equations, which involve the following integrals over the basis functions:

$$H_{ij} = \int \psi_i \hat{H}_v \psi_j d\tau . \tag{2.68}$$

If an exact vibrational Hamiltonian is used (e.g., either the internal coordinate, Eq. (2.25), or normal mode, Eq. (2.52) Hamiltonian), the accuracy of the calculated results only depends on the treatment of anharmonicity in the potential energy function. The variational calculation is not restricted to a particular type of basis functions; for example, either normal mode or internal coordinate bases can be used. However, one wants to choose a basis which simplifies the evaluation of the integrals in Eq. (2.68). Thus, if the calculations are performed with the normal mode Hamiltonian, a normal mode basis is preferred. Normal mode and internal coordinate basis sets are usually centered at the molecule's equilibrium geometry. In recent work Bačić and Light, 1986; Bačić and Light, 1989; Choi and Light, 1992) a discrete variable representation, DVR, basis has been advanced for calculating vibrational energy levels. This is a particularly useful approach for treating large-amplitude motion. Whatever basis set is used, its size is increased until the calculated vibrational energy levels converge to within a fixed criterion; for example, 0.01 cm^{-1}.

If there is a principal ψ_i in the linear combination, Eq. (2.66), which gives the wave function Ψ_n for the energy level, the quantum numbers for this ψ_i define the "almost good" quantum numbers for Ψ_n. For example, at low levels of excitation, where the vibrational Hamiltonian is nearly separable in normal mode coordinates and forms a good zero-order Hamiltonian, there will be only one principal normal mode basis function in the linear combination. However, at higher levels of excitation where anharmonic couplings between the normal mode coordinates become possible, there may not be one principal normal mode basis function in ψ_n. As a result, it may be impossible to assign normal mode quantum numbers to the energy levels. If this is the case it is useful to determine whether basis functions associated with another type of zero-order Hamiltonian, would give assignable wave functions. (Note, if bases of sufficient size are used, changing the basis functions for the calculation will not affect the calculated energy levels, but will alter the coefficients c_{in}, Eq. (2.66), for the energy level's wave function). If there is no zero-order Hamiltonian and associated basis functions which yield one principal ψ_i for each Ψ_n, the vibrational energy levels are said to be *intrinsically unassignable*. In the last section of this chapter the concept of intrinsic unassignability is discussed in terms of regular and irregular spectra.

The result of a variational calculation of the formyl radical HCO vibrational energy levels is given in table 2.5 (Cho et al., 1990). An internal coordinate Hamiltonian and internal coordinate basis functions were used in this calculations. The bases

Table 2.5. Variational Calculation of Vibrational States for the Formyl Radical.

Energy, cm^{-1}	Wave Function[a]
2815.13	$-0.9962\vert 000\rangle +0.0506\vert 110\rangle +0.0424\vert 100\rangle$
3896.22	$0.9808\vert 001\rangle +0.1462\vert 010\rangle -0.0784\vert 101\rangle$
4689.41	$0.9803\vert 010\rangle -0.1477\vert 001\rangle -0.0690\vert 120\rangle$
4966.72	$-0.9579\vert 002\rangle -0.1986\vert 011\rangle +0.1140\vert 102\rangle$
5227.70	$0.9670\vert 100\rangle -0.1531\vert 200\rangle +0.1134\vert 002\rangle$
5767.13	$-0.9411\vert 011\rangle -0.2085\vert 020\rangle +0.1982\vert 002\rangle$
6025.14	$0.9216\vert 003\rangle +0.2297\vert 012\rangle +0.2129\vert 101\rangle$
6245.86	$0.9196\vert 101\rangle -0.2167\vert 003\rangle -0.2123\vert 201\rangle$
6537.25	$0.9638\vert 020\rangle -0.2095\vert 011\rangle -0.0814\vert 130\rangle$
6833.82	$-0.8966\vert 012\rangle -0.2789\vert 021\rangle +0.2277\vert 003\rangle$
7068.20	$0.8479\vert 004\rangle +0.3427\vert 102\rangle +0.2389\vert 013\rangle$
7090.64	$0.9134\vert 110\rangle -0.1543\vert 210\rangle -0.1483\vert 200\rangle$
7220.80	$0.7504\vert 200\rangle +0.3868\vert 102\rangle -0.2676\vert 300\rangle$
7267.20	$0.7451\vert 102\rangle -0.4210\vert 200\rangle -0.2810\vert 004\rangle$
7611.67	$0.8996\vert 021\rangle -0.2767\vert 012\rangle +0.2575\vert 030\rangle$

[a]Results are for a calculation with a 10 × 9 × 10 = 900 basis set. The largest three coefficients in the eigenvector are listed. The notation $|n_1n_2n_3\rangle$ indicates n_1 quanta in the H—C stretch mode, n_2 quanta in the C═O stretch mode, and n_3 quanta in the H—C═O bending mode. Calculations by Cho et al. (1990).

consists of Morse stretch basis functions for the first 10 C—H stretch internal coordinate energy levels, Morse stretch basis functions for the first 9 C=O stretch internal coordinate energy levels, and harmonic basis functions for the first 10 H—C=O bend energy levels. For the lowest energy (zero-point) level the coefficient is -0.9962 for the $|000>$ basis function. The most highly mixed levels are at 7090.6 and 7267.2 cm^{-1}. Quantum numbers can be assigned to each of the levels. A survey of vibrational variational calculations for different molecules is given in table 2.6.

2.3.2.2 Rotational and vibrational/rotational energy levels

To determine the quantum mechanical rigid-rotor energy levels, the quantum mechanical Hamiltonian operator is formed from the classical Hamiltonian in Eq. (2.50) and the eigenvalue equation, Eq. (2.54), is solved. For a symmetric top rigid-rotor, which has two equal moments of inertia (i.e., $I_a = I_b \neq I_c$), the resulting energy levels are

$$E_{J,K} = [J(J+1) - K^2]\hbar^2/2I_a + K^2\hbar^2/2I_c, \tag{2.69}$$

where $J = 0,1,2, \ldots$ and $K = -J, -(J-1), \ldots 0, \ldots, J-1, J$. For "almost symmetric tops" which have $I_a \approx I_b$, one can often use the approximation (Townes and Schalow, 1955)

$$E_{J,K} = \left(\frac{1}{I_a} + \frac{1}{I_b}\right)\frac{[J(J+1) - K^2]\hbar^2}{4} + \frac{K^2\hbar^2}{2I_c}. \tag{2.70}$$

Approximate vibrational/rotational energy levels for symmetric top and "almost symmetric top" molecules are found by combining Eq. (2.65) with either Eq. (2.69) or Eq. (2.70). This approach is incomplete since it neglects vibrational/rotational coupling and centrifugal distortion, which for a diatomic are the fourth and fifth terms in Eq. (2.61). Using perturbation theory, Watson (1968a) has derived an analytic expression for the vibrational/rotational levels of a polyatomic molecule (Toselli and Barker, 1989). The expression includes anharmonic, coriolis, and centrifugal distortion terms and expresses the energy in terms of the vibrational quantum numbers n_i and the

Table 2.6. Variational Calculations of Vibrational and Vibrational-Rotational Energy Levels.

Molecule	Hamiltonian	Levels	Reference
H_2O	Radau	Vib.	Choi and Light (1992)
SO_2,H_2CO	Internal Coordinate[a]	Vib.-Rot.	McCoy et al. (1991)
LiCN/LiNC	Jacobi	Vib.	Bačić and Light (1986)
HCN/HNC	Jacobi	Vib.	Bačić and Light (1987)
KCN	Jacobi	Vib.-Rot.	Tennyson and Sutcliff (1982)
H_2CO	Watson	Vib.-Rot.	Maessen and Wolfsberg (1985)
CH_3F	Normal Mode	Vib.	Dunn et al. (1987)
H_2CO	Watson	Vib.	Romanowski et al. (1985)
H_2CO	Watson	Vib.	Maessen and Wolfsberg (1984)
HCO	Internal Coordinate	Vib.	Cho et al. (1990)
HCO	Watson	Vib.	Bowman et al. (1986)
C_2H_2	Internal Coordinate	Vib.	McCoy and Sibert (1991)
C_2H_2	Internal Coordinate	Vib.	Bramley and Handy (1993)
C_2H_2	Internal Coordinate	Vib.	Sibert and Mayrhofer (1993)

[a]Canonical Van Vleck perturbation theory and variational calculation.

rotational quantum numbers J and K. However, even this expression breaks down at sufficiently high energy and angular momentum, because it does not include all anharmonic and vibrational/rotational coupling terms.

Vibration/rotation energy levels can always be determined with the use of variational procedures. In this approach the wave function Ψ_J for a given vibration-rotation level is written as a linear combination of basis functions $\psi_i R_{JK}$

$$\Psi_J = \sum_i \sum_K c_{iK}^J \psi_i R_{JK} , \tag{2.71}$$

where ψ_i is a suitable vibrational wave function (e.g., the normal mode wave function in Eq. (2.64)), R_{JK} is a symmetric top wave function, and c_{iK}^J is a variational parameter. Vibrational/rotational coupling will mix the $(2J + 1)$ K-levels for a particular J. If this coupling is extensive so that all K-levels are mixed, K will no longer be a meaningful quantum number for an energy level. Variational calculations of vibrational/rotational energy levels for several molecules have provided insight into the importance of K-mixing and its relationship to anharmonic coupling (McCoy et al., 1991; Burleigh and Sibert, 1993).

As discussed above, anharmonic coupling in the potential can make it impossible to assign vibrational quantum numbers to an energy level. Thus, at levels of excitation where there is extensive anharmonic and vibrational/rotational coupling it may only be possible to identify an energy level by its energy (i.e., position in the spectrum) and its total angular momentum J. However, regardless of the assignability or unassignability of the energy levels, below the unimolecular threshold the spectrum is discrete, with individual lines only broadened by their coupling with the radiation field.

2.3.3 Semiclassical Calculation of Vibrational/Rotational Energy Levels

Another approach for finding vibrational energy levels is to apply the semiclassical quantization condition to classical trajectories determined by solving Hamilton's equations of motion, Eq. (2.9). This approach originates from critical insights by Einstein (1917) and is usually referred to as Einstein-Brilloum-Keller (EBK) semiclassical quantization (Percival, 1977; Noid et al., 1981). It can be derived by either applying quantization conditions to classical mechanics or by taking the Wentzel-Kramer-Brillouin (WKB) semiclassical limit of quantum mechanics (Schiff, 1968). For a diatomic molecule with no angular momentum, so that H in Eq. (2.15) becomes $H = p_r^2/2\mu + V(r)$, the semiclassical quantization condition is that $\int\int dp_r\, dr$ integrated over one vibrational period is $(n + \frac{1}{2})h$, where n is an integer and equals 0,1,2, ... A more convenient notation for this condition is

$$\oint p_r dr = (n + \tfrac{1}{2})h , \tag{2.72}$$

where the cyclic integral denotes integration over one period (i.e., orbit). The integral on the left side of Eq. (2.72) is called the action integral. The fraction $\frac{1}{2}$ in Eq. (2.72), is the Maslov index and is found by a detailed semiclassical study of the classical motion (Gutzwiller, 1990). Orbits for harmonic oscillator energy levels, Eq. (2.58), are shown in figure 2.3.

To find the semiclassical vibrational energy levels for a diatomic, Eq. (2.72) is written as

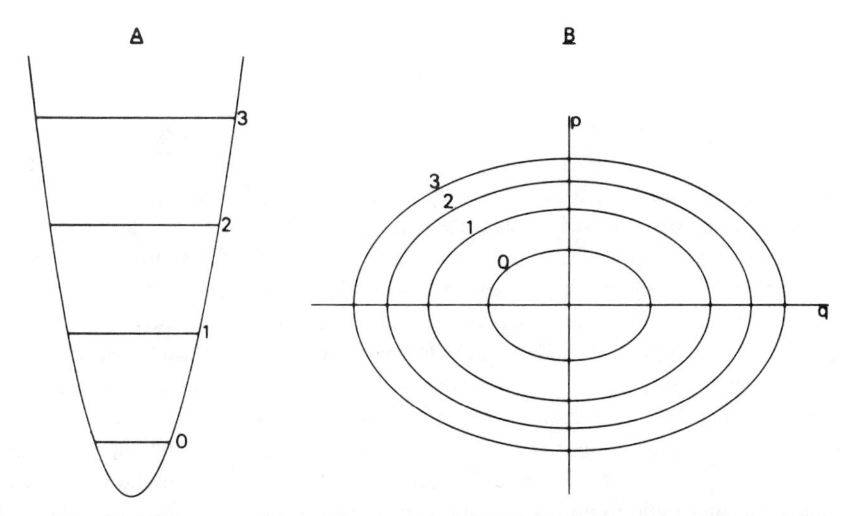

Figure 2.3 (a) The $n = 0, 1, 2$, and 3 energy levels for a harmonic oscillator. (b) Phase space volumes for the $n = 0, 1, 2$, and 3 harmonic oscillator energy levels.

$$(2\mu)^{1/2}\oint[E - V(r)]^{1/2} = (n + 1/2)h \tag{2.73}$$

and the energy E is varied until the equality in Eq. (2.73) is realized for integer values of n. For both a harmonic oscillator, Eq. (2.56), and Morse oscillator, Eq. (2.57), Eq. (2.72) is solvable analytically, and the resulting expressions for E are the same as those found quantum mechanically; Eqs. (2.58) and (2.59). It is straightforward to show that this is the case for a harmonic oscillator. By replacing the cyclic integral in Eq. (2.72) by the double integral over p_r and r, and using Hamilton's equations of motion to equate $dp_r dr$ to $dHdt$, Eq. (2.72) becomes

$$\int\int dHdt = (n + \tfrac{1}{2})h . \tag{2.74}$$

Since the integration limits are 0 to E for H and 0 to $1/\nu$ for t, Eq. (2.74) becomes $E = (n + \frac{1}{2})h\nu$. Analytic agreement between semiclassical and quantum vibrational energy levels for a diatomic molecule has only been found for the harmonic and Morse potentials. Though these energy levels are not in exact agreement for other potentials, they usually do agree to within experimental precision.

Vibrational/rotational levels for diatomic molecules are found by replacing p_θ^2 in Eq. (2.15) with its quantum equivalent $J(J + 1)\hbar^2$, so that Eq. (2.73) becomes

$$(2\mu)^{1/2}\oint[E - J(J + 1)\hbar^2/2\mu r^2 - V(r)]^{1/2} = (n + \tfrac{1}{2})h . \tag{2.75}$$

Solving this equation gives the energy for integer values of n and J. As discussed in chapter 3, inverting Eq. (2.75) is the Rydberg-Klein-Rees (RKR) procedure (Hirst, 1985) for determining diatomic potential energy functions from experimental energy levels.

Since the normal mode Hamiltonian for a polyatomic molecule, Eq. (2.47), is separable into a Hamiltonian for each normal mode, the semiclassical quantization in Eq. (2.72) may be applied independently to each normal mode. The resulting semi-classical expression for the normal mode energy levels is the same as that obtained from quantum mechanics, Eq. (2.63). If anharmonicity is included in the normal mode Hamiltonian, as in Eq. (2.48), the Hamiltonian remains separable if the anharmonicity does not couple the normal modes. Thus, the semiclassical energy is obtained by applying the condition of Eq. (2.72) to each anharmonic normal mode. For example, if the anharmonicity is given by the Morse function, Eq. (2.57), the semiclassical normal mode energy is a sum of Morse energy level terms, Eq. (2.59), for the normal modes.

Anharmonicity can also couple the modes of the molecule, so that the separable normal mode coordinates can no longer be used to obtain the semiclassical energy levels. Studies have shown that a molecule with coupled modes can have two types of classical vibrational motion: quasiperiodic and chaotic. Quasiperiodic motion is regular, like that of the normal-mode model. The vibrational motion can be represented by a superposition of individual vibrational modes each containing a fixed amount of energy. For some cases these modes resemble normal modes, but they may be identifiably different. Thus, although molecular Hamiltonians contain potential and kinetic energy coupling terms, they may still exhibit regular vibrational motion. Coexisting with the quasiperiodic motion is chaotic (i.e., irregular) motion, for which the molecule does not exhibit the regular vibrational motion as predicted by the separable normal mode model. Figure 2.4 depicts quasiperiodic and chaotic motion for a collinear HCC model molecule with H—C(R_1) and C—C(R_2) stretching degrees of freedom. For most molecular systems the vibrational motion becomes increasingly more chaotic as the energy is increased.

EBK semiclassical quantization is applicable if the classical motion is quasiperiodic. For N coupled modes, the quantum conditions are

$$\int_{C_i} \sum_{k=1}^{N} p_k dq_k = (n_i + \tfrac{1}{2})h \qquad (i = 1 \text{ to } N)\,, \tag{2.76}$$

where the n_i are integers, and the q_k and p_k are the coordinates and their conjugate momenta. The C_i are topologically independent closed paths, which need not be along actual trajectories. The first step in applying the quantum conditions in Eq. (2.76) is the identification of the C_i, which is illustrated here for a model two degree of freedom Hamiltonian. A two mode model with x,p_x and y,p_y has three independent coordinates and momenta at a fixed energy. Thus, if, during the course of a classical trajectory, a point is plotted in the x,p_x-plane every time $y = 0$ for $p_y > 0$, an "invariant curve" will result if the classical motion is quasiperiodic (i.e., regular). Such curves are called Poincare surfaces of section and are illustrated in figure 2.5 (in the figure x and y are identified by 1 and 2). Semiclassical eigenvalues are obtained by simultaneously satisfying the conditions

$$\begin{aligned} \oint p_x dx &= (n_x + \tfrac{1}{2})h \qquad (y = 0) \\ \oint p_y dy &= (n_y + \tfrac{1}{2})h \qquad (x = 0)\,. \end{aligned} \tag{2.77}$$

Figure 2.4 Two trajectories for a model HCC Hamiltonian described in Hase (1982). Top trajectory is for $n_{HC} = 0$ and $n_{CC} = 2$, and is quasiperiodic. Bottom trajectory is for $n_{HC} = 5$ and $n_{CC} = 0$, and is chaotic. R_1 is the HC bond length and R_2 the CC bond length. Distance is in angstroms (Å).

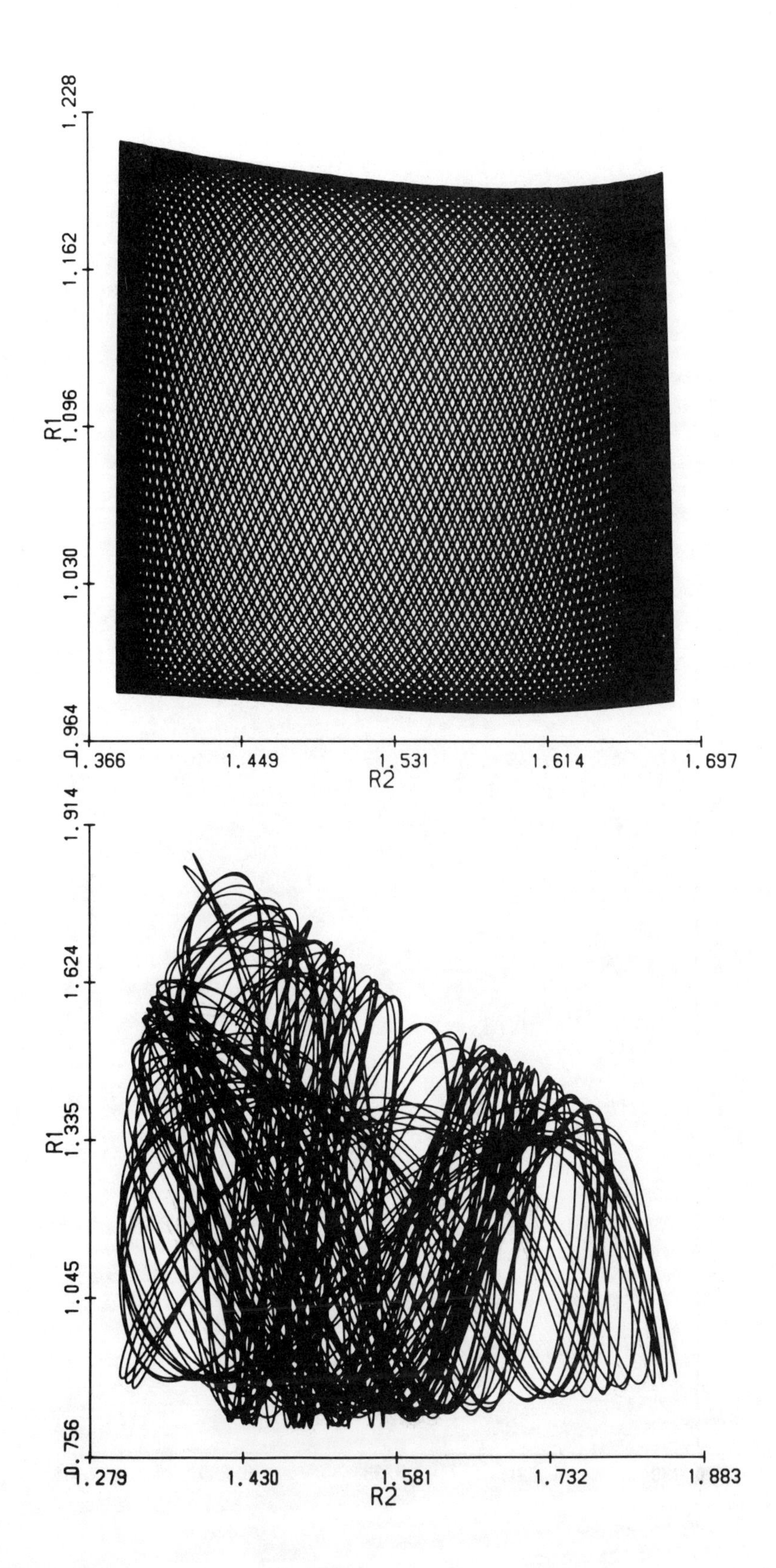

R1
1.228
1.162
1.096
1.030
0.964
1.366
1.449
1.531
1.614
1.697
R2
R1
1.914
1.624
1.335
1.045
0.756
1.279
1.430
1.581
1.732
1.883
R2

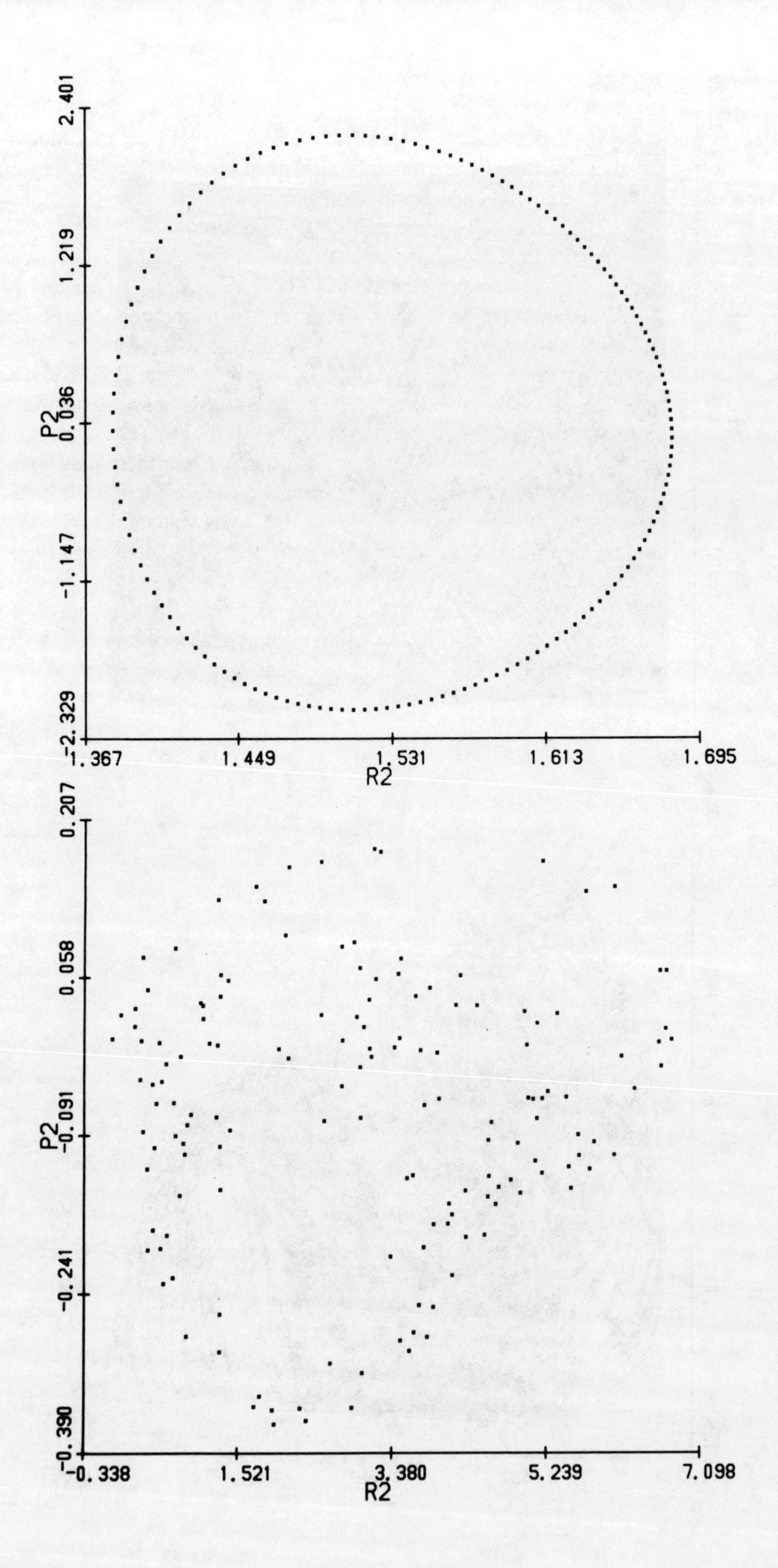

P2
2.401
1.219
0.036
-1.147
-2.329
1.367
1.449
1.531
1.613
1.695
R2
P2
0.207
0.058
-0.091
-0.241
-0.390
-0.338
1.521
3.380
5.239
7.098
R2

For actual molecules with three or more vibrational modes methods more sophisticated than the Poincare surface of section approach must be used to solve the semiclassical quantization conditions, and several approaches have been advanced (Eaker et al., 1984; Martens and Ezra, 1985; Skodje et al., 1985; Johnson, 1985; Duchovic and Schatz, 1981; Martens and Ezra, 1987; Pickett and Shirts, 1991). Semiclassical vibrational energy levels have been determined for SO_2, H_2O, H_3^+, CO_2, Ar_3, and I_2Ne_n. Semiclassical wave functions have also been determined for vibrational energy levels of molecules (DeLeon and Heller, 1984).

A surface of section for a chaotic trajectory is shown in figure 2.5. For such a case, the invariant curves C_i (i.e., tori) do not exist and EBK semiclassical quantization cannot be applied. Einstein was the first to point out this difficulty. Thus, though the energy levels can be obtained by pure quantum mechanical methods, they may be impossible to find by the semiclassical prescription if the classical motion is chaotic. As a result the classical-quantal correspondence becomes somewhat unclear. However, by use of new theoretical approaches, such as the trace formula (Gutzwiller, 1990), it may be possible to determine semiclassical energy levels and wave functions from chaotic motion. A qualitative correspondence which does exist is that the classical motion for an energy level is chaotic if the quantum mechanical wave function is highly mixed and unassignable. This is considered in more detail in the next section.

2.3.4 Regular and Irregular Spectra

The ability to assign a group of vibrational/rotational energy levels implies that the complete Hamiltonian for these states is well approximated by a zero-order Hamiltonian which has eigenfunctions $\psi_i(m)$. The ψ_i are product functions of a zero-order orthogonal basis for the molecule, or, more precisely, product functions in a natural basis representation of the molecular states, and the quantity m represents the quantum numbers defining ψ_i. The wave functions Ψ_n are given by

$$\Psi_n = \sum_i c_{in}\psi_i(m) . \tag{2.78}$$

Energy levels in the spectra, which are assignable in terms of the zero-order basis $\psi_i(m)$, will have a predominant expansion coefficient. Hose and Taylor (1982) have argued that for an assignable level there should be some representation for which $c_{in}^2 >$ 0.5 for one of the c_{in}. For some cases, more than one zero-order Hamiltonian may be necessary to assign energy levels. Spectra that can be assigned approximate quantum numbers are called *regular spectra*.

As the energy is increased, mode couplings come into play, and the spectrum may lose its simple regular structure (Nordholm and Rice, 1974). If the couplings are extensive, a zero-order Hamiltonian and basis cannot be found to represent the Ψ_n for these states. The spectrum for these states is irregular without patterns. For the most

---

Figure 2.5 Poincaré surfaces of section in the P_2, R_2-plane for the trajectories in figure 2.4. The area inside the closed curve of the periodic trajectory (top) is 5/2 h. The bottom trajectory is chaotic.

statistical (i.e., nonseparable) situation, the expansion coefficients in Eq. (2.78) are random variables, subject only to the normalization and orthogonality conditions

$$\sum_i c_{in}^2 = 1 \quad \text{and} \quad \sum_i c_{in} c_{im} = 0 . \tag{2.79}$$

The distributions of nearest neighbor energy level spacings differ for regular and irregular spectra (Buch et al., 1982; Matsushita and Terasaka, 1984; Terasaka and Matsushita, 1985; Zimmermann et al., 1987). For regular spectra, the energy level positions are related to a zero-order Hamiltonian and are well approximated by a total energy E which is a sum of individual mode energies, that is, $E = E_1 + E_2 + E_3. . . .$ For such a system it has been demonstrated (Brody, 1981; Berry and Tabor, 1977) that the energy levels behave as a random sequence, since there is no correlation between the individual mode energies. Such a spectrum may be constructed by ordering a sequence of uniformly distributed random numbers. For the levels one finds the Poisson, or random, distribution

$$P(s) = \exp\left(-s/\langle s\rangle\right)/\langle s\rangle, \tag{2.80}$$

where $\langle s\rangle$ is the average spacing between the energy levels. The Poisson distribution peaks at zero spacing, that is, the levels tend to cluster.

For an irregular spectrum there is strong mixing in any zero-order representation and the level energy is no longer a sum of individual mode energies. The energy levels may be calculated by diagonalizing a matrix whose elements are coupling integrals such as the one in Eq. (2.68). For any reasonable basis the elements of this matrix are correlated and not random. However, at high energies it is found (Burleigh and Sibert, 1993) that the distribution of energy level spacings $P(s)$ is the same as is found when diagonalizing a matrix of elements which are independent and truly random. A simple and widely studied matrix of this type is a Gaussian orthogonal ensemble (GOE), whose matrix elements are real Gaussian random variables. From numerical studies of such random matrices (Zimmermann et al., 1987; Buch et al., 1982), it was found that the energy level spacing is very closely approximated by the Wigner surmise (Wigner, 1957):

$$P(s) = \pi s \exp\left(-\pi s^2/4\langle s\rangle^2\right)/2\langle s\rangle^2. \tag{2.81}$$

Thus, for an irregular spectrum the distribution of energy level spacings no longer peaks at zero as for a regular system. Couplings between the levels of the zero-order Hamiltonian cause the levels to repel (Wigner, 1957). Spectra which are neither regular nor irregular, but intermediate between these two limiting cases, can be fit by a weighted superposition of the above Poisson and Wigner distributions, or by the distribution function proposed by Brody (1981).

Distributions of nearest neighbor energy level spacings have been determined from quantum mechanical variational calculations of vibrational energy levels. The energy spacing distributions are Wigner for KCN/KNC (Tennyson and Farantos, 1984) and LiCN/LiNC (Farantos and Tennyson, 1985; Henderson and Tennyson, 1990), while Poisson for HCN/HNC (Bačić and Light, 1987). The H_2O energy level spacing distribution shows a mild inclination toward the characteristics of the Wigner distribution as the energy range increases (Choi and Light, 1992). Vibrational energy levels for an HCC Hamiltonian are assignable, but the distribution of nearest neighbor energy spacings is not clearly Poisson (Hase et al., 1989). This is also the case for the uncoupled HCC Hamiltonian. For H_2CO the energy spacing distribution is intermediate between the Poisson and Wigner distributions (Burleigh and Sibert, 1993).

A number of studies have shown there is a qualitative correspondence between the classical vibrational/rotational motion of a molecule and the nature of its vibrational/rotational energy levels. In general, energy levels are assignable and the spectrum regular if the classical motion is quasiperiodic (i.e., regular). Energy levels apparently unassignable are associated with classical chaotic (i.e., irregular) motion. An exception to the above correspondence is the finding of chaotic classical trajectories for energy levels which are assignable with regular wave functions (Shaprio et al., 1984; Patterson 1985; Bai et al., 1985). This occurs when the trajectory traverses chaotic regions of phase space which have volumes smaller than $\hbar$ and, thus, are not relevant to quantum mechanics.

2.3.5 Experimental Studies

Recent experimental studies (Reisner et al., 1984; Abramson et al., 1985; Smith et al., 1987) have focused on the high-resolution spectroscopy of vibrationally excited molecules, and these studies have provided valuable insight into the nature of vibrational and vibrational/rotational energy levels as the energy is increased. Two of the most extensively studied molecules are acetylene and formaldehyde, and the properties of their spectra are summarized below.

For energies $\sim$9,500 cm^{-1} above the vibrational ground state, vibrational/rotational spectra for acetylene are assignable, and can be fit to an expression similar to Eq. (2.65) to determine potential parameters like ω_i and x_{ik} (Abramson et al., 1985). This type of behavior extends to 11,400–15,700 cm^{-1} above the zero-point level (Chen et al., 1990). Here there appears to be some mixing between zero-order vibrational levels, but without extensive vibrational/rotational coupling. However, spectra taken at higher energies near 27,900 cm^{-1} are much different, and appear to be intrinsically unassignable and cannot be fit by Eq. (2.65) (Sundberg et al., 1985). An analysis of spacings between adjacent lines at these higher energies shows that they approximately follow a Wigner distribution as expected for an irregular spectrum. From the number of lines observed in the spectrum it is concluded that about 70–80% of the accessible zero-order states mix in forming the exact eigenstates. It is also of interest that the energy level density at $\sim$27,900 cm^{-1} is approximately 5–6 times greater (Abramson et al., 1985) than that predicted from Eq. (2.65) using the parameters fit to the spectra at $\sim$9,500 cm^{-1}.

The transition from regular to completely irregular spectra for acetylene may occur over a small energy range. Though completely irregular spectra are found near 27,900 cm^{-1}, there is a transition to mixed regular/irregular spectra by only lowering the energy by $\sim$1,400 cm^{-1} to 26,500 cm^{-1}.

The rotationless vibrational levels of formaldehyde can each be assigned a set of normal mode quantum numbers and fit to Eq. (2.65) up to 9,300 cm^{-1} above the zero-point level (Reisner et al., 1984). At higher J and K values the spectra become more complex as a result of rotation-induced mixing of anharmonic vibrational basis functions, which compromises the "goodness" of both vibrational and K quantum numbers (Dai et al., 1985a). However, the mixing between zero-order basis functions is not complete (Dai et al., 1985b). At ~28,000 cm^{-1}, near the energy of the $H_2CO \rightarrow H_2 + CO$ potential energy barrier, mixing between the vibrational basis functions seems to be complete (Polik et al., 1990). However, there is still some question regarding the extent to which the $(2J + 1)$ K-levels are mixed (Hernandez et al., 1993).

REFERENCES

Abramson, E., Field, R.W., Imre, D., Innes, K.K., and Kinsey, J.L. (1985). *J. Chem. Phys.* **83**, 453.

Bačić, Z., and Light, J.C. (1986). *J. Chem. Phys.* **85**, 4594.

Bačić, Z., and Light, J.C. (1987). *J. Chem. Phys.* **86**, 3065.

Bačić, Z., and Light, J.C. (1989). *Annu. Rev. Phys. Chem.* **40**, 469.

Bai, Y.Y., Hose, G., Stefanski, K., Taylor, H.S. (1985). *Phys. Rev. A* **31**, 2821.

Berry, M.V., and Tabor, M. (1977). *Proc. R. Soc. London Ser. A.* **356**, 375.

Blais, N.C., and Bunker, D.L. (1962). *J. Chem. Phys.* **37**, 2713.

Bowman, J.M., Bittman, J.S., and Harding, L.B. (1986). *J. Chem. Phys.* **85**, 911.

Bramley, M.J., Green, W.H., and Handy, N.C. (1991). *Mol. Phys.* **73**, 1183.

Bramley, M.J., and Handy, N.C. (1993). *J. Chem. Phys.* **98**, 1378.

Brody, T.A. (1981). *Rev. Mod. Phys.* **53**, 385.

Buch, V., Gerber, R.B., and Ratner, M.A. (1982). *J. Chem. Phys.* **76**, 5397.

Burleigh, D.C., and Sibert, III, E.L. (1993). **98**, 8419.

Carney, G.D., Sprandel, L.L., and Kern, C.W. (1978). *Adv. Chem. Phys.* **37**, 305.

Chen, Y., Halle, S., Jonas, D.M., Kinsey, J.L., and Field, R.W. (1990). *J. Opt. Soc. Am. B* **7**, 1805.

Cho, S.-W., Hase, W.L., and Swamy, K.N. (1990). *J. Phys. Chem.* **94**, 7371.

Choi, S.E., and Light, J.C. (1992). *J. Chem. Phys.* **97**, 7031.

Califano, S. (1976). *Vibrational States*. Wiley, New York.

Dai, H.L., Korpa, C.L., Kinsey, J.L., and Field, R.W. (1985a). *J. Chem. Phys.* **82**, 1688.

Dai, H.L., Field, R.W., and Kinsey, J.L. (1985b). *J. Chem. Phys.* **82**, 2161.

Darling, B.T., and Dennison, D.M. (1940). *Phys. Rev.* **57**, 128.

Davidson, N. (1962). *Statistical Mechanics*. McGraw-Hill, New York.

DeLeon, N., and Heller, E.J. (1982). *J. Chem. Phys.* **81**, 5957.

Duchovic, R.J., and Schatz, G.C. (1986). *J. Chem. Phys.* **84**, 2239.

Duchovic, R.J., Hase, W.L., and Schlegel, H.B. (1984). *J. Phys. Chem.* **88**, 1839.

Duncan, J.L., McKean, D.C., and Mallinson, P.D. (1973). *J. Mol. Spectrosc.* **45**, 221.

Dunn, K.M., Boggs, J.E., and Pulay, P. (1987). *J. Chem. Phys.* **86**, 5088.

Eaker, C.W., Schatz, G.C., DeLeon, N., and Heller, E.J. (1984). *J. Chem. Phys.* **81**, 5913.

Einstein, A. (1917). *Verh. Dtsch. Phys. Ges.* **19**, 82.

Eyring, H., Walter, J., and Kimball, G.E. (1944). *Quantum Chemistry*. Wiley, New York.

Farantos, S.C., and Tennyson, J. (1985). *J. Chem. Phys.* **82**, 800.

Goldstein, H. (1950). *Classical Mechanics*. Addison-Wesley, London.

Gutzwiller, M.C. (1990). *Chaos in Classical and Quantum Mechanics*. Springer-Verlag, New York.

Hadder, J.E., and Frederick, J.H. (1992). *J. Chem. Phys.* **97**, 3500.

Hase, W.L., Cho, S.-W., Lu, D.-h., and Swamy, K.N. (1989). *Chem. Phys.* **139,** 1.

Hase, W.L. (1996) personal communication.

Hase, W.L. (1982). *J. Phys. Chem.* **86**, 2873.

Henderson, J.R., and Tennyson, J. (1990). *Mol. Phys.* **69**, 639.

Hose, G., and Taylor, H.S. (1982). *J. Chem. Phys.* **76**, 5356.

Hernandez, R., Miller, W.H., Moore, C.B., and Polik, W.F. (1993). *J. Chem. Phys.* **99**, 950.

Herzberg, G. (1945). *Molecular Spectra and Molecular Structure II. Infrared and Raman Spectra of Polyatomic Molecules*. Van Nostrand Reinhold, New York.
Hirst, D.M. (1985). *Potential Energy Surfaces. Molecular Structure and Reaction Dynamics*. Taylor and Francis, New York.
Johnson, B.R. (1985). *J. Chem. Phys.* **83**, 1204.
Maessen, B., and Wolfsberg, M. (1984). *J. Chem. Phys.* **80**, 4651.
Maessen, B., and Wolfsberg, M. (1985). *J. Phys. Chem.* **89**, 3876.
Martens, C.C., and Ezra, G.S. (1985). *J. Chem. Phys.* **83**, 2290.
Martens, C.C., and Ezra, G.S. (1987). *J. Chem. Phys.* **86**, 279.
Matsushita, T., and Terasaka, T. (1984). *Chem. Phys. Lett.* **105**, 511.
McCoy, A.B., Burleigh, D.C., and Sibert, III, E.L. (1991). *J. Chem. Phys.* **95**, 7449.
McCoy, A.B., and Sibert, III, E.L. (1991). *J. Chem. Phys.* **95**, 3476.
Noid, D.W., Koszykowski, M.L., and Marcus, R.A. (1981). *Ann. Rev. Phys. Chem.* **32**, 267.
Nordholm, K.S., and Rice, S.A. (1974). *J. Chem. Phys.* **61**, 203.
Patterson, C.W. (1985). *J. Chem. Phys.* **83**, 4618.
Pauling, L., and Wilson, Jr., E.B. (1935). *Introduction to Quantum Mechanics*. McGraw-Hill, New York.
Percival, I.C. (1977). *Adv. Chem. Phys.* **36**, 1.
Pickett, H.M. (1972). *J. Chem. Phys.* **56**, 1715.
Pickett, T.J., and Shirts, R.B. (1991). **94**, 6036.
Polik, W.F., Guyer, D.R., Miller, W.H., and Moore, C.B. (1990). *J. Chem. Phys.* **92**, 3471.
Reisner, D.E., Field, R.W., Kinsey, J.L., and Dai, H.-L. (1984). *J. Chem. Phys.* **80**, 5968.
Romanowski, H., Bowman, J.M., and Harding, L.B. (1985). *J. Chem. Phys.* **82**, 4155.
Rosenberg, B.J., Ermler, W.C., and Shavitt, I. (1976). *J. Chem. Phys.* **65**, 4072.
Schiff, L.I. (1968). *Quantum Mechanics*. McGraw-Hill, New York, p. 268.
Shapiro, M., Taylor, R.D., and Brumer, P (1984). *Chem. Phys. Lett.* **106**, 325.
Sibert, III, E.L., and Mayrhofer R.C. (1993). *J. Chem. Phys.* **99**, 937.
Skodje, R.T., Borondo, F., and Reinhardt, W.P. (1985). *J. Chem. Phys.* **82**, 4611.
Smith, A.M., Jørgensen, U.G., and Lehmann, K.K. (1987). *J. Chem. Phys.* **87**, 5649.
Steinfeld, J.I., Francisco, J.S., and Hase, W.L. (1989). *Chemical Kinetics and Dynamics*. Prentice-Hall, Englewood Cliffs, NJ.
Sundberg, R.L., Abramson, E., Kinsey, J.L., and Field, R.W. (1985). *J. Chem. Phys.* **83**, 466.
Tennyson, J., and Farantos, S.C. (1984). *Chem. Phys. Lett.* **109**, 160.
Tennyson, J., and Sutcliffe, B.T. (1982). *J. Chem. Phys.* **77**, 4061.
Terasaka, T., and Matsushita (1985). *Phys. Rev. A* **32**, 538.
Toselli, B.M., and Barker, J.R. (1989). *J. Chem. Phys.* **91**, 2239.
Townes, C.H., and Schalow, A.L. (1955). *Microwave Spectroscopy*. McGraw-Hill, New York.
Watson, J.K.G. (1968a). *J. Chem. Phys.* **48**, 4517.
Watson, J.K.G. (1968b). *Mol. Phys.* **15**, 479.
Watson, J.K.G. (1970). *Mol. Phys.* **19**, 465.
Wigner, E.P. (1957). Oak Ridge National Laboratory Report ORNL-2309, p. 59.
Wilson, E.B. Jr., Decius, J.C., and Cross, P.C. (1955). *Molecular Vibrations*. McGraw-Hill, New York.
Wilson, E.B. Jr., and Howard, J.B. (1936). *J. Chem. Phys.* **4**, 260.
Zimmermann, Th., Cederbaum, L.S., Meyer, H.-D., and Köppel, H. (1987). *J. Phys. Chem.* **91**, 4446.

3

Potential Energy Surfaces

Properties of potential energy surfaces are integral to understanding the dynamics of unimolecular reactions. As discussed in chapter 2, the concept of a potential energy surface arises from the Born-Oppenheimer approximation, which separates electronic motion from vibrational/rotational motion. Potential energy surfaces are calculated by solving Eq. (2.3) in chapter 2 at fixed values for the nuclear coordinates R. Solving this equation gives electronic energies $E_e^i(R)$ at the configuration R for the different electronic states of the molecule. Combining $E_e^i(R)$ with the nuclear repulsive potential energy $V_{NN}(R)$ gives the potential energy surface $V^i(R)$ for electronic state i (Hirst, 1985). Each state is identified by its spin angular momentum and orbital symmetry. Since the electronic density between nuclei is different for each electronic state, each state has its own equilibrium geometry, sets of vibrational frequencies, and bond dissociation energies. To illustrate this effect, vibrational frequencies for the ground singlet state (S_0) and first excited singlet state (S_1) of H_2CO are compared in table 3.1.

For a diatomic molecule, potential energy surfaces only depend on the internuclear separation, so that a potential energy curve results instead of a surface. Possible potential energy curves for a diatomic molecule are depicted in figure 3.1. Of particular interest in this figure are the different equilibrium bond lengths and dissociation energies for the different electronic states. The lowest potential curve is referred to as the *ground electronic state potential*. The primary focus of this chapter is the ground electronic state potential energy surface. In the last section potential energy surfaces are considered for excited electronic states.

3.1 PROPERTIES OF POTENTIAL ENERGY SURFACES

3.1.1 Potential Energy Contours

A unimolecular reactant molecule consisting of N atoms has a multidimensional potential energy surface which depends on $3N$-6 independent coordinates. For the smallest nondiatomic reactant, a triatomic molecule, the potential energy surface is four-dimensional (three independent coordinates plus the energy). Since it is difficult, if not impossible, to visualize surfaces with more than three dimensions, methods are used to reduce the dimensionality of the problem in portraying surfaces. In a graphical representation of a surface the potential energy is depicted as a function of two coordinates with constraints placed on the remaining $3N$-8 coordinates. The contour diagram in figure 3.2 for a triatomic ABC molecule, such as HOD, is an example of such a graphical representation. Potential energy contours are plotted versus the two bond lengths with the A—B—C angle fixed at a specific value.

A contour diagram for unimolecular dissociation of the ethyl radical, $C_2H_5 \rightarrow H$

Table 3.1. Vibrational Frequencies for the S_0 and S_1 States of Formaldehyde.[a]

	S_0	S_1
ν_1	2766	2847
ν_2	1746	1173
ν_3	1501	887
ν_4	1167	689
ν_5	2843	2968
ν_6	1251	904

[a]Frequencies are anharmonic values for $1 \leftarrow 0$ transitions and are given in cm^{-1} (Gelbart et al., 1980).

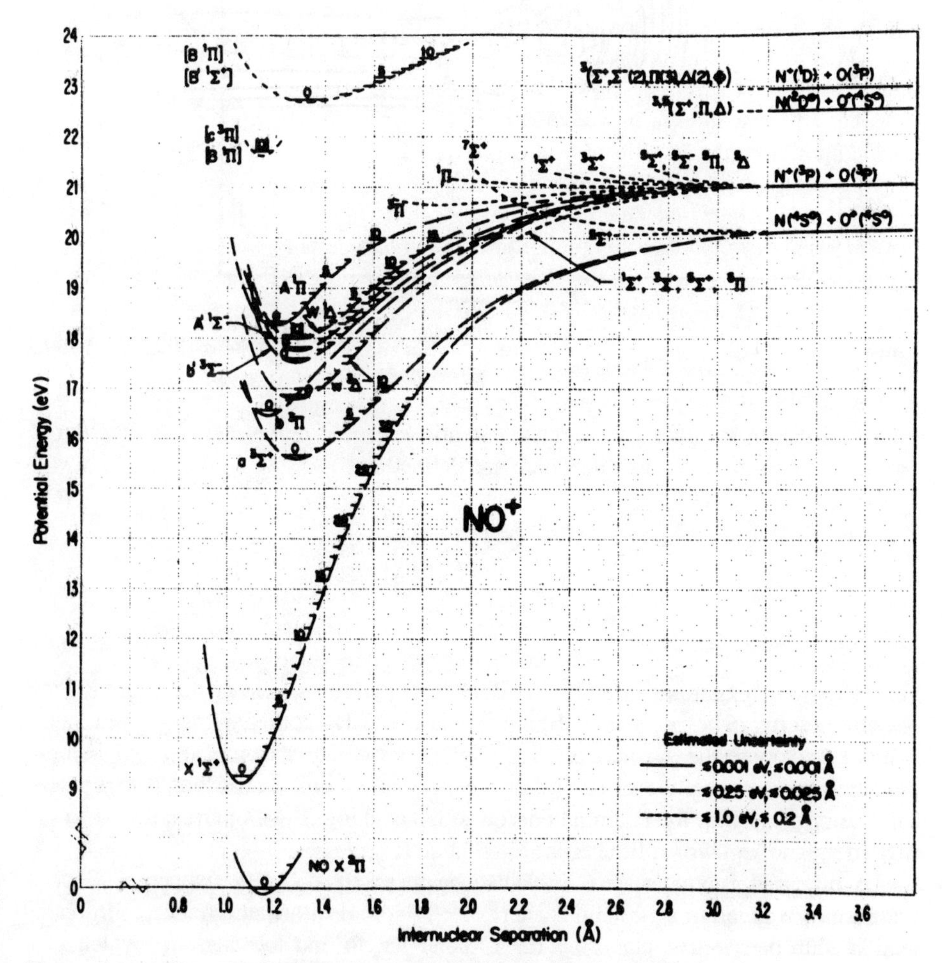

Figure 3.1 NO^+ potential energy curves (Albritton et al., 1979).

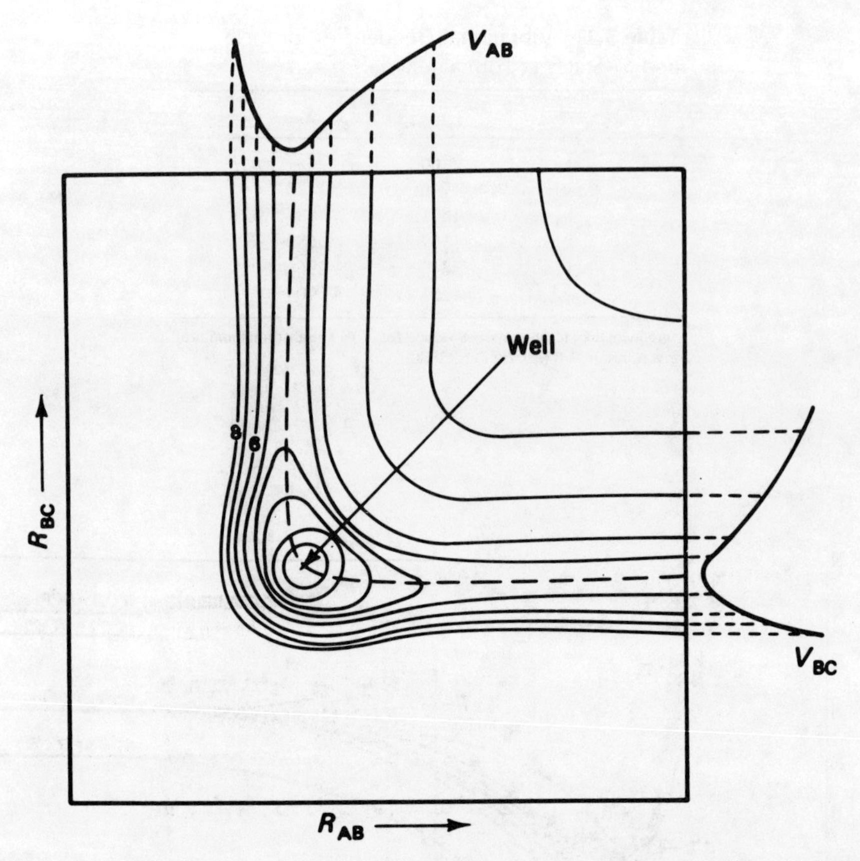

Figure 3.2 Depiction of a potential energy contour map for ABC (Steinfeld et al., 1989).

+ C_2H_4 is shown in figure 3.3. Here the potential energy is represented as a function of the internal coordinates r and θ, which are defined as

Values for the remaining internal coordinates of C_2H_5 are chosen to minimize the potential energy. The point w in figure 3.3 is the C_2H_5 potential energy minimum. Points z and x are the saddlepoints for H-atom dissociation from C_2H_5 and H-atom migration between the two carbon atoms, respectively. Point y corresponds to a potential energy maximum for H-atom addition to C_2H_4 along an axis perpendicular to the ethylene plane and which bisects the C═C bond.

A potential energy surface may also be represented by a perspective. Such a perspective is given in figure 3.4 for $CH_4 \rightarrow CH_3 + H$ dissociation. The coordinates used in this perspective are r the H—C bond length and the angle χ, which is a

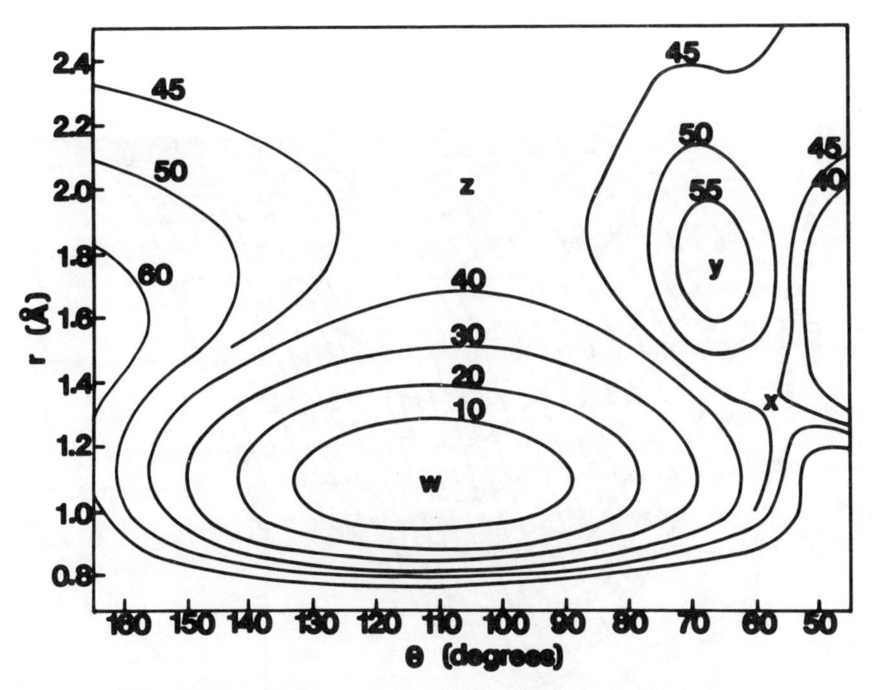

Figure 3.3 Potential energy contour map for $C_2H_5 \rightarrow H + C_2H_4$ dissociation; r, H—C distance; H—C—C angle; w, potential energy minimum; x, saddle point for H-atom migration; y, barrier for H-atom addition in C_{2v} symmetry; and z, saddle point for H-atom dissociation (Hase et al., 1978).

displacement from the C_{3v} symmetry axis in a plane that includes C and two H-atoms, and bisects the HCH angle of the remaining two H-atoms; that is,

The coordinates of the CH_3 group are varied to minimize the potential energy versus r, but not versus χ.

3.1.2 Reaction Path

An important property of a potential energy surface is the reaction path s which connects reactants and products by a path which may pass across a saddle point on the surface. The reaction path for $A + BC \rightarrow AB + C$ is depicted by the dashed line in figure 3.2. In evaluating reaction paths one finds that they depend on the coordinates used in the analysis. An *intrinsic* (i.e., unique) *reaction path* may be found by starting

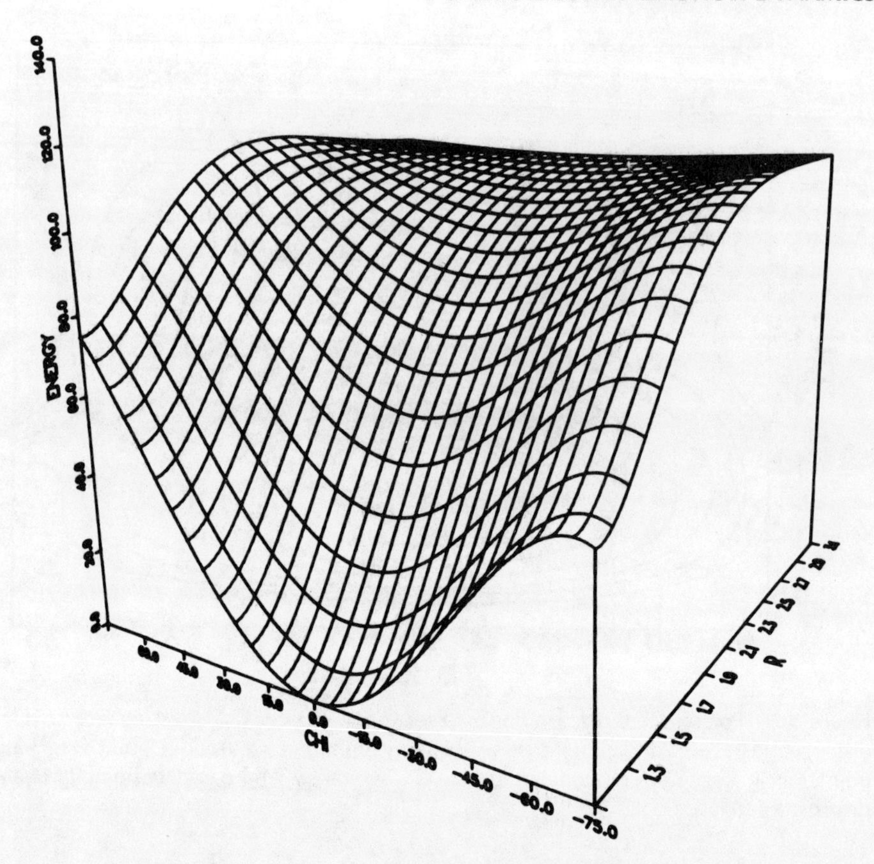

Figure 3.4 Plot of the $H + CH_3 \rightarrow CH_4$ potential energy surface, where r is the H—C distance (Å) and χ is the H—CH_3 angle (degrees). The geometry of the CH_3 group is optimized at each value of r. Energy is in kcal/mol (Hase et al., 1987).

at the saddlepoint and moving toward reactants and products in infinitesimal steps, with the kinetic energy removed after each step (Truhlar and Kupperman, 1970; Fukui, 1970). This motion traces out a path of steepest descent, which is obtained by following the negative gradient vector in mass-weighted Cartesian coordinates q_i (Eq. 2.27, p. 23) from the saddle point. If the surface does not have a saddle point, the reaction path is found by initializing the gradient vector at a sufficiently large separation between the unimolecular products. The negative gradient vector v_i is given by

$$v_i = \frac{dq_i}{ds} = -c\frac{\partial V}{\partial q_i} \qquad (i = 1 \text{ to } 3N)\,, \tag{3.1}$$

where the q_i are $3N$ mass-weighted Cartesian coordinates, V is the potential energy surface, and c is a constant to normalize the $3N$-dimensional vector v_i to unity;

$$c = \left[1/\sum_{i=1}^{3N} (\partial V/\partial q_i)^2 \right]^{1/2} . \tag{3.2}$$

Equation (3.1) comprises $3N$ first-order differential equations which are solved numerically.

By following the reaction path one determines the variation in the potential energy and in the Cartesian coordinates as the chemical system moves along the reaction path. Internal coordinates such as bond lengths, valence angle bends, and dihedral angles can be determined at any point on the reaction path from the Cartesian coordinates. A particular position along the reaction path is called the *reaction coordinate*, which can be written as a linear combination of the internal coordinates for the chemical system. This linear combination varies along the reaction path. For the reaction

$$\mathrm{H_2C{=}CHCl} \longrightarrow \mathrm{HCl} + \mathrm{H{-}C{\equiv}C{-}H}$$

the reaction coordinate is principally H-atom motion towards the Cl-atom at earlier stages of the reaction path. However, toward the end of the reaction path, as HCl is detached, the reaction is primarily C - - - Cl bond rupture.

Figure 3.5 depicts the three distinct shapes of reaction path potential energy curves for unimolecular reactions. Figure 3.5(a), shows the reaction path potential energy curve for an isomerization reaction such as

$$CH_3NC \rightleftharpoons CH_3CN.$$

Figure 3.5(b) shows the potential energy curve for a dissociation reaction like $C_2H_5 \rightarrow H + C_2H_4$ (figure 3.3), which has an energy barrier for the reverse association reaction. The final figure, 3.5(c), is the potential energy curve for a unimolecular dissociation reaction without a potential energy barrier for the reverse association reaction (figure 3.2).

An important component of the reaction path is a saddle point. As will be discussed later the difference between the saddle point's potential energy and the reactant molecule's potential energy minimum, that is, the classical activation energy E_o, is an important parameter and is used in a statistical calculation of the unimolecular rate constant. Also, the width of the potential energy barrier near the saddle point affects the rate at which the reactant molecule tunnels to products. Since the saddle point is a stationary point on the surface, the derivative of its potential energy is zero with respect to each Cartesian coordinate, that is, $\partial V/\partial q_i = 0$. Thus, the harmonic vibrational frequencies and eigenvectors at the saddle point are found in the same manner as for a potential energy minimum, viz, by use of Eqs. (2.27)–(2.45) in chapter 2. However, at the saddle point the potential energy surface curves downward along the reaction path, so that the corresponding force constant is negative for that one coordinate. This gives rise to an imaginary frequency for infinitesimal motion along the reaction path and $3N$-7 nonzero real frequencies, which represent harmonic vibrational motions orthogonal to the reaction path. In the simplest models for tunneling, the imaginary frequency determines the tunneling rate (see section 7.6).

As one moves along a reaction path connecting a saddle point and potential minimum, harmonic frequencies can be found for the $3N$-7 vibrational modes orthogonal to the reaction path motion (Miller et al., 1980; Kato and Morokuma, 1980). However, since all derivatives of the potential $\partial V/\partial q_i$ are not zero as one moves along the path, these vibrational modes cannot be found by diagonalizing the mass-weighted

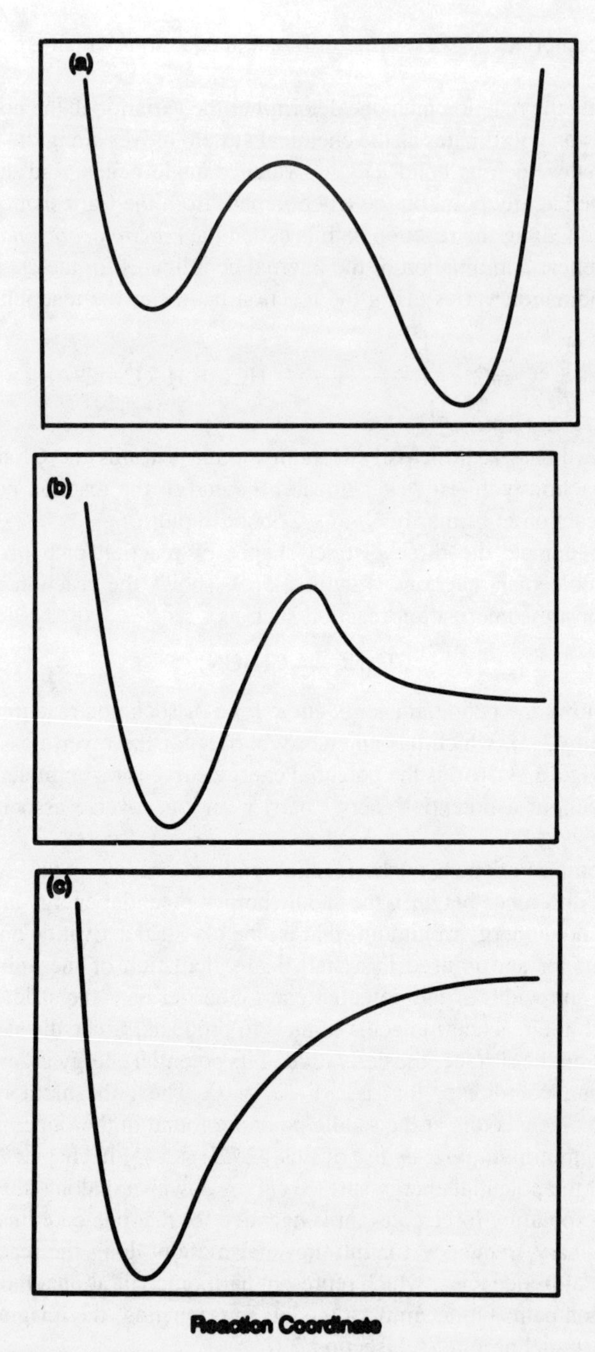

Figure 3.5 Schematic potential energy profiles for three types of unimolecular reaction. (a) Isomerization; (b) dissociation where there is a high energy barrier and a large activation energy for reaction in both the forward and reverse directions; (c) dissociation where the potential energy rises monotonically as for rotational ground state species, so that there is no barrier to the reverse association reaction (Steinfeld et al., 1989).

Cartesian force constant matrix **F** as is done for potential minima and saddle points; i.e., Eqs. (2.27) and (2.45) in chapter 2. Instead, a projected Cartesian force constant matrix $\mathbf{F^P}$ must be diagonalized. One forms $\mathbf{F^P}$ by projecting from **F** the direction along the reaction path and the directions corresponding to infinitesimal rotations and translations. When $\mathbf{F^P}$ is diagonalized there are seven zero eigenvalues which correspond to infinitesimal translations and rotations, and motion along the reaction path. The remaining $3N$-7 nonzero eigenvalues give harmonic frequencies for the vibrations orthogonal to the path. By evaluating and diagonalizing $\mathbf{F^P}$ as one moves along the reaction path, one determines the vibrational frequencies as a function of the path. Harmonic vibrational frequencies are plotted versus the reaction path in figure 3.6 for the $H_2CO \rightarrow H_2 + CO$ reaction, and in figure 3.7 for the $CH_4 \rightarrow CH_3 + H$ reaction. The potential energy along the reaction path s is called the vibrationally adiabatic ground state potential, which at the harmonic level is given by

$$V_a^G(s) = V(s) + \sum_{i=1}^{3N-7} h\nu_i(s)/2\,. \tag{3.3}$$

In this expression, $V(s)$ is the classical potential energy and the $\nu_i(s)$ are the vibrational frequencies.

3.1.3 Skewed and Scaled Coordinates

Insight about the dynamics of a unimolecular reaction can be obtained by examining the reaction's potential energy contour map. Usually this is at best only a qualitative analysis. However, it can be made quantitative for a linear triatomic *ABC* molecule by using skewed and scaled coordinates (Glasstone et al., 1941; Levine and Bernstein, 1987). The significance of these coordinates becomes readily apparent by considering the internal coordinate classical Hamiltonian for the linear ABC molecule: that is,

$$H = T(\dot{r}_1,\dot{r}_2) + V(r_1,r_2), \tag{3.4}$$

where

$$T = m_A(m_B + m_C)\dot{r}_1^2/2M + m_C(m_A + m_B)\dot{r}_2^2/2M + m_A m_C \dot{r}_1 \dot{r}_2/M. \tag{3.5}$$

In these equations r_1 and r_2 are the *AB* and *BC* internuclear separations, respectively, and M is the total mass. Because of the r_1r_2 coupling term in Eq. (3.5) and different values for the masses, the intramolecular motion of the linear *ABC* molecule cannot be studied by simply inspecting the molecule's (r_1,r_2) potential energy contour map. However, if the coordinates r_1 and r_2 are transformed, so that the kinetic energy is written as

$$T = (P_1^2 + P_2^2)/2M \tag{3.6}$$

the *ABC* intramolecular motion can be visualized in terms of a single hypothetical particle of mass M moving in two dimensions subject to the potential $V(Q_1,Q_2)$, where the coordinates Q_1 and Q_2 are conjugate to the momenta P_1 and P_2.

Transforming the kinetic energy from the expression in Eq. (3.5) to that in Eq. (3.6) is accomplished by skewing and scaling the axes r_1 and r_2 in the $V(r_1,r_2)$ contour map according to

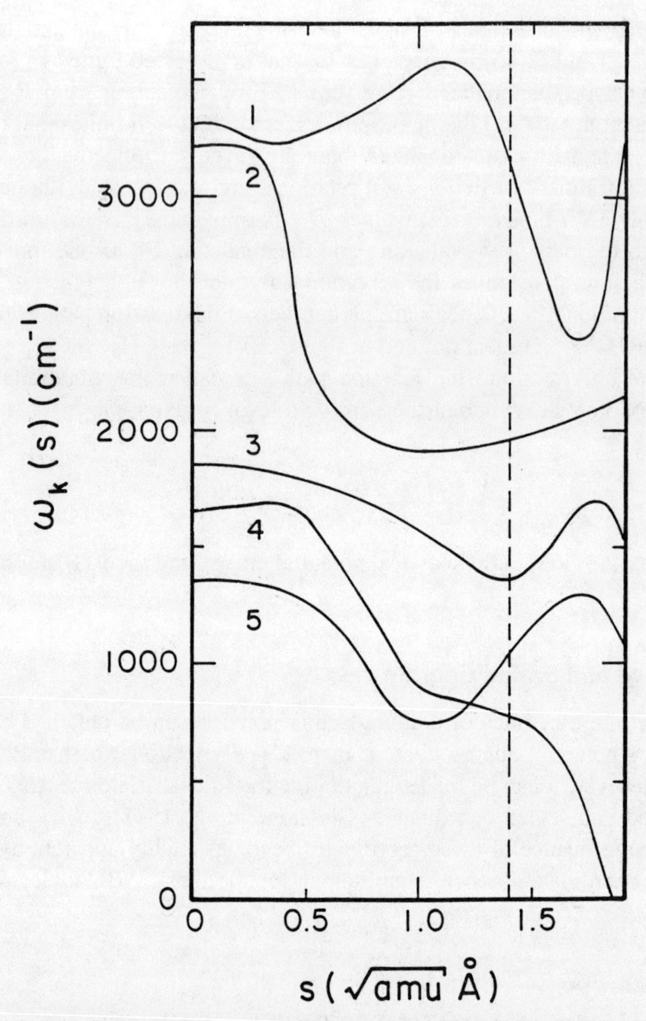

Figure 3.6 Vibrational frequencies for $H_2CO \rightarrow H_2 + CO$ versus the reaction path. Modes 1–5 are the asymmetric CH stretch, symmetric CH stretch, CO stretch, symmetric CH_2 bend, and out of plane bend, respectively, at the H_2CO equilibrium geometry (Waite et al., 1983).

$$Q_1 = r_1 + \beta\, r_2 \sin\phi,$$

$$Q_2 = \beta\, r_2 \cos\phi. \tag{3.7}$$

The reverse transformation is

$$r_1 = Q_1 - Q_2 \tan\phi\,, \tag{3.8}$$

$$r_2 = \frac{Q_2}{\beta} \sec\phi\,.$$

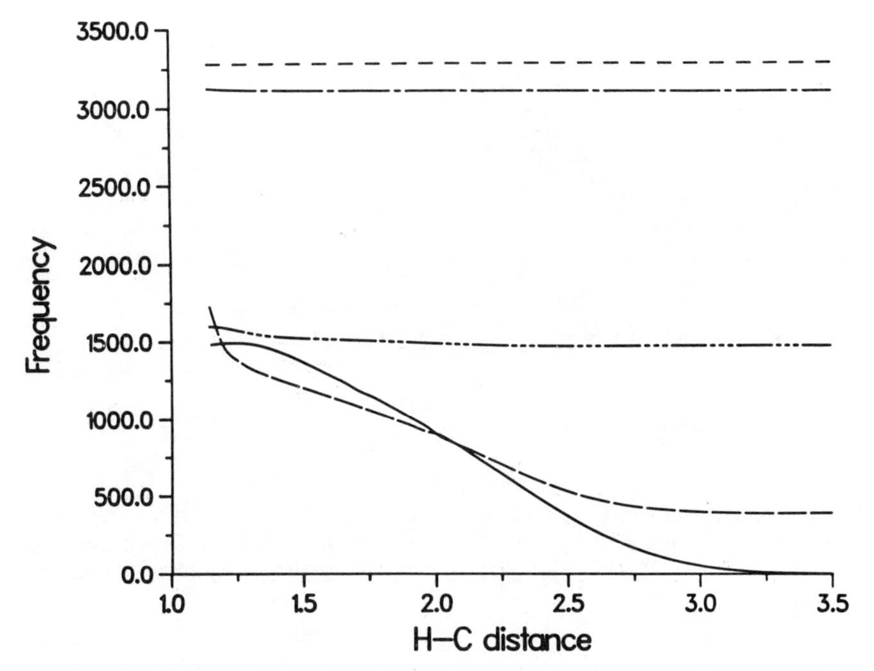

Figure 3.7 Vibrational frequencies (cm^{-1}) for $CH_4 \rightarrow CH_3 + H$ versus the reaction path. In descending order the modes at r_{HC} = 3.5 Å are the doubly degenerate CH stretch, symmetric CH stretch, doubly degenerate CH_3 deformation, out of plane CH_3 bend, and doubly degenerate H---CH_3 bend (Hase and Duchovic, 1985).

The transformation between r_1,r_2 and Q_1,Q_2 depends on two parameters β and ϕ, which are function of the masses of the three atoms, that is,

$$\sin\phi = \left[\frac{m_A m_C}{(m_A + m_B)(m_B + m_C)}\right]^{1/2}$$
$$\beta = \left[\frac{m_C(m_A + m_B)}{m_A(m_B + m_C)}\right]^{1/2} . \tag{3.9}$$

The ϕ parameter skews the rectangular r_1,r_2 axis system, while the β parameter is the scale factor for Q_2. A contour map for the $Cl_a^- + CH_3Cl_b \rightarrow Cl_aCH_3 + Cl_b^-$ reactive system in skewed and scaled coordinates is given in figure 3.8. (The CH_3 group is considered as a single atom in preparing this contour map.) More explicit details for constructing a contour map in skewed and scaled coordinates are given elsewhere (Glasstone et al., 1941; Levine and Bernstein, 1987).

By using the Q_1 and Q_2 coordinate system instead of the r_1 and r_2 internal coordinates, the kinetic energy (Eq. (3.6)) can be interpreted as that of a point particle (or system point) of mass M. The motion of this particle on the potential surface $V(Q_1,Q_2)$ can be simulated by letting a ball of mass M roll along the surface. Thus, quantitative aspects of the reaction dynamics can be understood by simply studying the $V(Q_1,Q_2)$ potential energy contour map.

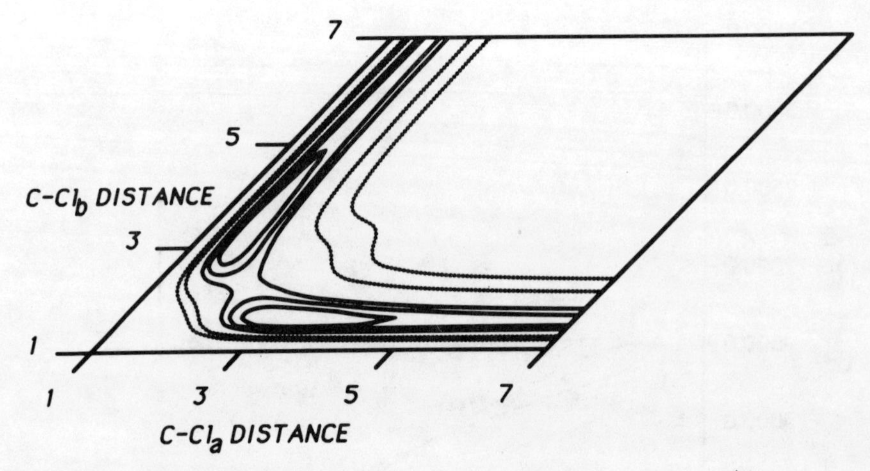

Figure 3.8 Potential energy contour diagram for the $Cl_a^- + CH_3Cl_b \rightarrow Cl_aCH_3 + Cl_b^-$ reaction in terms of the two C—Cl distances in a skewed axis representation. The remaining coordinates are set equal to their optimized values at each set of C—Cl_a and C—Cl_b distances. Solid and dotted contour lines are at 5 and 15 kcal/mol intervals, respectively (Cho et al., 1992).

3.2 DETERMINATION OF POTENTIAL ENERGY SURFACES

3.2.1 *Ab Initio* Calculations

Potential energy surfaces can be calculated by solving the electronic Schrödinger equation (Hirst, 1985). Because of the electronic–electronic repulsion terms in the electronic Hamiltonian, the electronic Schrödinger equation can only be solved in closed form for systems with very few electrons. Hence, various approximations must be made to solve the Schrödinger equation for most molecules. In an *ab initio* calculation the variational principle is used to give an upper bound to the electronic energy:

$$E_{\text{var}} = \frac{\langle \Psi | \hat{H} | \Psi \rangle}{\langle \Psi | \Psi \rangle} > E_{\text{exact}}\,. \tag{3.10}$$

Adjustable parameters in the approximate wave function Ψ are varied to minimize E_{var}, and obtain an improved energy and wave function.

There are excellent general descriptions of electronic structure calculations (Simons, 1991; Foresman and Frisch, 1993), and general computer programs for performing the calculations, such as GAUSSIAN (Frisch et al., 1992), have received wide use. Here, a brief review is given of the methodology of electronic structure calculations (Hehre et al., 1986; Schlegel and Frisch, 1991) and the results that can be obtained.

3.2.1.1 Methodology

In the Hartree-Fock (HF) model, the n electron wave function is written as a single Slater determinant of n orthonormal spin orbitals, ϕ_i:

$$\Psi_{\text{o}} = \frac{1}{\sqrt{n}} \,|\, \phi_1\phi_2 \cdots \phi_n \,|\,. \tag{3.11}$$

The spin orbitals [or molecular orbitals (MO)] are constructed from a linear combination of atomic orbitals (LCAO):

$$\phi_n = \sum_i c_{ni}\chi_i\alpha \quad \text{or} \quad \phi_n = \sum_i c_{ni}\chi_i\beta\,, \tag{3.12}$$

where α and β are spinfunctions and χ_i are basis functions for the spatial part of the spin orbitals. The basis functions χ_i are usually linear combinations of Gaussian functions. Minimizing the energy with respect to molecular orbital coefficients c_{ni} leads to the Hartree-Fock equations. These equations must be solved iteratively until self-consistency is achieved for the spin orbitals; thus, this is also known as the self consistent field (SCF) method (Roothan, 1951). In effect, each spin orbital is optimized in the average field of all the other orbitals. There are different types of Hartree-Fock calculations, depending on the constraints placed on the molecular orbitals. In restricted Hartree-Fock (RHF) calculations, the alpha (spin up) and the beta (spin down) MO coefficients are the same, whereas in the unrestricted Hartree-Fock (UHF) method, the alpha and beta MO coefficients are allowed to vary independently. The UHF method is particularly important for treating radicals (Hehre et al., 1986).

In the Hartree-Fock model each electron sees only the average field of the other electrons. In reality, the electrons must explicitly avoid each other because of their mutual coulombic repulsion; hence their motions are correlated. The difference between the Hartree-Fock energy and the exact energy is called the correlation energy. The Hartree-Fock wave function can be improved by taking a linear combination of Slater determinants, yielding a configuration interaction (CI) wave function (Boys, 1950):

$$\Psi = a_o\Psi_o + \sum_{ia} a_i^a\Psi_i^a + \sum_{i<j,a<b} a_{ij}^{ab}\Psi_{ij}^{ab} \cdots, \tag{3.13}$$

where Ψ_i^a, Ψ_{ij}^{ab}, etc. are Slater determinants in which occupied spin orbitals ϕ_i, ϕ_j, etc. in the reference determinant Ψ_o are replaced by unoccupied or virtual spin orbitals ϕ_a, ϕ_b, etc. (i.e., determinants that are singly excited, doubly excited, etc.). In a standard CI calculation Ψ_o is first determined by a Hartree-Fock calculation. The coefficients a_i^a, a_{ij}^{ab}, etc. for the excited Slater determinants are then determined variationally. In the multiconfiguration self consistent field (MCSCF) method (Werner and Shepard, 1987), a variational calculation is performed by simultaneously minimizing the energy with respect to the coefficients c_{ni} for the molecular orbitals, Eq. (3.12), and the a_i^a, a_{ij}^{ab}, etc. coefficients, Eq. (3.13). This must be done iteratively as in the Hartree-Fock approach.

Although the CI energy is variational (an upper bound to the exact energy given the form of the wave function), it is not size-consistent, that is, E_{CI} for X and Y at large separation is not the sum of E_{CI} for X and E_{CI} for Y computed individually. This is often a severe problem in computing reaction energetics and it is, thus, desirable to have a CI-like method that is size-consistent. This role is filled by the coupled-cluster (CC) method and the quadratic configuration interaction (QCI) approach. The MCSCF method is size-consistent.

The coefficients a_i^a, a_{ij}^{ab}, etc. in Eq. (3.13) can also be determined by perturbation theory. A convenient zero-order Hamiltonian operator $\hat{H}_o$ is the Fock operator, since the Hartree-Fock wave function is an eigenfunction of the Fock operator, as are excited configurations derived from it by replacing occupied orbitals with virtual orbitals. This choice of $\hat{H}_0$ yields Møller-Plesset (MP) perturbation theory (Bartlett and Silver, 1975; Krishnan and Pople, 1978), in which the correlated motion of the electrons is the perturbation. A calculation of this type is referred to as MP2, MP3, or MP4 if it is performed through second, third, or fourth order, respectively. The M*Pn* methods are size-consistent.

3.2.1.2 Stationary Points and Reaction Paths

Ab initio calculations are widely used to find structures, potential energies, and vibration frequencies for stationary points (i.e., potential minima and saddle points) on potential energy surfaces. The accuracy of the calculated result is determined by the number of basis functions and the treatment of electron correlation, both of which can be treated more completely the smaller the number of electrons in the molecular system. A widely used size-consistent computational approach is to determine stationary point geometries and vibrational frequencies at the MP2 level of theory, and then calculate the reaction energetics at the MP4 level using the MP2 geometries (Hehre et al., 1986). It is often desirable to compute reaction energetics with the size-consistent coupled-cluster (CC) method (Cizek, 1966; Purvis and Bartlett, 1982) or the quadratic-CI (QCI) method (Pople et al., 1987), since iterative methods can give better estimates of the correlation energy than perturbative methods, especially if the correlation energy is large. The MCSCF approach is often needed if the reaction involves considerable electron reorganization, as for a reaction in which the unimolecular reactant and products are closed shell systems, but there is extensive open shell character for the saddle point and nearby environs. For reactions which are composed of three or fewer second row atoms, activation energies and heats of reaction can often be calculated to within several kJ/mole.

For many reactions the calculated structures for potential energy minima are as accurate as those found experimentally. *Ab initio* and experimental harmonic vibrational frequencies usually agree to within 10–15% at the Hartree-Fock level and 5% at the MP2 level (Hehre et al., 1986). It has been found that Hartree-Fock harmonic frequencies computed with a medium-size basis set can be scaled by the factor 0.9 to give approximate anharmonic $n = 0 \rightarrow 1$ transition frequencies (Hehre et al., 1986). A detailed study has been made of how the computed *ab initio* frequencies for benzene depend on the size of the basis set and the treatment of electron correlation (Maslen et al., 1992).

In recent work, algorithms have been introduced into *ab initio* calculations so that reaction paths can be directly calculated from *ab initio* calculations (Schlegel, 1994). Also determined from these calculations are the frequencies for the $3N$-7 vibrational frequencies orthogonal to the reaction-path.

3.2.1.3 Potential Energy Surfaces

Ab initio calculations have also been used to study features of potential energy surfaces other than potential energy minima, saddle points, and reaction paths. For small reactive systems with three atoms it is possible to determine the complete potential energy surface from an *ab initio* calculation. A three-atom molecular systems such as H_2O has three independent coordinates and, if the potential energy is calculated for ten values of each coordinate, one thousand total potential energy points results. These points can then be fit by an analytic function. This approach has been used to determine an accurate analytic potential energy function for the $HCO \rightarrow H + CO$ unimolecular reaction (Bowman et al., 1986). In deriving analytic potential energy functions for unimolecular reactions with four or more atoms, each degree of freedom is not treated independently and model potential functions (e.g., Morse functions) are often used for degrees of freedom not directly involved in the unimolecular process.

3.2.2 Experimental Studies

Experimental measurements can provide information about potential energy surfaces for unimolecular reactions. As shown by Tolman, Fowler, and Guggenheim (Steinfeld et al., 1989) the thermal activation energy E_a for a chemical reaction is

$$E_a = \langle E_r(T) \rangle - \langle E(T) \rangle , \tag{3.14}$$

where $\langle E_r(T) \rangle$ is the average energy of molecules undergoing reaction and $\langle E(T) \rangle$ is the average energy of all reactant molecules. For thermal unimolecular reactions Eq. (3.14) is particularly useful at the high-pressure limit where the reactant molecules are represented by a canonical ensemble and $\langle E(T) \rangle$ is easily evaluated using statistical thermodynamics. Also, if transition state theory is valid for the unimolecular reaction and tunneling is unimportant, at high pressure $\langle E_r(T) \rangle$ is given by the average canonical energy of the transition state plus E_o. This latter energy is the difference between the zero point energy levels of the transition state and reactant.

Transition state theory, if valid, can also be used to determine the tightness or looseness of the transition state molecular configuration as compared to that of the reactants. When transition state theory is formulated in thermodynamic terms it is found that the high pressure Arrhenius A- factor is given by

$$A_\infty = e \frac{k_B T}{h} e^{\Delta S_o^\ddagger / R} , \tag{3.15}$$

where $\Delta S_o^\ddagger$ is the standard entropy of activation. For a unimolecular reaction with a very tight transition state it is possible that the transition state entropy is less than that of the reactant. However, for a unimolecular dissociation reaction like $C_2H_6 \rightarrow 2CH_3$ the transition state is loose, with an entropy considerably larger than that of the reactant. The relationship between the entropy of activation and the vibrational frequencies of the molecule and the transition state is given and discussed in chapter 7.

The force constants associated with a molecule's potential energy minimum are the harmonic values, which can be found from harmonic normal mode vibrational frequencies. For small polyatomic molecules it is possible (Duncan et al., 1973) to extract harmonic normal mode vibrational frequencies from the experimental anharmonic $n = 0 \rightarrow 1$ normal mode transition frequencies (the harmonic frequencies are usually approximately 5% larger than the anharmonic $0 \rightarrow 1$ transition frequencies). Using a normal mode analysis as described in chapter 2, internal coordinate force constants (e.g., table 2.4) may be determined for the molecule by fitting the harmonic frequencies.

For a diatomic molecule the Rydberg-Klein-Rees (RKR) method may be used to determine the potential energy curve $V(r)$ from the experimental vibrational/rotational energy levels (Hirst, 1985). This method is based on the Einstein-Brillouin-Keller

(EBK) semiclassical quantization condition given by Eq. (2.72). In contrast to the RKR method for diatomics, a direct method has not been developed for determining potential energy surfaces from experimental anharmonic vibrational/rotational energy levels of polyatomic molecules. Methods which have been used are based on an analytic representation of the potential energy surface (Bowman and Gazdy, 1991). At low levels of excitation the surface may be represented as a sum of quadratic, cubic, and quartic normal mode coordinates (or internal coordinate) terms, that is,

$$V = \frac{1}{2}\sum_{i,j} f_{ij}Q_iQ_j + \frac{1}{6}\sum_{i,j,k} a_{ijk}Q_iQ_jQ_k + \frac{1}{24}\sum_{i,j,k,l} b_{ijkl}Q_iQ_jQ_kQ_l . \tag{3.16}$$

A vibrational basis set is chosen for the molecule and the quantum mechanical variational method (Eqs. (2.66)–(2.68)) is used to calculate vibrational energy levels from the analytic potential. The parameters f_{ij}, a_{ijk}, b_{ijkl} in the potential are varied until the experimental vibrational energy levels are fit. This variational method is only practical at low levels of excitation.

Dynamical properties of chemical reactions are also used to determine potential energy surfaces. Dynamical calculations, such as classical trajectories, performed on a potential energy surface can be used to reproduce experimental dynamical results. By varying adjustable parameters in an analytic function, depicting the surface, it is often possible to determine what surface properties are required to fit the experimental results (Bunker, 1974). In unimolecular dissociation reactions, the partitioning of available energy to vibration, rotation and translation in the products may be used to deduce qualitative shapes of potential energy surfaces (Chang et al., 1992).

3.3 ELECTRONICALLY EXCITED STATES

3.3.1 Potential Energy Surfaces

One of the ramifications of the Born-Oppenheimer approximation is that potential energy surfaces for states with the same orbital symmetry and spin do not cross, that is, the noncrossing rule (Kauzmann, 1957; Hirst, 1985). However, this rule is only strictly applicable to the exact potential energy surfaces and energy curves derived from approximate molecular wave functions may cross. Approximate potential energy curves that cross are the ionic and covalent curves for NaCl, which have symmetry of $^1\Sigma^+$. Near the NaCl equilibrium geometry the ionic potential energy curve is lower in energy. However, at a large Na + Cl separation the covalent curve is lower and the curves cross near 10 Å (see figure 3.9).

As described above, to obtain accurate Born-Oppenheimer wave functions Ψ_i and potential energy surfaces one has to account for electron correlation. This can be done with a configuration interaction wave function, Eq. (3.13), by writing Ψ_i as a mixture of the Ψ_j° for the different Slater determinants, that is,

$$\Psi_i = \sum_j c_{ij}\Psi_j^\circ \tag{3.17}$$

In this linear combination, all states of a given symmetry are included. The result is a series of potential energy surfaces which do not cross. According to perturbation

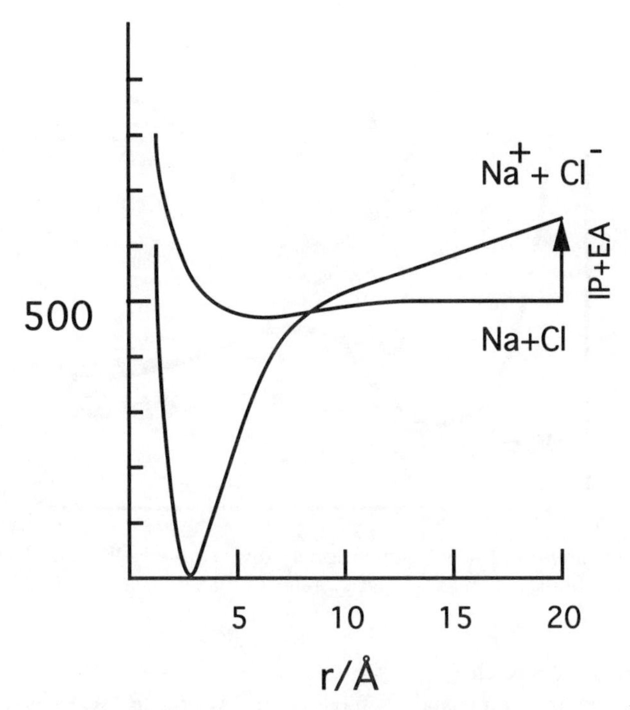

Figure 3.9 Covalent and ionic potential energy curves for NaCl.

theory the mixing is most pronounced at nuclear configurations where the potential surfaces for the Ψ_j^o have similar energies. As a result, the extent of mixing between the Ψ_j^o depends on the nuclear geometry.

An illustration of the noncrossing rule is given in figure 3.10 where potential energy curves are plotted for two electronic states with accurate wave functions Ψ_1 and Ψ_2. The dashed curves indicate the potential energy curves for the approximate wave functions χ_1 and χ_2. In the region of crossing between the approximate energies, the approximate electronic wave functions strongly mix in forming the accurate wave functions Ψ_i. The effect of this mixing is always to push the approximate potential energies apart, and this interaction is strongest at the *avoided crossing* point at $R = R_o$. The accurate potential energy curves with mixing are called *adiabatic* potentials. The term *diabatic* is used to describe the approximate potential curves. The electronic character of a wave function for an adiabatic potential can change significantly at an avoided crossing. For example, in figure 3.10, Ψ_1 is approximately equal to χ_1 before the crossing, but after the crossing becomes χ_2. Similarly, the accurate ground electronic state wave function for NaCl has considerable ionic character at small NaCl separations, but is covalent at large separations (see figure 3.9).

For a diatomic molecule, the crossing of two potential curves depends on only one coordinate, the internuclear separation. However, for a polyatomic molecule there are multiple coordinates to consider. For example, the radical CH_2 is of C_{2v} symmetry and has the triplet electronic states of $^3A_2$ and $^3B_1$ symmetry. The molecule CH_2 has three normal modes of vibration, namely, the symmetric stretch Q_1, the symmetric bend Q_2,

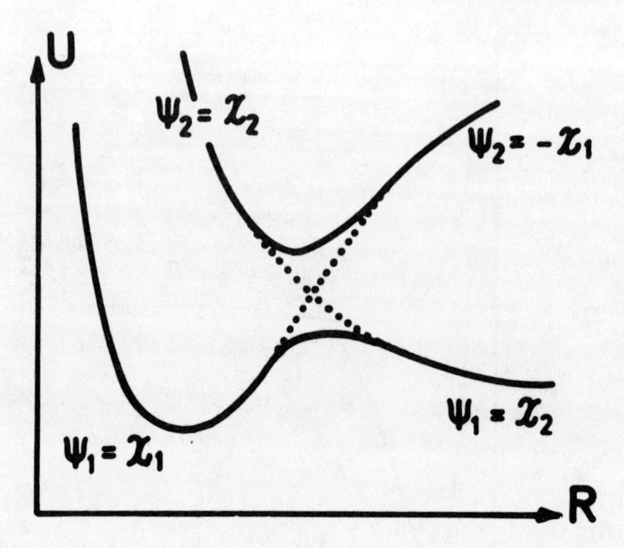

Figure 3.10 Illustration of a one-dimensional curve crossing. Ψ_1 and Ψ_2 designate adiabatic wave functions, whereas χ_1 and χ_2 refer to diabatic states (Lorquet et al., 1985).

and the asymmetric stretch Q_3. If the CH_2 geometry is maintained at C_{2v} by only displacing the symmetric Q_1 and Q_2 modes, the $^3A_2$ and $^3B_1$ surfaces are allowed to cross since they are of different symmetries. Thus, there is a line of intersection between the $^3A_2$ and $^3B_1$ surfaces which is a function of Q_1 and Q_2. However, if the molecule is distorted by displacing the asymmetric stretch mode Q_3, the molecular symmetry is lowered to C_s and the symmetry of each surface becomes $^3A''$. For this nonsymmetric displacement the two surfaces will not cross, but have an avoided crossing. If the potential energy surface is expressed as a function of one symmetric coordinate (Q_1 or Q_2) and the asymmetric coordinate, Q_3, in the vicinity of the crossing point the two surfaces have a double cone geometrical shape known as a *conical intersection*. This is illustrated in figure 3.11. A specific conical intersection is shown in figure 3.12 for the unimolecular dissociation of planar H_2CO^+. In the general polyatomic case the potential energy surfaces depend on n variables (3 for CH_2) and, thus, the intersection between the two surfaces is a surface of n-2 dimensions.

Spin-orbit coupling and a breakdown in the Born-Oppenheimer approximation can lead to a transition between adiabatic potentials. These two perturbations can also cause transitions between potential curves of different electronic symmetries and/or spin. In the *electronic adiabatic approximation* it is assumed that transitions between adiabatic curves are negligible, so that nuclear motion evolves on a single electronic potential surface.

3.3.2 Unimolecular Reactions

There are numerous situations in which molecules, initially prepared in electronically excited states, undergo unimolecular reaction. Figure 3.13 shows some possibilities. In

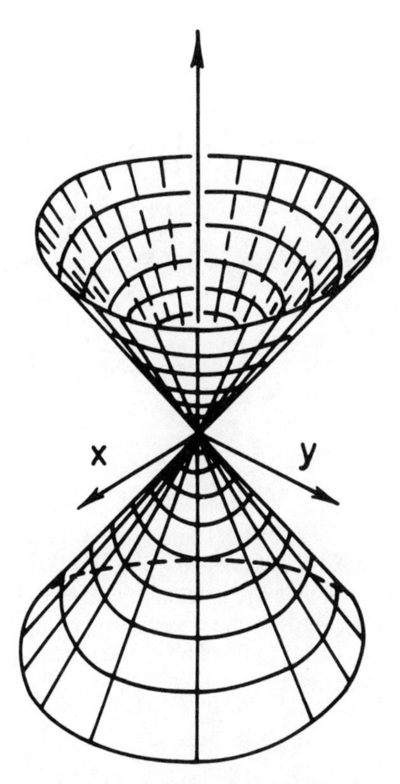

Figure 3.11 Schematic drawing of a canonical intersection between two potential energy surfaces (Hirst, 1985).

these figures absorption of electromagnetic radiation is the assumed excitation step. However, excitation could also occur by other processes such as electron impact. In Figure 3.13(a) the upper electronic state is repulsive so that the molecule, once excited to this surface, dissociates directly and immediately. This is termed direct dissociation, as opposed to predissociation. It is not a process that can be understood in terms of statistical unimolecular rate theory which is a theory of vibrational predissociation. However, it is a unimolecular reaction and some very interesting new results on the measurement of these rates have been recently reported (Schinke, 1993).

The second situation, figure 3.13(b), is an example of vibrational predissociation. The situation is the same as molecular decomposition on the ground electronic state, except now the decomposition occurs on an electronically excited surface. The process is electronically adiabatic since the reaction occurs on the electronic surface initially excited. The same theoretical models used to describe unimolecular reactions on the ground electronic surface are applicable to this situation.

The third situation, figure 3.13(c), is an example of electronic predissociation. That is, the molecule passes from one electronic state, which is bound, to a dissociative state. The reaction rate is then dependent on the strength of the coupling between the two surfaces, a problem that can be understood by curve crossing models

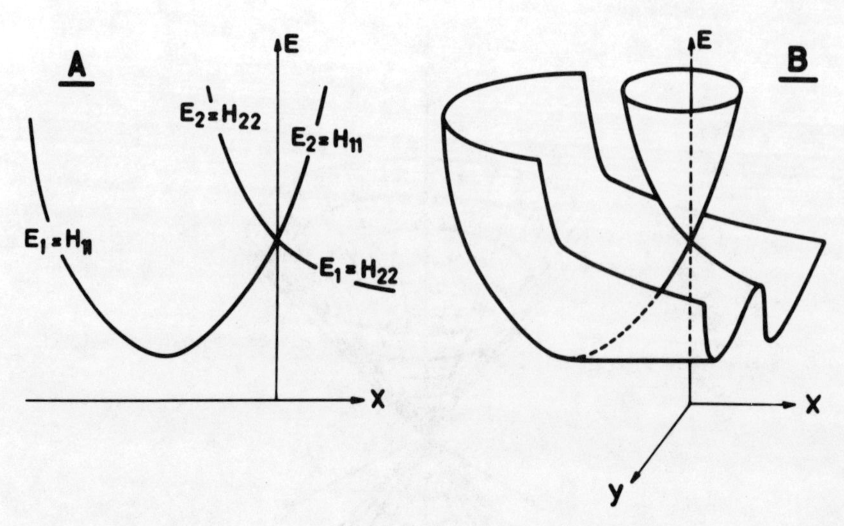

Figure 3.12 (a) Cross section of a conical intersection along the symmetry-conserving coordinate *x*. (b) Three-dimensional view when the symmetry-lowering coordinate *y* is added. The figure results from an *ab initio* calculation on H_2CO^+ described in Vaz Pires et al. (1978). Coordinate *x* corresponds to the dissociation path leading to H_2 + CO^+ fragments. Coordinate *y* is the out-of-plane bending mode of formaldehyde (i.e., the angle of pyramidalization ω) (Desouter-Lecomte et al., 1979).

as proposed by Landau and Zener many years ago (see chapter 8). As discussed in chapter 8, a statistical unimolecular rate theory can be combined with the curve crossing process to calculate the predissociation rate constant. The curve crossing is represented in one dimension while the remaining dimensions (i.e., coordinates) are treated statistically.

Finally, there is the situation (figure 3.13(d)) in which the molecule in an excited electronic state can either be stabilized by fluorescence (rate k_f) or undergo a nonradiative transition to the ground electronic state with a rate, k_{nr}. Once the molecule is in the ground electronic state, it can dissociate with a rate constant (k_d) that is often treated in terms of statistical unimolecular rate theory. If the excited electronic state is the first excited single state (S_1), one has the mechanism

$$S_1 \xrightarrow{k_f} S_0 + h\nu,$$

$$S_1 \xrightarrow{k_{nr}} S_0^*,$$

$$S_0^* \xrightarrow{k_d} \text{products},$$

The electronic nonadiabatic process, k_{nr}, is frequently used to prepare vibrationally excited ground electronic state molecules and is discussed in more detail in the next chapter.

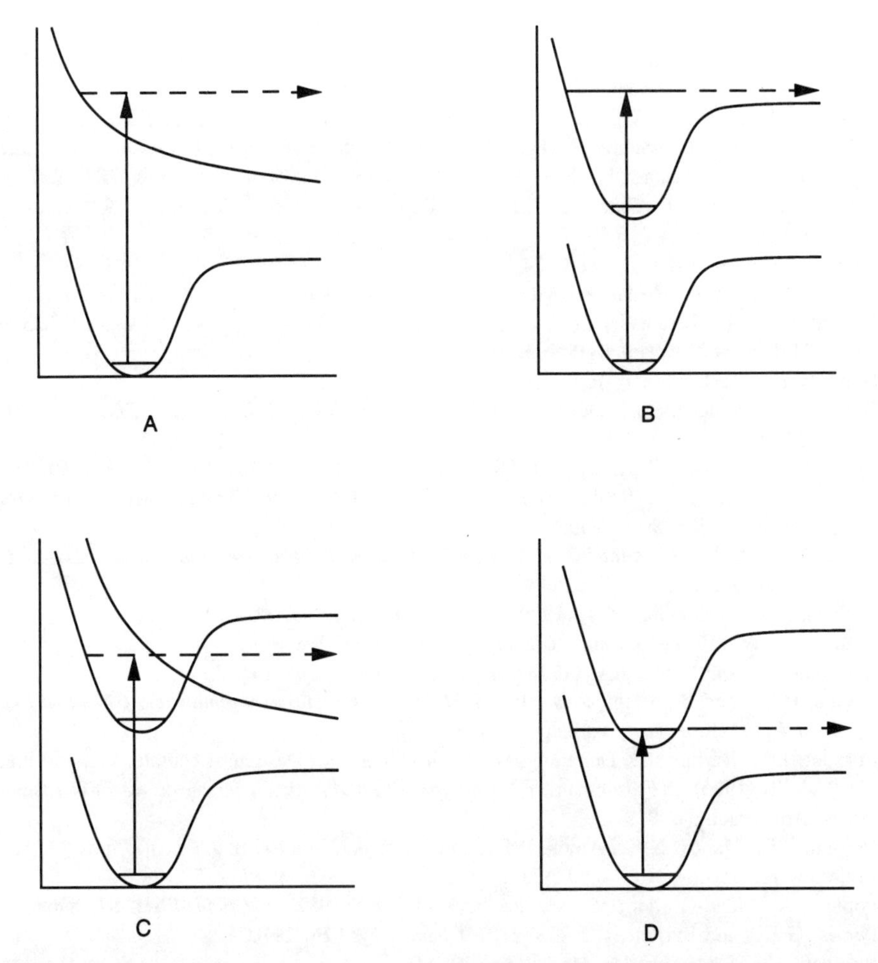

Figure 3.13 Examples of unimolecular decomposition from excited electronic states: (a) direct dissociation, (b) vibrational predissociation, (c) electronic predissociation, and (d) internal conversion.

REFERENCES

Albritton, D.L., Schmeltekopf, A.L., and Zare, R.N. (1979). *J. Chem. Phys.* **71**, 3271.
Bartlett, R.J., and Silver, D.M. (1975). *J. Chem. Phys.* **62**, 3258.
Bowman, J.M., and Gazdy, B. (1991). *J. Chem. Phys.* **94**, 816.
Bowman, J.M., Bittman, J.S., and Harding, L.B. (1986). *J. Chem. Phys.* **85**, 911.
Boys, S.F. (1950). *Proc. Roy. Soc. London A* **201**, 125.
Bunker, D.L. (1974). *Acc. Chem. Res.* **7**, 195.
Chang, Y.-T., Minichino, C., and Miller, W.H. (1992). *J. Chem. Phys.* **96**, 4341.
Cho, Y.J., Vande Linde, S.R., Zhu, L., and Hase, W.L. (1992). *J. Chem. Phys.* **96**, 8275.
Cizek, J. (1966). *J. Chem. Phys.* **45**, 4256.
Desouter-Lecomte, M., Galloy, C., Lorquet, J.C., and Vaz Pires, M. (1979). *J. Chem. Phys.* **71**, 3661.

Duncan, J.L., McKean, D.C., and Mallinson, P.D. (1973). *J. Mol. Spectrosc.* **45**, 221.

Foresman, J.B., and Frisch, Æ. (1993). *Exploring Chemistry with Electronic Structure Methods*. Gaussian, Pittsburgh.

Frisch, M.J., Trucks, G.W., Head-Gordon, M., Gill, P.M.W., Wong, M.W., Foresman, J.B., Johnson, B.G., Schlegel, H.B., Robb, M.A., Replogle, E.S., Gomperts, R., Andrés, J.L., Raghavachari, K., Binkley, J.S., Gonzalez, C., Martin, R.L., Fox, D.J., DeFrees, D.J., Baker, J., Stewart, J.J.P., and Pople, J.A. (1992). *GAUSSIAN 92*. Gaussian, Pittsburgh.

Fukui, K. (1970). *J. Phys. Chem.* **74**, 4161.

Gelbart, W.M., Elert, M.L., and Heller, D.F. (1980). *Chem. Rev.* **80**, 403.

Glasstone, S., Laidler, K.J., and Eyring, H. (1941). *The Theory of Rate Processes*. McGraw-Hill, New York, pp. 100–103.

Hase, W.L., and Duchovic, R.J. (1985). *J. Chem. Phys.* **83**, 3448.

Hase, W.L., Mondro, S.L., Duchovic, R.J., and Hirst, D.M. (1987). *J. Am. Chem. Soc.* **109**, 2916.

Hase, W.L., Mrowka, G., Brudzynski, R.J., and Sloane, C.S. (1978). *J. Chem. Phys.* **69**, 3548.

Hehre, W.J., Radom, L., Schleyer, P.v.R., and Pople, J.A. (1986). *Ab Initio Molecular Orbital Theory*. Wiley, New York.

Hirst, D.M. (1985). *Potential Energy Surfaces, Molecular Structure, and Reaction Dynamics*. Taylor and Francis, Philadelphia.

Kato, S., and Morokuma, K.J. (1980). *J. Chem. Phys.* **73**, 3900.

Kauzmann, W. (1957). *Quantum Chemistry*. Academic, New York.

Krishnan, R., and Pople, J.A. (1978). *Int. J. Quantum Chem.* **14**, 91.

Levine, R.D., and Bernstein, R.B. (1987). *Molecular Reaction Dynamics and Chemical Reactivity*. Oxford, New York, pp. 165–169.

Lorquet, J.C., Barbier, C., Dehareng, D., Leyh-Nihant, B., Desouter-Lecomte, M., and Praet, M.T. (1985). In F. Lahmani, Ed. *Photophysics and Photochemistry above 6 eV*. Elsevier, Amsterdam, p. 319.

Maslen, P.E., Handy, N.C., Amos, R.D., and Jayatilaka, D. (1992). *J. Chem. Phys.* **97**, 4233.

Miller, W.H., Handy, N.C., and Adams, J.E. (1980). *J. Chem. Phys.* **72**, 99.

Pople, J.A., Head-Gordon, M., and Raghavachari, K. (1987). *J. Chem. Phys.* **87**, 5968.

Purvis, G.D., and Bartlett, R.J. (1982). *J. Chem. Phys.* **76**, 1910.

Roothan, C.C.J. (1951). *Rev. Mod. Phys.* **23**, 69.

Schinke, R. (1993). *Photodissociation Dynamics*. Cambridge University Press, Cambridge.

Schlegel, H.B. (1994). In *Modern Electronic Structure Theory*, Yarkony, D. R., Ed. World Scientific, Singapore.

Schlegel, H.B., and Frisch, M.J. (1991). *Theoretical and Computational Models for Organic Chemistry*. Formosinho, J.S., Csizmadia, I.G., and Armaut, L.G., Eds., NATO-ASI series C 339, Kluwer Academic, the Netherlands, pp. 5–33.

Shepherd, R. (1987). "The Multiconfiguration self-consistent field method." In K.P. Lawley ed., Ab initio Methods in Quantum Chemistry II, Adv. Chem. Phys. vol 69. Wiley, New York, pp. 63–200.

Simons, J. (1991). *J. Phys. Chem.* **95**, 1017.

Steinfeld, J.I., Francisco, J.S., and Hase, W.L. (1989). *Chemical Kinetics and Dynamics*. Prentice-Hall, Engelwood Cliffs, NJ.

Truhlar, D.G., and Kuppermann, A. (1970). *J. Am. Chem. Soc.* **93,** 1840.

Vaz Pires, M., Galloy, C., and Lorquet, J.C. (1978). *J. Chem. Phys.* **69,** 3242.

Waite, B.A., Gray, S.K., and Miller, W.H. (1983). *J. Chem. Phys.* **78**, 259.

Werner, H.-J. (1987). Matrix-formulated direct multiconfiguration self-consistent field and multiconfiguration reference configuration-interaction methods. In K.P. Lawley ed., *Ab initio Methods in Quantum Chemistry II*, Adv. Chem. Phys. Vol 69, pp. 1–62. Wiley, New York.

4

State Preparation and Intramolecular Vibrational Energy Redistribution

The first step in a unimolecular reaction involves energizing the reactant molecule above its decomposition threshold. An accurate description of the ensuing unimolecular reaction requires an understanding of the state prepared by this energization process. In the first part of this chapter experimental procedures for energizing a reactant molecule are reviewed. This is followed by a description of the vibrational/rotational states prepared for both small and large molecules. For many experimental situations a superposition state is prepared, so that intramolecular vibrational energy redistribution (IVR) may occur (Parmenter, 1982). IVR is first discussed quantum mechanically from both time-dependent and time-independent perspectives. The chapter ends with a discussion of classical trajectory studies of IVR.

4.1 FORMATION OF ENERGIZED MOLECULES

A number of different experimental methods have been used to energize a unimolecular reactant. Energization can take place by transfer of energy in a bimolecular collision, as in

$$C_2H_6 + Ar \rightarrow C_2H_6^* + Ar \tag{4.1}$$

Another method which involves molecular collisions is chemical activation. Here the excited unimolecular reactant is prepared by the potential energy released in a reactive collision such as

$$F + C_2H_4 \rightarrow C_2H_4F^* \tag{4.2}$$

The excited C_2H_4F molecule can redissociate to the reactants F + C_2H_4 or form the new products H + C_2H_3F.

Vibrationally excited molecules can also be prepared by absorption of electromagnetic radiation. A widely used method involves initial electronic excitation by absorption of one photon of visible or ultraviolet radiation. After this excitation, many molecules undergo rapid radiationless transitions (i.e., intersystem crossing or internal conversion) to the ground electronic state, which converts the energy of the absorbed photon into vibrational energy. Such an energization scheme is depicted in figure 4.1 for formaldehyde, where the complete excitation/decomposition mechanism is

$$H_2CO(S_0) + h\nu \rightarrow H_2CO(S_1) \rightarrow H_2CO^*(S_0) \rightarrow H_2 + CO. \tag{4.3}$$

Here, S_0 and S_1 represent the ground and first excited singlet states.

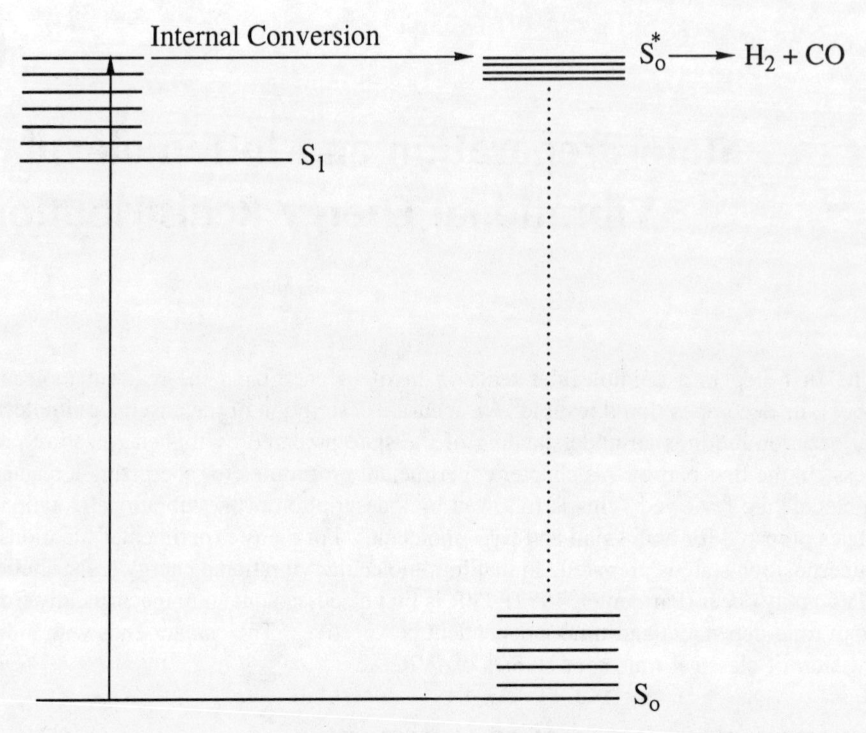

Figure 4.1 Preparation and decomposition of a vibrationally excited ground-state H_2CO molecule (S_o^*) by internal conversion from a vibrational level of S_1.

The formation of energized molecules by radiationless transitions does not necessarily require that the initial electronic excitation occur via absorption of electromagnetic radiation. For example, in the formation of ions by electron impact, that is,

$$e^- + A \rightarrow A^+ + 2e^-, \tag{4.4}$$

the product ion A^+ is often prepared in electronically excited states (Forst, 1973). Rapid radiationless transitions may then lead to a vibrationally excited A^+ ion in the ground electronic state.

In recent experiments lasers have been used to excite unimolecular reactants. In overtone excitation, states in which an MH or MD bond contains n quanta of vibrational energy are prepared by direct single-photon absorption. Here, M is a massive atom in contrast to H or D. For states with large n, the energy in the bond may exceed the molecule's unimolecular threshold. Overtone excitation of CH and OH bonds has been used to decompose molecules (Crim, 1984, 1990). One limitation of this technique is the very weak oscillator strengths of overtone absorptions.

Molecules also acquire vibrational energy by the absorption of infrared photons. In contrast to visible and ultraviolet radiation, the energy provided by a single infrared photon is usually much less than the unimolecular threshold energy. However, by using a sufficiently intense monochromatic radiation source such as an infrared laser, a molecule may be compelled to absorb multiple infrared photons so that its total vibrational energy is in excess of the threshold.

Two additional excitation methods, both of which employ two lasers, have been developed for highly selective state preparation: stimulated emission pumping (SEP) (Kittrell et al., 1981) and double resonance excitation (Luo and Rizzo, 1990). The excitation mechanism for SEP is depicted in figure 4.2. The first laser prepares a vibrational/rotational level of an excited electronic state. Before this state undergoes a spontaneous radiative or nonradiative transition, a second laser with a lower energy photon stimulates emission of light. The stimulated emission forms a highly excited vibrational/rotational level of S_0. Because of the high resolution of the lasers, individual rotation/vibration states of S_0 may often be prepared.

The energy level diagram for an infrared-optical double resonance experiment is given in figure 4.3. An infrared laser promotes molecules from the ground vibrational state to a state with one quantum in a particular mode, for example, an OH stretch. After a delay, these molecules absorb a visible photon from a second laser to put them in a final state with energy in excess of the unimolecular threshold. Because the first transition prepares a single J,K state, this excitation scheme may resolve individual rotational transitions to the final state. The infrared optical double resonance, stimulated emission pumping and overtone excitation experimental techniques are discussed in considerable detail in chapter 5.

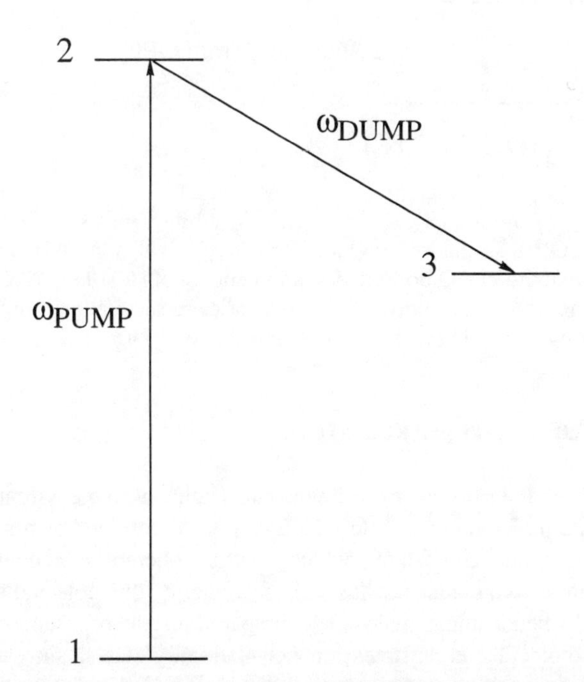

Figure 4.2 Three-level stimulated emission pumping (SEP) scheme, showing PUMP and DUMP transitions. The PUMP laser excites a vibrational/rotational level of an electronic state. The DUMP laser then stimulates emission from this state to an excited vibrational/rotational level of the ground electronic state (Kittrell et al., 1981).

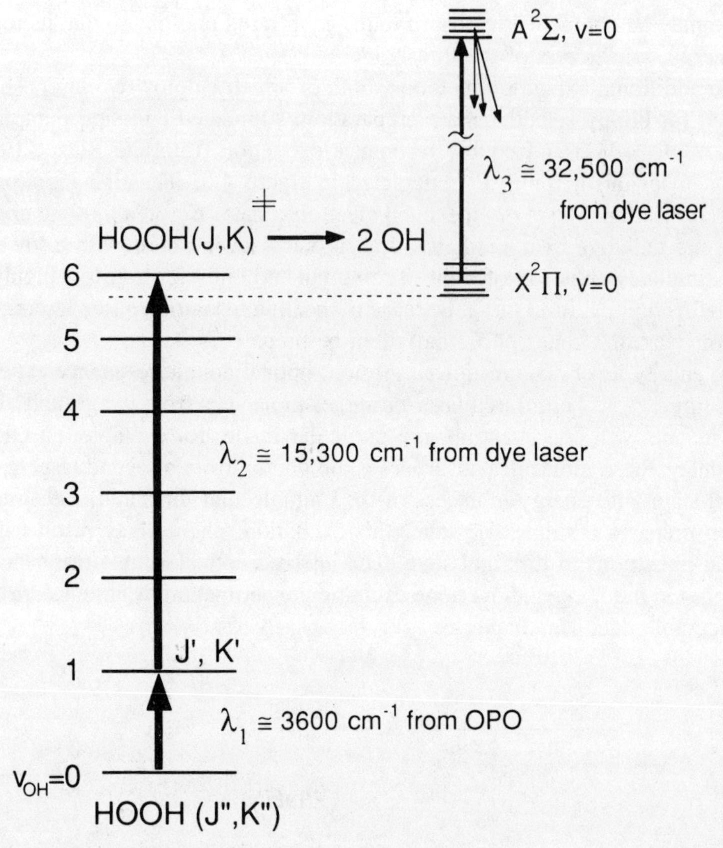

Figure 4.3 Energy level diagram for infrared-optical double resonance excitation of HOOH to the $6\nu_{OH}$ vibrational overtone level and subsequent laser-induced fluorescence detection of the OH dissociation products. The $6\nu_{OH}$ level is at 18,943 cm^{-1} above the ground state. Since the O—O bond dissociation energy of HOOH is 17,035 cm^{-1}, molecules dissociating from $6\nu_{OH}$ have 1913 cm^{-1} of excess vibrational energy to be partitioned between the two OH fragments (Luo and Rizzo, 1990).

4.2 NATURE OF THE PREPARED STATE

In the most general terms an experiment can excite a single vibrational/rotational eigenstate of the molecule or a state which can be represented as a superposition (or collection) of individual eigenstates, which is either coherent or incoherent (see discussion below). Since each eigenstate has a discrete energy, the molecule's energy is more highly resolved when a single eigenstate is prepared instead of a superposition. Strictly speaking, the molecular eigenstates are not stationary states, since the true system eigenstates have contributions from both the molecule and radiation field. This is the origin of radiative lifetimes for the molecular energy levels (Steinfeld, 1985). However, for the presentation in this chapter, the molecular eigenstates will be treated as stationary with respect to coherent superposition states which are much shorter-lived.

The nature of the prepared state is affected by the: (1) energy resolution of the excitation process; (2) the size of the molecule; and (3) the properties of the vibrational/rotational energy levels.

4.2.1 Energy Resolution of the Excitation Process

The origin of the energy resolution is a very important issue and should be considered very carefully, since it affects the intramolecular and/or unimolecular dynamics. A transform limited pulsed laser (i.e., a laser for which the energy resolution is only limited by the time-energy uncertainty principle) has an energy resolution given by the Heisenberg uncertainty principle $\Delta E \Delta t \geq \hbar$. Such a laser prepares, in a single molecule, a coherent superposition of states

$$\Psi = \Sigma c_n \psi_n , \tag{4.5}$$

where the ψ_n are the wave functions for the individual eigenstates and the c_n are determined by the interaction between the molecule and the laser. By increasing the laser's pulse width and, thus, its energy resolution, there are fewer eigenstates in the superposition. Figures 4.4(a) and 4.4(b) illustrate a single eigenstate and a superposition state prepared by a pulsed laser.

The same energy resolution as that for a pulsed laser can arise from a cw laser running at many discrete modes. However, there are no phase relations between the excited molecules, and the prepared state can be viewed as an incoherent superposition. It is simply a collection of different molecules each excited to an individual eigenstate by one of the discrete modes of the cw laser. If the cw laser has a high resolution so that it runs at only one mode a single eigenstate would be prepared. This assumes that spacings between the energies for the eigenstates is larger than the band width of the laser.

States of unimolecular reactants prepared by collisional energization and chemical activation, reactions (1) and (2), can also be viewed as incoherent superposition states. The most specific excitation will occur when the collision partners are in specific vibrational/rotational states and the relative translational energy is highly resolved. However, even for this situation it is difficult to avoid preparing a superposition state since the collisions have a distribution of orbital angular momentum.

Selective excitation becomes easier as fewer initial states are populated by the molecule to be excited. The most desirable situation is when the molecule is initially in only one vibrational/rotational state, for example, the zero point level. Then a single mode cw laser or a pulsed laser with a sufficiently long pulse width can excite the molecule from this state to a single vibrational/rotational eigenstate. Figure 4.4(c) illustrates that the effect of an initial thermal energy distribution may be to allow many more transitions between initial and final states, and, thus, excite many states for the same laser resolution. This is because it is very likely that there will be an accessible final state for most (if not all) of the thermally populated initial states, given the laser excitation energy and its resolution. More selective excitation is possible, from an initial thermal distribution, as the excitation energy is decreased and the final state density (see below) becomes smaller. With two lasers, one of low frequency and a second at a higher frequency (e.g., an infrared-optical double resonance experiment, as in figure 4.3), it is possible to excite a single vibrational/rotational eigenstate from a

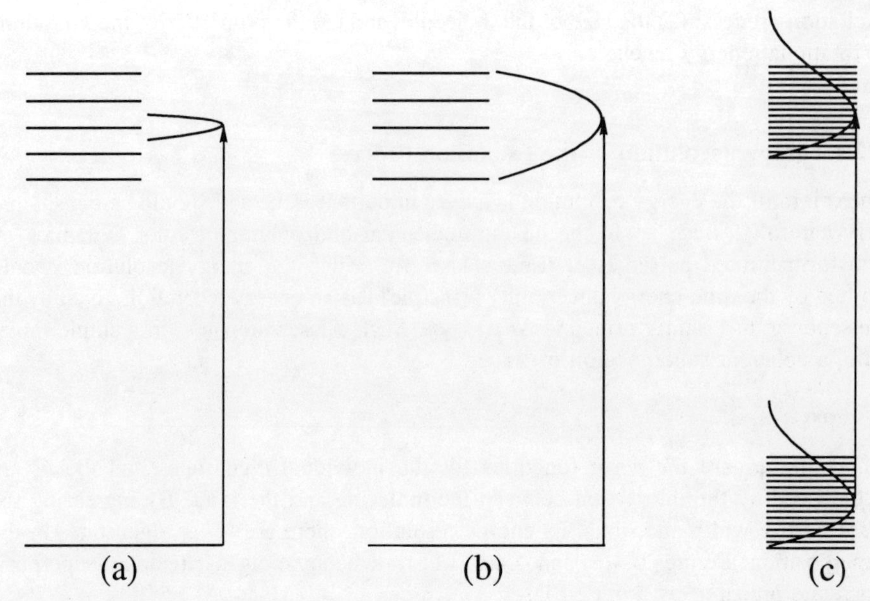

Figure 4.4 Illustration of different state preparations by laser excitation: (a) transition from a single initial state to one final state, (b) transition from a single initial state to a superposition of final states, (c) exciting a Boltzmann distribution of initial states.

thermal distribution for a laser resolution that would excite many final eigenstates in a one photon excitation process.

4.2.2 Small versus Large Molecules

Another important factor in determining the type of state prepared when energizing a molecule is the magnitude of the molecule's density of states. The density of states denoted by $\rho(E)$ is defined as the number of states per energy interval (e.g., states/cm^{-1}) and prescriptions for evaluating $\rho(E)$ are given in chapter 6. Harmonic vibrational densities of state are plotted in figure 4.5 for CH_4 and C_6H_6. At low energies $\rho(E)$ is very small for CH_4 and a particularly high energy resolution is not needed to excite an individual vibrational state. For example, with $\rho(E) = 2$ cm^{-1} a resolution of 0.5 cm^{-1} or less is required. However, as E is increased higher resolution in the excitation process is required to excite individual states.

At the same energy $\rho(E)$ is much larger for C_6H_6 than CH_4. This illustrates the general property that $\rho(E)$ tends to increase as the size of the molecule increases. The state density is also affected by the frequencies of the vibrational modes, since a mode with a low vibrational frequency will give rise to a larger $\rho(E)$ than a mode with a high vibrational frequency. Thus, since the vibrational frequencies for CCl_4 are lower than those for CH_4, at the same energy CCl_4 will have a larger $\rho(E)$ than CH_4. In summarizing the above factors that influence $\rho(E)$, one finds that it is easiest to resolve individual states at low energies for small molecules with high vibrational frequencies.

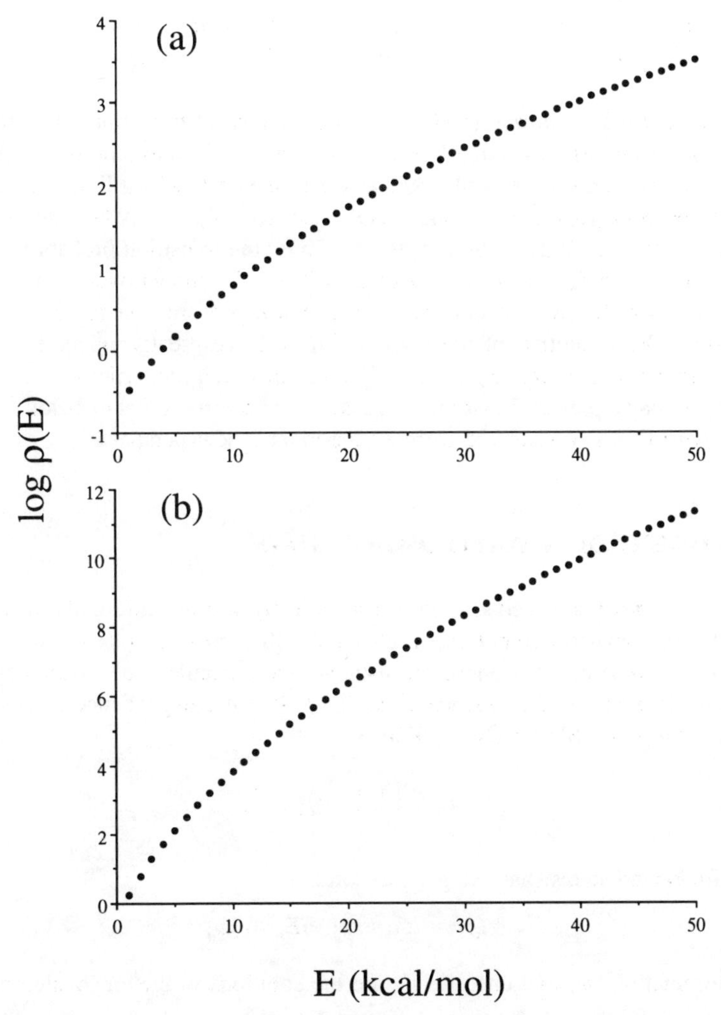

Figure 4.5 Harmonic densities of state for methane (a) and benzene (b).

4.2.3 Properties of the Vibrational/Rotational Eigenstates

It is worthwhile considering how properties of the wave functions ψ_n for the vibrational/rotational eigenstates (see section 2.3.2) affect the prepared state. Consider a symmetric top molecule, initially in a specific energy level with angular momentum J, absorbing laser radiation which has insufficient resolution to excite an individual energy eigenstate during the excitation step. Thus, the prepared state may contain many eigenstates. Assume the wave function for the initial state is characterized by good vibrational (n) and rotational (J,K) quantum numbers. If K is also a good quantum number for the final states, only states which obey the selection rule $\Delta K = 0, \pm 1$ are included in the superposition state. Thus, the ψ_n included in the prepared state are restricted by this selection rule. However, as discussed in chapter 2, coriolis coupling can mix the $(2J + 1)$ K-states and destroy K as a good quantum number. Therefore, the transition probability which is restricted to $\Delta K = 0, \pm 1$ when K is a good quantum number will now be spread over all the $(2J + 1)$ K-states. As a result there will be many more ψ_n in the prepared state Ψ, Eq. (4.5), when K is not a good quantum number.

Assume that in a normal mode zero-order picture all transitions from the ground state to states in the energy range $E \rightarrow E + \Delta E$ are forbidden except one. Thus, even if the laser's resolution is of the order ΔE, a single eigenstate will still be prepared by the excitation process. However, for the actual eigenstates there may be extensive mixing between the normal mode zero-order levels. Thus, the transition probability, which is to only one state in the absence of mixing, will now be spread over many eigenstates and a laser with the same resolution as above will prepare a state containing many eigenstates. The resolution of the laser will have to be greatly enhanced to excite a single vibrational/rotational eigenstate. The way in which properties of the eigenstates affect the initially prepared state is discussed in much more detail below, in section 4.4.2, where IVR is considered from a time-independent perspective.

4.3 TIME-DEPENDENT SUPERPOSITION STATE

As described above, a pulsed laser with a sufficiently narrow pulse width will overlap a group of vibrational/rotational eigenstates and will prepare a coherent superposition, Eq. (4.5), of some or all of these eigenstates in a molecule. This superposition state evolves in time and is thus identified as $\Psi(t)$. Its time dependence is given by the equation of motion (Merzbacher, 1970):

$$i\hbar \frac{d\Psi(t)}{dt} = \hat{H}\Psi(t) \tag{4.6}$$

which, for bound states, has the general solution

$$\Psi(t) = \Sigma c_n \exp[-iE_n t/\hbar]\psi_n . \tag{4.7}$$

The eigenfunctions ψ_n and energies E_n are the solutions to the eigenvalue problem

$$\hat{H}\psi_n = \mathrm{E}_n\psi_n, \tag{4.8}$$

where $\hat{H}$ is the vibrational/rotational Hamiltonian for the molecule. The ψ_n and E_n are the wave functions and energies for the vibrational/rotational eigenstates.

4.3.1 Zero-Order Picture

In chapter 2 it was shown that the eigenstate wave functions ψ_n and energies E_n in Eq. (4.8) can be obtained by diagonalizing the molecular Hamiltonian with a zero-order product basis set ϕ_ℓ. The ψ_n and ϕ_ℓ are related by the matrix equation

$$\phi = \mathbf{C}\psi \tag{4.9}$$

where $\mathbf{C}$ is the eigenvector matrix which results from the diagonalization of the Hamiltonian. If the Hamiltonain matrix is real symmetric, the $\mathbf{C}$ matrix is unitary and

$$\psi = \tilde{\mathbf{C}}\phi \tag{4.10}$$

The exact (i.e., complete) Hamiltonian $\hat{H}$ is related to the zero-order Hamiltonian $\hat{H}_o$ by

$$\hat{H} = \hat{H}_0 + V, \tag{4.11}$$

where V is the perturbation. The perturbation may arise from either potential or kinetic energy coupling, depending upon what coordinate system is used to represent the Hamiltonian, for example, normal mode or internal coordinates (see chapter 2). The operator $\hat{H}_o$ is often called the unperturbed Hamiltonian. Its eigenvalue equation is $\hat{H}_o\phi_l = E_l^o\phi_l$.

If the perturbation V were absent, the superposition state $\Psi(t)$ in Eq. (4.7) would be

$$\Psi(t) = \Sigma c_l exp[-iE_l^o t/\hbar]\phi_l . \tag{4.12}$$

When the perturbation is present this expression is no longer a solution to the equation of motion, Eq. (4.6), but it is still correct to expand $\Psi(t)$ in terms of the ϕ_l of the unperturbed problem, provided that the expansion coefficients become time-dependent (Merzbacher, 1970):

$$\Psi(t) = \sum_l c_l(t)\exp[-iE_l^o t/\hbar]\phi_l , \tag{4.13}$$

If this equation is substituted into Eq. (4.6) one obtains the following equations of motion for the $c_l(t)$:

$$i\hbar \frac{dc_l}{dt} = \sum_k V_{lk}c_k\exp[-i(E_k^o - E_l^o)t/\hbar], \tag{4.14}$$

where

$$V_{lk} = \langle\phi_l|V|\phi_k\rangle \tag{4.15}$$

is the matrix element of the perturbation between the zero-order states. The term $|c_l(t)|^2$ represents the probability $P_l(t)$ of being in the zero-order state ϕ_ℓ at time t.

The probability amplitude versus time of the initially prepared superposition state is given by

$$C(t) = \langle\Psi(0)|\Psi(t)\rangle. \tag{4.16}$$

The interpretation of the time evolution of $\Psi(t)$ is facilitated (Baggott et al., 1986) by using a zero-order product basis which includes the initially prepared state $\Psi(0)$. This product basis function can be identified as ϕ_s. Then, according to Eq. (4.16), the probability amplitude of finding the system in $\Psi(0) = \phi_s$ versus time is given by

$$C_s(t) = c_s(t)\ \exp\{-iE_s^o t/\hbar] \tag{4.17}$$

where it has been assumed the ϕ_l constitute an orthonormal basis. The term $|C_s(t)|^2$ represents the probability $P_s(t)$ of being in the zero-order state ϕ_s at time t.

4.3.2 Two Examples of $\Psi(t)$

An example of the time evolution of a nonstationary state is the well-documented double-well problem, for instance, the NH_3 inversion. The stationary eigenstates of the system are symmetric and antisymmetric linear combinations of the eigenstates for each well:

$$\psi_s = \frac{1}{2^{\frac{1}{2}}} (\phi_I + \phi_{II}), \qquad \psi_a = \frac{1}{2^{\frac{1}{2}}} (\phi_I - \phi_{II}). \tag{4.18}$$

For these stationary states the system cannot be defined as being in either well, and there is no time evolution; that is, the wave function is spread over both wells. If the excitation is sufficiently broad to excite both ψ_s and ψ_a, which are separated by an amount ΔE, a nonstationary superposition (or zero-order) state will be prepared. At $t = 0$ this state is found by inverting Eq. (4.18) and is given by

$$\Psi(0) = \phi_I = \frac{1}{2^{\frac{1}{2}}} (\psi_s + \psi_a). \tag{4.19}$$

The combinations $\psi_s + \psi_a$ and $\psi_s - \psi_a$ represent localized wave functions in the two different wells. According to Eq. (4.7), the time evolution of the initial superposition $\psi_s + \psi_a$ is given by

$$\Psi(t) = \exp(-iE_s t/\hbar)\, \psi_s + \exp(-iE_a t/\hbar)\, \psi_a. \tag{4.20}$$

Using $\Delta E = E_a - E_s$, $\nu = \Delta E/h$, and $\omega = 2\pi\nu$, $\Psi(t)$ becomes

$$\begin{aligned}\Psi(t) &= 1/2 \exp(-iE_s t/\hbar)\{(\psi_s + \psi_a)[1 + \exp(-i\omega t)] + (\psi_s - \psi_a)[1 - \exp(-i\omega t)]\} \\ &= \exp[-i(E_s + E_a)t/2\hbar][(\psi_s + \psi_a)\cos\pi\nu t + i(\psi_s - \psi_a)\sin \pi\nu t].\end{aligned} \tag{4.21}$$

This last form shows that the system shuttles between $\psi_s + \psi_a$ and $\psi_s - \psi_a$ with frequency $\nu/2$. The physical process is tunneling back and forth between the wells. Since the splitting ΔE between ψ_s and ψ_a increases with energy (Merzbacher, 1970), the tunneling rate increases with energy.

Since each well is identified with a particular conformation, one sees that the time dependence of $\Psi(t)$ is akin to the concept of molecular structure. For example, consider the isomers ethyl alcohol and dimethyl ether. At an internal energy of ~10 kcal/mol neither is a stationary state. The reason that structural attributes may be associated with each is because tunneling is very slow. Thus, a $\Psi(t)$ that evolves slowly is similar to a stable molecular conformation.

An analog to the above double-well problem are the local mode O—H vibrations found in water. Extensive studies have been performed (Lawton and Child, 1979, 1980, 1981; Sibert et al., 1982a,b) for a water model which accurately represents the coupling between O—H stretches but neglects the bending degree of freedom. A zero-order picture for the O—H stretch energy levels consists of pairs of local-mode states ϕ_{nm} in which there are n quanta in one bond and m in the other or vice versa. However, local-mode states are not the eigenstates for the system. As a result of couplings in the Hamiltonian, the eigenstates are symmetry-adapted combinations of the (n,m) and (m,n) states like the combinations for the double-well problem, Eq. (4.18); that is, $\psi_s = 1/2^{\frac{1}{2}}(\phi_{nm} + \phi_{mn})$ and $\psi_a = 1/2^{\frac{1}{2}}(\phi_{nm} - \phi_{mn})$. (In a more complete analysis small amounts of local-mode states higher and lower in energy will be included in forming ψ_s and ψ_a. However, the model used here incorporates all the important dynamics of the energy transfer.) Plots of ψ_s and ψ_a for the (5,0) and (0,5) local-mode states are shown in figure 4.6(a) and 4.6(b). The expressions for ψ_s and ψ_a may be inverted to obtain the superpositions

$$\phi_{nm} = \frac{1}{2^{\frac{1}{2}}}(\psi_s + \psi_a) \qquad \phi_{mn} = \frac{1}{2^{\frac{1}{2}}}(\psi_s - \psi_a)\,. \tag{4.22}$$

A plot of ϕ_{mn} for the (5,0) local-mode state is given in figure 4.6(c).

If either of the superpositions ϕ_{nm} or ϕ_{mn} is prepared experimentally, the system will oscillate back and forth between these two local-mode states, as above for the double-well problem. The energy splitting between the ψ_s and ψ_a eigenstates gives the rate of intramolecular vibrational energy transfer (or tunneling) between the OH bonds. However, in contrast to the double-well problem, here the tunneling rate decreases with energy. For the $(0,m)$ local-mode progression the rate decreases by five orders of magnitude between $m = 1$ and $m = 8$. In classical terms, a water molecule excited in the (0,8) state may execute 10^6 vibrations in one bond before transferring energy to the other. The energy splitting between ψ_s and ψ_a for the $(0,m)$ and $(m,0)$ local-mode states is given in table 4.1. The energy splitting, and, thus tunneling rate, is related to the coupling between the two local modes. This coupling decreases as m is increased.

4.3.3 Relationship between $\Psi(t)$ and Absorption Spectra

The fundamental equation which relates the time dependence of the superposition state $\Psi(t)$ and the absorption spectrum is, for the weak-field limit, the Fourier transform

$$I(\omega) = \frac{1}{2\pi}\int_{-\infty}^{\infty} dt\,\langle \Psi(0) \mid \Psi(t)\rangle\, e^{i\omega t}\,, \tag{4.23}$$

where, according to Eq. (4.16), $\langle\Psi(0)|\Psi(t)\rangle$ defines the probability amplitude of the superposition state prepared at $t = 0$. This equation provides the connection between time-independent spectroscopy in the frequency domain and time-dependent spectroscopy in the time domain. If $\Psi(t)$ is evaluated out to infinite time, it is straight forward to show that $I(\omega)$ becomes a stick spectrum of the energies for the individual eigenstates. For $\Psi(t)$ in Eq. (4.7) evaluated to infinite time, the absorption spectrum becomes

$$I(\omega) = \sum_n c_n^2 \delta(\omega - E_n/\hbar)\,. \tag{4.24}$$

The probability amplitude $\langle\Psi(0)|\Psi(t)\rangle$ can be found from the absorption spectrum $I(\omega)$ by inverting Eq. (4.23) and taking the Fourier transform of $I(\omega)$. The integration is now over ω instead of t.

If the superposition state is written as a zero-order state instead of a linear combination of eigenstates, the probability amplitude is given by Eq. (4.17), so that the absorption spectrum becomes

$$I(\omega) = \frac{1}{2\pi}\int_{-\infty}^{\infty} c_s(t)\exp[i(\omega - E_s^{\circ}/\hbar)t]dt\,, \tag{4.25}$$

where E_s° is the transition energy to the zero-order state ϕ_s. If $c_s(t)$ is evaluated to infinite time, $I(\omega)$ in Eq. (4.25) becomes a stick spectrum and is given by

$$I(\omega) = \sum_n |\langle\phi_s \mid \psi_n\rangle|^2 \delta(\omega - E_n/\hbar)\,. \tag{4.26}$$

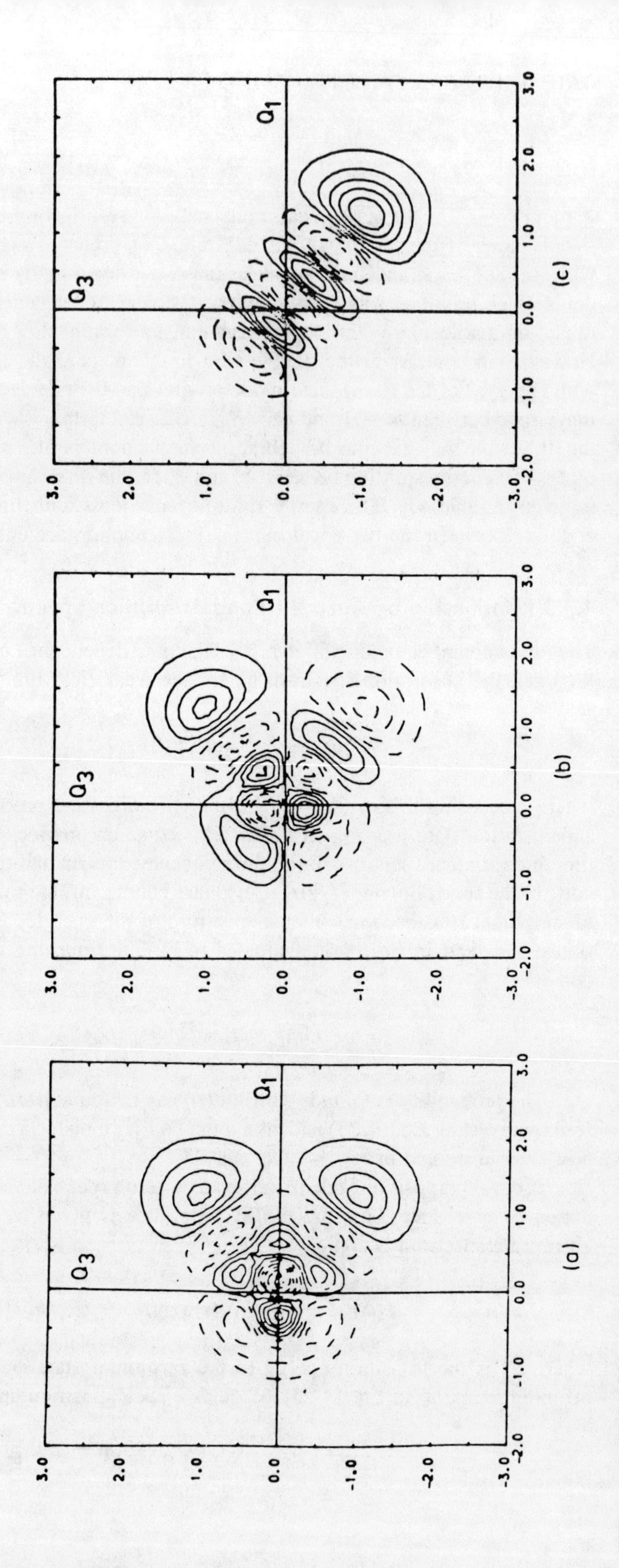
Q3
Q1
(a)
(b)
(c)
3.0
2.0
1.0
0.0
-1.0
-2.0
-3.0

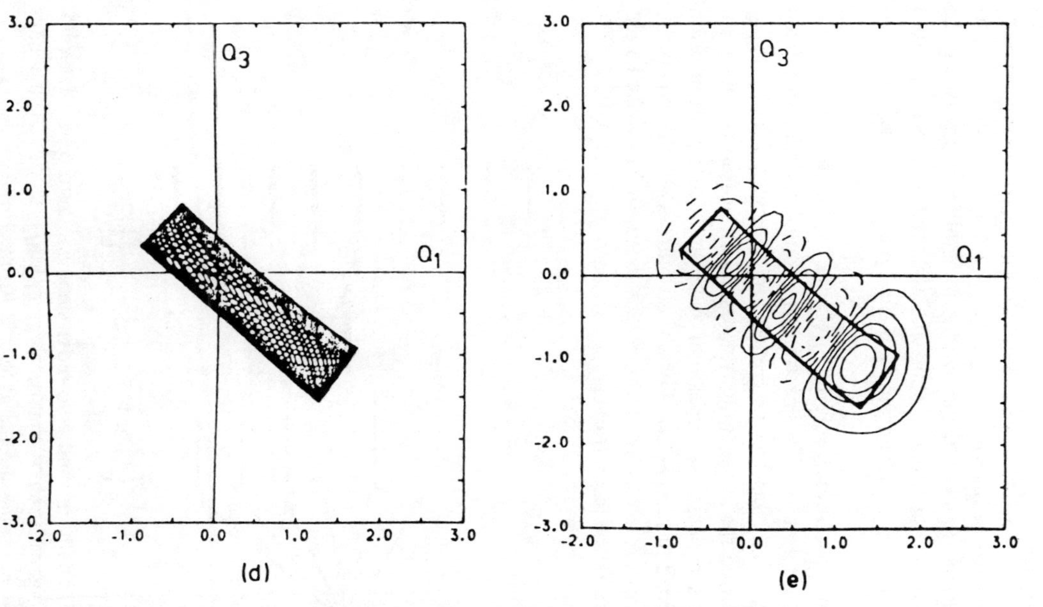

Figure 4.6 Wave functions and local trajectories, with continuous and dashed contours used to denote positive and negative phases of the wave functions. Parts (a) and (b) show the stationary-state wave functions for the symmetric and antisymmetric linear combinations of the $\phi_{5,0}$ and $\phi_{0,5}$ local-mode states, respectively. Part (c) shows the localized superposition $\phi_{5,0}$, (d) the corresponding local-mode trajectory, and (e) the caustics of the classical motion in relation to the localized superposition. (Reprinted with permission from Lawton and Child, 1981. Copyright 1981, Taylor and Francis.)

Table 4.1. Energy Splitting between Symmetric and Antisymmetric H_2O Eigenstates Corresponding to the $(0,m)$ Local-Mode State[a]

Local-Mode State	Energy Splitting (cm^{-1})
0,1	106.362
0,2	57.659
0,3	19.796
0,4	4.231
0,5	0.653
0,6	0.077
0,7	0.007
0,8	<0.000

[a]See Lawton and Child (1981).

This spectrum, figure 4.7, consists of lines at the eigenstate energies E_n whose heights are proportional to the magnitude squared of the overlap integral $\langle\phi_s|\psi_n\rangle$ between the superposition state and molecular eigenstates. Thus, the eigenstates seen in the $I(\omega)$ spectrum are just those which carry oscillator strength from the initial unexcited molecular state to ϕ_s.

An unresolved absorption spectrum $I(\omega)$ results from the evolution of $\Psi(t)$ in Eq. (4.6) and, thus, $C(t)$ for a finite time T. The spectrum $I(\omega)$ and $\Psi(t)$ are still related according to Eqs. (4.16) and (4.23), but the integration limits become T and $-T$. As $\Psi(t)$ and, thus, $C(t)$ are followed for longer times the resolution of $I(\omega)$ is improved. Consider the preparation of a coherent superposition state prepared by a transform limited pulsed laser as described in section 2.1. For a short pulse width the superposition will contain many eigenstates; that is, the number of n in Eq. (4.5) will be larger. If $\Psi(t)$ is followed in time by a probe laser, $I(\omega)$ can be calculated from Eq. (4.23). The

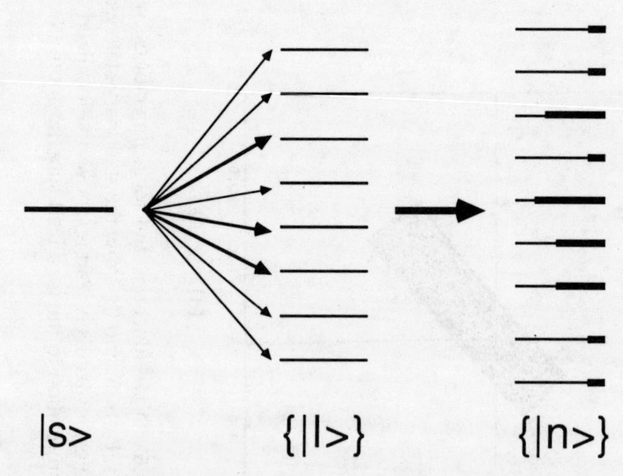

Figure 4.7 Depiction of the initially excited zero-order state $|s\rangle$, zero-order states $|l\rangle$ coupled to $|s\rangle$, and the exact eigenstates $|n\rangle$, that is, $|n\rangle = c_s^n|s\rangle + \Sigma_l c_l^n|l\rangle$. The heights of the bars associated with the $|n\rangle$ states are equal to $|c_s^n|^2$ (Crim, 1995).

resolution of $I(\omega)$ will improve as the time is increased. In the long time limit, an eigenstate spectrum will result, and one will know the c_n^2 in Eq. (4.24), which give the probability the eigenstate is present in $\Psi(t)$. The relationship between the frequency (energy) and time domain is that a highly resolved spectrum requires a long-time profile, while a highly unresolved spectrum is the result of a short-time profile (Heller, 1978; Lee and Heller, 1979; Lorquet et al., 1982).

4.4 INTRAMOLECULAR VIBRATIONAL ENERGY REDISTRIBUTION

The concept of intramolecular vibrational energy redistribution (IVR) can be formulated from both time-dependent and time-independent viewpoints (Li et al., 1992; Sibert et al., 1984a). IVR is often viewed as an explicitly time-dependent phenomenon, in which a nonstationary superposition state, as described above, is initially prepared and evolves in time. Energy flows out of the initially excited zero-order mode, which may be localized in one part of the molecule, to other zero-order modes and, consequently, other parts of the molecule. However, delocalized zero-order modes are also possible. The nonstationary state initially prepared is often referred to as the "bright state," as it carries oscillator strength for the spectroscopic transition of interest, and IVR results in the flow of amplitude into the manifold of so-called "dark states" that are not excited directly. It is of interest to understand what physical interactions couple different zero-order modes, allowing energy to flow between them. A particular type of superposition state that has received considerable study are M–H local modes (overtones), where M is a heavy atom (Child and Halonen, 1984; Hayward and Henry, 1975; Watson et al., 1981).

In contrast to time-domain measurements, experiments in the frequency domain excite eigenstates of the molecule, which do not evolve with time (Stuchebrukhov et al., 1989). Splittings and perturbations observable in high-resolution spectra can yield detailed information on the amount of mode coupling in a given molecule. At the most fundamental level one is interested in determining the amount of mixing between the zero-order states in forming the eigenstates. In the following both the time-independent and time-dependent pictures of IVR are described.

4.4.1 Time-Dependent IVR

4.4.1.1 Zero-Order Description

The zero-order picture of IVR follows the presentation given above in section 4.3.1 (p. 74) (Freed and Nitzan, 1980; Uzer, 1991). A superposition or zero-order state is formed by a pulsed laser. This state then evolves in time by couplings with other zero-order states. Figure 4.8 illustrates a situation, outlined by Stannard and Gelbart (1981), where there are three types of zero-order states. The zero-order state $|s\rangle$ is the superposition prepared by the excitation process. The zero-order states $\{|l\rangle\}$ are strongly coupled to $|s\rangle$; that is, V_{sl} from Eq. (4.15) is large. The remaining states comprise a "quasi-continuum", $\{|q\rangle\}$, which is weakly coupled to both $|s\rangle$ and $|l\rangle$. The molecular eigenstates $|n\rangle$ (i.e., ψ_n) are related to the zero order states through expansion:

$$|n\rangle = c_s^n|s\rangle + \sum_l c_l^n|l\rangle + \sum_q c_q^n|q\rangle . \tag{4.27}$$

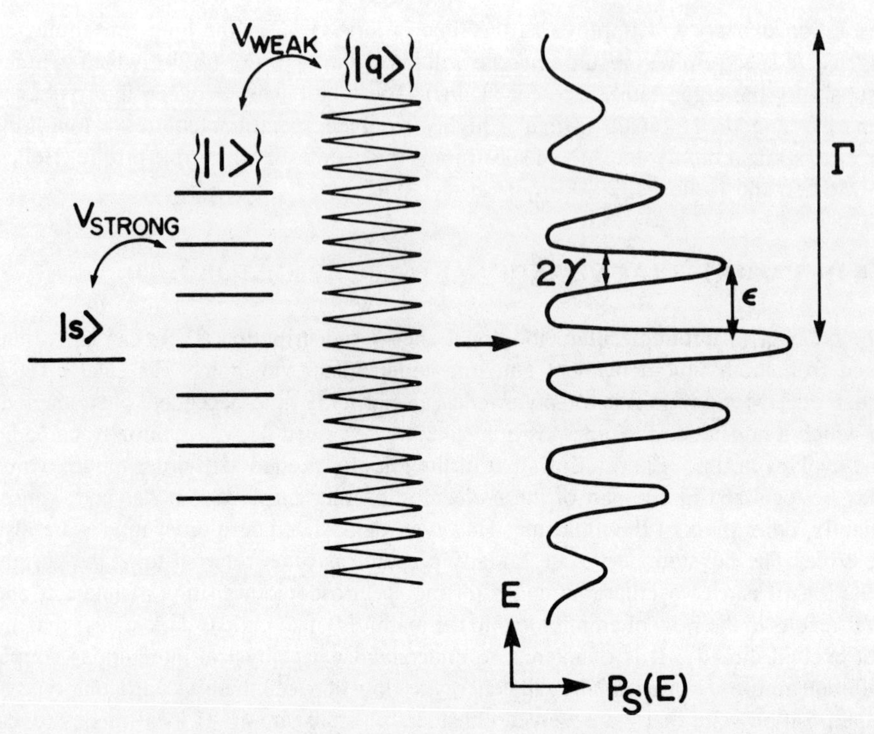

Figure 4.8 Coupling and resultant line shape for an initially prepared zero-order state |s (Stannard and Gelbart, 1981).

The relationship between the product basis ϕ_l described in section 4.3.1 and the states in figure 4.8 is straightforward. The initially prepared state $|s\rangle$ is the same as $\phi_s = \Psi(0)$. The remaining $\{|l\rangle\}$ and $\{|q\rangle\}$ states are the ϕ_l states, with ϕ_s removed.

Other scenarios for couplings between zero-order states are possible besides the one illustrated in figure 4.8. For example, $|s\rangle$ could be strongly coupled to sets of states, $\{|j\rangle\}$ and $\{|l\rangle\}$, with the coupling the strongest for the $\{|j\rangle\}$ states. In addition, there could be weak coupling to the quasi-continuum $\{|q\rangle\}$ and weak coupling between the $\{|l\rangle\}$ and $\{|j\rangle\}$ states.

Figure 4.9 depicts a possible time evolution $P_s(t)$, section 4.3.1, for the initially prepared state $|s\rangle$ in figure 4.8. The absorption spectrum $I(\omega)$ that results from this $P_s(t)$ is shown on the right side of figure 4.8. There are three principal features to the spectrum: (a) The strong coupling between the $|s\rangle$ and $\{|l\rangle\}$ states gives rise to a new set of molecular states within the $|s\rangle$ and $\{|l\rangle\}$ subspace; that is,

$$|m\rangle = c_s^m |s\rangle + \sum_l c_l^m |l\rangle . \tag{4.28}$$

In the absence of the continuum these molecular states would be the exact molecular eigenstates [see Eq. (4.27)]. (b) Each $|m\rangle$ state can be considered a "resonance" since it is broadened into a Lorentzian-like envelope of width γ as a result of the weak coupling between $|s\rangle$ and $\{|l\rangle\}$ to the quasi-continuum. The spacing between the resonance states,

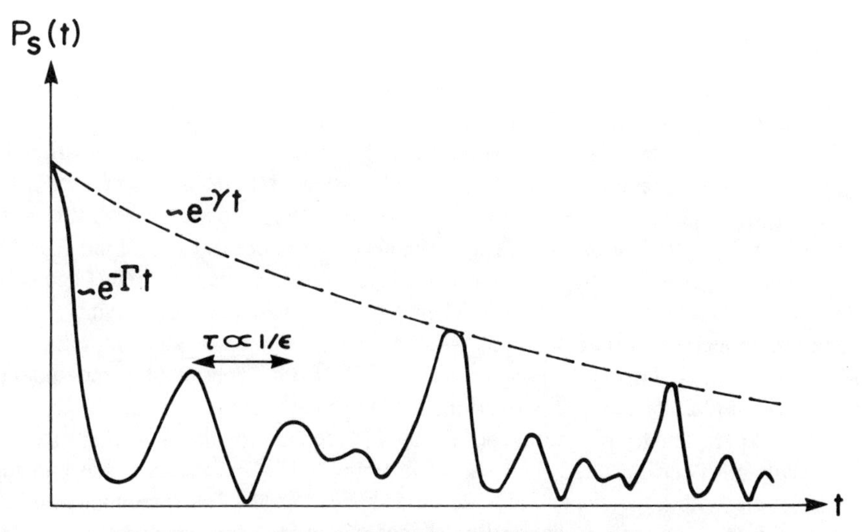

Figure 4.9 Time evolution of an initially prepared zero-order state |s> associated with the line shape in Figure 4.8 (Stannard and Gelbart, 1981).

ϵ, is on the order of the inverse density of $|l\rangle$ levels. (c) The $|s\rangle$–$|l\rangle$ interaction leads to a distribution of $|s\rangle$ probability over the energy range Γ. Each of the molecular states $|m\rangle$ in Eq. (4.28) has a different energy E_m. A plot of the $|c_s^m|^2$ [see Eq. (4.28)] versus E_m defines the spreading out (width $\sim \Gamma$) of $|s\rangle$ due to its interaction with $\{|l\rangle\}$.

As shown in figure 4.9, the time evolution $P_s(t)$ shows three distinct time dependencies, each characterized by either γ, Γ, or ϵ: (1) The Lorentzian-like envelope, with width Γ for the complete absorption spectrum transforms into an exponential decay with rate constant $\Gamma/\hbar$, which is for the short-time decay of $P_s(t)$. (2) The set of resonances, separated on average by an energy ϵ, transform into a set of oscillations (i.e., recurrences) whose periods are approximately $\epsilon/\hbar$. (3) The envelope of each individual resonance also transforms into an exponential-like decay, characterized by the rate $\gamma/\hbar$, which corresponds to "leakage from the sparse $|s\rangle - \{|l\rangle\}$ subspace into the quasi-continuum $\{|q\rangle\}$. The recurrences described above in (2) are damped out by this slow decay.

The above analysis pertains to a particular time-scale for $P_s(t)$. If only the initial decay of $P_s(t)$ is measured with a rate $\Gamma/\hbar$, the resulting absorption spectrum will be broad and featureless with a width Γ. Thus, a low-resolution spectrum corresponds to only observing the short-time decay of $P_s(t)$. On the other hand, if $P_s(t)$ were followed for a sufficiently long time, structure in the individual resonance envelopes will begin to appear in the spectrum. Finally, as discussed above, if $P_s(t)$ is evaluated to infinite time $I(\omega)$ will become a stick spectrum of the individual eigenstates $|n\rangle$ in Eq. (4.27) for which $c_s^n \neq 0$. A projection of $|s\rangle$ onto eigenstates is illustrated in figure 4.7.

In figure 4.8 only one particular type of superposition state $|s\rangle$ and couplings are considered. With higher resolution, the individual resonances $|m\rangle$, Eq. (4.28), in the spectrum could be excited. The time evolution for each $|m\rangle$ state would be well-described by a single exponential with rate $k_m = \gamma_m/\hbar$, where γ_m is the width of the resonance envelope.

4.4.1.2 Nonstatistical IVR: Benzene Overtones

If only a small number of states are coupled to the initially prepared zero-order state $|s\rangle$, as depicted in figure 4.8, and/or if there is a hierarchy of couplings between $|s\rangle$ and other zero-order states, the time evolution of $|s\rangle$ is nonexponential as illustrated in figure 4.9. Such behavior is often called nonstatistical IVR. Experimental studies of the absorption spectra of CH overtones in benzene show nonstatistical IVR with a nonexponential $P_s(t)$. The studies most relevant to discussions of IVR are those of the $n = 3$ and $n = 4$ overtones in which rotationally cold benzene is prepared (Page et al., 1987, 1988; Scoton et al., 1991a,b). There is still considerable uncertainty to what extent inhomogeneous broadening plagued the earlier room temperature experiments of the $n = 1$ to 9 overtones (Reddy et al., 1982). Both experimental and quantum mechanical studies of the $n = 3$ overtone are presented here.

The experimental spectrum reported by Page et al. for the $n = 3$ overtone of rotationally cold benzene (~5 K) is shown in figure 4.10. The spectrum has considerable structure with a major peak at 8827 cm^{-1}. Wyatt et al. (1992) reconstructed this spectrum with a sum of Lorentzian line shapes centered at frequencies ω_j,

$$I(\omega) = \sum_{j=1}^{N} \frac{A_j \Gamma^2}{(\omega - \omega_j)^2 + \Gamma^2}, \tag{4.29}$$

where Γ is the common half-width at half-maximum (hwhm) of each Lorentzian. Figure 4.11 shows reconstruction of the Page et al. spectrum with $N = 26$ and $\Gamma = 1$, 3, and 6 cm^{-1}.

As described at the beginning of section 4.3.3 (p. 77), the Fourier transform of the frequency spectrum gives the survival probability for the initially prepared zero-order (i.e., superposition) state, $|s\rangle$. A limited Fourier transform of $I(\omega)$ in Eq. (4.29) can be defined as

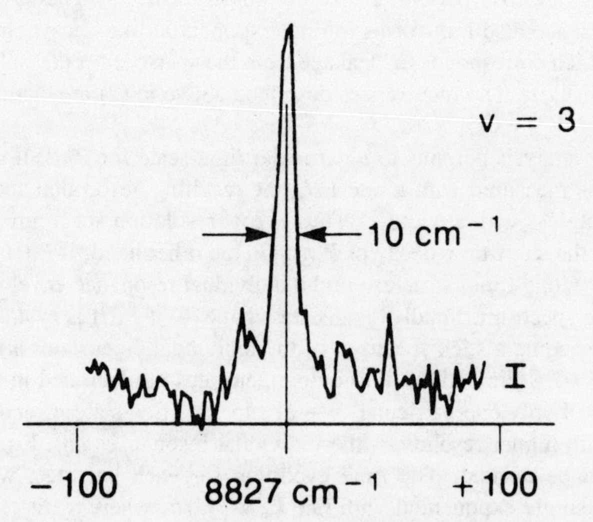

Figure 4.10 Spectrum of the $n = 3$ benzene overtone transition; $T_{rot} \approx 5$ K (Page et al., 1987).

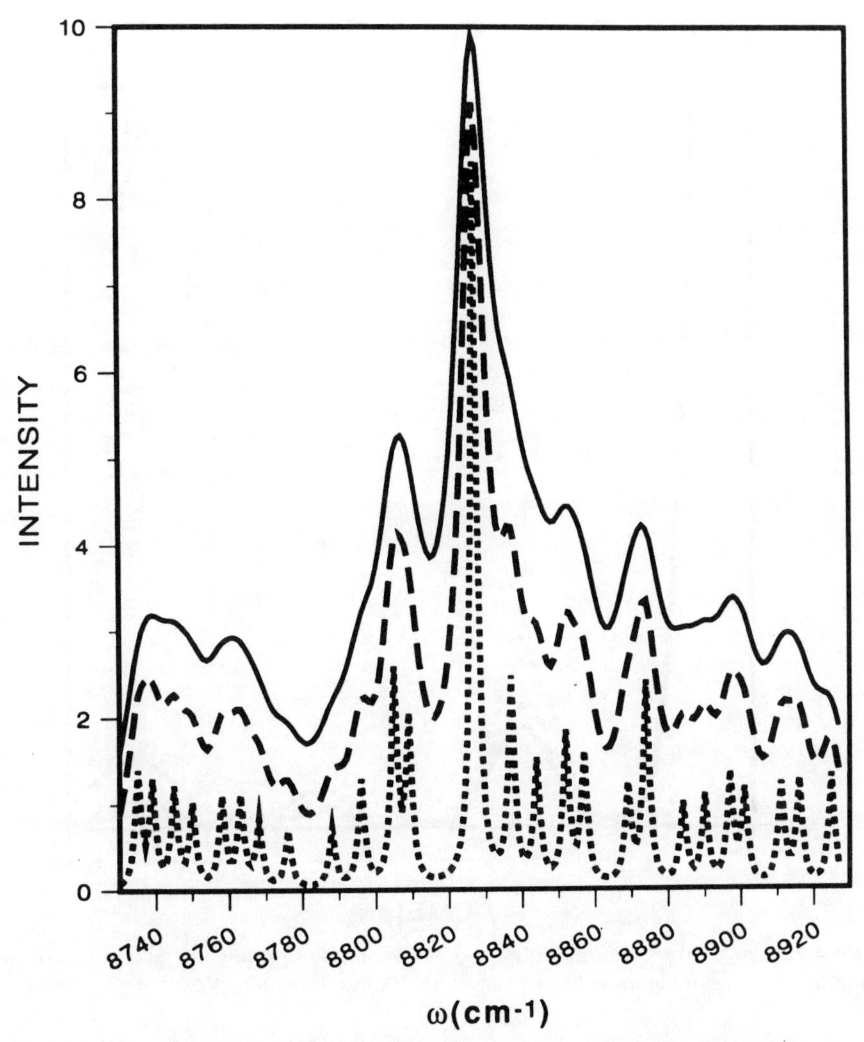

Figure 4.11 Reconstruction of the spectrum in Figure 4.10 with 26 Lorentzians with $\Gamma = 6\ \mathrm{cm}^{-1}$ (——), $\Gamma = 4\ \mathrm{cm}^{-1}$ (– – –), and $\Gamma = 1\ \mathrm{cm}^{-1}$ (----) (Wyatt et al., 1992).

$$p(t) = \left| \int_{\omega_1}^{\omega_2} I(\omega) e^{i\omega t} d\omega \right|^2, \tag{4.30}$$

where ω_1 and ω_2 define the frequency range of the experimental spectrum. The survival probability, normalized so that $P_s(0) = 1$, is

$$P(t) = p(t)/p(0). \tag{4.31}$$

The survival probability computed in this way is displayed in figure 4.12 for the three reconstructed spectra in figure 4.11. The survival plot shows rapid fall-off by $t = 10{,}000$ au ($\sim$0.25 psec), and small recurrences near $t = 17{,}000$ au, 32,000 au, and

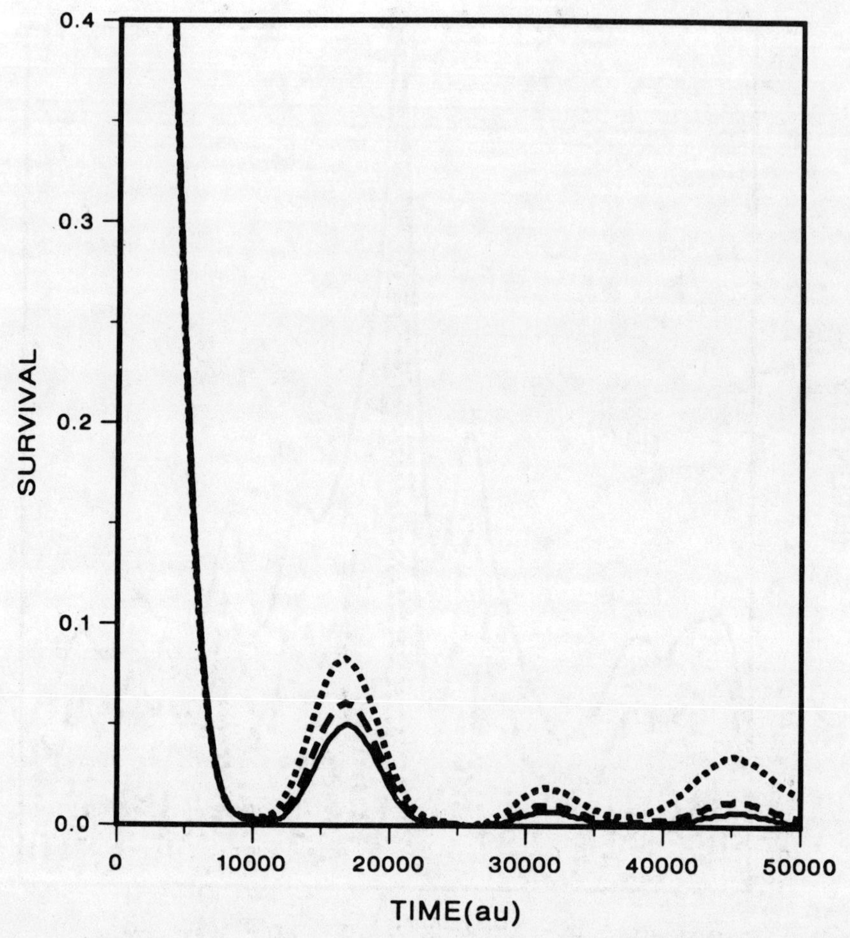

Figure 4.12 Survival probabilities obtained by Fourier transforming the reconstructed experimental spectra in Figure 4.11. 1 psec = 41,300 au time units (Wyatt et al., 1992).

45,000 au. The positions of the peaks are not sensitive to the value of Γ, but the peak heights increase as Γ decreases.

The rate of the initial decay of $P_s(t)$ is related to the width of the overall absorption envelope for the 8727 to 8927 cm^{-1} frequency range of the experimental spectrum. If a Fourier transform were taken of a Lorentzian $I(\omega)$ for only the principal peak in the spectrum, the decay rate would be significantly smaller.

The survival probability for the $n = 3$ overtone of benzene has been determined by a time-dependent quantum mechanical calculation based on a 16-mode planar benzene model (5 CH stretch modes are inactive out of the 21 planar modes). All states for these modes are treated in the calculation. The $n = 3$ zero-order overtone state is assumed to be a state with three quanta in one C—H internal coordinate (i.e., a local mode). The quantum survival probability is plotted in figure 4.13. It is similar to the experimental $P_s(t)$ in that there is a rapid fall-off at short times, followed by multiple small recurrences. This agreement supports the interpretation that the broad absorption

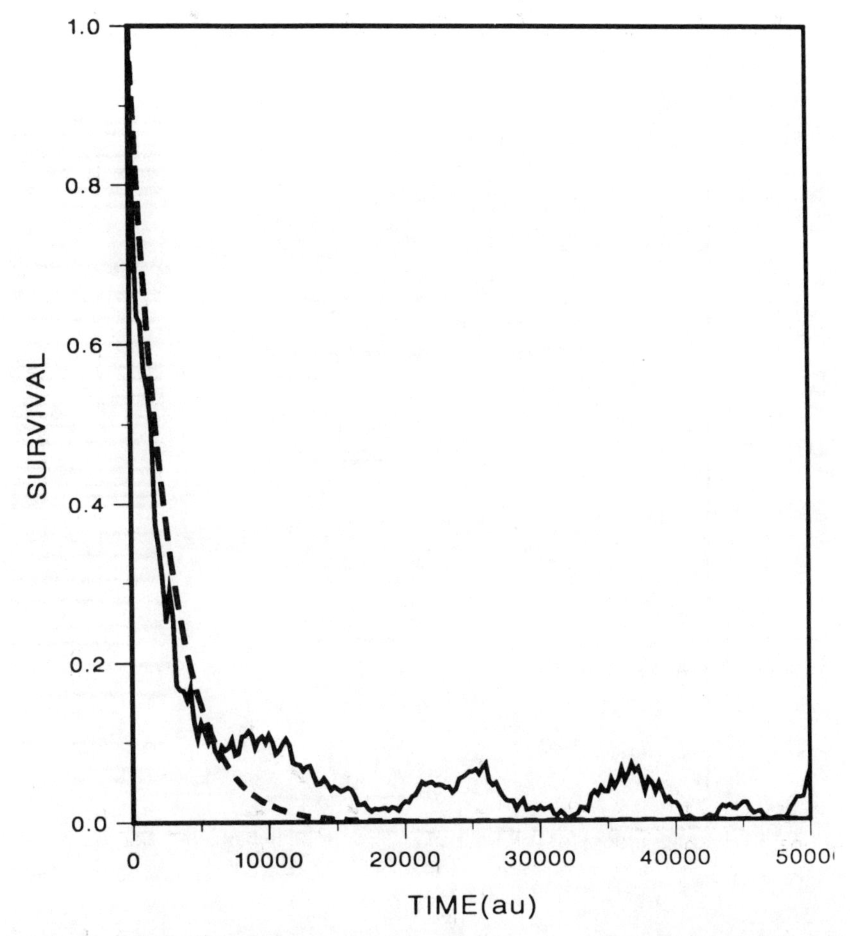

Figure 4.13 Comparison of quantum (——) and quasi-classical (---) survival probabilities for the $v = 3$ benzene overtone state. 1 psec = 41,300 au time units (Wyatt et al., 1992).

envelope of the experimental spectrum corresponds to a zero-order state with three quanta in only one C—H bond.

It was also found, from the quantum calculations, that the short time dynamics ($t < 0.5$ psec) of the $n = 3$ overtone is accurately reproduced by reduced four- and five-mode models. In addition, only a small subset of states for these modes are significantly populated during the early decay of the $n = 3$ overtone, even though the total density of states is very large. In the next section a Fermi resonance model is presented for the flow of energy from the excited C—H bond.

4.4.1.3 Multitiered and Fermi Resonance Model

A hierarchy of couplings are possible in a zero-order state model for IVR. In such a model only a subset of the zero-order states may be coupled to the initially prepared zero-order state ϕ_s. This subset is then strongly coupled to another subset and so forth. Each subset of zero-order states is called a tier. Figure 4.14 depicts a two tier model

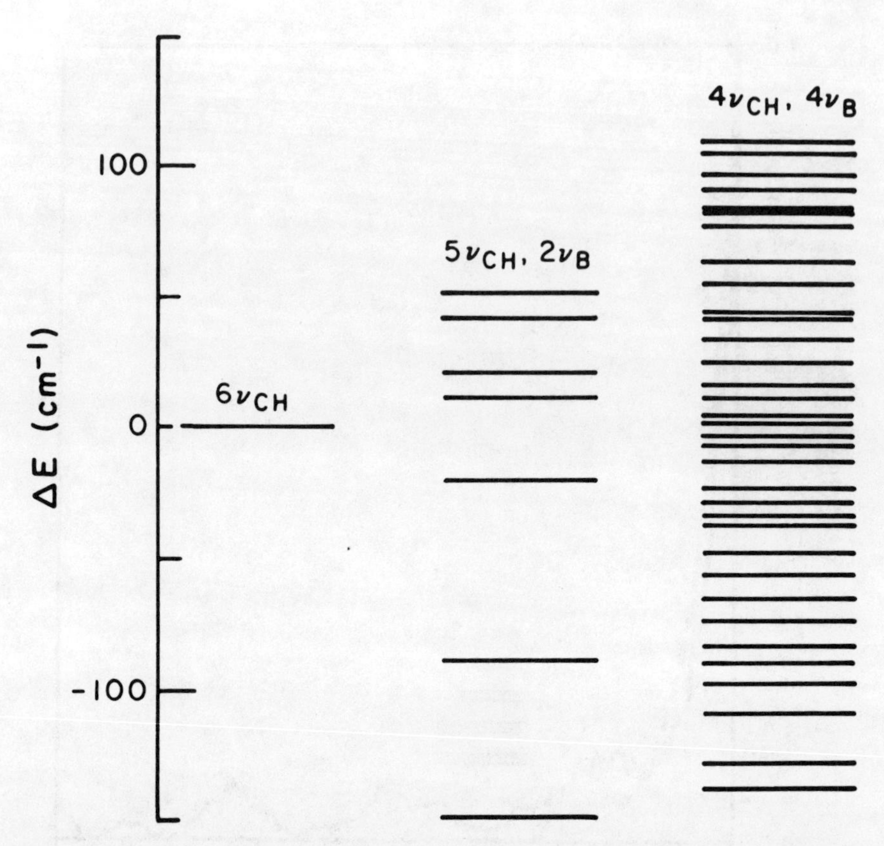

Figure 4.14 Two-tier model for IVR from a $6n_{CH}$ overtone in benzene. $2n_B$ represents two quanta in a HCC bending mode (Sibert et al., 1982c).

proposed to explain IVR from a zero-order state of benzene with six quanta in one C—H bond.

The concept of quantum mechanical resonance coupling is often used to explain IVR. An anharmonic resonance between two zero-order modes occurs when the ratio of frequencies for the modes is approximately given by a ratio of integers. From studies of vibrational spectra it is well known that low order resonances with small integer ratios are particularly effective for coupling modes. The splitting of almost degenerate vibrational energy levels in CO_2 was correctly interpreted as a 2:1 frequency resonance (i.e., "Fermi resonance") between the bending and stretching modes. Another named resonance is the 2:2 "Darling-Dennison" resonance. Other anharmonic resonances such as 1:3, 1:4, etc. are well known but are not named. In a multitiered model for IVR, each tier may be linked to a specific anharmonic resonance.

The model in figure 4.14 results from a 2:1 bend–stretch Fermi resonance model for IVR from benzene overtone states (Sibert et al., 1984a). The first tier is a group of states for which one quantum of CH stretch, n_{CH}, has been transferred to two quanta of CH bend, n_B. In the second tier an additional n_{CH} quantum has been transferred to two more n_B quanta. Though this model is based on strong physical arguments, it is still not known whether this Fermi resonance and tier model is the complete picture for IVR in benzene.

Quack and co-workers (Segall et al., 1987; Baggott et al., 1985; Marquardt and Quack, 1991) have solved Eqs. (4.13)–(4.17) for CH overtone states in a variety of molecules including CF_3H, CD_3H, and $(CF_3)_3CH$. They used the tridiagonal zero-order model $|v_s, v_b, \ell\rangle$, where v_s is the number of CH-stretch quanta, n_b the number of CH-bend quanta, and $\ell\hbar$ the vibrational angular momentum. Each overtone transition can be characterized by the quantum number $N = n_s + \frac{1}{2} n_b$. In these studies they found strong evidence for a 2:1 bend–stretch Fermi-resonance coupling in IVR for the CH overtones. Terms such as $\lambda q_s q_b^2$, where q_s is the stretch coordinate and q_b the bend coordinate, are important since these terms couple nearly degenerate stretch and bend states. For the $N = 6$ overtone of $(CF_3)_3CH$ efficient coupling and essentially complete redistribution between stretching and bending excitation occurs via the tridiagonal chain of $|n_s, n_b, \ell\rangle$ couplings $|6,0,0\rangle \rightleftharpoons |5,2,0\rangle \rightleftharpoons |4,4,0\rangle \rightleftharpoons |3,6,0\rangle \rightleftharpoons |2,8,0\rangle \rightleftharpoons |1,10,0\rangle \rightleftharpoons |0,12,0\rangle$. Time-dependent populations of the $|6,0,0\rangle$, $|5,2,0\rangle$, and $|0,12,0\rangle$ zero-order states are given in figure 4.15.

4.4.1.4 Statistical Limit

The statistical limit for IVR is attained when the initially prepared zero-order state $|s\rangle$ is coupled to a sufficiently large number of other zero-order states, $|l\rangle$, so that $|s\rangle$ appears to decay irreversibly and practically to zero, within the experimental time scale. A recent theoretical treatment of this model has been given by Voth (1987, 1988). For this situation the probability $|c_s(t)|^2$, Eq. (4.17), of being in the initially prepared state $|s\rangle$ is given by the exponential

$$|c_s(t)|^2 = \exp(-kt). \tag{4.32}$$

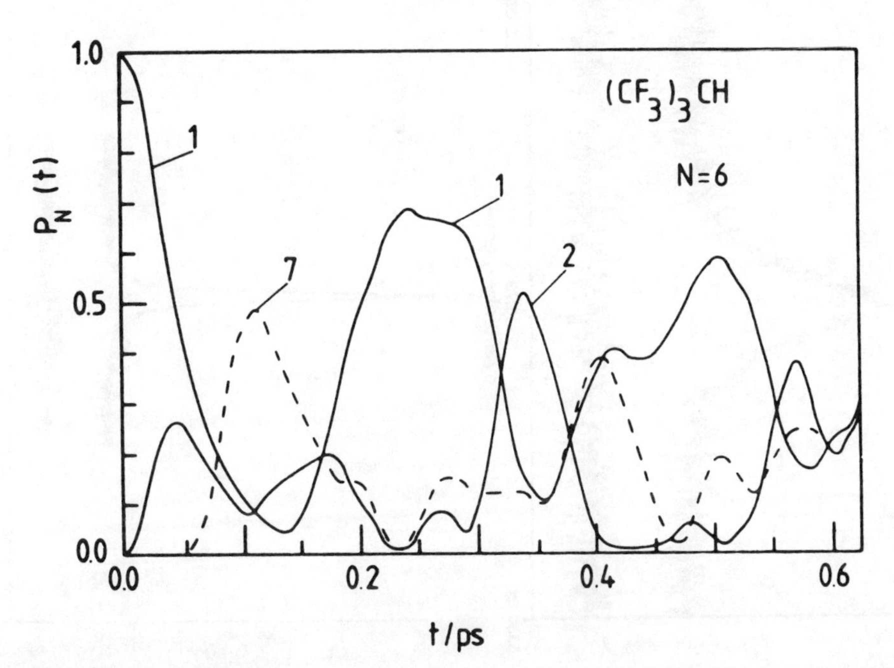

Figure 4.15 Time-dependent populations of some vibrational states $|n_s, n_b, l\rangle$ with $|6,0,0\rangle$ initially populated (labeled 1). The label 2 refers to $|5,2,0\rangle$ and 7 refers to $|0,12,0\rangle$ (Baggott et al., 1985).

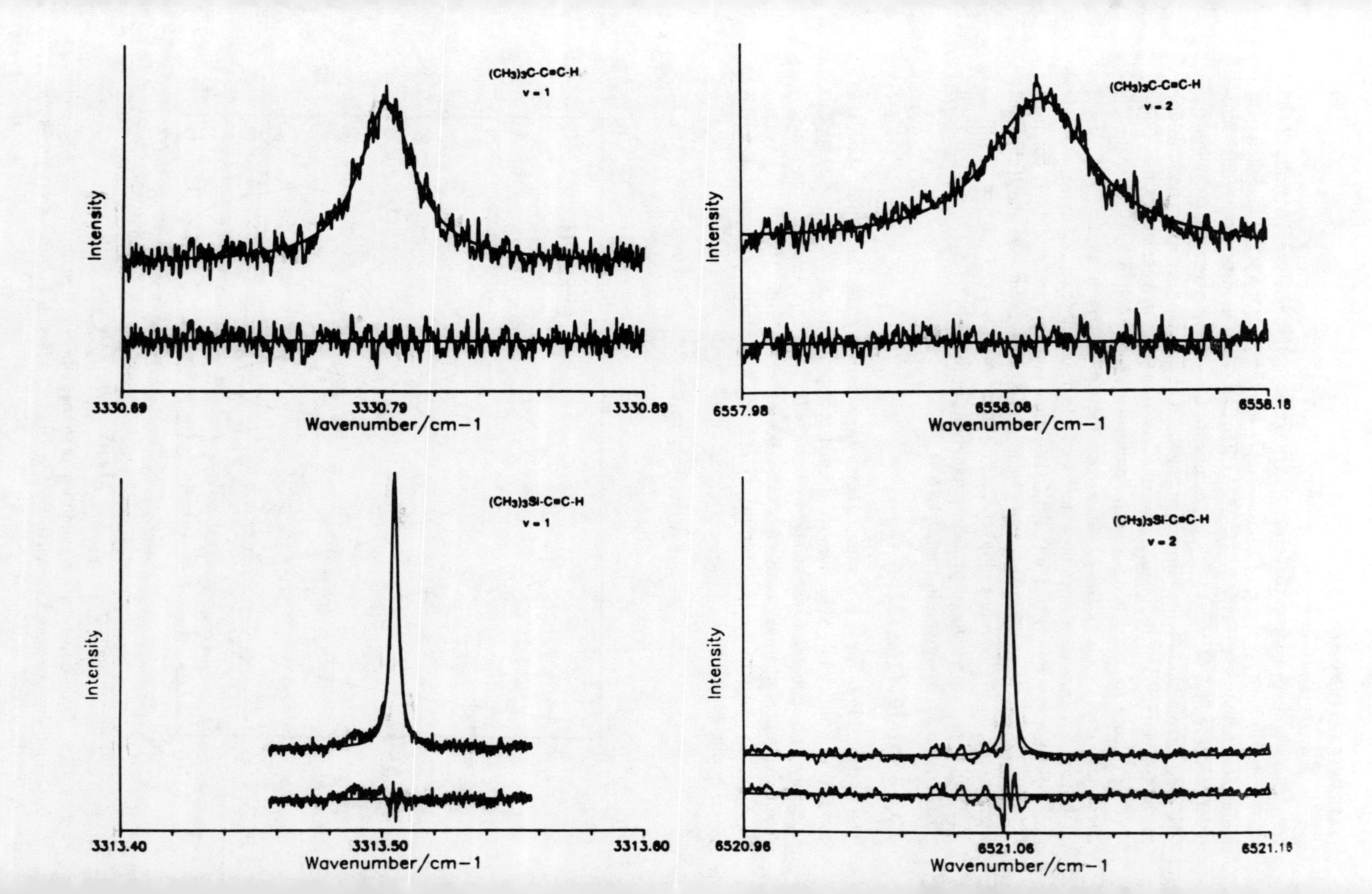
(CH3)3C-C≡C-H
v = 1
Intensity
3330.69
3330.79
3330.89
Wavenumber/cm−1
(CH3)3C-C≡C-H
v = 2
Intensity
6557.98
6558.08
6558.18
Wavenumber/cm−1
(CH3)3Si-C≡C-H
v = 1
Intensity
3313.40
3313.50
3313.60
Wavenumber/cm−1
(CH3)3Si-C≡C-H
v = 2
Intensity
6520.96
6521.06
6521.16
Wavenumber/cm−1

The parameter k is the rate constant for intramolecular vibrational energy redistribution and, according to the Fermi golden rule is given by

$$k = \frac{2\pi}{\hbar} \bar{V}_{sl}^2 \rho_l(E), \tag{4.33}$$

where $\bar{V}_{sl}$ is the rms coupling between $|s\rangle$ and the $|l\rangle$ zero-order states and $\rho_l(E)$ is the density of the $|l\rangle$ states.

If $c_s(t)$ is found from Eq. (4.32) and inserted into Eq. (4.25), the absorption spectrum becomes the structureless Lorentzian band envelope:

$$I(\omega) = \frac{c_o k/2}{(\omega - \omega_o)^2 + k^2/4}, \tag{4.34}$$

where c_o is a normalization constant and $\omega_o = E_s^o/\hbar$ is the peak of the absorption band. An analysis of the width of the absorption envelope shows that the full-width-half-maximum (fwhm) of the envelope $\Gamma(\mathrm{cm}^{-1})$ is related to the IVR rate constant via the relationship

$$k = 2\pi c\Gamma. \tag{4.35}$$

The statistical limit does *not* require that $|s\rangle$ be coupled to *all* the other zero-order states. All that is required for application of the statistical limit is that a sufficiently large number of the $|\ell\rangle$ states are coupled to $|s\rangle$, so that IVR appears to be irreversible. Thus, the experimental observation of a Lorentzian band envelope in an absorption spectrum does not necessarily imply that the initially prepared zero-order state is coupled to all the remaining zero-order states so that a microcanonical ensemble is formed. What is required is that the effective density of states multiplied by the resolution be much larger than one.

It has been found that IVR is in the statistical limit for a series of molecules with the general formula $(CX_3)_3Y{-}C{\equiv}C{-}H$, where Y is C or Si and X is H, D, or F (Kerstel et al., 1991; Gambogi et al., 1993). The initially excited state is a fundamental or first overtone of the acetylenic C—H stretch. The spectra for the $R(7)$ transitions of the fundamentals and the $R(5)$ transitions of the overtones for 3,3-dimethylbutane, $(CH_3)_3CC{\equiv}CH$, and (trimethylsilyl) acetylene $(CH_3)_3SiC{\equiv}CH$ are shown in figure 4.16. The solid lines are Lorentzian fits, Eq. (4.34), to the spectra. In the statistical limit of intramolecular vibrational energy redistribution a Lorentzian line shape is

Figure 4.16 Left-hand side: $R(7)$ of the fundamental acetylenic C—H stretch rovibrational spectra of $(CH_3)_3CC{\equiv}CH$ (above) and $(CH_3)_3SiC{\equiv}CH$ (below). Right-hand side: $R(5)$ of the overtone acetylenic C—H stretch rovibrational spectra of $(CH_3)_3CC{\equiv}CH$ (above) and $(CH_3)_3SiC{\equiv}CH$ (below). In all four cases the measured rotational line and a nonlinear least-squares fit to a single Lorentzian are shown in the upper traces, while residual of the Lorentzian fit and the zero line are shown in the lower traces. (For the sake of clarity upper and lower traces are staggered.) The residuals indicate a true Lorentzian line shape for the carbon compound as expected for the statistical regime of IVR. For the silicon compound the fit to a single Lorentzian is not as exact. The small residuals at the low-frequency side for Si compound (below) in both the fundamental and the overtone are likely due to two isotopes of Si with 4.67% and 3.1% natural abundance or to a hot-band transition (Kerstel et al., 1991).

predicted and, according to Eq. (4.35), the line width of the profile provides a relaxation rate.

The spectra in figure 4.16 show that a single atom substitution may have an enormous effect on the intramolecular dynamics, since there is a remarkable difference in the rotationally resolved spectra of the fundamental and first overtone of these two molecules. The line width of the silicon-substituted compound is significantly narrower than that of tert-butylacetylene in both the fundamental and first overtone. There is also a striking different behavior of the two molecules in going from fundamental to overtone excitation. Tert-butylacetylene shows a decreased IVR lifetime in the overtone, dropping by almost a factor of 2. In contrast, the silicon substituted compound shows exactly the opposite behavior since the lifetime in the fundamental is decreased by almost a factor of 2 compared to the overtone.

The line widths of the tert-butylacetylene overtone transition correspond to an overtone relaxation lifetime of ~110 psec. The corresponding lifetime for the fundamental is ~200 psec. For $(CH_3)_3Si{\equiv}CH$ the estimated lifetimes are 2 and 4 nsec for the fundamental and first overtone, respectively. The origin of the different IVR lifetimes for $(CH_3)_3CC{\equiv}CH$ and $(CH_3)_3SiC{\equiv}CH$ has not been identified. As shown in table 4.2, it cannot be associated with the density of states, since the $(CH_3)_3SiC{\equiv}CH$ density of states is significantly larger. The densities of state do not explain why the fundamental and overtone line widths only differ by a small factor. It is also not clear why the overtone lifetime is shorter for $(CH_3)_3CC{\equiv}CH$, but longer for $(CH_3)_3SiC{\equiv}CH$. The tentative interpretation of IVR for these two molecules is that only a minimum density of vibrational states is required for statistical relaxation to occur, yielding a homogeneous Lorentzian line width. Increasing the density beyond this minimum does not substantially increase the IVR rate. It has been suggested that the states strongly coupled to the acetylenic C—H stretch are those which give rise to low-order resonance transitions with small changes in quantum numbers for the zero-order states (Gambogi et al., 1993). Trajectory studies, which are discussed below in section 4.3, indicate that the role of Si in restricting IVR is not a simple mass effect (Swamy and Hase, 1985).

To conclude this section on statistical IVR, it is instructive to return to the zero-

Table 4.2. Density of A_1 States for $(CH_3)_3CC{\equiv}CH$ and $(CH_3)_3SiC{\equiv}CH$

Level of Excitation[a]	ρA_1[b]
$(CH_3)_3CC{\equiv}CH$	
$v = 1$	704
$v = 2$	1.06×10^6
$(CH_3)_3SiC{\equiv}CH$	
$v = 1$	2.09×10^4
$v = 2$	4.99×10^7

[a]This is the vibrational excitation level of the acetylenic C—H stretch.

[b]This is the number of states/cm^{-1} of A_1 symmetry calculated by using the molecular symmetry designations of G_{162}. The full density of states is 162 times the density of A_1 states.

order model outlined above in section 4.4.1.1 (p. 81). Resonance states, identified by $|m\rangle$, Eq. (4.28), are formed by strong couplings between $|s\rangle$ and a small number of $|l\rangle$ states. The Lorentzian-like line widths of the $|m\rangle$ states, figure 4.8, result from couplings with the quasi-continuum $|q\rangle$. For the coupling scheme in figure 4.8, the Fermi golden rule in Eq. (4.33) may give an accurate representation of the IVR rates k_m for the resonance states. In using this equation to calculate k_m, the coupling term becomes $\bar{V}_{mq}$, the rms coupling between the state $|m\rangle$ and the continuum $\{|q\rangle\}$, and the density becomes $\rho_q(E)$, the density of the continuum states. Finally, the same type of model can be used to interpret the Lorentzian-like lines in the $\nu = 3$ absorption spectrum of benzene, figure 4.10.

4.4.1.5 Quantum Beats

The recurrences in $P_s(t)$ for a superposition state, as shown in figure 4.9, are often called quantum beats. A three-level model for quantum beats is depicted in figure 4.17. For this example the states do not decay by coupling with a continuum, but by fluorescence to the ground state. Two zero-order states $|s\rangle$ and $|r\rangle$ are formed by linear combinations of the exact eigenstates $|l\rangle$ and $|j\rangle$. The $|s\rangle$ and $|r\rangle$ states are coupled by the element V_{sr}. The zero-order state $|s\rangle$ carries oscillator strength from the ground state and is excited by a laser pulse. After $|s\rangle$ is excited, the coupling between $|s\rangle$ and $|r\rangle$ will result in oscillatory flow of amplitude $\Psi(t)^2$, Eq. (4.7), between $|s\rangle$ and $|r\rangle$. This will

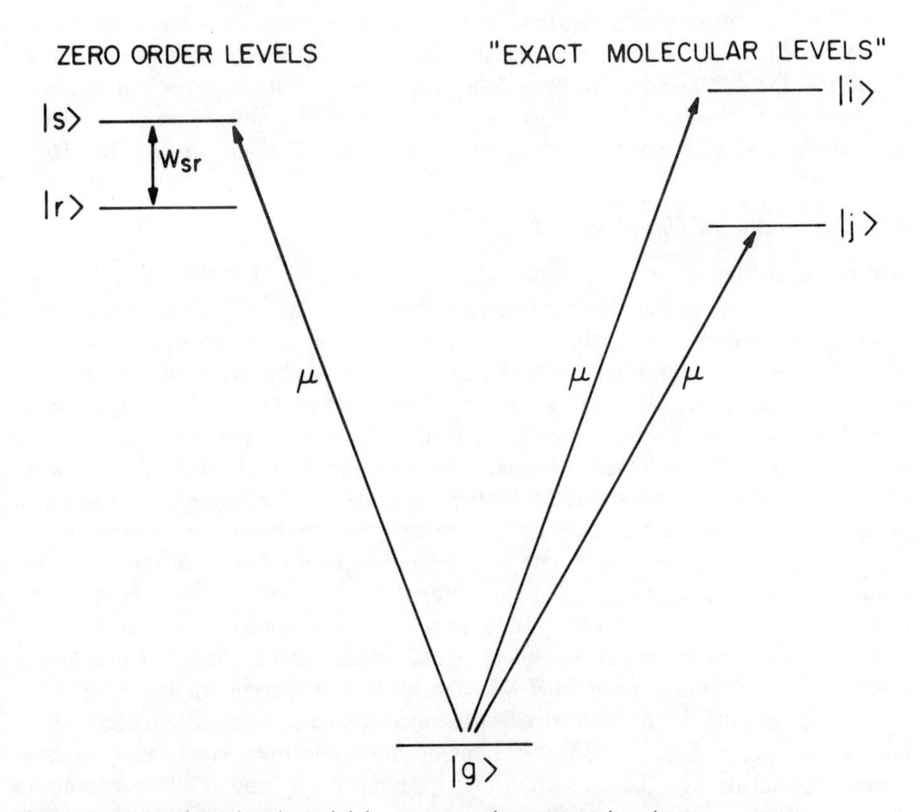

Figure 4.17 A three-level model for quantum beats (Freed and Nitzan, 1980).

manifest itself by fluorescence decays from $|s\rangle$ and $|r\rangle$ which are modulated by a $\cos\omega_{lj}t$ interference term, where $\omega_{lj} = (E_l - E_j)/\hbar$ and $E_l - E_j$ is the energy difference between the two eigenstates $|l\rangle$ and $|j\rangle$ which combine to form the zero-order states $|s\rangle$ and $|r\rangle$; see Eq. (4.21). Since $|s\rangle$ and $|r\rangle$ have different emission Franck-Condon factors this modulation (a quantum beat) may be observed by measuring the time-resolved dispersed fluorescence. The quantum beat or recurrence is damped by fluorescence (i.e., transitions to the ground state). It should be clear that this quantum beat is the spectroscopic analog of the double-well tunneling and OH local-mode IVR discussed above in section 4.3.2 (p. 75). The existence of quantum beats requires the preparation of a coherent superposition state, since if either of the exact eigenstates $|l\rangle$ or $|j\rangle$ were prepared there would simply be exponential decay due to fluorescence from the eigenstate. If an incoherent superposition of $|l\rangle$ and $|j\rangle$ were prepared, nonexponential decay would be observed as a result of different fluorescence rates for the two eigenstates. Finally, to observe a quantum beat, one must have a detection method which is able to distinguish between the two zero-order states $|s\rangle$ and $|r\rangle$.

Quantum beats are observed (Felker and Zewail, 1985) when jet-cooled anthracene is excited with 1420 cm^{-1} excess energy in S_1 (resolution of 2 cm^{-1}). Modulated fluorescence from three different types of emission bands is shown in figure 4.18. The bands at $\bar{\nu}_d$ = 1750 and 1500 cm^{-1} are in phase and out of phase with the band at $\bar{\nu}_d$ = 390 cm^{-1}. From Fourier analyses of the modulated fluorescence decay for the different emission bands, the number of eigenstates in $\Psi(t)$ may be determined as well as the transformation matrix **C** between the zero-order states ϕ and eigenstates ψ, that is, Eq. (4.9). For E_{vib} of S_1 equal to 1420 cm^{-1} four eigenstates contribute to $\Psi(t)$. From the radiation bandwidth and the density of states this number is found to be less than the total number of eigenstates energetically accessible. Results similar to these are observed for other S_1 excitations of jet-cooled anthracene in the 1300–1600-cm^{-1} range.

4.4.1.6 Critique of Time-Dependent IVR

One of the ambiguous features of time-dependent studies of IVR is the identity of $|s\rangle$ the zero-order state prepared experimentally (the bright state) and the remaining zero-order states (the dark states), which define the basic set and the nature of the couplings. As described in section 4.4.1.1 (p. 81), since $|s\rangle$ is a superposition of transitions to individual eigenstates, Eq. (4.5), the nature of $|s\rangle$ is determined by the energy resolution of the experiment. Figure 4.8 shows that in a low-resolution experiment $|s\rangle$ is a superposition of all the transitions under the absorption band. However, with higher resolution it becomes possible to excite the individual resonances $|m\rangle$, Eq. (4.28), in the absorption band. Such a resonance would then become the zero-order state $|s\rangle$. As the initially prepared zero-order state $|s\rangle$ is altered by the experimental resolution one changes the zero-order basis set and concomitant couplings used to describe the spectrum. One is left to wonder if there is a fundamental zero-order basis for a molecule, or is the zero-order model simply dependent on the experimental resolution? Consider the different $P_s(t)$ and, thus, spectra that would result from preparing different $\Psi(0) = |s\rangle$ by varying the pulse width of a transform-limited pulsed laser from femtoseconds to milliseconds. One expects to observe a smooth transition from exponential decay to nonexponential decays (i.e., recurrences), quantum beats, and resolved eigenstate spectra as the pulse width is lengthened.

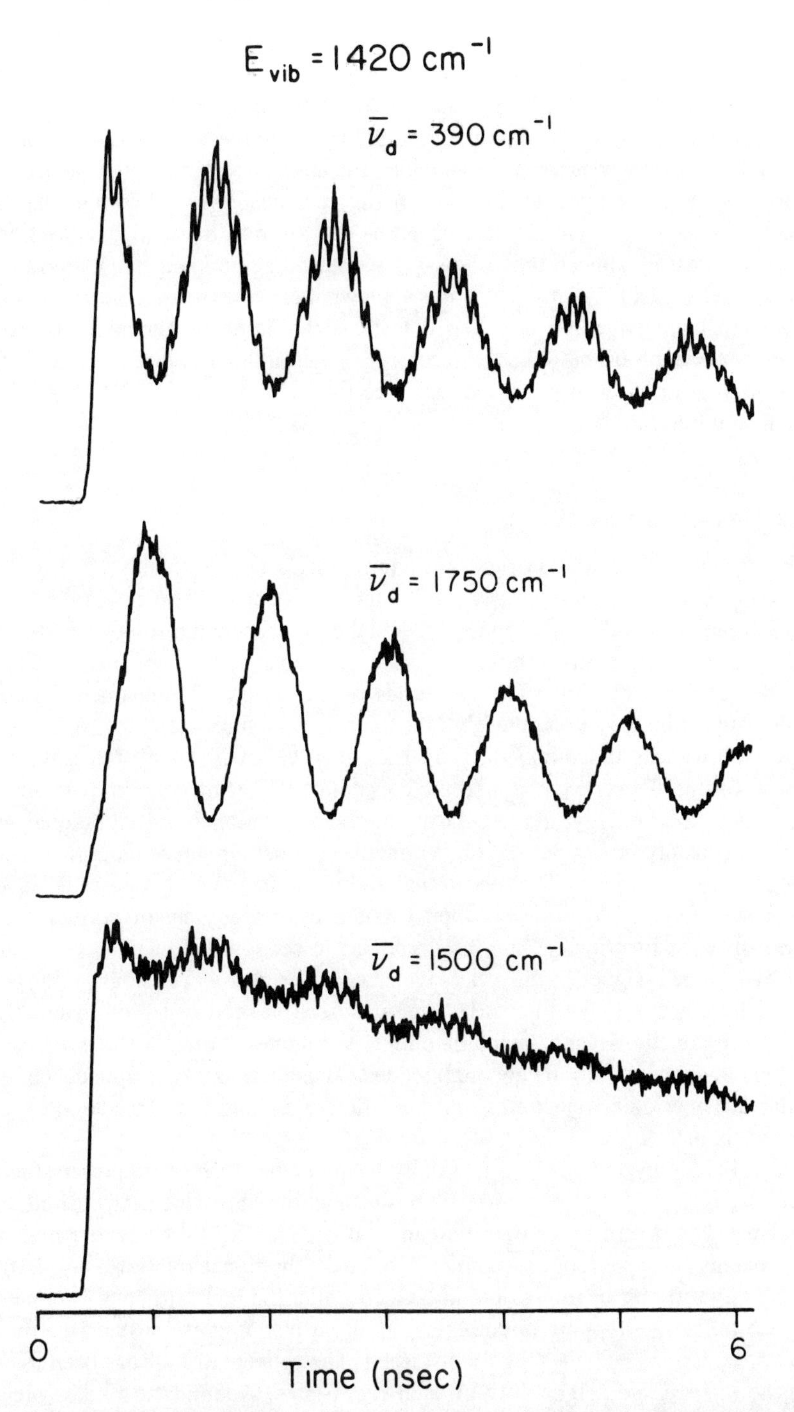

Figure 4.18 Representative decay types for fluorescence bands in the E_{vib} = 1420 cm^{-1} spectrum of anthracene. The wavenumber shifts of the bands from the excitation energy are given in the figure (Felker and Zewail, 1985).

As will be discussed in chapter 6, of fundamental importance in the theory of unimolecular reactions is the concept of a microcanonical ensemble, for which every zero-order state within an energy interval ΔE is populated with an equal probability. Thus, it is relevant to know the time required for an initially prepared zero-order state $|s\rangle$ to relax to a microcanonical ensemble. Because of low resolution and/or a large number of states coupled to $|s\rangle$, an experimental absorption spectrum may have a Lorentzian-like band envelope. However, as discussed in the preceding sections, this does not necessarily mean that all zero-order states are coupled to $|s\rangle$ within the time scale given by the line width. Thus, it is somewhat unfortunate that the observation of a Lorentzian band envelope is called the statistical limit. In general, one expects a hierarchy of couplings between the zero-order states and it may be exceedingly difficult to identify from an absorption spectrum the time required for IVR to form a microcanonical ensemble.

4.4.2 Time-Independent IVR

In the absence of a radiation field, the vibrational/rotational manifold of a molecule consists of discrete eigenstates, ψ_n, with energy E_n. For an ideal experiment in which inhomogeneous broadening is negligible and the experimental resolution is sufficient to resolve each of these eigenstates, ψ_n, the excitation radiation prepares a state which does not evolve with time. The properties of the ψ_n are of fundamental interest to unimolecular kineticists and closely tied to an understanding of IVR. In section 3 of chapter 2 it was discussed how the properties of ψ_n vary with energy. Here we consider the relationship between the ψ_n and the concept of IVR (Stuchebrukhov et al., 1989).

At low total energy and angular momentum, rotational and vibrational motions are approximately separable. (At higher angular momentum these motions are coupled through coriolis effects.) The vibrational Hamiltonian is nearly separable in normal coordinates and the ψ_n are well approximated by the product of normal-mode and rotational wave functions. Thus, as described in section 3.2 of chapter 2, quantum numbers can be assigned to the exact eigenstates. The classical analogs of this behavior are regular quasiperiodic trajectories. As discussed in section 3.3 of chapter 2, energies, wave functions, and quantum numbers obtained from the trajectories are in excellent agreement with the quantum values. Thus there is a correspondence between regular quasiperiodic trajectories and eigenfunctions, and the name *regular* is often attached to such Ψ_n.

As the energy is increased it usually becomes necessary to represent the eigenfunctions by a linear combination of basis functions ϕ_i, which, if desired, can be associated with a zero-order Hamiltonian; that is, Eqs. (4.9)–(4.11). For higher energies, making spectroscopic assignments for the vibrational/rotational levels may be exceedingly difficult or impossible in the sense that the level structure becomes complex due to the overlapping or combination of various progressions and the regularity of the progressions becomes highly disturbed. The concept of IVR has been associated with the *intrinsic* unassignability of such eigenstates. Attempts to find semiclassical eigenvalues for these states in general fail, since the trajectories are chaotic instead of quasiperiodic (section 3.3 of chapter 2). As a result, these unassignable states are often called nonregular or chaotic. In addition to the intrinsic unassignability of chaotic eigenstates as compared to regular ones, IVR may also be identified spectroscopically by an analysis of nearest-neighbor spacings of energy levels (section 3.4 of chapter 2). Regular levels should follow a Poisson distribution. From numerical experiments with random matrices it is suggested that chaotic spectra should obey a Wigner surmise (Eq. (2.81) of chapter 2).

For chaotic eigenstates ψ_n it has been conjectured that it is impossible to find a separable zero-order Hamiltonian with eigenfunctions ϕ_i for which there is only one principal ϕ_i contributing to each ψ_n. Instead, many different ϕ_i's will be included in representing ψ_n. It is possible that chaotic wave functions have similar global amplitudes so that average values for an observable a are nearly the same for two eigenstates ψ_i and ψ_j with similar energies. That is,

$$\bar{a}_i = \langle\psi_i|\hat{a}|\psi_i\rangle \approx \bar{a}_j = \langle\psi_j|\hat{a}|\psi_j\rangle . \tag{4.36}$$

Much different average values for a are expected for regular wave functions.

The extent of mixing between zero-order states ϕ_i in forming the ψ_n can be systematically evaluated using the natural orbital expansion (NE) analysis (Certain, 1985; Choi and Light, 1992). The NE analysis gives the most rapidly convergent set of zero-order basis functions. The natural orbitals $\bar{\phi}_i$ are a linear combination of an arbitrarily chosen zero-order product basis ϕ_i, so that the eigenfunctions become

$$\psi_n = \sum_i d_{ni}\bar{\phi}_i . \tag{4.37}$$

In an NE analysis, the mode mixing is quantified in terms of the population probability coefficients, $\{d_{ni}^2\}$. The following "entropy of mixing" S is a measure of the mode mixing for a state ψ_n:

$$S_n = \sum_i - d_{ni}^2 \ln d_{ni}^2 . \tag{4.38}$$

If there exists only one dominant configuration, for example, $\bar{\phi}_{n1}$, then $d_{n1}^2 = 1$, giving $S = 0$. This will correspond to a case in which the modes are perfectly separated.

The regular eigenfunctions are usually localized in configuration space in contrast to the chaotic ones which are delocalized. Thus, the transition moments $\mu_{if} = \langle\psi_f|\hat{\mu}|\psi_i\rangle$ for the two types of eigenfunctions have much different behavior. For regular eigenfunctions it is possible that only a small number of the ψ_n in an energy interval will have an appreciable transition moment so that they can be seen spectroscopically. If the eigenfunctions are chaotic all ψ_n should be observed in the spectra. However, the transition probability, which was localized on only a few regular ψ_n, will be spread, with appreciable fluctuations, over all the chaotic ψ_n.

Heller (1983) has quantified the relationship between spectral intensities in an absorption spectrum and IVR. From Eq. (4.26) the spectral intensities for an excited zero-order state ϕ_1 are determined by

$$P_n^1 = |\langle\phi_1|\psi_n\rangle|^2. \tag{4.39}$$

The key to understanding the importance of spectral intensities in elucidating IVR lies in the two formulae

$$P(1|1) = \sum_n p_n^1 p_n^1 , \tag{4.40}$$

and

$$P(1|i) = \sum_n p_n^1 p_n^i , \tag{4.41}$$

where $P(1|i) = P(i|1)$ is the time-averaged probability of starting in state ϕ_1 and being found later in the state ϕ_i; that is,

$$P(1|i) = \lim_{T\to\infty} \frac{1}{T} \int_o^T |\langle\phi_i|\Psi(t)\rangle|^2 \, dt , \tag{4.42}$$

where $\Psi(t)$ is the time dependence of the initially prepared zero-order state, that is, $\Psi(0) = \phi_1$. Thus, by measuring a spectrum to high resolution one can say how much time $\Psi(t)$ spends in the vicinity of its birthplace (for example, $P(1|1)$, Eq. (4.40)].

There is a very simple interpretation for $P(1|1)$ (McDonald, 1979; Chaiken et al., 1981; Kim et al., 1988). Suppose ϕ_1 is an eigenstate. Then $P(1|1) = 1$. Suppose ϕ_1 is an equal-amplitude sum of N_o eigenstates. Then

$$\phi_1 = \frac{1}{\sqrt{N_o}} \sum_{n=1}^{N_o} \psi_n ,$$

$$\begin{aligned} P(1|1) &= \sum_{n=1}^{N_o} (p_n^1)^2 \\ &= \sum_{n=1}^{N_o} \left(\frac{1}{N_o}\right)^2 \\ &= \frac{1}{N_o} . \end{aligned} \tag{4.43}$$

Suppose ϕ_i is one of the eigenstates composing ϕ_1. Then also

$$P(1|i) = 1/N_o. \tag{4.44}$$

Suppose ϕ_j is in a normalized state constructed out of the N_o ψ_n, but otherwise arbitrary. Then

$$P(1|j) = 1/N_o. \tag{4.45}$$

Thus ϕ_1 spends $(1/N_o)$th of its time in the state ϕ_1, or $(1/N_o)$th of its time in any state ϕ_i or ϕ_j, which is made up of one or more of the N_o states that have intensity in the spectrum. The spectrum is

$$I(\omega) = \sum_{n=1}^{N_o} \frac{1}{N_o} \delta(\omega - \omega_n) , \tag{4.46}$$

where $\omega_n = E_n/\hbar$. It is as if ϕ_1 has N_o zero-order states to visit, and it spends equal time in each state. The point of Eqs. (4.43)–(4.46) is that

$$P(1|1) = (\text{no. of zero-order states visited by } \phi_1)^{-1} = \sum_n (p_n^i)^2 . \tag{4.47}$$

Equation (4.47) is paramount to the realization that measurable spectra contain information about energy flow in molecules. Note that assignments of the spectra are not necessary. We take a spectrum, normalize the line intensities p_n^1 so that

$$\sum_n p_n^1 = 1 \tag{4.48}$$

and then use Eq. (4.45) to determine the number of zero-order states visited by $\Psi(t)$.

This number becomes all the more significant if we know how many states N exist within a range ΔE of the mean energy of E of the state ϕ_1 (where ΔE is the energy dispersion of a given ϕ_1). That is, if the N_o states seen in the spectrum are only a subset of those states available in the same energy regime, then we can say that ϕ_1 has visited only a fraction of the available states. This fraction is simply

$$\begin{aligned} F &= N_o/N \\ &= \frac{P^{ch}(1|1)}{P(1|1)}, \end{aligned} \tag{4.49}$$

where

$$P^{ch}(1|1) = 1/N; \tag{4.50}$$

ch stands for *chaotic*, because if $P(1|1) = P^{ch}(1|1)$, the state ϕ_1 visits all N available states. All we need to determine N is the energy uncertainty ΔE of ϕ_1, and the density of states $\rho(E)$ of the system so that

$$N = \rho(E)\Delta E. \tag{4.51}$$

Note that in a given spectrum a lot of missing lines (i.e., low or zero spectral intensity in certain eigenstates) implies $N_o < N$. Crudely speaking, if we look at a spectrum and see N_o lines in a range ΔE, but there are N eigenstates in ΔE, then F is just the ratio given by Eq. (4.49). These arguments lead to the conclusion that a completely chaotic system in quantum mechanics ought to have every line present in a spectrum, as dense a spectrum as the density of states permits. The sparser the spectrum, the fewer states sampled by the initial state.

The study of IVR in the time-independent domain becomes complicated when one deals with unresolved spectra (Gruner and Brumer, 1991a,b). However, if the individual eigenstates are resolved, an unambiguous analysis of IVR is possible, since a Fourier transform of the spectra $I(\omega)$, Section 3.3, gives the time profile of an initial zero-order state. A set of experiments in which individual eigenstates are resolved involve molecules in which the acetylenic C—H stretch mode ν_1 is excited with one or two quanta; that is, $n_1 = 1 \leftarrow 0$ and $n_1 = 2 \leftarrow 0$. An infrared laser spectrometer with a 10^{-4} cm^{-1} resolution was used to study the acetylenic C—H overtones of propyne, 1-butyne, and 1-pentyne; that is, $CH_3—C{\equiv}C—H$, $C_2H_2—C{\equiv}C—H$, and n–$C_3H_7—C{\equiv}C—H$, respectively (McIlroy and Nesbitt, 1989, 1990). Propyne has a low density of states of ~1 state/cm^{-1} at the 3300 cm^{-1} energy for the $n_1 = 1 \leftarrow 0$ transition. Thus, the laser is capable of resolving individual lines in the spectrum and no time-dependent IVR is observed. However, the intensity and structure of the spectrum of the acetylenic CH overtone is consistent with negligible coupling between the state corresponding to the CH normal mode motion and any other states. Individual eigenstates are also observed in the $n_1 = 1 \leftarrow 0$ spectrum of 1-butyne, since the density of states of 114 ± 30 states/cm^{-1} for this excitation gives rise to spectral lines spaced by ~0.01 cm^{-1}, which is much larger than the laser resolution of 10^{-4} cm^{-1}. However, that every eigenstate is observed in the spectrum suggests that there is coupling between a zero-order state corresponding to excitation of the acetylenic C—H stretch and *all* states that have a similar energy. For 1-pentyne the $n_1 = 1 \leftarrow 0$ spectrum is consistent with coupling between the acetylenic C—H stretch and only 1/3 of the remaining zero-order states. Though IVR does not actually occur in these experiments, one can calculate $C_s(t)$, Eq. (4.17), for an acetylenic C—H stretch by taking a Fourier transform of $I(\omega)$ in the vicinity of the C—H stretch excitation (see discussion at the beginning of section 3.3). Such a calculation mimics an experiment in which a pulsed laser prepares a C—H stretch overtone. These calculations show that the C—H stretch overtones of 1-butyne and 1-pentyne have an IVR time of ~500 psec. In related experiments Perry and co-workers have used high-resolution infrared spectroscopy and a novel high-resolution infrared double-resonance technique to study molecular eigenstate spectroscopy in *trans*-ethanol (Bethardy and Perry, 1993), 1-butyne (Perry, 1993), and propyne (Go and Perry, 1992).

4.4.3 Classical Trajectory Studies of IVR

There have been two general directions for classical trajectory studies of IVR (Gomez and Pollak, 1992). One involves the analysis of classical Hamiltonians for molecules and is concerned with the structure of the multidimensional phase space for excited molecules and the mechanism(s) for intramolecular energy transfer (Lichtenberg and Lieberman, 1991). The other is concerned with determining how to use classical mechanics to represent the initially excited zero-order state ϕ_1 and then propagate $\Psi(t)$. There is also an interest in the correspondence between classical and quantal descriptions of IVR (Brumer and Shapiro, 1988).

4.4.3.1 Phase Space Structure and Dynamics

Consider the molecular Hamiltonian

$$H = \sum_i h_i(P_i, Q_i) + V \,, \tag{4.52}$$

where each $h_i(P_i, Q_i)$ is a Hamiltonian for an individual mode, depending on the mode's momentum and coordinate, and V is a perturbation which couples the h_i. The sum of the h_i's is a separable zero-order Hamiltonian. For example, if the individual h_i are harmonic oscillators, this zero-order Hamiltonian is the normal mode Hamiltonian (Eq. (2.47) of chapter 2). Numerical solutions of Hamilton's equations of motion for the Hamiltonian in Eq. (4.52) have identified two distinct types of classical motion: quasiperiodic (regular) and chaotic (irregular) as in figure 2.4. Quasiperiodic motion can be represented by a linear combination of separable vibrational modes. A Fourier analysis of coordinate displacements for a molecule executing quasiperiodic motion identifies a specific frequency for each mode. For chaotic motion, energy flows freely between the modes in the molecule and, thus, the molecule does not exhibit regular vibrational motion. A Fourier analysis of chaotic vibrational motion results in a multitude of vibrational frequencies, without identifiable frequencies for particular modes.

The nature of the intramolecular motion of a molecule is intimately related to the concept of ergodicity. To quote Maxwell, the ergodic hypothesis ". . . is that the system (i.e., molecule), if left to itself in its actual state of motion, will, sooner or later, pass through every phase which is consistent with the equation of energy." A molecule undergoing chaotic motion has the possibility of satisfying the ergodic hypothesis, but one with quasiperiodic motion does not. For a molecule with n degrees of freedom there are $2n$ momenta and coordinates and, thus, the molecular phase space has a dimension $2n - 1$, since constant energy provides a constraint. For quasiperiodic motion the trajectory moves in a restricted region of phase space which, according to a theorem in topology, has the shape of a n-dimensional torus. For a two-mode Hamiltonian, this torus has the shape of a tube (fig. 4.19). The two motions giving rise to the frequencies ω_1 and ω_2 are depicted on the tube.

At low levels of excitation, the motion associated with molecular Hamiltonians is quasiperiodic. However, as the energy is increased, in most cases there is a gradual destruction of the tori and a transition from quasiperiodic to chaotic motion, with both types present at intermediate energies. If all the tori are destroyed, all regions of phase space become accessible to a trajectory, and the ergodic hypothesis becomes valid.

Numerical simulations have shown that the presence of the perturbation V in Eq. (4.52) is not sufficient for ergodic behavior. It was discovered that a system of N weakly coupled harmonic oscillators will not freely exchange energy as long as there is no collection of integers $\{n_k\}$ for which

$$\sum_{k=1}^{N-1} n_k \omega_k \cong 0 \tag{4.53}$$

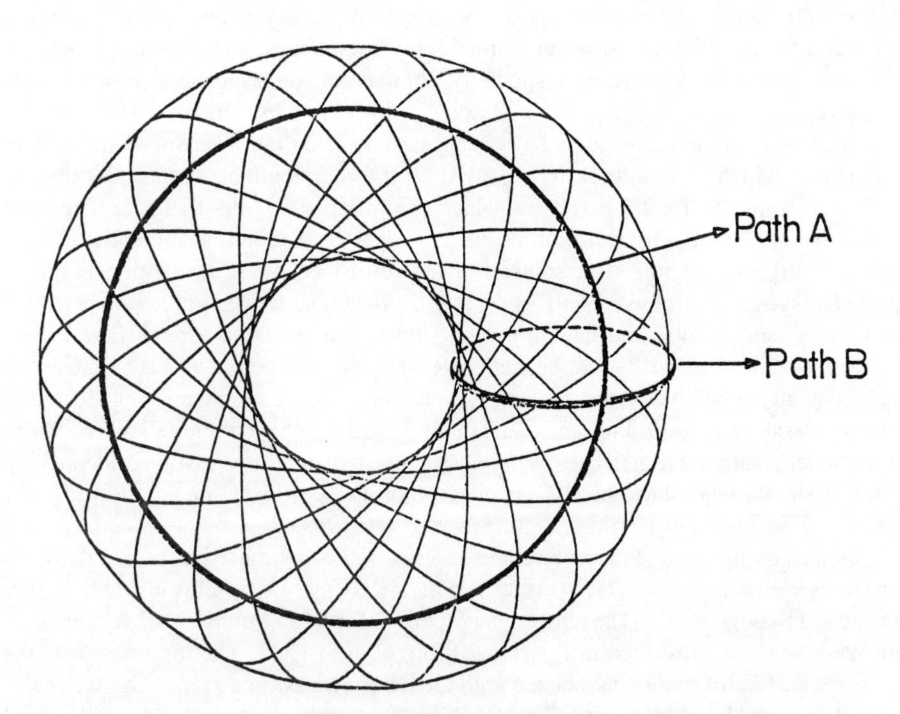

Figure 4.19 Two topologically independent contours (closed paths A and B) on an invariant torus (Noid and Marcus, 1977).

other than $\{n_k\} = 0$. These frequency matchings are known as "internal resonances," and the lack of these resonances precludes appreciable energy sharing in the weak coupling limit. The reason for this result is explained by the Kolmogorov-Arnold-Moser (KAM) theorem, which states that if there are no resonances among a set of harmonic oscillators, the addition of an anharmonic perturbation that is sufficiently small compared to the total energy does not make the system ergodic.

The theorem can be explained by considering the effect of the perturbation on the phase space tori. Tori of the unperturbed Hamiltonian with incommensurate frequencies continue to exist, being only slightly distorted by the perturbation. On the other hand, the tori with commensurate frequencies, or with incommensurate frequencies whose ratio is well approximated by (r/s), where r and s are relatively small integers, are grossly deformed by the perturbation. The KAM theorem shows that, when V is sufficiently small, the majority of initial conditions for Hamiltonian (4.52) lie on the preserved tori and nonergodic motion results. This justifies the view that the perturbation V largely serves only to slightly change the frequencies. Nevertheless, the relatively small set of initial conditions leading to motion not on preserved tori are intermingled in phase space with the preserved tori. It is the existence of this relatively small set of destroyed tori which ultimately lead to ergodic behavior as the energy is increased.

An internal resonance, for example, between ω_1 and ω_2, has a width defined by $\delta\omega_1$ and $\delta\omega_2$. Thus, the resonance condition of Eq. (4.53) is satisfied by a band of ω_1 and ω_2 frequencies. As the coupling is increased the resonance bands broaden and it becomes more likely that a resonance band between ω_1 and ω_2 overlaps another resonance band between ω_2 and ω_3. Efficient IVR between modes 1, 2, and 3 is expected for such a resonance overlap. Tori are destroyed in the overlapping resonance zone and the resulting classical motion is highly complicated and perhaps ergodic. A single resonance does not lead to ergodicity, but overlapping resonances do (Martens et al., 1987; Atkins and Logan, 1992; Oxtoby and Rice, 1976; Chirikov, 1960, 1979).

Classical dynamical studies have been used to study the mechanism for IVR in molecules. Much of this work has focused on the role nonlinear resonances play in IVR; for example, the 2:1 Fermi resonance (Kellman, 1985, 1990). There is interest in the role of overlapping nonlinear resonances when the motion is chaotic (Oxtoby and Rice, 1976) and the role of a single isolated resonance when the motion is quasi-periodic (Jaffe and Brumer, 1980; Sibert et al., 1982). The former studies are pertinent to time-dependent IVR and have provided support for the bend–stretch Fermi resonance mechanism for IVR from benzene overtone states (Sibert et al., 1984b; Garcia-Ayllón et al., 1988). The latter studies have provided insight into identifying the nature of the classical motion and, thus, quantum probability in configuration space for vibrational/rotational eigenstates. The classic study of this type involved identifying the OH stretch eigenstates of H_2O as either "normal" or "local" modes (Lawton and Child, 1979; Jaffé and Brumer, 1980; Sibert et al., 1982a).

Rates of intramolecular vibrational energy redistribution have been related to phase space bottlenecks (Davis, 1985; Davis and Gray, 1986; Gibson et al., 1987; Davis and Skodje, 1992). The effect of a bottleneck for energy transfer between regions of phase space is illustrated in figure 4.20 for collinear OCS. This figure depicts the classical surface of section associated with the CS stretch normal mode. The surface of section is found by plotting the CS stretch normal mode momentum P and coordinate Q

every time the CO stretch normal mode coordinate becomes zero with a positive momentum; see section 2.3.3 (p. 36). The outer curve A in Figure 4.20 is the energy boundary and the inner curve D marks the outer edge of a region of quasiperiodic motion. Trajectories in both regions I and II are chaotic; however, they differ by the resonance condition between the CO and CS stretches. In region I, ω_{CO}:ω_{CS} is 3:1, but in region II, it is 5:2. The groups of three and five islands (B and C) mark the 3:1 and 5:2 resonances. (In the CO stretch surface of section there are groups of one and two islands.) The reason for the different frequency ratios is that the frequency depends upon the energy in the mode, and regions I and II are, respectively, associated with high and low energies in the CS mode. Thus, a transition between regions I and II leads to a change in the CS mode energy. In going from region I to region II, the trajectory leaves one *resonance zone* and enters another.

The solid curve between regions I and II is the boundary bottleneck between these two regions. Davis (1985) has suggested that the trajectory which defines this boundary has a frequency ratio with the "worst" irrational number. Such numbers are well known (Berry, 1978) and the golden mean,

$$\gamma = \frac{1}{2}(1 + \sqrt{5}) \tag{4.54}$$

is the most difficult to express with rationals. Numerical studies by Davis (1985)

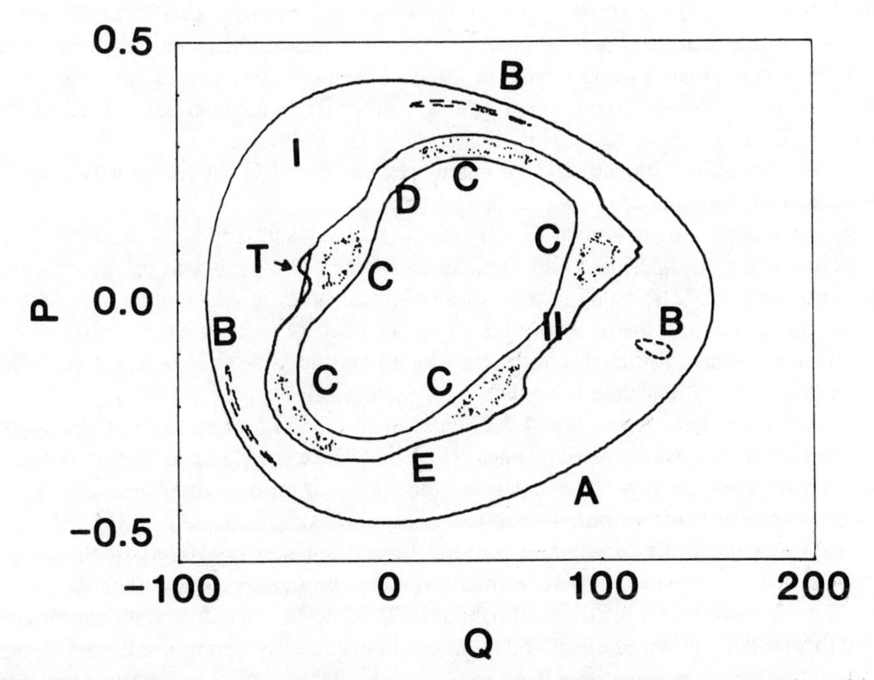

Figure 4.20 Classical surface of section associated with the CS stretch normal mode of OCS showing the following regions: (A) total energy contour, (B) 3:1 resonance islands, (C) 5:2 resonance islands, (D) outer boundary of quasi-periodic region, (E) golden mean torus between chaotic regions labeled I and II, (T) turnstile (in golden mean torus) between regions I and II (located by arrow) (Gibson et al., 1987).

suggest that the golden mean frequency ratio and any frequency ratio by adding an integer to the golden mean may act as a bottleneck to energy transfer. It has been speculated (Davis and Skodje, 1992) that the phase space boundary between two resonance tones may be associated with a cantorus, which results from the breakup of a torus whose frequency ratio is far enough from rational.

The region marked T in figure 4.20, that is, the figure eight, is known as the *turnstile* through which classical flux passes in flowing between regions I and II. The turnstile is determined by further propogation of the curve identifying the boundary between the two regions (Davis, 1985). Part of the turnstile lies inside the boundary curve and part lies outside. These two regions are called "flux in" and "flux out," respectively, because they are the pathways for the trajectories to enter and leave the two regions. Due to Liouville's theorem the areas bounded by the original and propagated curves are equal, so that both halves of the turnstile have equal areas. If the motion inside regions I and II is chaotic, a statistical value for the rate through the turnstile (for either direction) is proportional to the area of a lobe of the turnstile divided by the area of the region being exited.

4.4.3.2 Absorption Spectra and IVR Rates

Classical trajectories may either under- or overestimate the rate of IVR. Some relaxation processes are not allowed classically; for example, quantum mechanical tunneling through potential energy barriers. Also, in the absence of such a barrier classical mechanics may still not allow an initial zero-order state to relax, even though the state is quantum-mechanically nonstationary. In the other extreme classical mechanics may be more chaotic than quantum dynamics. Quantum mechanics often gives more structured motion with more recurrences among zero-order states than does classical mechanics. Each of these extremes is illustrated in the following.

When classical trajectories are calculated for the H_2O model with two O—H stretches discussed in section 4.3.2 (p. 76), local-mode type motion is found for which energy is trapped in individual O—H bonds (Lawton and Child, 1979, 1981). The trajectories are quasiperiodic and application of the EBK semiclassical quantization condition [Eq. (2.72)] results in pairs of local-mode states in which there are n quanta in one bond and m in the other or vice versa. The pair of local-mode states (n,m) and (m,n) have symmetry-related trajectories which have the same energy. The local-mode trajectory for the (5,0) state is depicted in Figure 4.6d.

As discussed above (section 4.3.2) these local-mode states are *not* the eigenstates for the system, but superposition states. However, since the classical motion is quasiperiodic for these local mode states (i.e., the state is a torus in the phase space), the system is trapped in the initially excited local-mode state and the quantum periodic oscillation between the (n,m) and (m,n) local mode states is not observed classically. Thus, classical mechanics severely underestimates the rate of energy transfer.

The opposite effect, an overestimate of IVR, is seen when classical trajectories are used to simulate the intramolecular dynamics of benzene. By definition, the zero-point level of benzene is an eigenstate (i.e., a stationary state) and there is no IVR. However, when zero-point energy is added to a benzene Hamiltonian which includes appropriate couplings between the normal modes, energy flows from high-frequency modes with large zero-point energies to low-frequency modes with smaller zero-point energies (Lu and Hase, 1988a). It is possible the zero-point level has ergodic classical behavior so that, if a trajectory were followed for a sufficiently long time, a classical microcanonical ensemble would be formed in which each mode has the same average energy. Certainly the observation of IVR for the benzene zero-point level is not correct! This is an extreme example of the situation where classical mechanics overestimates IVR. The incorrect flow of zero-point energy in classical trajectories is a serious problem, for which a general solution has not been found.

Though trajectories do not correctly describe the benzene zero-point level, they appear to adequately represent the initial IVR from benzene overtone local mode (i.e., zero-order) states in which n quanta are added to a particular CH bond (Sibert et al., 1984b; Clarke and Collins, 1987; Garcia-Ayllón, 1988; Gomez Llorente et al., 1990; Lu and Hase, 1988b, 1989). In the trajectory simulations of this overtone state the initial step is to add zero-point energy to the benzene normal modes. A CH bond is then excited to the local-mode state $|n\rangle$ by adding more energy to the bond so that the bond energy, as defined by

$$E = \mathrm{p}^2/2\mu + D[(1 - e^{-\beta(r-r_o)})]^2 \qquad (4.55)$$

is the same as the Morse eigenvalue $E(n)$, Eq. (2.59). The probability that the molecule remains in the initial overtone state is evaluated versus time by phase averaging over an ensemble of trajectories. This quantity, $P(n,t)$, is the quasi-classical trajectory approximation to the probability $|c_1(t)|^2$ discussed in section 4.3.1 (p. 75). The local mode is identified as being in state n if the bond energy, Eq. (4.55), is within the Morse eigenvalue energy interval $E(n - \frac{1}{2})$ to $E(n + \frac{1}{2})$.

$P(n,t)$ calculated in this manner for the $n = 3$ overtone of benzene (Lu and Hase, 1988b, 1989) is plotted in Figure 4.13, where it is compared with the quantum result. The classical $P(n,t)$ decays exponentially to zero without the recurrences seen in the quantum calculation. It is these recurrences which give the structure in the absorption spectrum. The spectrum calculated from the classical exponential $P(n,t)$ is a smooth Lorentzian band envelope with fwhm of 85 cm^{-1}.

This analysis for the benzene $n = 3$ overtone spectrum suggest that though trajectories may be adequate for calculating a low-resolution absorption spectrum, it is not clear that they may be used to determine fine details in the spectrum. To evaluate details in the absorption spectrum with a time-independent formalism requires knowing the energy for each of the transitions in the absorption envelope; that is, Eq. (4.24). Given the current understanding of classical and semiclassical mechanics, it may be fundamentally impossible to calculate the transition energies from a classical trajectory study. In general, the determination of semiclassical transition energies for polyatomic systems (i.e., more than three dimensions) has required quasi-periodic classical motion. For highly excited overtone states the classical intramolecular motion is usually chaotic, which seems to be the case for the benzene overtone states.

From a time-dependent point of view, recurrences in the probability of occupying the initially prepared state give rise to the fine structure in the overtone absorption spectrum. Though rudiments of these recurrences may be present in the short-time trajectory $P(n,t)$, chaotic classical motion destroys the longer time recurrences, which occur quantum mechanically. It is these latter recurrences which are needed to evaluate fine details in the absorption spectrum. Thus, the classical trajectory method may be limited to the evaluation of low-resolution absorption spectra. However, it should be pointed out that progress is being made in extracting information from systems with classical chaotic motion (Heller, 1984; Taylor, 1989; Farantos and Taylor, 1991; Heller, 1991).

REFERENCES

Atkins, K.M., and Logan, D.E. (1992). *J. Chem. Phys.* **97**, 2438.
Baggott, J.E., Chuang, M.-C., Zare, R.N., Dübal, H.R., and Quack, M. (1985). *J. Chem. Phys.* **82**, 1186.
Baggott, J.E., Law, D.W., Lightfoot, P.D., and Mills, I.M. (1986). *J. Chem. Phys.* **85**, 5414.
Berry, M.V. (1978) in *Topics in Nonlinear Dynamics*, AIP Conference Proceedings No. 46. AIP, New York.
Bethardy, G.A., and Perry, D.S. (1993). *J. Chem. Phys.* **99,** 9400.
Brumer, P., and Shapiro, M. (1988). *Adv. Chem. Phys.* **70**, 365.
Certain, P.R. (1985). *J. Chem. Phys.* **89**, 4464.
Chaiken, J., Gurnick, M., and McDonald, J.D. (1981). *J. Chem. Phys.* **74**, 117.
Child, M.S., and Halonen, L. (1984). *Adv. Chem. Phys.* **57**, 1.
Chirikov, B.V. (1960). *Plasma Phys.* (*J.N.E. Pt. C*) **1**, 253.
Chirikov, B.V. (1979). *Phys. Rep.* **52**, 265.
Choi, S.E., and Light, J.C. (1992). *J. Chem. Phys.* **97**, 7031.
Clarke, D.L., and Collins, M.A. (1987). *J. Chem. Phys.* **87**, 5312.
Crim, F.F. (1984). *Ann. Rev. Phys. Chem.* **35**, 657.
Crim, F.F. (1990). *Science* **249**, 1387.
Crim, F.F. (1995). Personal communication.
Davis, M.J. (1985). *J. Chem. Phys.* **83**, 1016.
Davis, M.J., and Gray, S.K. (1986). *J. Chem. Phys.* **84**, 5389.
Davis, M.J., and Skodje, R.T. (1992). In *Advances in Classical Trajectory Methods, Vol. 1, Intramolecular and Nonlinear Dynamics*. Hase, W.L., Ed., Plenum, New York, p. 77.
Farantos, S.C., and Taylor, H.S. (1991). *J. Chem. Phys.* **94**, 4887.
Felker, P.M., and Zewail, A.H. (1985). *J. Chem. Phys.* **82**, 2975.
Forst, W. (1973). *Theory of Unimolecular Reactions*. Academic Press, New York, p. 278.
Freed, K.F., and Nitzan, A. (1980). *J. Chem. Phys.* **73**, 4765.
Gambogi, J.E., Lehmann, K.K., Pate, B.H., Scoles, G., and Yang, X. (1993). *J. Chem. Phys.* **98**, 1748.
Garcia-Ayllón, A., Santamaria, J., and Ezra, G.S. (1988). *J. Chem. Phys.* **89**, 801.
Gibson, L.L., Schatz, G.C., Ratner, M.A., and Davis, M.J. (1987). *J. Chem. Phys.* **86**, 3263.
Go, J., and Perry, D.S. (1992). *J. Chem. Phys.* **97**, 6994.
Gomez Llorente, J.M., Hahn, O., and Taylor, H.S. (1990). *J. Chem. Phys.* **92**, 2762.
Gomez Llorente, J.M., and Pollak, E. (1992). *Ann. Rev. Phys. Chem.* **43**, 91.
Gruner, D., and Brumer, P. (1991a). *J. Chem. Phys.* **94**, 2848.
Gruner, D., and Brumer, P. (1991b). *J. Chem. Phys.* **94**, 2862.
Hayward, R.J., and Henry, B.R. (1975). *J. Mol. Spectrosc.* **57**, 221.
Heller, E.J. (1978). *J. Chem. Phys.* **68**, 3891.
Heller, E.J. (1983). *Faraday Discuss. Chem. Soc.* **75**, 141.
Heller, E.J. (1984). *Phys. Rev. Lett.* **53**, 1515.
Heller, E.J. (1991). *J. Chem. Phys.* **94**, 2723.
Jaffé, C., and Brumer, P. (1980). *J. Chem. Phys.* **73**, 5646.
Kellman, M.E. (1985). *J. Chem. Phys.* **83**, 3843.
Kellman, M.E. (1990). *J. Chem. Phys.* **93**, 6630.
Kerstel, E.R.Th., Lehmann, K.K., Mentel, T.F., Pate, B.H., and Scoles, G. (1991). *J. Phys. Chem.* **95**, 8282.

Kim, H.L., Minton, T.K., Ruoff, R.S., Kulp, T.J., and McDonald, J.D. (1988). *J. Chem. Phys.* **89**, 3955.
Kittrell, C., Abramson, E., Kinsey, J.L., McDonald, S.A., Reisner, D.E., Field, R.W., and Katayama, D.H. (1981). *J. Chem. Phys.* **75,** 2056.
Lawton, R.T., and Child, M.S. (1979). *Mol. Phys.* **37**, 1799.
Lawton, R.T., and Child, M.S. (1980). *Mol. Phys.* **40**, 773.
Lawton, R.T., and Child, M.S. (1981). *Mol. Phys.* **44**, 709.
Lee, S.-Y., and Heller, E.J. (1979). *J. Chem. Phys.* **71**, 4177.
Li, H., Ezra, G.S., and Philips, L.A. (1992). *J. Chem. Phys.* **97**, 5956.
Lichtenberg, A.J., and Lieberman, M.A. (1991). *Regular and Chaotic Dynamics, 2nd ed.* Springer-Verlag, New York.
Lorquet, A.J., Delwiche, J., and Hubin-Franskin, M.J. (1982). *J. Chem. Phys.* **76**, 4692.
Lu, D.-h., and Hase, W.L. (1988a). *J. Chem. Phys.* **89**, 6723.
Lu, D.-h., and Hase, W.L. (1988b). *J. Phys. Chem.* **92**, 3217.
Lu, D.-h., and Hase, W.L. (1989). *J. Chem. Phys.* **91**, 7490.
Luo, X., and Rizzo, T.R. (1990). *J. Chem. Phys.* **93**, 8620.
Marquardt, R., and Quack, M. (1991). *J. Chem. Phys.* **95**, 4854.
Martens, G.C., Davis, M.J., and Ezra, G.S. (1987). *Chem. Phys. Lett.* **142**, 519.
McDonald, J.D. (1979). *Ann. Rev. Phys. Chem.* **30**, 29.
McIlroy, A., and Nesbitt, D.J. (1989). *J. Chem. Phys.* **91**, 104.
McIlroy, A., and Nesbitt, D.J. (1990). *J. Chem. Phys.* **92,** 2229.
Merzbacher, E. (1970). *Quantum Mechanics*. Wiley, New York.
Noid, D.W., and Marcus, R.A. (1977). *J. Chem. Phys.* **67**, 559.
Oxtoby, D.W., and Rice, S.A. (1976). *J. Chem. Phys.* **65**, 1676.
Page, R.H., Shen, Y.R., and Lee, Y.T. (1987). *Phys. Rev. Lett.* **59**, 1293.
Page, R.H., Shen, Y.R., and Lee, Y.T. (1988). *J. Chem. Phys.* **88**, 4621.
Parmenter, C.S. (1982). *J. Phys. Chem.* **86**, 1736.
Perry, D.S. (1993). *J. Chem. Phys.* **98**, 6665.
Reddy, K.V., Heller, D.F., and Berry, M.J. (1982). *J. Chem. Phys.* **76**, 2814.
Scotoni, M., Boschetti, A., Oberhofer, N., and Bassi, D. (1991a). *J. Chem. Phys.* **94**, 971.
Scotoni, M., Leonardi, C., and Bassi, D. (1991b). *J. Chem. Phys.* **95**, 8655.
Segall, J., Zare, R.N., Dubäl, H.R., Lewerenz, M., and Quack, M. (1987). *J. Chem. Phys.* **86**, 634.
Sibert, III, E.L., Reinhardt, W.P., and Hynes, J.T. (1982a). *J. Chem. Phys.* **77**, 3583.
Sibert, III, E.L., Hynes, J.T., and Reinhardt, W.P. (1982b). *J. Chem. Phys.* **77**, 3595.
Sibert, III, E.L., Reinhardt, W.P., and Hynes, J.T. (1984a). *J. Chem. Phys.* **81**, 1115.
Sibert, III, E.L., Hynes, J.T., and Reinhardt, W.P. (1984b). *J. Chem. Phys.* **81**, 1135.
Sibert, III, E.L., Hynes, J.T., and Reinhardt, W.P. (1982c). *Chem. Phys. Lett.* **92**, 455.
Stannard, P.R., and Gelbart, W.M. (1981). *J. Phys. Chem.* **85**, 3592.
Steinfeld, J.I. (1985). *Molecules and Radiation, 2nd Edition*. MIT Press, London.
Stuchebrukhov, A., Ionov, S., and Letokhov, V. (1989). *J. Phys. Chem.* **93**, 5357.
Swamy, K.N., and Hase, W.L. (1985). *J. Chem. Phys.* **82**, 123.
Taylor, H.S. (1989). *Acc. Chem. Res.* **22**, 263.
Uzer, T. (1991). Phys. Rep. **199,** 73.
Voth, G.A. (1987). *J. Chem. Phys.* **87**, 5272.
Voth, G.A. (1988). *J. Chem. Phys.* **88**, 5547.
Watson, I.A., Henry, B.R., and Ross, I.G. (1981). *Spectrochimica Acta* **37A**, 857.
Wyatt, R.E., Iung, C., and Leforestier, C. (1992). *J. Chem. Phys.* **97**, 3477.

5

Experimental Methods in Unimolecular Dissociation Studies

5.1 INTRODUCTION

The experimental aspects of kinetic studies with state-selected reactants have played a critical role in the advancement of our understanding of unimolecular processes. The ultimate goal of these studies is the determination of the dissociation rate and final products of the following reaction:

$$AB(e, v, j) \xrightarrow{k(e,\, v,\, j)} A(e', v', j', m_j') + B(e'', v'', j'', m_j'') + KE, \quad (5.1)$$

where e, v, j, and m_j are the electronic, vibrational, rotational, and rotational projection quantum numbers, and KE is the kinetic energy released in the dissociation. Depending on the size of the products A and B, e, v, and j may be each composed of several quantum numbers. The experiment must thus be designed in such a fashion that as many of the quantum numbers as possible are measured. This is no easy task. The determination of each quantum number often requires a separate experiment. Furthermore, it is often of interest to obtain not only the distribution of final states in a given product, but also the correlations among several quantum numbers. For instance, the distribution of translational energies for products in a given rotational state can be measured.

The three major aspects in the study of unimolecular reactions are (a) the preparation of the excited molecule or ion, (b) the methods for determining the dissociation rates, $k(e, v, j)$; and (c) the determination of the products and their internal energy states or distributions.

5.1.1 State or Energy Selection of Reactant

If the experiment is designed so that e, v, and j of the reactant species are known, then the reactant can be claimed to be state-selected. If however, the resolution is insufficient to resolve individual quantum states, then it is best to claim simply that the reactant is energy-selected, with a given resolution. The term energy-selected is unambiguous. On the other hand, as we saw in the previous chapter, the term-state selected is not clear. Any system above the dissociation limit is not in a stationary state and is evolving in time, so that some of the quantum numbers are not conserved. It is ultimately the excited state lifetime, or the energy width, which determines whether state selection is possible. For all but the smallest molecules, the energy width of the excited states is larger than the spacing between adjacent quantum states so that state selection is generally not possible.

Even though it may not be possible to excite selected quantum states above the dissociation limit, it is always possible in principle to select the total angular momentum J because this quantity must be conserved.

5.1.2 Dissociation Rate Measurement

The measurement of the reaction rate constant is the heart of any dynamical experiment. To achieve this, a time base appropriate for the rate of the reaction in question must be available. Because unimolecular rates can vary over more than 10 orders of magnitude, no single approach is possible.

Dissociation rates can be determined by three methods which are based on (a) the measurement of relative rates, that is, the ratio of two rates one of whose rate constant is known; (b) direct real-time rate measurements; and (c) rate extraction from homogeneously broadened absorption line widths. The first two methods can be employed over very large time scales, whereas the last method is limited to rates in excess of about 10^9 sec^{-1}.

5.1.3 Product State Determination

While the reactants can often be state selected with an excitation source of sufficient resolution (with considerable effort expended in a full analysis of the molecule's spectrum), the full determination of the product states requires a far greater effort. This is because the products are formed in a distribution of states, some in the ground state and others in excited states. An electronically excited state which fluoresces can be analyzed by fluorescence spectroscopy. On the other hand, infrared fluorescence rates are usually much smaller than collisional stabilization rates so that the fluorescence analysis of vibrationally excited products in the ground electronic state is generally not feasible except with ultrasensitive detection schemes (Stewart and McDonald, 1981; Moss et al., 1981). Thus, other methods, such as laser induced fluorescence (LIF) or resonance enhanced multiphoton ionization (REMPI) must be used. If products are formed in several different electronic states, whose analysis requires a laser of a different color, it is apparent that a full determination of the product state distribution is almost impossible. However, the conservation of energy and momentum can often be used to advantage in order to determine quantities that are not directly measured. For instance, if the translational and vibrational energy of a product diatomic and atomic fragments are determined, than the rotational energy can be obtained from the difference between the measured product energies and the total energy. Similarly, conservation of linear momentum implies that measurement of the translational energy of only one of the fragments is necessary in order to determine the total energy released.

5.2 THE STATE OF THE SAMPLE

Before discussing the various methods for product detection and rate measurements, we consider first the state of the gas phase sample. The sample can be either in a bulb with a canonical distribution of internal energy states at the bulb temperature, or it can be entrained in a molecular beam where the temperature is considerably lower. Both have well-defined distributions of vibrational, rotational, and translation energies.

A supersonic jet obtained by the expansion of a gas through a nozzle differs from the traditional bulb sample in two important respects. First, the gas flow is directed in space with a velocity distribution that can be quite narrow. Second, the sample translational temperature is cooled to a few degrees Kelvin and many of the internal degrees of freedom are also cooled substantially. This is especially important for large molecules which have many vibrational modes excited at room temperature. The spectroscopy of even large molecules is thus simplified to the point where it is possible to excite single vibrational states and in some cases rotational levels as well.

The advantage of a jet-cooled sample is at the same time also a disadvantage. If the role of the rotational states in unimolecular reactions is of interest, a jet-cooled sample in which most of the molecules are in the very low J levels is clearly not suitable. For such investigations, it is necessary to study the sample in a bulb and use a laser to excite molecules in selected rotational states. This is extremely difficult for all but the smallest molecules because of the very large density of states and the difficulty in assigning all but the lowest energy levels. Double-resonance approaches help overcome this problem.

5.2.1 Molecules in a Bulb

As already pointed out, a room-temperature sample provides a rich source of molecules in high rotational states. However, in order to rotationally select a molecule in a photodissociation reaction, it is necessary to have assignable lower and upper states. Aside from the considerable problem associated with extreme anharmonicity near the dissociation limit, many molecules are in the ergodic limit and have intrinsically unassignable eigenstates. That is, the eigenstates consist of random admixtures of zero-order basis states as in the case of H_2CO (Polik et al., 1990).

A fundamental requirement for one photon state selection in thermal samples is that the spacing of adjacent absorption lines must be wider than the Doppler width of the lines. This line width arises from the distribution of colors which an ensemble of molecules absorbs as a result of the distribution of molecular speeds. The shift in the absorption energy due to a molecule traveling, for instance, toward the laser beam will be $\nu - \nu_o = \nu_o(u/c)$, where c is the speed of light, and ν_o and ν are the frequencies of the laser light as seen by the molecule at rest and one traveling with speed u. The average velocity (at 298 K) of a typical molecule (N_2) is 500 m/sec. Thus at a photon energy of 50,000 cm^{-1}, the difference in the measured transition energy between molecules traveling toward and away from the laser beam would be 0.17 cm^{-1}. When the Maxwell-Boltzmann distribution of velocities is taken into account, the absorption line (assumed here to be a delta function) becomes a Gaussian function:

$$F(\nu) = \exp \frac{-mc^2(\nu - \nu_o)^2}{2RT\nu_o^2}, \tag{5.2}$$

which has a full width at half maximum of

$$fwhm = \sqrt{\frac{8\ln(2)RT}{mc^2}}\,\nu_o, \tag{5.3}$$

This leads to a room temperature Doppler width of 0.11 cm^{-1} for the N_2 molecule at a transition energy of 50,000 cm^{-1}.

A circumvention of this problem is through the use of two-color excitation, or double resonance. An infrared laser can excite a selected rotational state (with a greatly reduced Doppler width because ν_o is much less). As discussed in a following section on overtone excitation, the absorption of the second (visible) photon leads to an excited state in a much more precisely defined energy.

5.2.2 Cold Molecules in a Molecular Beam

The cooling of molecules in supersonic expansions has been discussed in many excellent papers and chapters (Moore et al., 1989; Lubman et al., 1982; Smalley et al., 1977; Levy, 1980; Anderson, 1974; D.R. Miller, 1988; Scoles, 1988; Randeniya and Smith, 1990; DePaul et al., 1993). The treatments by DePaul et al. (1993) and D.R. Miller (1988) are particularly useful and concise. Figure 5.1 shows a molecular beam produced by a skimmed free jet expansion. At room temperature the molecules in the high-pressure reservoir (typically P_o = 0.5–10 atm) travel in random directions with a Maxwell-Boltzmann distribution of speeds. If the reservoir has a small orifice through which the molecules can escape into a vacuum, the molecules are forced to pass through the orifice by the collective effect of a directed flow. At high pressures (low mean free path) collisions during the expansion force the molecules to pass from the high-pressure region into the vacuum with a narrow speed distribution. The angular distribution can be approximated by a $\cos^n\theta$ function in which θ is the velocity angle relative to the beam direction and n is an exponent that depends upon the gas pressure and the heat capacity of the gas. A typical value for a neat beam of Ar or He is $n = 4$ (R.E. Miller, 1988). However, this value increases with the mass for molecules seeded in a low mass diluent.

The translational temperature, which is defined in terms of the width of the velocity distribution, decreases with the distance from the nozzle. As the temperature drops, collisions become more gentle and internal energy of the molecules (vibrational and rotational) flows to the translational modes. The translational temperature continues to cool as long as collisions are possible. Eventually, as the gas expands in the x and y directions (the molecular beam flow is in the z direction), the density is reduced

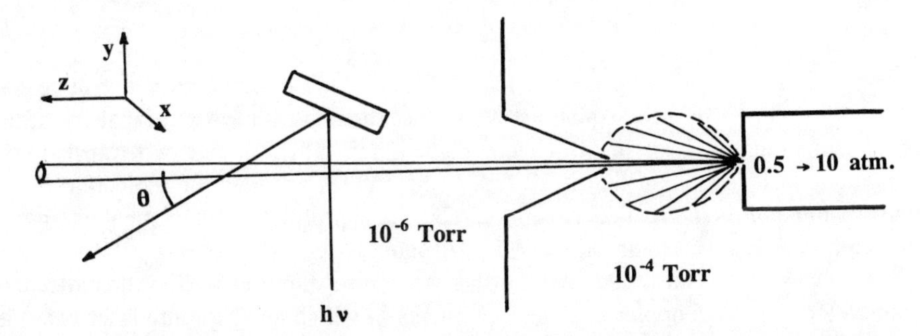

Figure 5.1 Schematic diagram of a nozzle and skimmer with representative pressures in the three regions. The laser intersects the beam at two different angles which permits measurement of the parallel and perpendicular translational temperatures as well as the beam velocity.

so that collisions effectively stop, and cooling of the internal degrees of freedom stops. Because vibrational, rotational, and translational relaxation rates differ, cooling of the various internal energies stops at different distances from the nozzle and therefore at different temperatures. It is thus important to distinguish the cooling for the various degrees of freedom.

5.2.2.1 Translational Beam Temperatures

The translational temperature of a beam can be derived from basic principles of gas dynamics which treat the expansion as an adiabatic, or isentropic, process. For a monatomic gas which has a kinetic energy of $E_o = 3/2\ RT_o$, and an enthalpy of $E_o + R$, conservation of the enthalpy requires that

$$H_o = 5/2\ RT_o = 1/2\ mu_b^2 + 3/2\ RT_{\parallel} + RT_{\perp}, \tag{5.4}$$

where T_o is the temperature of the gas prior to expansion (the stagnation temperature), u_b is the center of mass beam velocity, and $T_{\parallel}$ and $T_{\perp}$ are the final beam translational temperatures parallel and perpendicular to the molecular beam. The molecular beam defines the z direction along which the beam kinetic energy is $\frac{1}{2}\ RT_{\parallel}$. To this must be added $RT_{\parallel}$ which is the PV work associated with the enthalpy. The two degrees of freedom perpendicular to the molecular beam give rise to the $RT_{\perp}$ term. The lower of these temperatures is $T_{\perp}$ because molecules with perpendicular velocity vectors leave the skimmed beam. No such geometrical selection takes place for parallel velocities so that this is a more appropriate direction for characterizing the translational temperature of the molecular beam. That is, an effusive skimmed beam could have a low perpendicular translational temperature even though very little cooling has taken place. The maximum beam velocity is achieved when $T_{\parallel}$ and $T_{\perp}$ are reduced to 0. Thus the upper limit of the beam velocity for a monatomic gas is

$$u_b \leqq \sqrt{5RT_o/m}\,. \tag{5.5}$$

For the case of a neat beam of a polyatomic molecule with rotational and vibrational heat capacities, c_r and c_v, the above equation is written as

$$H_o = 1/2\ mu_b^2 + 3/2\ RT_{\parallel} + RT_{\perp} + \int_0^{T_r} c_r dT + \int_0^{T_v} c_v dT \tag{5.6}$$

$$= 1/2 mu_b^2 + 3/2\ RT_{\parallel} + RT_{\perp} + c_r T_r + c_v T_v\,, \tag{5.7}$$

where in Eq. (5.7) we have assumed that c_r and c_v are constant over the indicated temperature range. In these equations we further distinguish the vibrational and rotational temperatures, both of which are higher than either of the two translational temperatures. It is evident that if all final temperatures are low, the molecular beam converts most of its internal energy into directed center of mass translational energy of the beam [$\frac{1}{2}\ mu_b^2$]. The total enthalpy, H_o, is equal to $(5/2\ R + c_r + c_v)T_o$.

One way of measuring the two translational temperatures as well as the molecular beam velocity is by Doppler shift measurements in which an absorbing laser beam is passed through the molecular beam at two different angles (Bergmann et al., 1975; Hefter and Bergmann, 1988) The experimental arrangement is shown in Figure 5.1. The perpendicular laser beam will be absorbed at ν_o while the laser passing through the beam at the angle θ will be shifted to shorter wavelengths (higher frequencies). Figure

5.2 shows the resulting P_1 absorption peak (unshifted $\nu_o \approx 3920 \text{ cm}^{-1}$) for the expansion of 6% HF in a He beam (Bohac et al., 1992). The width of the narrow line (10 MHz or 0.00033 cm^{-1}) is limited by the resolution of the F-center IR laser, which means that according to Equation (5.3), $T_\perp < 0.1$ K. The width of the broad peak (36 MHz) is a measure of the parallel temperature. In order to convert the width measured at an angle of 30° to what it would be when $\theta = 0$, the measured width must be divided by cos(30), thereby increasing it from 36 MHz to 41.6 MHz (.00139 cm^{-1}). This yields a parallel temperature of 0.94 K for helium. The beam velocity can be calculated from the shift of 515 MHz between the two peaks. The beam velocity given by $u_b = \Delta\nu c/[\nu_o \cos\theta]$, yields a beam velocity of 1520 m/sec. This can be compared to the theoretical ultimate beam velocity (Eq. (5.5)) of 1580 m/sec. In the latter calculation, the effective mass of the 6% HF beam was taken to be 4.96 amu.

As the gas exits the nozzle, the temperature and gas densities drop rapidly. The rate of decrease depends only upon γ, the ratio of the constant pressure to constant volume heat capacity. For an ideal gas $\gamma = (C_v + R)/C_v$. Its value thus ranges from a maximum of $5/3 = 1.67$ for monatomic gases to $\gamma = 1$ for a large molecule with a very large vibrational heat capacity. The beam and stagnation temperatures and gas densities are related by the equations (D.R. Miller, 1988):

$$T/T_o = [1 + \tfrac{1}{2}(\gamma - 1)M^2)]^{-1} \tag{5.8}$$

$$n/n_o = [1 + \tfrac{1}{2}(\gamma - 1)M^2)]^{-1/(\gamma-1)} \tag{5.9}$$

The parameter M is the Mach number, which is the ratio of the molecular beam center of mass speed (u_b) to the speed of sound in the beam. The speed of sound is itself related to the temperature by $u_s = (\gamma RT/m)^{1/2}$, where m is the molecular weight. The

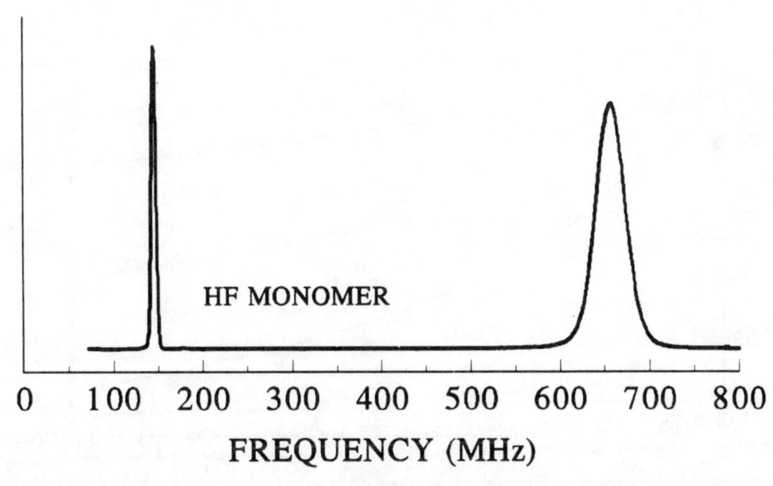

Figure 5.2 Doppler profiles for IR absorption of 6% HF in a He beam obtained with a laser arrangement shown in Figure 5.1. The sharp narrow peak is obtained by crossing the laser beam at an angle of 90° with respect to the molecular beam, while the broad peak is the result of a 30° angle between the laser and molecular beams. The center line absorption for the P_1 line is 3920 cm^{-1}. The shift in the two peaks is a measure of the beam velocity. A 40-μm-diameter nozzle was used with a stagnation pressure of 6.5 atm. Adapted with permission from Bohac et al. (1992).

temperature, density, and Mach number are plotted as a function of the distance from the nozzle in Figures 5.3 and 5.4 (D.R. Miller, 1988).

Equations (5.8) and (5.9) have been derived under the assumption that γ is constant throughout the expansion. Although this is true for a monatomic gas, it is not true for the vibrational heat capacity of polyatomic gases which become smaller as the temperature drops.

The Mach number varies with the reduced distance as $A(\gamma)[z/d]^{\gamma-1}$, where $A(\gamma)$ is a constant that depends upon γ. For Ar or He, $A = 3.23$ (DePaul et al., 1993). Even though the center of mass beam velocity does not change after about three nozzle diameters, M continues to increase with distance from the nozzle. This is a result of the decreasing sound velocity as the temperature drops. According to Figure 5.4, the Mach number increases much more rapidly for higher values of γ. As a result, the translational cooling is much more efficient if molecules are diluted in a rare gas such as He or Ar. An additional advantage of heavy dilution in a rare gas is that the equations derived for constant heat capacity can be employed.

None of the parameters plotted in Figures 5.3 and 5.4 depends upon the stagnation pressure. This is because their functional dependence was derived under the assumption that collisions continue to cool the gas, even up to very high z/d values. However, this does not happen in a real system in which collisions effectively cease after a few nozzle diameters. The precise distance depends on the stagnation pressure and the nozzle diameter. Thus, the greater the pressure, the greater the distance along which collisions cool the gas. By taking into account the gas density and the collision cross

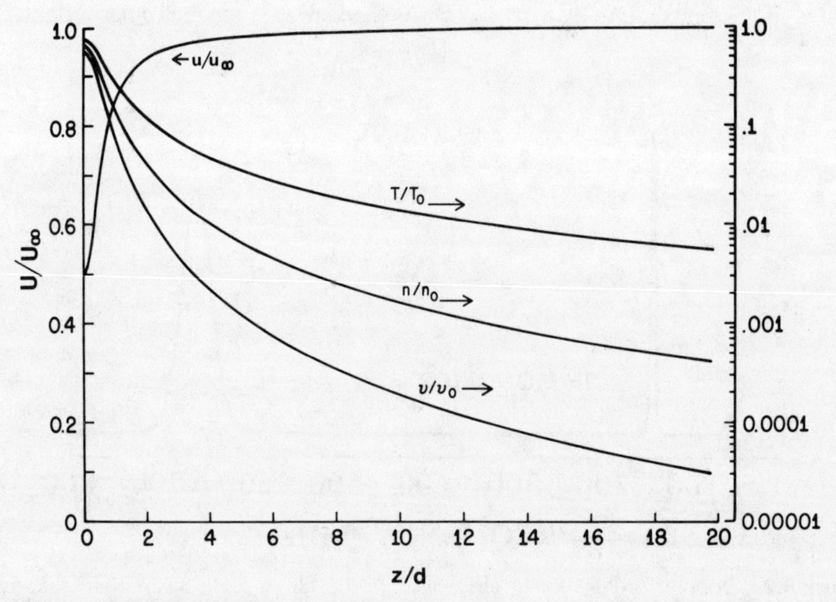

Figure 5.3 Free jet center line beam velocity (u/u_∞), temperature (T/T_o), gas density (n/n_o), and hard sphere collision frequency (ν/ν_o) as a function of the distance from the nozzle, in units of nozzle diameters for the case of $\gamma = 5/3$. Note that none of these parameters depends upon the stagnation pressure. Taken with permission from D.R. Miller (1988).

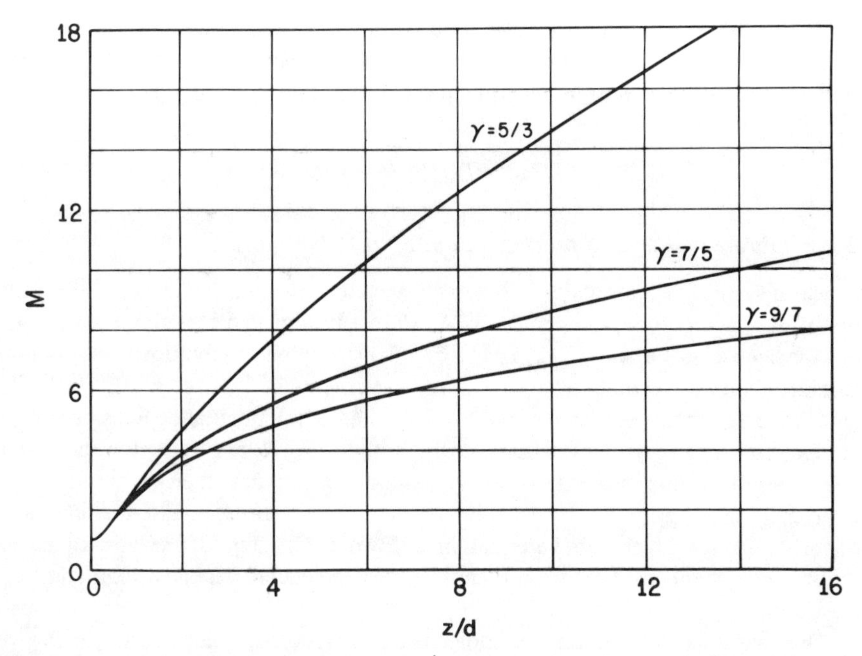

Figure 5.4 The Mach number as a function of distance from the nozzle for various values of γ, the ratio of the constant pressure to constant volume heat capacity. Note that the Mach number does not depend upon the source pressure. Taken with permission from D.R. Miller (1988).

section, and making the assumption that collisions stop abruptly at a certain point, the sudden freeze model predicts that the Mach number reaches a terminal value which is given by (Anderson, 1974):

$$M_T = 1.18[\lambda_o/d]^{[-(\gamma-1)/\gamma]}, \tag{5.10}$$

where λ_o is the mean free path of the gas in the source, and d is the nozzle diameter. When this is expressed in terms of the stagnation pressure in atmospheres at 298 K and the nozzle diameter in μm, the results for Ar and He are

$$\mathrm{M_T(Ar)} = 2.8(P_o d)^{0.4} \quad \text{and} \quad \mathrm{M_T(He)} = 2.1(P_o d)^{0.4}. \tag{5.11}$$

The terminal Mach number can be inserted into Equation (5.8) to determine the final translational temperature. For the case of the experiment shown in Figure 5.2, in which the stagnation pressure is 6.5 atm and the nozzle diameter is 50 μm, the final parallel translational temperature is calculated to be 2.4 K (compared to the experimentally determined value of 1 K).

If we approximate Equations (5.11) by changing the exponent from 0.4 to 0.5, it is possible to derive a simple relationship between the pressure and the nozzle diameter. Since $T_o/T \gg 1$, and $M_T^2 \propto P_o d$,

$$T/T_o \propto (P_o d)^{-1}. \tag{5.12}$$

Apparently, the cooling improves with increasing pressure and nozzle diameter. On the other hand, the gas flow Q increases with the pressure and the square of the nozzle diameter. Because most experiments are limited by the pumping speed, the gas flow is generally considered a constant for a given apparatus. Thus for a given Q, the degree of cooling is given by

$$T/T_o \propto (P_oQ)^{-1/2} \propto d/Q. \tag{5.13}$$

5.2.2.2 Vibrational and Rotational Cooling of Polyatomic Gases

Of central importance in state or energy selected dynamical studies are the initial vibrational and rotational temperatures. As shown in Figure 5.5, a room-temperature molecule such as isobutane (C_4H_{10}) has a broad distribution of vibrational and rotational energies with an average value of 120 meV (970 cm^{-1}) (Weitzel et al., 1991). The major interest in the use of beams is thus the cooling of the internal degrees of freedom of polyatomic molecules. Equations (5.8) through (5.13) are valid for monatomic gases. If a polyatomic molecule is highly diluted in a rare gas, then the gas dynamics will be determined primarily by the diluent gas, and the translational temperatures can be approximated with the above equations. However, the fate of the vibrational and rotational energy in the expansion of polyatomic molecules depends upon additional parameters.

The rotational temperature of molecular-beam-cooled molecules has been well documented by spectroscopic means. McClelland et al. (1979) have shown that I_2 molecules cooled in a molecular beam reach rotational temperatures that are within a factor of two (e.g., 3 K vs. 1.6 K or 23 K vs. 13 K) of the translational temperature if

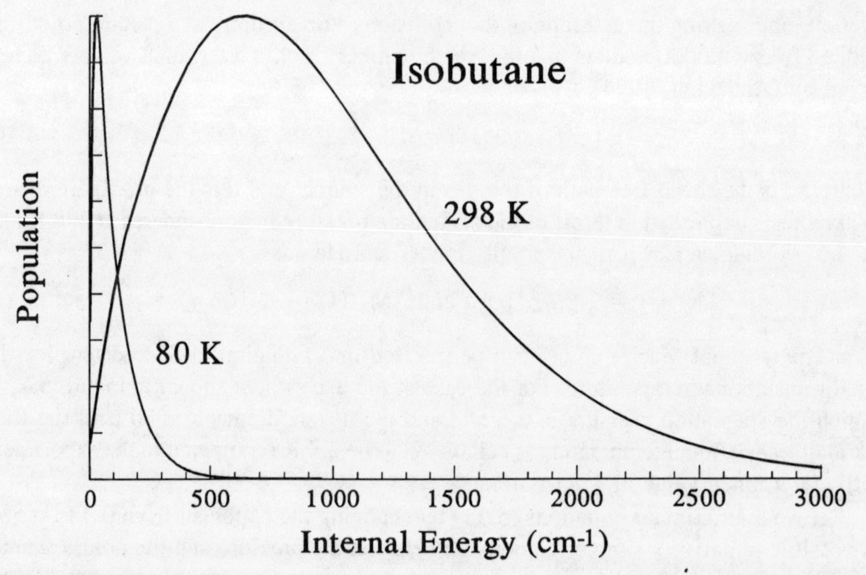

Figure 5.5 The calculated Boltzmann ro-vibrational energy distribution for isobutane at the two indicated temperatures. At 80 K, the average vibrational energy is reduced to just 17 cm^{-1}, while the rotational energy is 83 cm^{-1}. The vibrational frequencies were taken from Weitzel et al. (1991).

the diluent gas is Ar, CO_2, CH_4, or other massive molecules. On the other hand, if the diluent is He or H_2, then the rotational temperature is not as efficiently cooled (e.g., 27 K vs. 0.25 K) Figure 5.6 contrasts the room-temperature and molecular-beam-cooled spectra of NO by resonance enhanced multiphoton ionization (Miller and Compton, 1986). The facile cooling of the rotational degrees of freedom is consistent with rotational relaxation data which indicate that fewer than 10 collisions are required to transfer a rotational energy quantum to the translational degrees of freedom (Stevens, 1967).

In contrast to rotations, collisions are inefficient at transferring vibrational energy

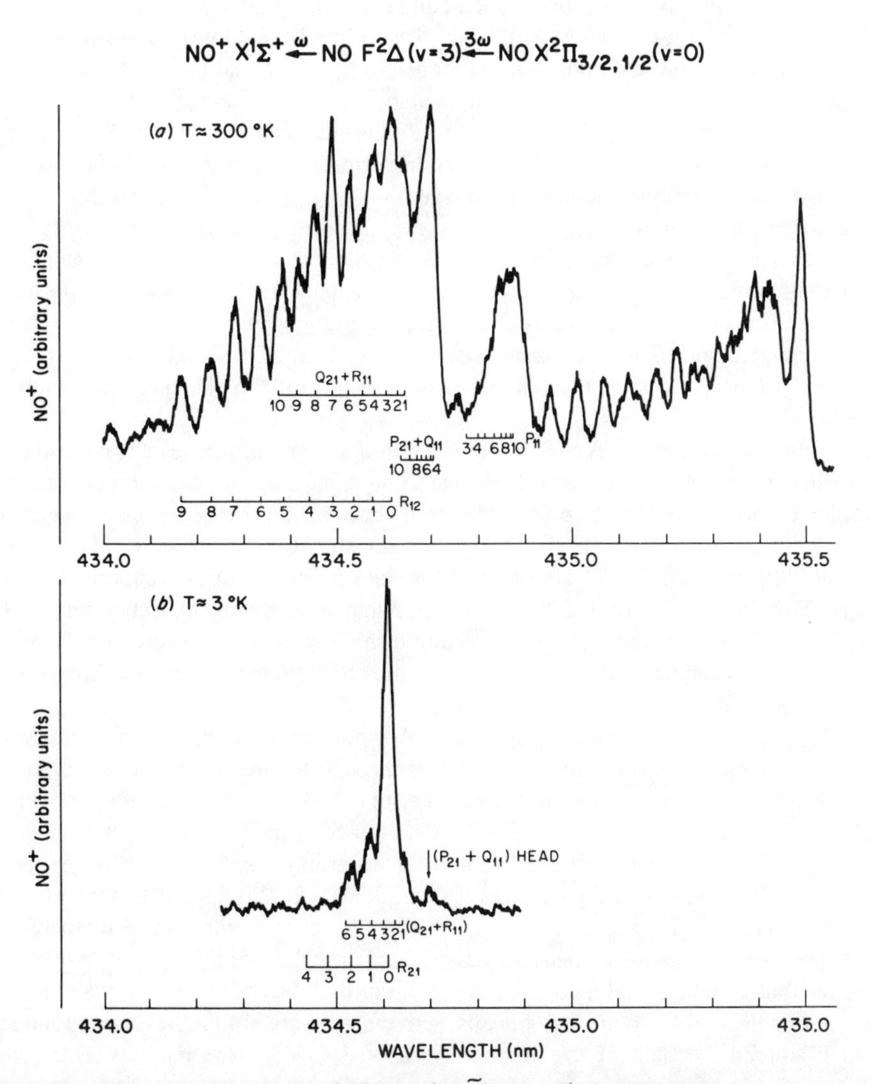

Figure 5.6 Two REMPI spectra of NO in the $\tilde{A}$ state. The room-temperature spectrum shows a broad rotational distribution while the cold spectrum shows just a few of the rotations. Taken with permission from Miller and Compton (1986).

to translational degrees of freedom (Stevens, 1967; Levine and Bernstein, 1987). For small molecules such as N_2 as many as 10^5 collisions are required to deactivate the v = 1 level (Cottrell and McCoubrey, 1961). According to the Schwartz-Slawsky-Herzfeld (SSH) theory, (Schwartz et al., 1952), the probability for vibrational energy transfer scales as $T^{-1/3}$. Thus, the relaxation rate drops as the temperature in the beam drops, so that vibrational cooling may stop at temperatures considerably higher than the translational temperature.

The vibrational relaxation rate also depends upon the total energy in the molecule and the vibrational energy spacing. At high internal energies (e.g., 50 kcal/mol or 18,000 cm^{-1}), significant amounts of energy (ca. 1000 cm^{-1}) are transferred per collision when the collision gas is itself a complex molecule (Gilbert and Smith, 1990). However, at room temperature and below, it is primarily the lowest-frequency vibrational modes which are responsible for transferring vibrational energy to the translational heat bath. According to the adiabaticity argument (Landau and Teller, 1936; Levine and Bernstein, 1987), energy transfer is maximized if the collision period matches the vibrational period. The collision period for thermal systems is generally less than the vibrational period so that the V–T energy transfer probability is low. Detailed analysis shows that it varies as $\exp(-2\pi\nu/\alpha u)$, where ν is the vibrational frequency, α is a parameter that characterizes the repulsive part of the interaction potential, and u is the center of mass collision velocity. This means that molecules with low vibrational frequencies relax much more rapidly than molecules in which the lowest frequency is high. For instance benzene, in which the lowest frequency is rather high (350 cm^{-1}), appears to cool rather inefficiently in a molecular beam (Sulkes, 1985).

Another aspect that affects vibrational cooling in a molecular beam is the nature of the diluent gas. Although Ar and He have the same translational temperature, the former is more efficient at cooling vibrations because the better match between the masses allows for more efficient momentum transfer. Even more efficient is a polyatomic diluent, such as CO_2, because it provides a V–V path in addition to a V–T path. This is in spite of the higher translational temperature expected for a CO_2 expansion. However, there is a point of diminishing returns. Both Ar and polyatomic species tend to condense at low temperatures. Hence, helium is necessary to reach very low temperatures.

How can the vibrational temperature of a polyatomic molecule in a molecular beam be assessed? A traditional approach is through the measurement of vibrational hot bands. If it is assumed that the vibrations are in a Boltzmann distribution of states, the temperature of the molecule can be determined from the ratio of the hot band $(1 \rightarrow 0)$ intensity to the $(0 \rightarrow 1)$ intensity according to: $I(1 \rightarrow 0)/I(0 \rightarrow 1) = \exp(-\Delta E/RT)$. However, there are several problems with this approach. First of all, it is not certain that the molecule can be described by a single vibrational temperature. The lack of experimental evidence concerning this question generally leads workers to assume that all vibrational modes are in equilibrium. A second, more serious problem, is one of sensitivity. Often, the optically active modes are not the low-energy modes, which contain the bulk of the thermal vibrational energy. Hence, the sensitivity in determining the temperature is considerably diminished.

A more direct method for determining the average vibrational energy in molecules cooled in molecular beams is by determining the dissociative ionization onset by photoionization (Weitzel et al., 1991). Because photoionization is a bound to continuum process, the whole molecular thermal energy distribution (vibrations and rotations) is transferred to the ion. That is, it can reasonably be assumed that at a given photon energy, all molecules are ionized with equal probability and that their energy as ions is given simply as $h\nu + E_{th}$. This assumes that the electrons are ejected with the same kinetic energy, independent of the precursor molecule's thermal energy. The dissociation onset (here defined as the energy at which 50% of the ions are dissociated) for room-temperature and cold molecules should be shifted by the difference in the median thermal energy between the cold and warm molecules. The amount of the shift is thus directly related to the thermal energy removed from the molecule. Such an experiment on isobutane (Fig. 5.7) shows that even with a modest expansion of 20% isobutane in Ar at a stagnation pressure of 400 Torr, the cooling is sufficient to reduce the vibrational temperature to below 80 K, a temperature sufficiently low to remove virtually all of the vibrational energy. Figure 5.5 shows the Boltzmann ro-vibrational distribution at 80 K for which the average energy is 12 meV (100 cm^{-1}), 80% of which is due to the rotational energy. A similar experiment with a neat sample of isobutane reduced the vibrational temperature to only 150 K (Weitzel et al., 1991).

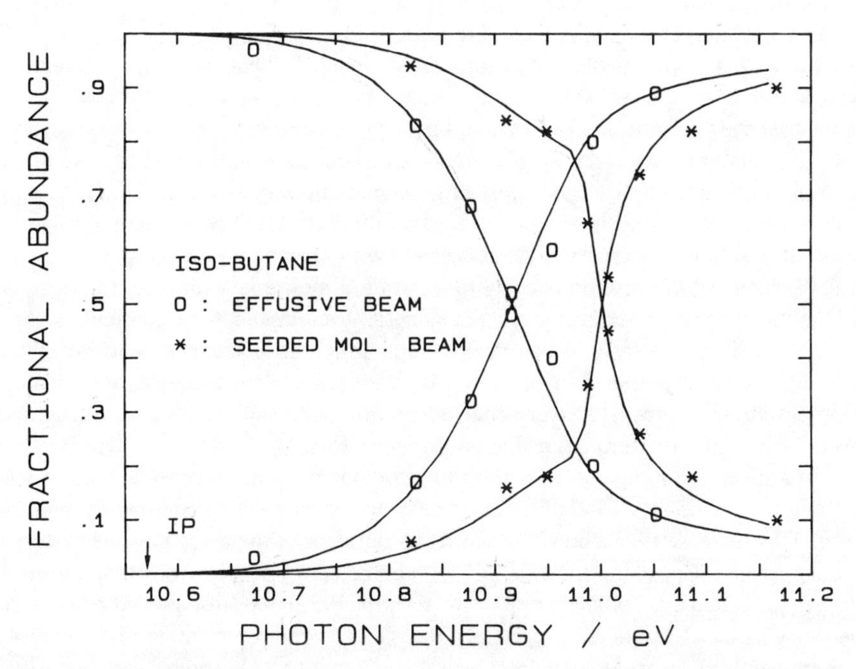

Figure 5.7 The dissociative ionization onsets for CH_4 loss from energy selected *iso*-butane ions obtained by PEPICO. The shift in the crossover energy is the median thermal energy removed when the room-temperature sample (effusive beam) is replaced by a seeded beam (20% C_4H_{10} in 400 Torr Ar). The cw nozzle had a diameter of 180 μm. Taken with permission from Weitzel et al. (1991).

5.2.2.3 Sample Density in Molecular Beams

In studies involving only the detection of fluorescence, the experiments can be carried out without a skimmer very close to the nozzle because photomultipliers can be operated at high pressures. The gas densities can thus be substantial. According to Figure 5.3, the gas density within 10 nozzle diameters from the nozzle are 0.001 P_o, or close to 1 Torr for a 1 atm stagnation pressure. However, for experiments involving the detection of charged particles, the gas pressure must be below 10^{-5} Torr. This is accomplished by placing the experiment in a separate and differentially pumped chamber. The separation of the two chambers by a skimmer permits operation of the source and experimental chambers with pressure differentials of two orders of magnitude. The price for this lower pressure is a vastly reduced gas density. After the molecular beam is established, the density drops as $(z/d)^2$. Thus, operating the experiment 50 mm from an 0.1 mm skimmer means that the gas density drops from $10^{-3}P_o$ to $4 \times 10^{-7}P_o$.

5.2.2.4 Pulsed versus Continuous Molecular Beams

Molecular beams come in two sorts: continuous and pulsed. The former are used with continuous light sources, while the latter are generally used with pulsed light sources. Although this difference seems trivial, it is actually profound because the two types of experiments differ in nearly every respect, from the size of the experimental chamber, the size of the pumps, the types of lasers or other light sources, to the electronics used to detect and analyze the experimental results. A continuous system uses a great deal of sample so that experiments with isotopically labeled substances can often not be carried out. The relatively low instantaneous power densities of continuous lasers means that they cannot be used for nonlinear processes, such as multiphoton ionization. The data production rate in continuous experiments is uniform in time so that the electronics are not required to digest large signals in very short times. Thus, single photon particle counting is an approach generally associated with continuous experiments. It is seldom necessary to be concerned with bleaching of a sample by the laser light. Because of the low probability of exciting a molecule with a continuous laser, two or three color experiments are not generally attempted with continuous lasers. Finally, a continuous system has no inherent timing associated with it, so that the derivation of rates is less evident than it is in pulsed systems. On the other hand, in favorable circumstances, the very high resolution achievable with continuous lasers permits extraction of rates from line width measurements.

At a given pumping speed, a pulsed nozzle operating at 10 Hz with a pulse width of 100 μsec can deliver 1000 times more peak gas density than a continuous molecular beam. This increase is related simply to the duty factor of the pulsed nozzle system. In practice, it is not possible to scale all the parameters by a factor of 1000 so that the intensity advantage is somewhat reduced (Gentry, 1988). Because pulsed and continuous working (cw) lasers have typically the same average power, the advantages of operating pulsed lasers with pulsed beams are obvious. The high laser intensity of pulsed lasers makes it possible to carry out multiphoton experiments in which an interaction volume is interrogated with several different laser pulses, either simultaneously or delayed with respect to each other. The rapid rate of data production (many data points must be collected within a few nanoseconds), requires the use of very fast electronics.

5.3 METHODS FOR STATE OR ENERGY SELECTION

5.3.1 Direct Electronic Excitation

Pulsed laser excitation of electronic states is one of the simplest means for measuring dissociation rates in real time. Once excited above the dissociation limit, the molecule evolves in one of two ways (see section 3.3, p. 60):

$$AB + h\nu \rightarrow AB^*(S_1) \begin{array}{l} \rightarrow A + B \\ \rightarrow AB^{\#}(S_0) \rightarrow A + B\,. \end{array}$$

If the excited S_1 state is dissociative, the reaction generally takes place on the upper potential energy surface and is extremely fast. This class of reactions is often referred to as photodissociation, a topic discussed in a recent book (Schinke, 1992). On the other hand, if the excited state decays first by internal conversion to the ground electronic state, the reaction rate is much slower and more readily measured.

Among the first laser photodissociation experiments carried out were those by Riley and Wilson (1972), in which the fragmentation of alkyl halides was investigated. Because the upper state is dissociative, the fragments are produced with considerable kinetic energy. Pulsed excitation was used to provide a start signal for measuring the product time of flight distribution and thereby the translational energy release. However, the direct dissociation rates, which are extremely fast, were not measured.

The precision in energy with which the molecule is prepared depends upon the laser resolution and its related pulse width. The pulse width of a dye laser pumped by the standard Nd-YAG, Excimer, or nitrogen laser is 5–10 nsec. Thus, its transform limited (maximum) resolution is given by the uncertainty principle as $\Delta E = \hbar/\Delta t = 0.001\ \text{cm}^{-1}$. This resolution is sufficient to resolve the rotational states of most molecules. The resolution of a pulsed dye lasers with just a grating as a dispersing element is about $0.5\ \text{cm}^{-1}$. In order to reach the ultimate transform-limited resolution, it is necessary to install an intracavity etalon (e.g., a Fabry Perot interferometer) which typically consists of two accurately spaced parallel plates which provide surfaces for repeated reflections of the laser light (Moore et al., 1989). The resulting interferences convert the laser output into a series of highly monochromatic spikes shifted in wavelength by $1/2d$, where d is the etalon spacing. The wavelength can be adjusted either by a physical change of the plate spacings, or by an effective spacing change achieved by varying the pressure of a gas which is contained between the two plates (so-called pressure tuning) which changes the refractive index of the medium.

If the time resolution is now increased to 5 psec, the ultimate energy resolution is reduced to $1\ \text{cm}^{-1}$. This is clearly insufficient to resolve rotational levels in all but a few hydrogen containing molecules such as H_2O, HCl, and H_2. On the other hand, the ability to study fast reactions has increased by three orders of magnitude (Zewail, 1983).

Another three orders of magnitude decrease in the laser pulse width brings us to femtosecond pulses with which rate constants for very fast reactions, such as direct dissociations on repulsive surfaces, can be measured (Zewail, 1991). However, the resolution now degrades to $1000\ \text{cm}^{-1}$. The complex experimental set-up required for this work has been described by Zewail and co-workers (Felker and Zewail, 1988; Khundkar and Zewail, 1990).

5.3.2 Direct Excitation of High Vibrational Levels

Most dissociation reactions, especially those activated by thermal means, take place on the ground potential energy surface. It is thus of great interest to have available methods for preparing molecules directly in these high vibrational levels with a controlled total energy close to the dissociation limit. Such an approach is necessary for molecules in which the excited electronic state is not at a convenient energy. The excitation of specific configurations of the vibrational energy would also appear to be an interesting capability of direct vibrational excitation. This is not possible when the molecule is prepared by electronic excitation followed by a nonradiative transition because control over which vibrational modes are excited is lost. In practice, control over the vibrational modes is seldom possible because of rapid intramolecular vibrational energy redistribution (IVR) which is discussed in chapter 4.

Two methods have been extensively used for pumping large amounts of energy in a controlled fashion directly into the vibrational modes of molecules. These are the direct excitation of vibrational overtones, and stimulated emission pumping (SEP). A third approach, stimulated raman adiabatic passage (STIRAP), is less well developed. However it appears to have some advantages over the other method in that a significant fraction of the molecules can be excited to the upper state (Gaubatz et al., 1990).

5.3.2.1 The Excitation of Overtone Vibrations

The potential energy surface of a polyatomic molecule made up of harmonic oscillators is given by

$$V(q_1, q_2, \cdots, q_n) = \sum_i k_i q_i^2 \tag{5.14}$$

in which the q's are the normal vibrational modes (see chapter 2). In this expression, involving only the square terms in the potential energy, the normal modes are independent of each other, which leads to the $\Delta v = +1$ infrared absorption selection rule. However, to the extent that the potential energy surfaces in real molecules are anharmonic, this selection rule breaks down. It is thus possible to excite combination as well as overtone bands at the expense of a greatly reduced transition probability which drops as the number of overtones increases. For instance in the case of HOOH, a well-studied molecule by overtone excitation, the cross section for the production of $v = 5$ relative to $v = 1$ is about 10^{-6} (Luo and Rizzo, 1990; Scherer and Zewail, 1987). With the power available from visible lasers, it is possible to excite a small fraction of the molecules to the fifth and even the sixth overtone. In the excitation of the fifth O—H overtone, as much as 18,000 cm^{-1} (= 2.2 eV = 51 kcal/mol) can be deposited into a molecule. This is sufficient energy to break bonds in a number of molecules.

The overtone excitation tends to excite molecules in local modes (chapter 4). Thus, the excitation would appear to be highly localized. Furthermore, by the appropriate choice of laser wavelength, it is possible to excite different parts of a molecule. For instance, in the isomerization of allyl isocyanide to allyl cyanide,

$$H_2C{=}CH{-}CH_2{-}N{\equiv}C \longrightarrow H_2C{=}CH{-}CH_2{-}C{\equiv}N$$

the reactant can be investigated by exciting three different C—H groups. Since each one can be excited to the fifth, sixth, and seventh overtone, nine different excitation modes or energies can be accessed (Reddy and Berry, 1979a). Figure 5.8 shows the three different C—H absorptions in the photoacoustic spectrum of the sixth overtone region (Segall and Zare, 1988).

Although it appears that overtone excitation can selectively place the energy in certain parts of the molecule, this conclusion is not entirely correct. As pointed out by Jasinski et al. (1983), the preparation of true local modes requires the coherent excitation of the state over the whole transition width (see chapter 4). Thus the pulse width of the laser must match the homogeneously broadened absorption line. This requires typically picosecond pulse widths. At one time it was thought that the overtone absorption peaks in Figure 5.8 were homogeneously broadened. However, later experiments in which various portions of each band were excited, showed that the room-temperature spectrum in figure 5.8 consists of multiple transitions, many of them arising from hot bands (Segall and Zare, 1988). A similar conclusion was reached for the case of benzene. When this molecule was investigated in a molecular beam, the observed overtone spectra were very sharp with homogeneous bandwidths of the order of 10 cm^{-1} (Page et al., 1988).

An interesting application of overtone excitation is vibrationally mediated photodissociation, in which a molecule excited to high vibrational levels is photodissociated to a repulsive state (Likar et al., 1988; Crim, 1990). Figure 5.9 shows schematically the process in question. The excited repulsive surface can be reached by either one

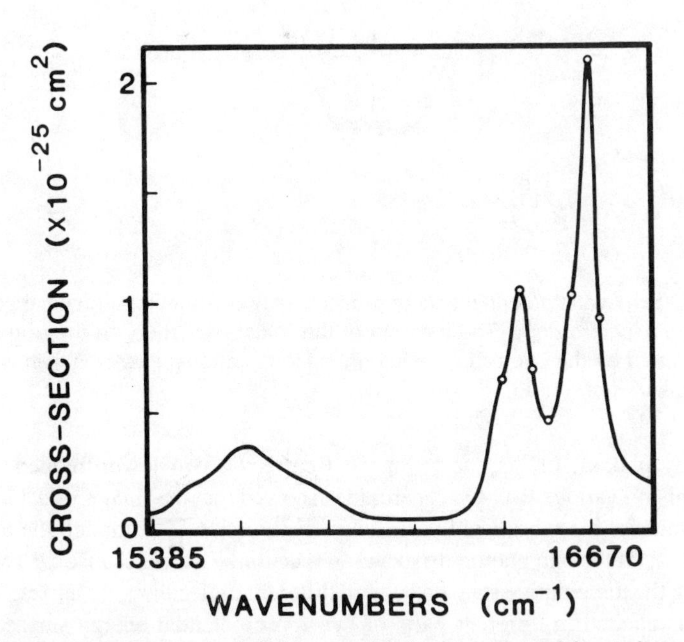

Figure 5.8 Overtone spectrum of allyl isocyanide in the region of the sixth overtone. The three peaks from low to high energies correspond to C—H overtones at the following positions: the methylenic CH_2, nonterminal olefinic CH, and the terminal olefinic CH groups, respectively. Adapted with permission from Segall and Zare (1988).

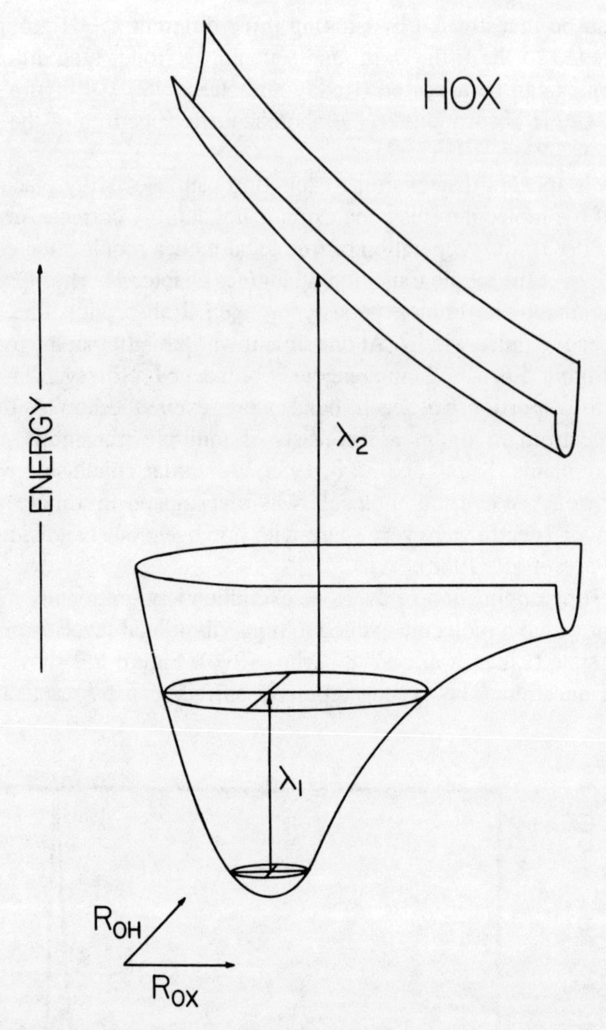

Figure 5.9 Schematic potential energy surfaces for two-color vibrationally mediated photodissociation spectroscopy. The location of the initial excitation on the upper repulsive surface is varied by the choice of the intermediate vibrational energy. Taken with permission from Likar et al. (1988).

high-energy photon, or, as shown in the figure, via two lower-frequency photons. Experiments on various R-O-O-R′ peroxides showed that the final product energies are very different for the vibrationally mediated dissociation even though the total energy is the same as in the one photon dissociation reaction. This is because the two methods of reaching the dissociative state do so with different molecular geometries. That is, the molecule is excited to different parts of the upper potential energy surface. Because molecules dissociating on repulsive surfaces tend to fragment without redistribution of their internal energy (Schinke, 1992), the final product state distribution reflects the geometry of the intermediate state. It should be emphasized that in most molecules, rapid IVR in the intermediate state does not permit mode selective chemistry. The variation in the observed products is simply a result of the average geometry of the intermediate states which vary with vibrational excitation.

Because overtone excitation is a low-probability process, it is necessary to carry out the experiment with considerable sample density. Thus, most of the studies on overtone spectroscopy have been with sample pressures in excess of 0.1 Torr (gas density $= n > 3 \times 10^{15}$ molec./cm^3). An estimate of the number of molecules excited can be made with use of Beer's law which is $N/N_o = \exp(-\sigma\ell n) \approx 1 - \sigma\ell n$, the latter approximation being valid for low-probability processes. In this expression, N and N_o are the intensities of the transmitted and incident light, respectively. Suppose that the cross section (σ) for excitation of the fifth overtone in HOOH is 10^{-23} cm^2 (Scherer and Zewail, 1987), that the path length $\ell = 1$ cm, and that the density, n is as given above. The probability for photon absorption will be 3×10^{-8}. If the laser delivers 10^{16} photons/second, as many as 10^8 molecules can be excited per second.

Because overtone excitation is a linear process, these high vibrations can be excited either by cw or pulsed lasers. However, most of the early work on overtone excitation was carried out with cw lasers (Reddy and Berry, 1977, Reddy and Berry, 1979a, Reddy and Berry, 1979b; Jasinski et al., 1983). Among the first such studies was one by Reddy and Berry (1977) in which the isomerization of $CH_3NC \rightarrow CH_3CN$ was investigated as a function of the internal energy. An intracavity circulating power of 80 W was used to irradiate the neat CH_3NC sample at pressures from 1 to 10 Torr for up to 30 min. Although the cross section at 726.5 nm is only 3×10^{-24} cm^2, conversions of up to 50% to the lower energy CH_3CN could be achieved. The experimental set-up for this study is shown in Figure 5.10.

The above estimates of signal level might suggest that a skimmed molecular beam, with a gas density of typically 10^{13} molecules/cm^3, would not be sufficiently

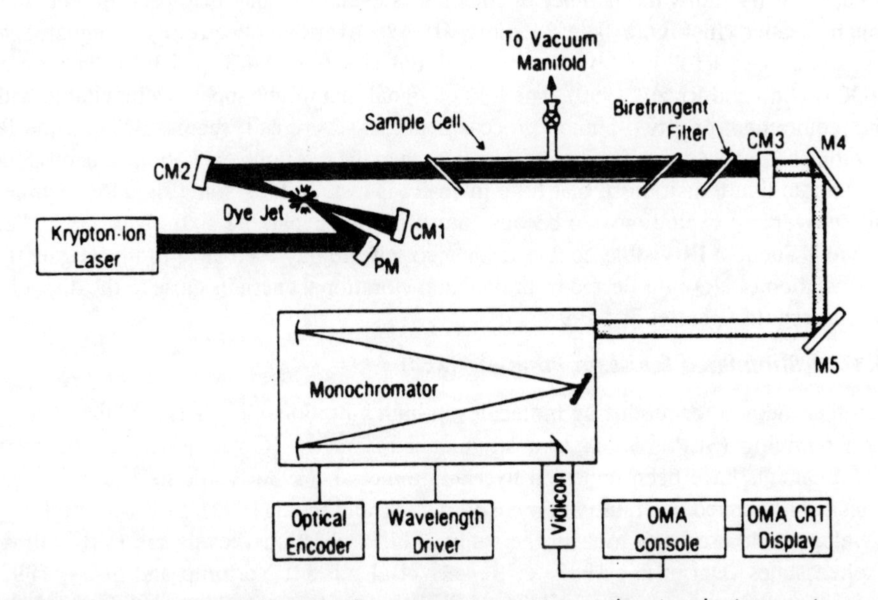

Figure 5.10 Experimental set-up of an overtone experiment showing the intracavity sample cell with a cw dye laser. Reaction products are analyzed by gas chromatography. Taken with permission from Reddy and Berry (1977).

dense to serve as an effective sample source. However, these estimates of signal must be made with some care. Because a room-temperature sample contains molecules in a large distribution of internal energy states, only a small fraction will be in the correct energy level to absorb the photon. The larger the molecule, the broader is this distribution of internal states. As a result, the measured absorption cross section is a strong function of the sample temperature, as well as the laser bandwidth. If the laser line width is as narrow as the natural line width of a single ro-vibrational absorption line, the phenomenological absorption cross section will increase as the gas is cooled and more molecules find themselves in fewer and fewer energy levels. Thus the gas density is only part of the story. What is really required is a knowledge of the density of sample gas in the *correct* energy level. This can be readily calculated from a knowledge of the molecule's density of states. In fact, overtone excitation with pulsed molecular beams and pulsed lasers has been reported (Butler et al., 1986; McGinley and Crim, 1986). It is thus apparent that a molecular beam source, although lower in gas density by more than two orders of magnitude relative to the thermal source, contains a higher density of molecules in a given energy level. The above observations underscore the fact that a phenomenological absorption cross section is not an inherently interesting number because it cannot be readily transferred from one experimental arrangement to another.

The drawback in the use of thermal samples becomes increasingly evident as the molecule increases in size. Single-state selection is nearly impossible because of the congested absorption lines in thermal samples. However, state selection (especially the rotational states) is possible in two-color experiments in which an infrared laser first excites a ro-vibrational mode of the molecule, followed by the visible laser that excites the overtone (Luo and Rizzo, 1990, 1991; Fleming et al., 1991; Fleming and Rizzo, 1991; Luo et al., 1992) The optical–optical double-resonance (OODR) approach greatly limits the number of final states excited so that the spectrum becomes much cleaner. In Figure 5.11, a direct $0 \rightarrow 6$ overtone spectrum is compared to sequential $0 \rightarrow 1$ followed by $1 \rightarrow 6$ transitions (see Figure 4.3, p. 70) for the case of HOOH (Luo and Rizzo, 1990). The loss of signal that might appear as inevitable with the requirement of a two-photon process is not as severe as it seems. Because the IR photon absorption cross section is far greater than the overtone excitation, it is possible to saturate that transition so that little intensity is lost. In fact, some signal is regained in the overtone excitation step because the fifth rather than the sixth overtone is then excited. Such an IR-visible double-resonance scheme has been used to investigate the dissociation of NO_2 in selected rotational and vibrational energies close to the dissociation threshold (Reid et al., 1993).

5.3.2.2 Stimulated Emission Pumping (SEP)

Another method for producing molecules in high vibrational levels is stimulated emission pumping (SEP), which is a stimulated resonance Raman process. The early developments have been reviewed by Hamilton et al. (1986) while its many applications are illustrated in a multivolume work by Dai and Field (1994). SEP was originally developed and used as a tool for the study of high vibrational levels and IVR in those excited states (Dai et al., 1985a,c; Reisner et al., 1984; Northrup and Sears, 1992; Stanley and Castleman, 1991). Because SEP can excite molecules to levels above the dissociation limit, it is highly useful for the study of unimolecular reactions as well (Dai et al., 1985b; Berry et al., 1992; Choi and Moore, 1989; Choi et al., 1992; Neyer

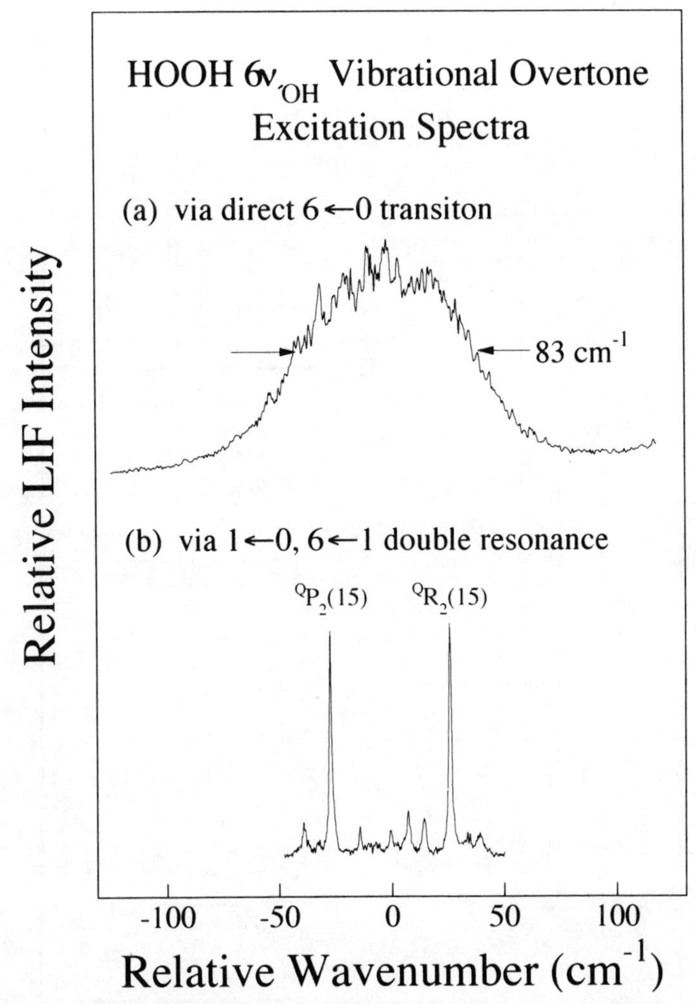

Figure 5.11 Comparison of the sixth OH overtone spectra in HOOH obtained by direct $6 \leftarrow 0$ excitation and by double-resonance absorption (see Fig. 4.3). In the latter process, the IR laser was held fixed on a $1 \leftarrow 0$ $P(16)$ transition, while the visible laser was scanned over the $6 \leftarrow 1$ transition. A third laser was used to monitor the overall absorption by FE of the OH dissociation product. Taken with permission from Luo and Rizzo (1990).

et al., 1993). An energy diagram for state preparation of HFCO by SEP is shown in Figure 5.12. Laser I (the pump laser) excites the molecule to an electronic state, while laser II (the dump laser) stimulates an emission back down to the ground electronic state. The final energy that remains in vibrations is just the difference in the energy of the pump and dump lasers plus the original energy of the molecule. The set-up for this SEP experiment is shown in Figure 5.13. The pulsed molecular beam which is oriented perpendicular to the page is intersected by the two collinear lasers about 4 cm from the nozzle. The detection scheme for the SEP process (see Fig. 5.14) is based on monitoring the fluorescence from the S_1 state before and after the dump laser pulse. Other

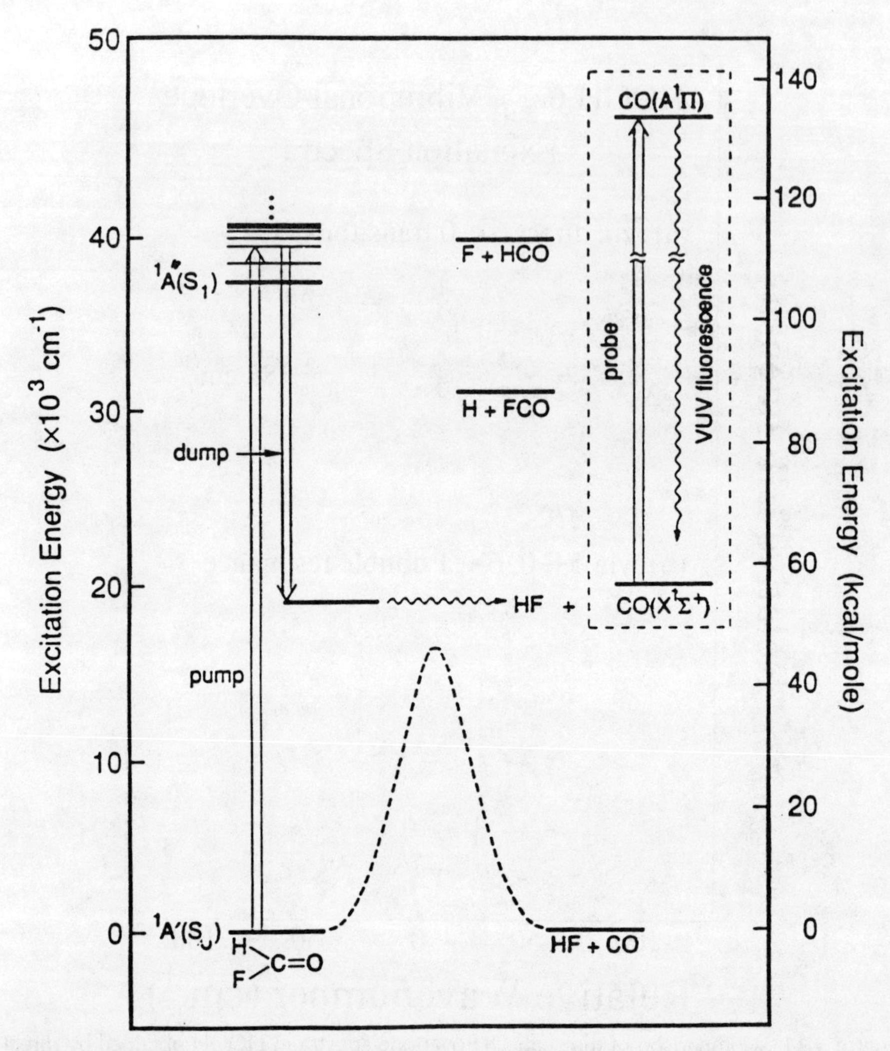

Figure 5.12 The potential energy diagram of HFCO showing the preparation of high vibrational levels of the molecule by stimulated emission pumping (SEP). The product CO molecules are monitored by fluorescence excitation. Taken with permission from Choi et al. (1992).

detection schemes can also be used. For instance, if the excited electronic state is further excited by a third laser to the ionization limit, the decrease in the ionization signal is a measure of the SEP process (Hamilton et al., 1986).

One of the important features of SEP, in common with other optical–optical double-resonance techniques, is the simplification of the spectra because laser I excites a single upper state. Thus, selection rules between the intermediate and final states greatly reduce the final allowed states. Of particular interest for the study of dissociation phenomena is that SEP allows for considerable control of the final states that are

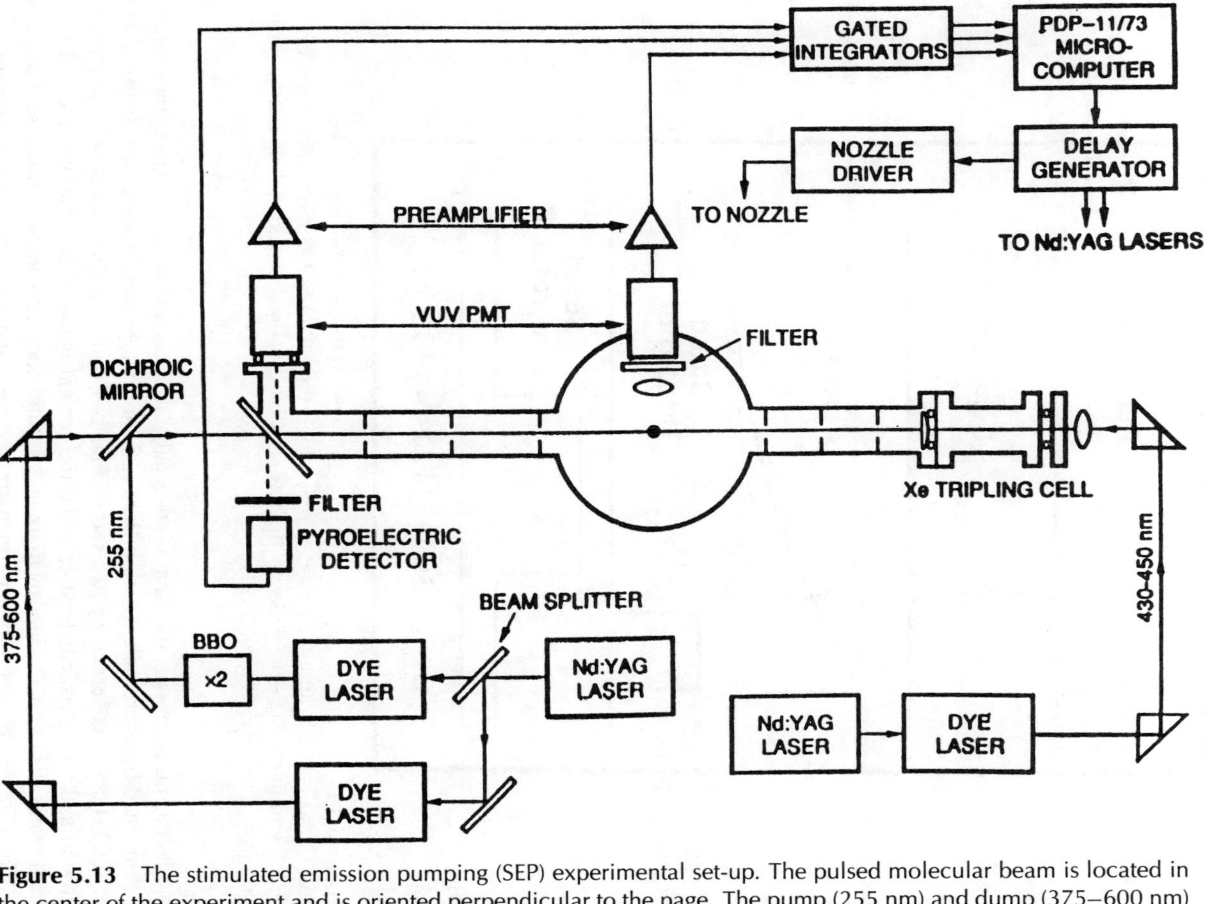

Figure 5.13 The stimulated emission pumping (SEP) experimental set-up. The pulsed molecular beam is located in the center of the experiment and is oriented perpendicular to the page. The pump (255 nm) and dump (375–600 nm) beams enter from the left, while the VUV fluorescence excitation (FE) laser enters from the right. Taken with permission from Choi et al. (1992).

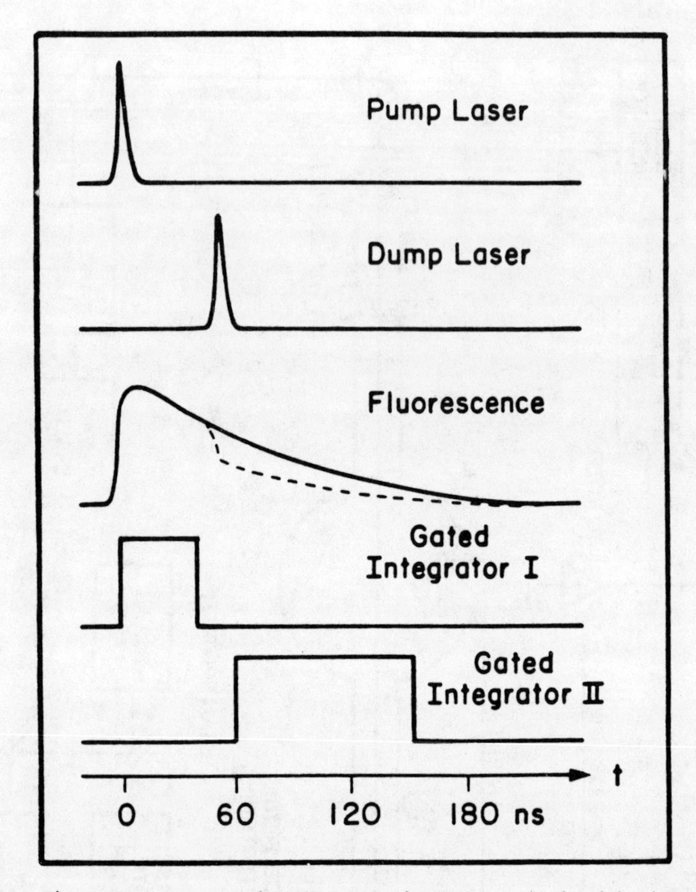

Figure 5.14 The timing sequence for measuring the SEP signal. The delay between pump and dump lasers is about 50 nsec during which time the excited molecules travel only 0.005 cm. Dumping efficiencies as high as 60% have been achieved. Taken with permission from Choi and Moore (1991).

produced. Furthermore, the final vibrational states reached are often ones that cannot be obtained by overtone excitation. In a resonant two-photon process, the distribution of final states is governed by two sets of Franck-Condon factors, the first between the ground and the intermediate states, and the second between the intermediate and the final states. Thus, if the intermediate and ground state geometries differ, the Franck-Condon factors between the intermediate and final states can be used to prepare the excited S_0 molecule in vibrational states not easily reached with one photon.

5.4 METHODS FOR ENERGY SELECTION OF IONS

The problems associated with the preparation of ions in selected energies or states differ from those of neutrals in that the species to be state selected lie in the ionization continuum. Energy selection cannot be accomplished simply by photoexcitation of a ground-state molecule because the energy removed by the departing electron must be accounted for. Two methods have been used to overcome this problem. In the photoelectron photoion coincidence (PEPICO) experiment, the electron kinetic energy is measured and the ions are collected in coincidence with these energy analyzed electrons. (Baer et al., 1991; Baer, 1986). In the other method, ground-state ions are formed by whatever means is most convenient and the ion is then photoexcited to dissociative states (Durant et al., 1984). The photoexcitation of a ground-state ion is a bound-to-bound transition and is thus similar to the excitation of a neutral reactant.

The main difference between neutral and ionic species lies in the density of the sample. Ions cannot be stored in great concentrations because of their mutual electrostatic repulsion (the space charge effect). Another difference is that ions can be readily accelerated by electric and magnetic fields. This ability to steer and to accelerate ions can be used to measure the dissociation rates.

5.4.1 Photoelectron Photoion Coincidence (PEPICO)

In the PEPICO method, an ion is produced by one-photon vacuum UV ionization (140 to 50 nm). In this bound-to-continuum process, the ions are produced in a distribution of internal energies, E_i, given by

$$E_i = h\nu - IE + E_{\text{th}} - KE_{\text{e}}, \tag{5.15}$$

where $h\nu$ is the photon energy, IE the ionization energy, E_{th}, the initial thermal energy of the neutral molecule, and KE_{e} the energy of the departing electron. Since $h\nu$ and IE are known and E_{th} is either known or made negligibly small, the ion internal energy distribution is a function mainly of the distribution of the electron kinetic energies. The ions can thus be energy selected by measuring them in delayed coincidence with energy-selected electrons. In the absence of thermal energy in the precursor molecule, the internal energy resolution will be determined by the electron energy resolution which can be as high as 1 to 5 meV but is more typically 25 meV (200 cm^{-1}).

A PEPICO set-up is shown in Figure 5.15. Ions and electrons are formed in a small electric field in which they are accelerated in opposite directions. The smaller the electric field, the higher the electron energy resolution. The electron signal not only establishes the internal energy of the ion, but it also provides a start signal for measuring the ion time of flight (TOF). This timing is the basis for mass analysis and ion lifetime measurement. One of the unique features of PEPICO is the ability to determine absolute collection efficiencies and therefore the total ionization rate. If we assume that the detected electrons and ions are derived from the same photoionization volume, then the collection efficiencies for electrons, E_{e}, and ions, E_{i}, are given by

$$E_{\text{e}} = C_{\text{c}}/C_{\text{i}} \qquad E_{\text{i}} = C_{\text{c}}/C_{\text{e}} \tag{5.16}$$

where C_{c}, C_{i}, and C_e, are the coincidence, total ion, and total electron counts per second, respectively (Baer, 1979). The total ionization rate is then $C_{\text{c}}/(E_{\text{i}}E_e)$. The ability to measure the absolute collection efficiencies is useful in a number of ways, among them in the collection of fragment ions that have considerable kinetic energy. A drop in the collection efficiency indicates that energetic ions are not detected because they hit apertures or the walls of the TOF region.

The first coincidence experiments were carried out with fixed energy line sources and the ions measured in coincidence with energetic electrons (Brehm and Puttkamer,

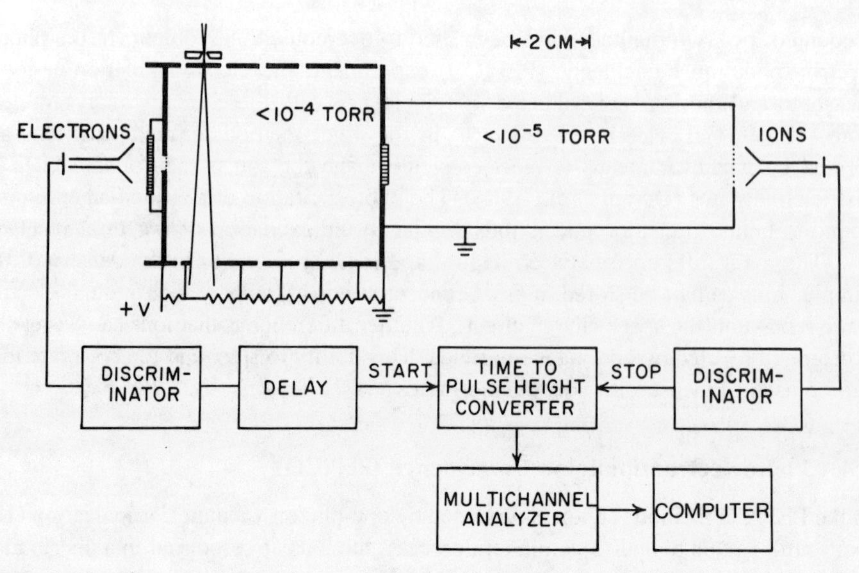

Figure 5.15 The experimental arrangement of a photoelectron photoion coincidence (PEPICO) set-up. Electron energy analysis is accomplished by an electrostatic energy analyzer, by electron TOF when the light source is pulsed, or by angular discrimination against energetic electrons. Taken with permission from Baer (1986).

1967). More recently, dispersed continuum photon sources have been used in which the ions are measured in coincidence with initially zero, or threshold, electrons (Stockbauer, 1973; Werner and Baer, 1975; Dutuit et al., 1991; Das et al., 1985; Rosenstock et al., 1980; Weitzel et al., 1993; Norwood et al., 1989). The latter method, often called threshold PEPICO or TPEPICO, has two major advantages relative to PEPICO. First, threshold electrons can be collected with near 100% collection efficiency precisely because they have no initial velocity and thus pass through small apertures with only minor losses. On the other hand, energetic electrons are formed with a distribution of initial velocity vectors, resulting in a collection efficiency of about 0.1%. This reduced collection efficiency cannot be compensated by the higher intensity of the line source (ca. 10^{13} photons/sec) compared to the dispersed continuum source (ca. 10^9 photons/sec) because false coincidences between unrelated electrons and ions are proportional to the total ionization rate. For a total ionization rate of C_{tot} the mean time between ion formation events is $1/C_{tot}$. If the time window for collection of coincidence signal is τ, the number of false coincidences in that time interval will be $C_{tot}\ \tau$. Thus, for an ionization rate of 10^5 counts/sec and a time window of 10μsec, there will be on the average one false coincidence event per second. The goal in a coincidence experiment is to keep the total ionization rate low, while optimizing the efficiencies with which the ions and electrons are collected. The factor of 1000 in the collection efficiency between PEPICO and TPEPICO results in a 1000-fold increase in the false coincidence rate in the former. It is also the false coincidence rate that prevents the use of pulsed VUV lasers in PEPICO experiments. Each 5-nsec laser pulse generates many electron–ion pairs so that it is not possible to distinguish the true from the false coincidences. Pulsed laser PEPICO would only be possible with high-repetition-rate lasers that can be operated at sufficiently low pulse energies to prevent multiple ionization events. Pulse rates in excess of 10 kHz, a pulse width of 5 nsec, and a total photon flux of 10^{10} photons/sec would be ideal parameters for a laser based PEPICO experiment.

The second advantage of TPEPICO is in the electron energy resolution. This resolution is determined by an angular discrimination against energetic electrons (Baer et al., 1969; Tsai et al., 1974; Spohr et al., 1971). The analyzer, consisting of a series of small apertures, allows threshold electrons to pass while rejecting most energetic electrons. It is more akin to a filter than a dispersive analyzer that is required for the analysis of energetic electrons. With sufficiently small apertures, and long drift tubes, the electron resolution can be made as high as desired. A major problem in the TPEPICO energy analysis, based only on angular discrimination of energetic electrons, is that a small fraction of the energetic electrons have their initial velocity vector in the direction of the detector. These can be distinguished from the real threshold electrons only by a true energy analysis. The most effective method is by TOF, which is possible when the light source is pulsed as in, for example, synchrotron radiation (Baer et al., 1979). With this approach, large apertures and drift regions of only one or two centimeters can be used which give collection efficiencies of 50% and resolutions of 5 meV (Merkt et al., 1993). The combination of angular discrimination and TOF analysis is also the basis of the laser based zero kinetic energy (ZEKE) electron spectroscopy in which the electron resolution is limited only by the bandwidth of the laser (Müller-Dethlefs and Schlag, 1991).

5.4.2 Photodissociation of Ions

The major advantages of laser photodissociation of ions relative to the PEPICO approach is the higher-energy resolution that can be achieved and the ability to measure faster rates. The higher resolution is possible because of the use of a laser light source and because electron energy analysis is no longer necessary. The faster rate measurements are possible because the ion internal energy resolution is not adversely affected by the application of large draw-out fields. On the other hand, the photodissociation of ions involves more steps. The first and most difficult one is the preparation of an ion in a well defined state. If electron impact ionization is used, the ions must first be cooled either radiatively by storing them in a trap, or by collisions in a supersonic expansion. However, the latter approach is unproven and risks the formation of cluster ions (Feinberg et al., 1992; Johnson and Lineberger, 1988). Radiative cooling of the excited ions down to the wall temperature of the apparatus has been successfully employed to prepare ions for photodissociation (So and Dunbar, 1988; Dunbar et al., 1987). However, the wall temperature used so far has been 298 K which does not sufficiently reduce the internal energy of large molecules.

Another approach to the production of ground-state ions is by laser ionization. The aim here is to prepare the ion in the ground state or in a single excited state. The former is generally easier to achieve. The simplest approach, which is the use of a single VUV laser photon to ionize the molecule to the ground state followed by a second dissociating laser pulse, has not been reported yet. Instead, various schemes leading to approximate ion energy selection via two photon (one color) absorption have been used (Durant et al., 1984; Stanley et al., 1990; Kuhlewind et al., 1986). These schemes lead to approximate energy selection because they depend upon favorable Franck-Condon factors between the intermediate neutral and the final ion states for pure state preparation of the ion. In benzene, the intermediate S_1 state fortuitously lies just halfway to the ionization energy, so that a single color resonant two-photon absorption produces the ion mostly in the ground vibrational levels of the electronic ground state.

A more general method for ion state preparation with a single-color laser is shown in Figure 5.16. If the intermediate state is a Rydberg state with a geometry similar to that of the ion, then the ionization step will produce an ion in that same vibrational level. Only a few molecules have been energy selected in this manner because it is not always possible to find a convenient and unperturbed Rydberg state (Conway et al., 1985; Orlando et al., 1989; Yang et al., 1991). Another general method for preparing the ion ground state is with two laser colors in which the first laser excites the molecule to a long-lived intermediate state while the second laser excites this state to the ion ground state (Lemaire et al., 1987). However, this approach requires a total of three laser colors because a final laser color is needed to prepare the ion in selected energies above the dissociation limit.

Once the ion has been energy selected, the subsequent photodissociation is straightforward. The major experimental difficulty involves overlapping in both space and time of laser pulses in order to maximize the probability of ion photon absorption.

5.5 RATE DETERMINATION METHODS

5.5.1 Relative Rates from Stern-Volmer Plots and Fluorescence Rates

The approach used in many early rate studies (by collisional and chemical activation) is one in which the decay rate is measured in competition with another reaction, whose rate is presumably known. The "known" reaction is often the collisional stabilization

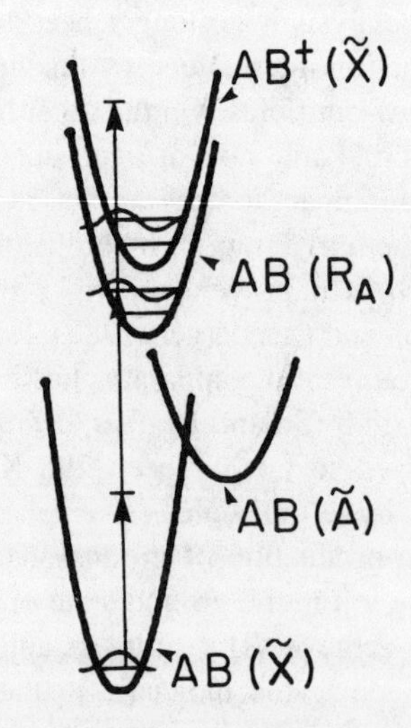

Figure 5.16 Scheme for state selection of ions by resonance enhanced multiphoton ionization. The Rydberg (R_A) and ion ($\tilde{X}$) states have the same geometry so that $\Delta v = 0$ transitions dominate in the ionization step. Taken with permission from Baer (1986).

reaction, $AB^* + M \rightarrow AB + M$, whose rate can be varied by changing the pressure. The energy-selected version of this involves optical excitation of the reactive species. Consider the following scheme utilized by Reddy and Berry (1979a):

$$AB + h\nu \xrightarrow{k_a} AB^* \qquad \text{photo-activation}, \tag{5.17a}$$

$$AB^* + M \xrightarrow{k_d} AB + M \qquad \text{collisional deactivation}, \tag{5.17b}$$

$$AB^* \xrightarrow{k(E)} \text{Products} \qquad \text{unimolecular reaction}. \tag{5.17c}$$

By applying a steady-state condition on the AB^* species $[d(AB^*)/dt = 0]$, we find that the rate of reaction, given by $dP/dt = k(E)[AB^*]$ is

$$\text{Rate} = \frac{k_a[h\nu]k(E)}{k(E) + k_d[M]}[AB] = k_{app}[AB] \tag{5.18}$$

where the quantities in square brackets are the concentrations of the species. The $[h\nu]$ is the photon number density. The apparent first-order rate constant, k_{app}, can be inverted to give

$$\frac{1}{k_{app}} = \frac{1}{k_a[h\nu]} + \frac{k_d}{k_a[h\nu]k(E)}[M]. \tag{5.19}$$

Thus, a plot of $(k_{app})^{-1}$ versus the gas pressure $[M]$ yields a straight line with intercept $(k_a[h\nu])^{-1}$ and slope $k_d/(k_a[h\nu]k(E))$. Since the photoactivation rate, $k_a[h\nu]$ is known from the intercept, the slope permits the determination of the ratio $k_d/k(E)$. An example of such a Stern-Volmer plot is shown in Figure 5.17 for the isomerization of the previously mentioned allyl isocyanide reaction, $C_3H_5NC \rightarrow C_3H_5CN$, in which $k_a[h\nu]/k_{app}$ is plotted. This results in an intercept of 1.0 and a slope of $k_d/k(E)$. The quantity of interest, $k(E)$, can be extracted if we know the deactivation rate, k_d. This is generally taken to be equal to the gas kinetic collision rate constant (strong collision assumption), which is typically about 10^{-9} cm^3/(molec·sec).

In order to have precision in the rate measurements, it is necessary to carry out experiments at pressures such that the deactivation rate, $k_d[M]$, is comparable to $k(E)$. At a pressure of 1 atm ($n = 2.5 \times 10^{19}$ molec./cm^3) the collision rate will be 2.5×10^{10} sec^{-1}. By increasing the pressure, faster rates can be measured. However, there is a practical upper limit imposed by the deactivation step, which reduces the product signal as the pressure is raised. At low pressures, the practical limit is imposed by the diffuse or thin sample which reduces the signal level. Typical lower limits are about 1 Torr where the deactivation rate is 3×10^7 sec^{-1}. Thus the range over which rates can be measured in such experiments is between 10^7 and 10^{10} sec^{-1}.

The rate results from Stern-Volmer plots depend directly upon the assumption that k_d is equal to the collision rate. This can be checked by diluting the sample gas with different inert collision partners. Unfortunately, because the product yield drops as the pressure is increased, these experiments have generally been done with neat samples. However, the Stern-Volmer plot also contains an internal check on the strong collision assumption. Suppose that a single collision does not completely stabilize the excited molecule so that it can continue to react, but at a lower rate. Residual reactivity at the

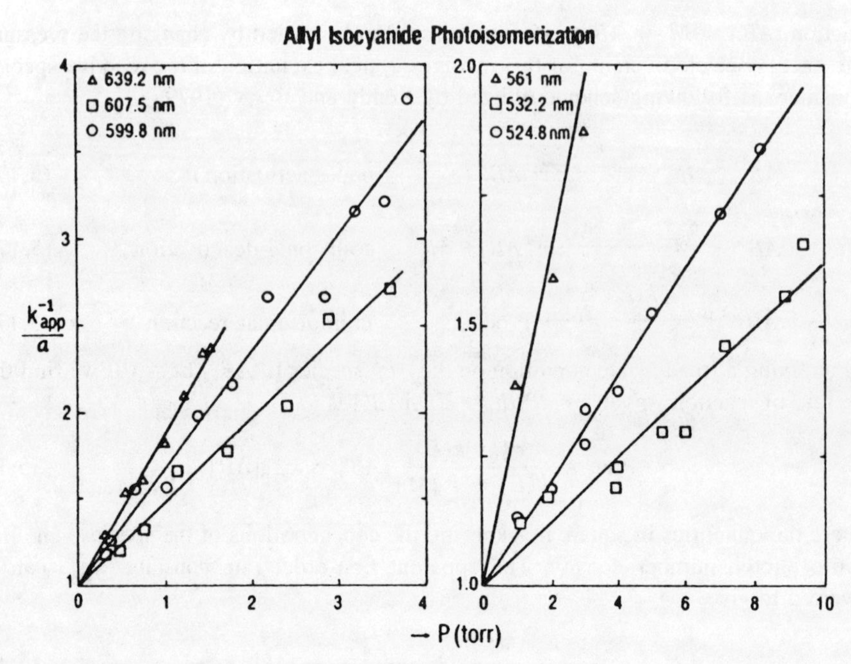

Figure 5.17 Stern-Volmer plots for the isomerization reaction of allyl isocyanide at several laser energies. The slope $[k_d/k(E)]$ is inversely proportional to the reaction rate constant. Adapted with permission from Reddy and Berry (1979a).

lower rate should become evident at low pressures, where the Stern-Volmer plot will be curved. Hence, complete stabilization is not necessary for the technique to work. What is necessary is that the deactivation step reduces the unimolecular rate to negligible levels, in which case the straight-line portion at high pressures can be used. However, curvature in straight lines is often difficult to quantify, or to distinguish from the noise. Several studies have in fact concluded that the strong collision assumption is not generally valid, and corrections for it have been shown to be necessary (Snavely et al., 1986; Baggott, 1985; Troe, 1983).

Another indirect method for rate measurements involves measuring the fluorescence decay rate and the quantum efficiency from which the rate of nonradiative decay can be calculated. The principle is based on the following mechanism:

$$AB + h\nu \longrightarrow AB^* \xrightarrow{k_r} AB + h\nu' \quad (5.20a)$$

$$AB^* \xrightarrow{k_{nr}} A + B \text{ (and other processes)}. \quad (5.20b)$$

The quantum efficiency, Φ, is proportional to the ratio of the fluorescence signal to the number of photons absorbed. In order to determine the absolute quantum efficiency, it is necessary to calibrate the experimental set-up with a molecule of known quantum efficiency.

The lifetime of the excited AB^* is generally measured by determining the fluorescence lifetime τ. The total decay rate of the excited AB^* is given by $k_T = k_r + k_{nr}$, where $\tau = 1/k_T$. Since $\Phi = k_r/k_T$, the nonradiative rate is given by $k_{nr} = k_T(1 - \Phi)$. Although this appears to be a simple experiment for systems in which the excited state decays via both fluorescence and dissociation, it is in fact not a simple matter to establish that k_{nr} can be attributed exclusively to dissociation. The excited state can also decay nonradiatively by internal conversion to the ground electronic state, by intersystem crossing to a triplet state followed by collisional deactivation, etc. That is, dissociation is only one of several nonradiative paths that may be important in the decay of an excited state. Thus, considerable care must be exercised in the interpretation of such data. However, these data can always be used to place an upper limit on the dissociation rate constant.

Figure 5.18 shows the fluorescence decay rate of *trans*-stilbene, a molecule which isomerizes to *cis*-stilbene on the excited potential energy surface (Syage et al., 1982). At low energies the *trans*-stilbene lifetime as measured by fluorescence is independent

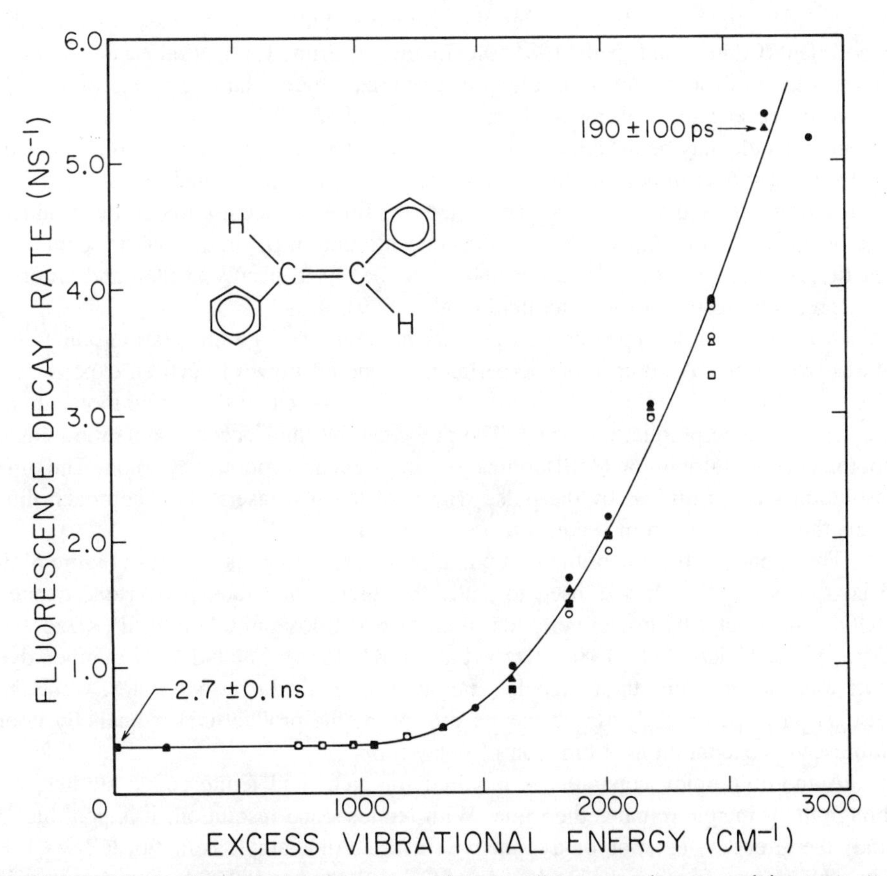

Figure 5.18 The fluorescence decay rate of *trans*-stilbene as a function of the excitation energy. The rate increases abruptly above the barrier for isomerization at 1200 cm^{-1}, an increase that is attributed to the isomerization reaction. The rate is extracted under the assumption that the fluorescence rate remains constant at (1/2.7 nsec^{-1}). Taken with permission from Syage et al. (1982).

of the excitation energy. However, above about 1200 cm^{-1}, the decay rate increases rapidly with a corresponding decrease in the quantum yield. The total decay rate curve could be fitted by including an isomerization rate obtained with the RRKM model (Khundkar et al., 1983). The sharp onset at 1200 cm^{-1} and the appearance of the product *cis*-stilbene at the same energy, appears to establish that the increase in k_{nr} is due exclusively to the isomerization reaction.

5.5.2 Direct Real-Time Measurements

The most reliable method for measuring dissociation rates is through real-time determination of the time evolution of a reacting system. Real-time experiments require a start pulse which signals the initiation of the reaction, and a means for monitoring the disappearance of reactant, or the appearance of products as a function of the time after excitation. If the reaction is very slow (on the order of microseconds) and the reaction products are produced in an excited electronic state, the natural fluorescence of the product as a function of time after the excitation pulse can be used to time the experiment (Cannon and Crim, 1981; McGinley and Crim, 1986; Nimlos et al., 1989). Fluorescence of the parent molecule is characterized by a total decay rate of $k_f + k_r$ (fluorescence and reaction rates). If $k_r \gg k_f$, the fluorescence quantum yield for the parent molecule may be too small to be measurable. On the other hand, fluorescence of the excited product molecule will be characterized by a rise in signal that is related to the reaction rate, and a decay that is related to the fluorescence lifetime of the product. The time profile is identical to that of the concentration of $[B]$ in the classic sequential reaction $A \rightarrow B \rightarrow C$ (in this case B and C are the electronically excited and ground-state products, respectively) (Steinfeld et al., 1989).

A more versatile approach that permits monitoring of ground-state products requires two lasers. A pump-probe experiment is one example of such an experiment. The second laser, time delayed from the first, can probe either the parent molecule or the newly formed product molecule. The probe can be fluorescence excitation, single photon (PI) or multiphoton (MPI) ionization, transient absorption, and so on. The time resolution will be limited by the pulse widths of the two lasers. This approach thus spans the time range from seconds to femtoseconds.

The apparatus for one of the first pump probe experiment is shown in Figure 5.19 (Rizzo et al., 1984). It was used to study the dissociation rate of overtone excited HOOH and *t*-butyl hydrogen peroxide, both of which lose an OH radical (Rizzo and Crim, 1982; Ticich et al., 1986; Rizzo et al., 1983, 1984). The Nd:YAG-pumped dye laser was used to excite the molecule (in a bulb) to selected overtone states, while the nitrogen laser pumped dye laser was used to probe the product OH radicals by laser fluorescence excitation as a function of delay time.

A more complex apparatus is required for pico- and femtosecond studies, although the principle remains the same. With femtosecond resolution, it is possible to study the direct dissociation on a repulsive surface. An example is that of ICN $\rightarrow$ I + CN. The time profiles for the production of CN with the use of 100 fsec-wide pulses is shown in Figure 5.20. Because the CN group in the transition state region absorbs to the red of "free" CN, it is possible to monitor the time profile of the CN unit in the transition state and the appearance of "free" CN (Knee and Zewail, 1988).

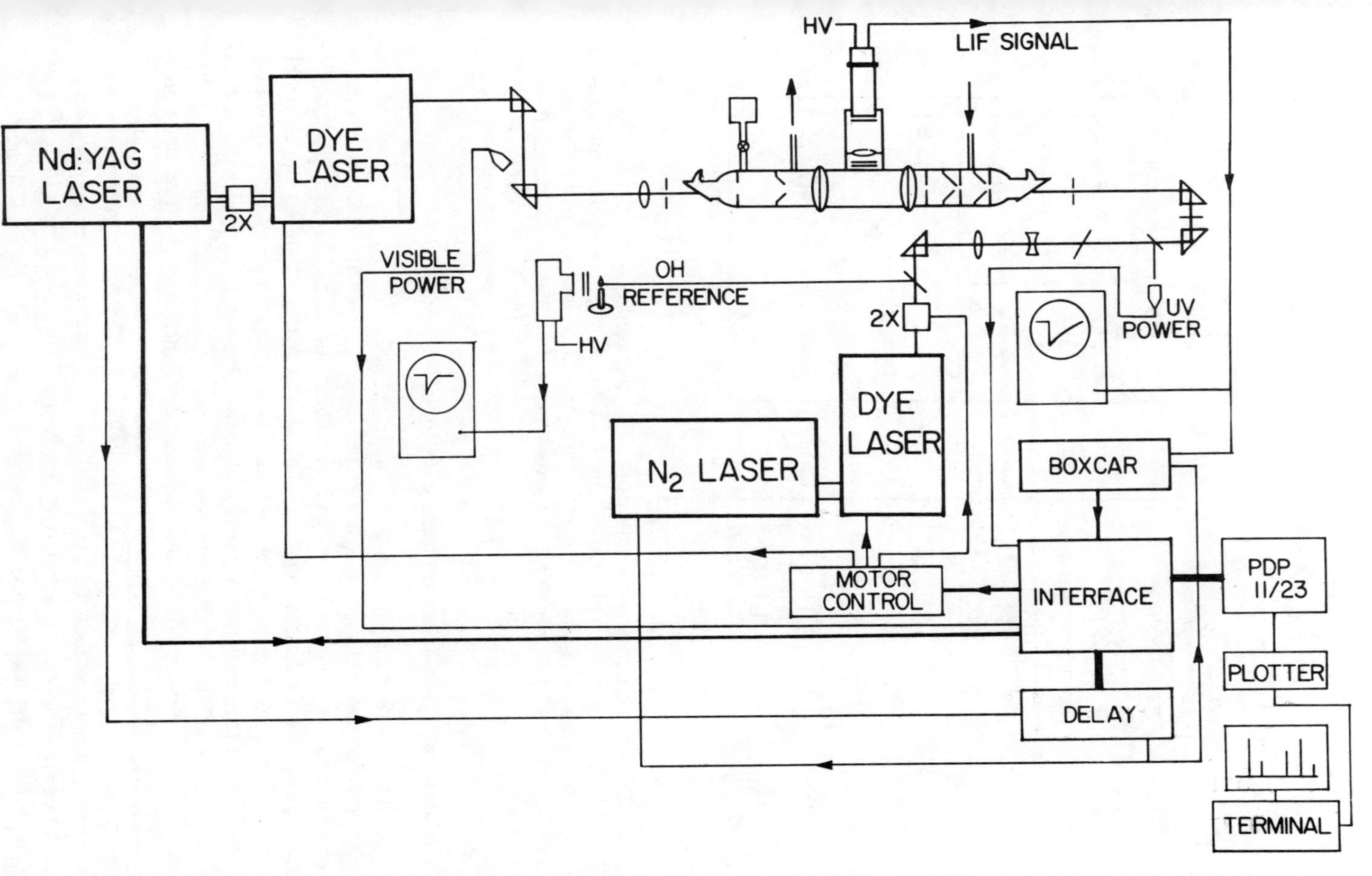

Figure 5.19 Two color apparatus for the study of an overtone-induced reaction followed by time-delayed LIF or FE analysis of products. The two laser beams counterpropagate in the pressurized sample cell. Taken with permission from Rizzo et al. (1984).

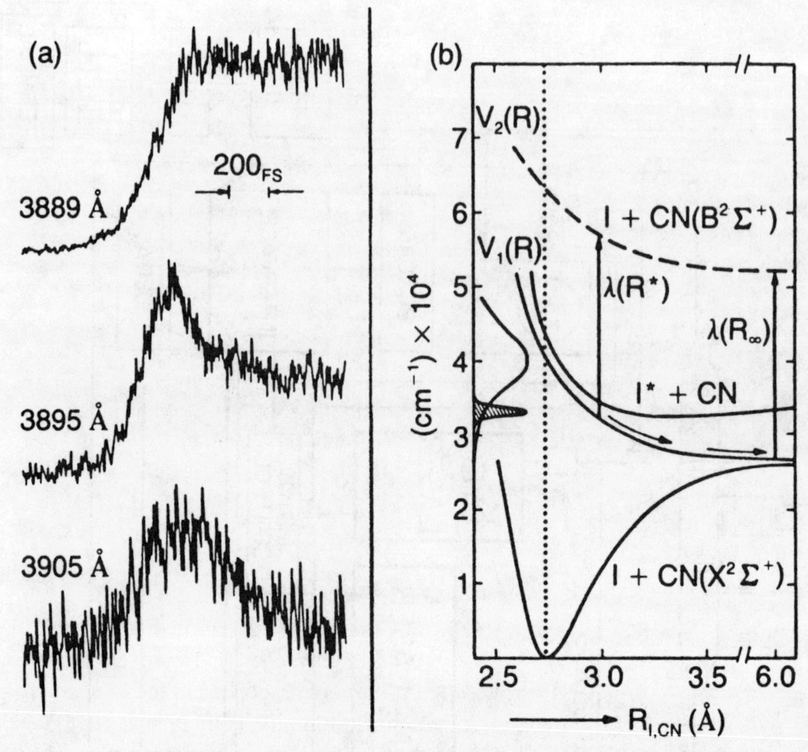

Figure 5.20 Femtosecond laser studies of the ICN dissociation rates. The 100 fsec wide laser pulse with its resolution of about 50 cm^{-1} is shown along the vertical energy scale of the potential energy diagram in Figure 5.20(b). The upper trace in part (a) shows the production of "free" CN while the two lower traces show the appearance and disappearance of the transition state CN. Taken with permission from Knee and Zewail (1988).

5.5.3 Real-Time Rate Measurements with Ions

Although the above pump probe methods can be applied to ions as well as neutrals, the low density of state-selected ions makes the application of fluorescence excitation somewhat tenuous. On the other hand, the ability to accelerate ions permits rate measurements by a different technique. The kinetic energy of an ion can be readily measured either by time of flight methods or by a direct energy analysis with an electrostatic energy analyzer. Suppose that an ion dissociates in the course of being accelerated by an electric field. An ion gains energy during acceleration, but loses some of it upon dissociation because the neutral fragment carries off its share of the kinetic energy which is proportional to its mass (recall that $KE = \frac{1}{2}mu^2$). Thus, the product ions' kinetic energy or TOF distribution can be converted directly into a distribution of ion lifetimes. Figure 5.21 shows a simple two-stage acceleration plus drift experiment for measuring ion dissociation rates. The electric field in the first and second acceleration regions is $\epsilon_1 = (V_2 - V_1)/a_1$, and $\epsilon_2 = V_2/a_2$, respectively. For nondissociating ions, the ion TOF and kinetic energy is given by

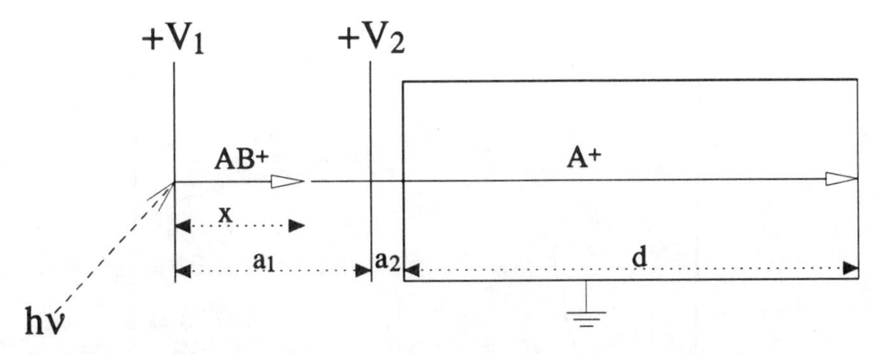

Figure 5.21 Schematic diagram for measuring ion dissociation rates in a two-stage acceleration region. The rates of ions which dissociate in the acceleration region can be determined by measuring either the A^+ product ion kinetic energies, or their times of flight.

$$\text{TOF(parent ions)} = T_{a1} + T_{a2} + T_{d} \tag{5.21a}$$

$$\text{KE(parent ions)} = \epsilon_1 a_1 + \epsilon_2 a_2, \tag{5.21b}$$

where the subscripts a and d refer to the dimensions of the acceleration and drift regions. The kinetic energy of daughter ions that dissociate immediately upon photon absorption will be identical to the parent ion kinetic energy. However, their time of flight will be less than the parent ion TOF. If the ion dissociates in the first acceleration region at a distance x from the point of ionization, its TOF in that region will be composed of two terms, the first as a heavy parent ion, the second as a lighter daughter ion.

$$\text{TOF}(A^+) = T_{0\to x}(AB^+) + \text{T}_{x\to a1}(A^+) + T_{a2}(A^+) + T_{d}(A^+), \tag{5.22a}$$

where $T_{0\to x}$ and $T_{x\to a1}$ are the flight times of the parent and daughter ions in the first acceleration region. Similarly, the kinetic energy of the slowly formed daughter ions will be given by

$$KE(A^+) = \epsilon_1 xm/M + \epsilon_1(a_1 - x) + \epsilon_2 a_2, \tag{5.22b}$$

where m and M are the masses of the daughter and parent ions, respectively. A distribution of decay times in the first acceleration region results in an asymmetrically broadened TOF as shown in Figure 5.22, from which the decay rate can be extracted (Baer and Carney, 1982; Baer, 1986). An analogous data set is obtained if, instead of the TOF, the product kinetic energies are measured (Andlauer and Ottinger, 1971; Stanley et al., 1990).

The ion lifetime range that can be investigated in these experiments depends upon the applied electric field. In field ionization experiments, lifetimes in the picosecond range have been measured by applying fields as high as 20 keV/cm (Derrick and Burlingame, 1974; Kluft et al., 1988; Darcy et al., 1978). These high fields are achieved by replacing the left electrode in Figure 5.21 with a razor blade and moving the blade to within a fraction of a millimeter of the $+V_2$ plate. However, no ion internal energy information is obtained in field ionization. In PEPICO experiments, where it is

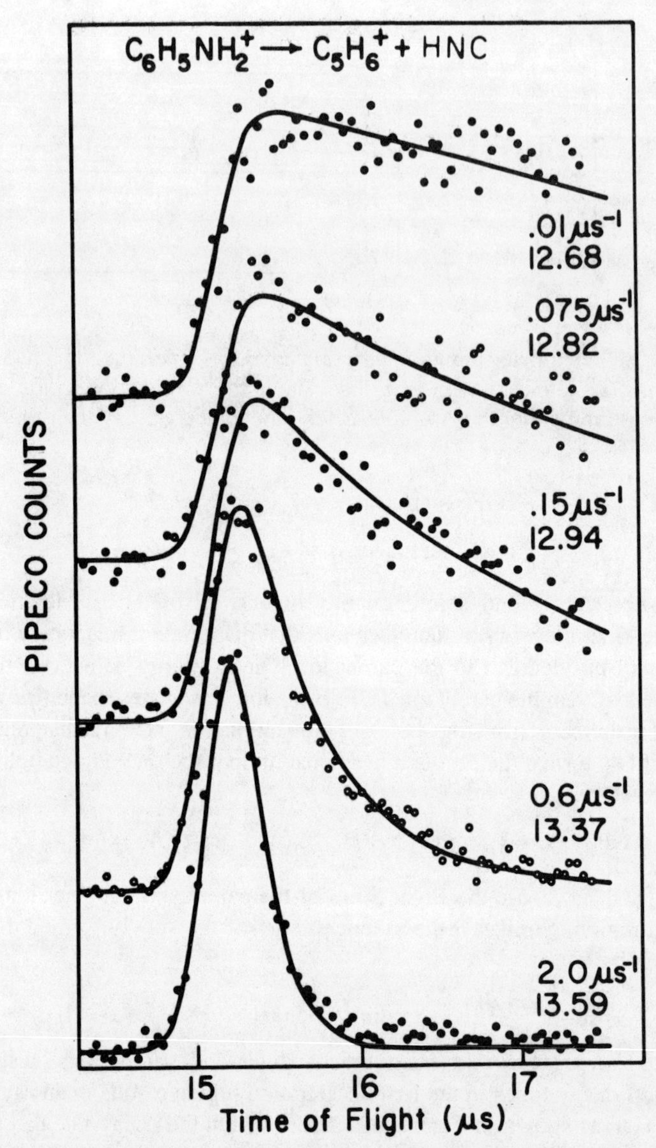

Figure 5.22 The product time of flight distribution for the dissociation of aniline ions at various photon energies ranging from 12.68 to 13.59 eV. The solid lines through the experimental points are calculated TOF distributions in which the indicated dissociation rates were assumed. Taken with permission from Baer (1986), and Baer and Carney (1982).

necessary to energy analyze the coincident electrons, rather low fields (typically 10 V/cm) must be maintained. The range of ion lifetimes under these conditions is from .1 to 100 μsec. If the ions are energy selected by ion absorption of a laser photon, the acceleration field can be considerably increased so that lifetimes down to the sub-nanosecond range can be measured (Choe and Kim, 1991; Boesl et al., 1992).

Very slowly dissociating (meta-stable) ions can also decay in the drift region.

Unless additional acceleration regions are applied after this region, these ions (let's call them $A^{\#+}$) are all detected at the same time as the parent ions because they have the same nominal velocity. On the other hand, their kinetic energy $[KE(AB^+)m/M]$ will differ from that of the undissociated parent ions. Information about these ions can thus be extracted by adding a post acceleration or retardation field. All $A^{\#+}$ ions formed in the drift region will have a single kinetic energy, independent of where in this region they were produced. Although a simple retarding field would suffice to separate the undissociated parents from the daughter ions, the most common method used has been a reflectron which turns the ion trajectory by nearly 180° (Fig. 5.23). By proper adjustment of the reflectron voltages, it is possible to shift the peak due to the daughter ions that were formed in the drift region anywhere between the fast daughter ions and the parent ion peak. Consider the two ions, AB^+ and $A^{\#+}$ which arrive at the first grid of the reflectron at precisely the same time. They both penetrate into the deceleration region until their kinetic energy is zero, after which they retrace their path and exit the reflectron at the same energy with which they entered. They thus complete their journey through the second drift with the same time of flight. The AB^+ and $A^{\#+}$ TOF differ only in the time spent in the reflectron. Thus, a soft reflection will permit these two ions to penetrate more deeply into the reflectron which results in a larger TOF difference. The parent ion will penetrate more deeply because of its greater energy and thus will lag behind the daughter ion. It is even possible to reflect only the lower energy $A^{\#+}$ ions while the parent ions pass through the reflectron without reflection. On the other hand, if the reflection is very hard, achieved with the application of a very strong electric field, the penetration is very shallow and relatively little time is spent in the reflectron. In the limit of a sudden reflection, the AB^+ and $A^{\#+}$ ions will spend no time in the reflectron and thus have precisely the same TOF.

The dissociation rate can be derived from the ratio of the $A^{\#+}$ and the undissociated AB^+ signals. Under the assumption of a single exponential decay rate, the rate can be extracted from the relation

$$\frac{A^{\#+}}{AB^+} = \frac{\displaystyle\int_{t_1}^{t_2} \exp[-k(E)t]dt}{\displaystyle\int_{t_2}^{\infty} \exp[-k(E)t]dt\ .} \tag{5.23}$$

The rate can also be obtained from a similar expression which relates the daughter signals in the acceleration and drift regions.

Even slower dissociation rates can be measured by storing ions in an ion trap such as a pulsed ion cyclotron resonance (ICR) cavity (So and Dunbar, 1988) or a quadrupole ion trap (March et al., 1992), both of which can trap the ions up to several seconds. In the ICR, ions are trapped by a combination of DC electric and magnetic fields, while in the quadrupole trap, they are stored by a combination of RF and DC electric fields. Analysis of either the depleted parent ions or the newly formed product ions is carried out by pulsed extraction of mass selected ions. Thus, the timing with respect to the photodissociation pulse is achieved by delayed ion extraction. The long time limit in this experiment is determined not by the trapping time of the instrument, but by the IR fluorescence rate of ions which is typically 10^2 to 10^3 sec^{-1} (Dunbar, 1990; Dunbar et al., 1987).

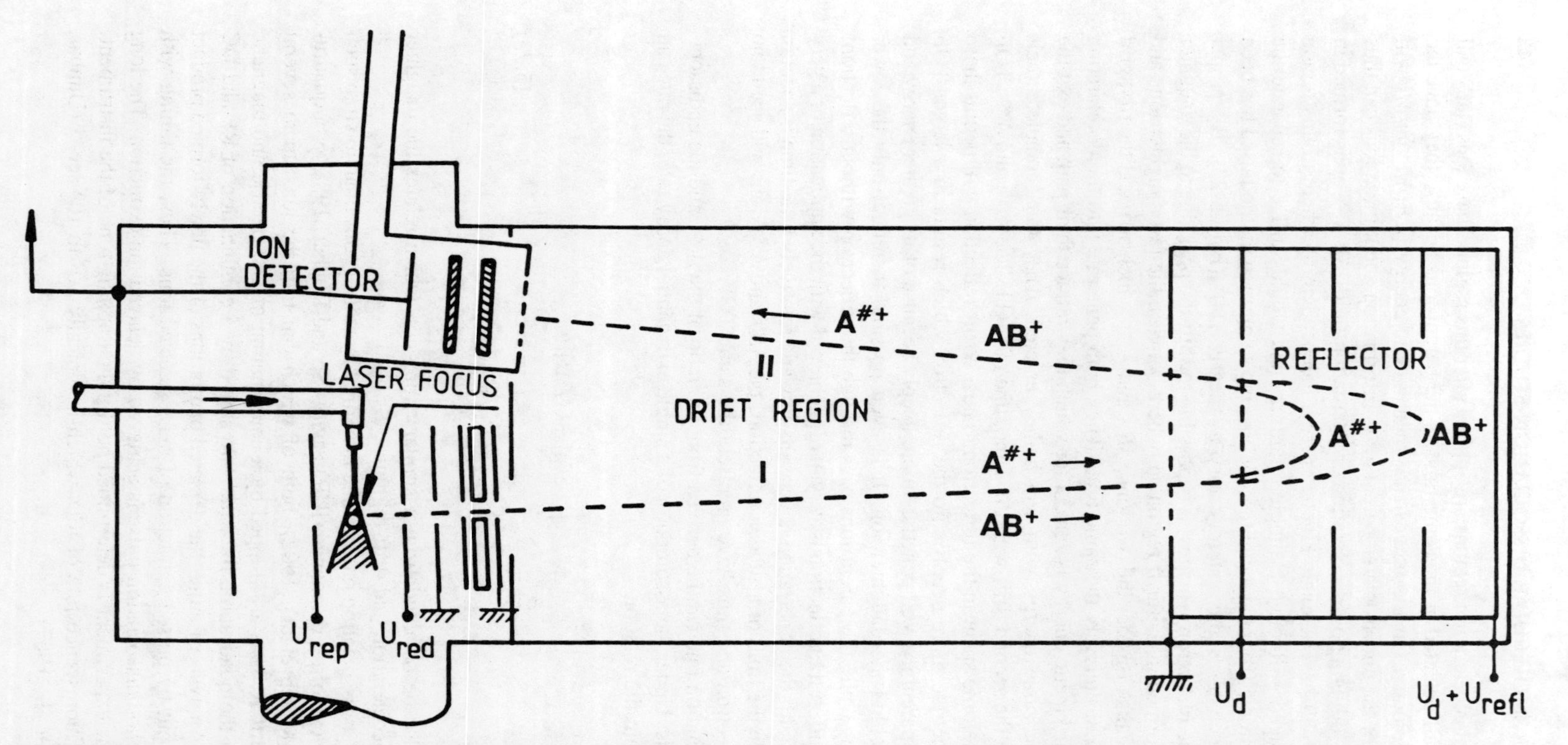

Figure 5.23 Experimental set-up of a reflectron TOF mass spectrometer. Ions are reflected by the retarding electric field. Fragment ions ($A^{\#+}$) formed in drift region I have velocities similar to their parent ions (AB^+). Hence they have less kinetic energy than their parent ions. The reflecting field thus separates the $A^{\#+}$ ions from the undissociated AB^+ ions. The times of flight of $A^{\#+}$ and AB^+ in drift region II are identical. Adapted from Neusser (1987).

A related approach is time-resolved photoionization mass spectrometry (TPIMS) or PEPICO in which the ions are trapped for specified periods of time prior to mass analysis. Fragmentation onsets shift to lower energies as the delay time is increased. Such experiments have been performed with microsecond delays in simple PEPICO ion sources, (Dannacher et al., 1983; Rosenstock et al., 1979) or with delays up to several milliseconds in ion storage traps by TPIMS (Gefen and Lifshitz, 1984; Faulk et al., 1990). An example of this energy shift is shown in Figure 5.24 for the loss of iodine from energy-selected iodobenzene ions (Dannacher et al., 1983). The derived rates range from 10^3 to 10^6 sec^{-1}.

5.5.4 Inversion of Absorption Peak Widths

The third method for determining the dissociation rate is by measuring peak widths in absorption spectra. According to the uncertainty principle, the excited state lifetime (Δt) and the uncertainty in establishing the energy of that state (ΔE) are related by $\Delta t \cdot \Delta E = \hbar$. When this is expressed in terms of time in picoseconds and resolution in cm^{-1}, it takes the form

$$\Delta t(\text{psec}) \cdot \Delta E(\text{cm}^{-1}) = 1/(2\pi c) = 5.3 \text{ psec} \cdot \text{cm}^{-1}, \tag{5.24}$$

in which ΔE is the full width at half maximum of a Lorentzian absorption peak. A Lorentzian peak is obtained in the limit when the laser resolution is much smaller than the peak width. In most experiments, the Gaussian-shaped resolution function of the

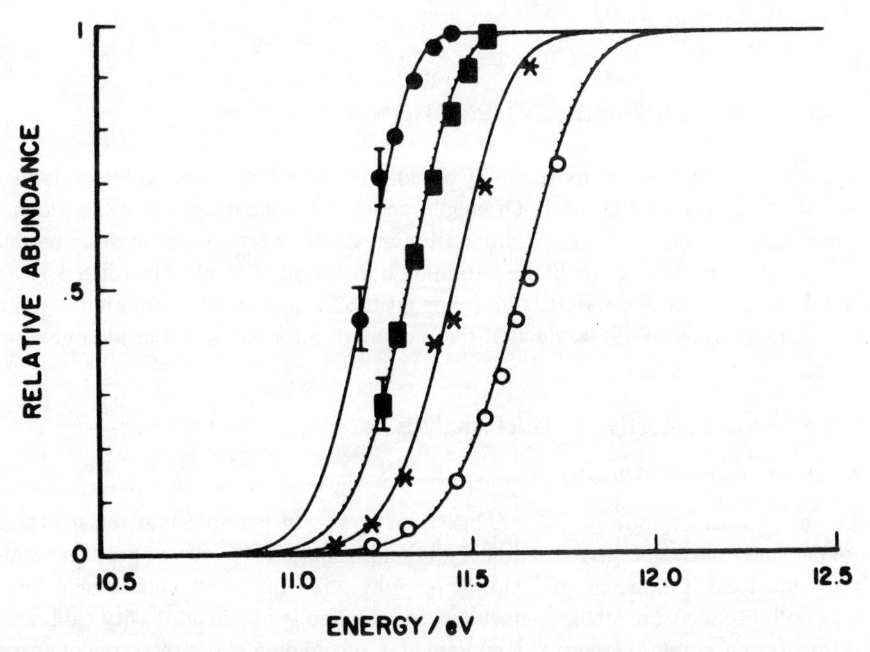

Figure 5.24 The relative abundance of $C_6H_5^+$ PEPICO signal from iodobenzene as a function of the ion internal energy at four delayed ion extraction times. The four different curves (from right to left) were taken with ion residence times of 1.0, 5.9, 21, and 57 μsec, respectively. Taken with permission from Dannacher et al. (1983).

laser must be convoluted with the Lorentzian peak shape. The resulting Voight profile can then be compared to the experimentally obtained absorption line.

Although high-laser-resolution experiments permit measurements of rates down to 10^8 sec^{-1}, the typical laser resolution of 0.1 cm^{-1} limits this method to reactions with rate constants greater than 2×10^{10} sec^{-1}. However, this approach is fraught with pitfalls. First, determining that a peak is homogeneously broadened rather than simply broad because of many closely spaced absorption lines which cannot be resolved by the available resolution is no easy task (Luo and Rizzo, 1990; Semmes et al., 1987). One test for homogeneity of the line is that excitation at any energy within the broadened absorption line results in identical dynamical measurements (i.e., rate constants or product energy distributions). Another test is that bleaching the transition at one energy serves to deplete the whole broad line. However, even if it is established that the line is homogeneously broadened, it is not necessarily the case that the broadening mechanism is dissociation. In fact, in most cases it probably is not. A particular absorption peak can be broadened by a number of processes, the most common one being IVR. Miller (Uzer and Miller, 1991) has suggested that as long as the number of dark states within the homogeneously broadened line is greater than two or three, it is impossible to determine whether the broadening mechanism is dissociation or IVR. This severely limits this approach to very small molecules or to molecules at very low energies at which the density of states is still small (R.E. Miller, 1990). Systems which satisfy the above requirements are D_2CO (Polik et al., 1990) and HCO (Tobiason et al., 1995; Neyer et al., 1995; Adamson et al., 1993) as well as small dimers such as HCCH—HF (Huang and Miller, 1989) and $(HF)_2$ (Huang et al., 1986) which can be dissociated by high-resolution IR lasers.

5.6 METHODS FOR PRODUCT DETECTION

The excitation of a reactant molecule by photons is the primary method for initiating a state- or energy-selected reaction. Of equal importance is the detection of the products because this provides the critical information about the reaction rate and the product state distribution. The method used for product detection is closely related to the method used to initiate the reaction. For instance, picosecond initiation requires picosecond analysis of the products if the maximum information about the rates is of interest.

5.6.1 Non-State-Selective Product Analysis

5.6.1.1 GC Product Analysis

When the primary rate information consists of a ratio of unimolecular rates (derived from the study of competitive reactions), all that is required is a ratio of product yields. This is most easily done by gas chromatography as in the Stern-Volmer analysis in which both products are stable molecules. The method is universal, easily calibrated, and therefore accurate. However, it contains no information about the product energy states. Also the reaction yield must be significant in order to be detectable.

5.6.1.2 The Universal Detectors - EI, PI, and MPI Mass Spectrometry

5.6.1.2.a Electron impact ionization. A universal detector is one that detects all species, independent of their internal energy, their spectroscopic structure, or their identity. The simplest of all such detectors is the electron impact (EI) mass spectrometer. Electrons are produced from a hot filament and are accelerated to about 70 V. They can be contained either in an ionization source so that the sample beam passes through this source, or they can be produced in the form of an electron beam which can be directed toward the molecular beam. The beam size can be varied by the use of appropriate electrostatic lenses (Moore et al., 1989). The advantage of such a scheme is the simplicity of the EI source. The disadvantage is that it is difficult to specify and control the energy of the electron beam. As a result, the ions generated from a given neutral molecule will include not only the parent ion mass, but frequently various fragmentation product masses as well. If the product of the neutral dissociation reaction under investigation happens to result in a product neutral that has the same mass as a fragment ion formed in the ion source from the parent ion, then it is difficult to distinguish the reactants from products without additional effort. On the other hand, if angular distributions or kinetic energy measurements are part of the detection scheme, it may be possible to distinguish a given mass ion as originating from the neutral reaction product from one which is merely the fragment produced in the ion source.

Electrons are generated in a continuous fashion from a hot filament. Although electrons can be produced in copious quantities, electron impact ionization is not a very efficient process so that a large electron flux is necessary. An efficient gun can generate as much as 10 mA of current, which corresponds to nearly 10^{17} electrons/sec. With such an intense source, ionization efficiencies as high as 10^{-3} can be achieved. This is the case when electron production and sample ionization are carried out in a single, self-contained ionization region. If instead, the system is constructed so that the electrons are transported in the form of a focused beam to the reaction interaction zone, the space charge in the electron beam substantially diminishes the usable electron beam flux. Typical currents in electron beams are 10 to 100 μA which corresponds to 10^{14}–10^{15} electrons/sec.

The EI ionization cross section starts at zero at the ionization threshold, increases approximately linearly (Wannier, 1953), and finally peaks at an electron energy of about 70 eV. The latter is the standard electron energy used in mass spectrometers. The vanishing cross section at the ionization threshold is a particularly unfortunate property of electron ionization.

5.6.1.2.b One photon photoionization (PI). One photon photoionization (PI) is a nearly universal detector. It ionizes any molecule with an ionization energy less than the photon energy. This property is extremely useful because it is possible to be partly selective. More important is the fact that PI has a large cross section even at the ionization energy of the molecule (Berkowitz, 1979). This means that the photon energy can be kept low so as to minimize fragmentation of the sample that is being analyzed. Often dissociation products are free radicals which have ionization energies that are lower than that of the reactant molecules. This means that the products can be selectively detected with little background signal.

There are several continuously working VUV sources, none of which is very intense. The weakest ones are the laboratory-based continuum sources such as the "hydrogen many line discharge," the He Hopfield continuum, and the Ar continuum, with intensities of about 10^{9}–10^{10} photons/sec when the VUV monochromator is set for 1 Å bandpass. With good optics and very clean mirrors, dispersed radiation from a synchrotron can be as intense as 10^{12}–10^{13} photons/sec·Å (Samson, 1967; Baer, 1989). These intensities are weak so that the fraction of molecules that can be ionized is very small. If we assume a typical ionization cross section of 10^{-17} cm^2 and an interaction distance of 0.1 cm, we can expect to ionize no more than about 1 in 10^8 molecules. In addition, the continuum sources require a VUV monochromator.

Considerably more intensity can be obtained from a low-pressure rare gas discharge which gives line emissions at mostly a single wavelength, thereby obviating the need for a monochromator. Since all of the discharge energy is concentrated in a few emission lines, the intensities can be higher (ca. 10^{14} photons/sec). The available lines are Xe (9.57, 8.44 eV), Kr (10.64,10.03 eV), Ar (11.83,11.62 eV), Ne (16.85,16.67 eV), and He (21.22 eV) (Samson, 1967).

VUV lasers offer the most effective means for PI product detection. Although their average intensity is not particularly high (ca. 10^{11} photons/sec), when their pulsed structure is combined with pulsed molecular beam experiments, the photons concentrated in a few nanoseconds increase the effective intensity by orders of magnitude. That is, the peak intensity of a VUV laser in a typical 5-nsec pulse is on the order of 10^{19} photons/sec, which means that a sample can be completely ionized. However, only those molecules which happen to be in the laser interaction region during those 5 nsec can be detected.

A particularly simple, and very intense, VUV source is one obtained from the third harmonic generation of the 355 nm Nd-YAG laser line, which gives a VUV line at 118 nm (10.5 eV). This is accomplished by passing the 355 nm light through a cell containing Xe gas at an adjustable pressure of about 25 Torr. Both LiF and MgF are partly transparent to light of this energy so that the tripling cell can have windows. Tunable VUV photons can be generated down to the windowless region by tripling the light at the throat of a Xe free jet (Page et al., 1987; Cromwell et al., 1991; Tonkyn and White, 1989). Because third harmonic generation is a highly nonlinear process, the power drops rapidly as the pump laser intensity is reduced. The most intense tunable VUV sources use 4-wave mixing in which two laser colors are combined in a nonlinear medium such as Hg or Mg vapor (Hepburn, 1991; Hilbig et al., 1986; Vidal, 1986). Photons with energies up to 17.5 eV have been reported by this technique (Kong *et al.* 1993a,b).

5.6.1.2.c Nonresonant MPI. Nonresonant multiphoton ionization (MPI) has many of the same characteristics as PI. Although the cross section is low, the visible or UV laser intensity is orders of magnitude greater than the VUV intensity. However, even this is not sufficient for ionization without tightly focusing the laser to a cross section of about 100 μm^2. Thus, the two lasers (pump and probe) must be carefully overlapped. MPI is not as selective as PI because it is difficult to control the ionization process. Even more difficult is the control of additional photons absorbed by the ions once formed. The ion tends to absorb many photons by a complex series of absorption and dissociation steps, ultimately producing many small fragments which are often difficult to associate with a particular neutral precursor (Dietz et al., 1982).

5.6.2 State-Selective Methods

5.6.2.1 Fluorescence Excitation (FE) or Laser-Induced Fluorescence (LIF)

Fluorescence excitation (FE) [often called laser-induced fluorescence (LIF)] is one of the principle techniques for detecting reaction products which are produced in the electronic ground state (Zare, 1984; Dagdigian, 1988). As implied by its name, a reaction product is induced to fluoresce (spontaneously) by the absorption of a photon. No monochromator is needed in FE because it is the totality of the fluorescence signal that is proportional to the absorption by the product species. A typical experimental arrangement involving two pulsed lasers, the dissociation laser and the time-delayed FE laser, is shown in Figure 5.19. Another scheme is illustrated in Figures 5.12 and 5.13.

Because the absorption of a photon is state specific, that is, only one photon energy can excite a molecule from some ground state $\langle v,J|$ to some excited state $\langle v',J'|$, the FE signal is a measure of the population density in states v and J. Because the product state detection is done without dispersing the fluorescence and because it is a zero background experiment, it is a highly sensitive technique. Of equal importance is the capability to extract time information. If the time delay of the FE probe laser is scanned, the product fluorescence signal rises exponentially with delay time. The time resolution is limited only by the laser pulse width and not by the often slower fluorescence time of the excited product. Since most reactions form products in their ground electronic states, the FE technique is considerably more useful than plain fluorescence. Thus the three major features are the (a) sensitivity to ground electronic state products, (b) the ability to determine the product state without loss of sensitivity, and (c) the ability to extract time information down to the femtosecond level.

The FE technique is not universally applicable because not all molecules fluoresce. In particular, large molecules often decay by internal conversion (radiationless transition) to vibrationally excited levels of the ground electronic state from which they fluoresce slowly in the infrared or undergo collisional deactivation. The excited state may also decay by dissociation. The quantum yield, which is defined as the fraction of the excited states that fluoresce, may change with excited state vibrational and rotational level so that the comparison of the FE signal for different products needs to be corrected for these effects. Even if the quantum yield remains constant, extracting populations from the FE signal is not a straightforward procedure because the rotational line strengths depend upon the coupling schemes, which themselves may change with J or v (Hefter and Bergmann, 1988; Dagdigian, 1988). Finally, it is evident that for this technique to work, the spectroscopy of the ground and excited electronic state must be very well known. Because of these limitations the FE method has been applied mainly to small molecules such as CN, OH, NO, CO, NO_2, and CH_3O.

5.6.2.2 Resonance Enhanced Multiphoton Ionization (REMPI)

Two of the three laser ionization methods have already been discussed, namely one-photon PI and multiphoton MPI. The third type is resonance enhanced MPI, or REMPI. In the latter method the laser is tuned so that an intermediate state of the molecule is excited with one, two, or perhaps three photons. The excitation of the intermediate state determines the overall cross section for the process because the absorption of additional photons to reach the ionization continuum is generally rapid. In contrast to PI and MPI, REMPI is state selective if the absorption process is resonant between two bound and reasonably long-lived states of the molecule. It is an extremely sensitive method for product detection because the result of the REMPI process is an ion which can be detected with near 100% efficiency. Not only is the ion collection efficiency of the detector (e.g., by channeltron electron multiplier or a multichannel plate detector) extremely high (ca. 50%), but all ions regardless of their initial velocity vector can be collected by the application of appropriate electric fields. This is a major advantage over FE, in which only a fraction of the photons can be collected and for which the detectors are generally of lower efficiency (ca. 20%).

As with FE, not all molecules can be ionized by REMPI. In fact, nearly the same requirements hold for FE and REMPI. That is, the excited state must have a sufficiently long lifetime with respect to dissociation or interconversion so that ionization (or fluorescence in the case of FE) will take place. As with FE, a knowledge of the ground and intermediate state spectroscopy is essential in order to determine the internal energy of the reaction products. Finally, the extraction of sample populations from the REMPI signal requires the same careful analysis as it does with FE.

If the REMPI process involves the absorption of two or more nonresonant photons to reach the intermediate state, the process becomes highly nonlinear and the cross section drops significantly. In that case the laser must be tightly focused in order to drive the transition to the intermediate state. The cross section for a two-photon absorption is typically about 10^{-52} cm^{4}sec and the rate of absorption is given by $dN/dt = \sigma I^2 n$, where I is the photon flux in photons/cm^2·sec, and n is the gas density. The fraction of molecules excited is then $dn/n = \sigma I^2 dt$. If a typical pulsed laser delivers 10^{16} photons per pulse and the pulse length is about 5 nsec, then the peak flux is 2×10^{24} photons/sec. If this laser is then focused to a 50-μm diameter spot, it will have a density flux of 10^{29} photons/cm^2·sec. Thus the fraction of molecules excited is 0.005. Although this is a good efficiency, which can be increased even further by more tightly focusing the REMPI laser, the interaction volume becomes very small.

5.6.3 The Bolometer as a Detector for Neutral Molecules

The detection of molecules in a molecular beam by a bolometer is based on the bolometer's response to the total beam energy, including the center of mass translational energy (Zen, 1988). The bolometer consists of a liquid-helium-cooled thermocouple whose electrical response varies rapidly with the energy of the bolometer. The low temperature is necessary in order to reduce the heat capacity of the thermocouple, thereby increasing its sensitivity, as well as to minimize the thermal detector noise.

The bolometer can be used in a variety of ways, the simplest one being a detector for spectroscopy. Suppose a molecular beam, which is directed at the bolometer and thus raises its temperature to some level, is intersected by an infrared laser beam. As the laser is scanned through the IR spectrum the molecules absorb energy. This energy is retained by the molecules for times characteristic of vibrationally excited states (typically several milliseconds) so that they impart this additional internal energy to the bolometer which then records a further rise in the temperature. The success of the method depends upon a very stable continuous molecular beam because the signal has associated with it a large background due to the translational energy of the beam. Nevertheless, it is estimated that as few as 10^6 molecules with an energy of 4000 cm^{-1} are needed to record a signal (R.E. Miller, 1988). The absorption spectrum of HF in Figure 5.2 was obtained with a bolometer.

Of interest for the study of dissociation reactions is the product translational energy and the spacial resolution of the bolometer. Suppose that a molecular beam consisting of monomers and dimers is directed onto a bolometer. If a loosely bound dimer absorbs an IR photon with an energy in excess of its dissociation limit, it will dissociate with some of the excess energy released as translational energy of the two products. If the dissociation imparts sufficient translational energy normal to the beam axis, the products will be scattered out of the beam so that the bolometer will record a decrease in the number of molecules. For a drift distance to the bolometer of 10 cm and a beam velocity of 8×10^4 cm/sec, a perpendicular velocity of about 2×10^3 cm/sec is sufficient to avoid a 2-mm-diameter bolometer. Thus, absorption of an IR photon by a monomer will increase the bolometer temperature, while the absorption of an IR photon by a dimer (which dissociates) will reduce the temperature of the bolometer. In this fashion spectra of dimers and higher-order van der Waals clusters can be obtained. However, the identity of the species is determined solely by the spectrum. The bolometer is incapable of aiding in the identification of the molecule. Because this is a continuous molecular beam and laser experiment, rate information comes only from an analysis of the absorption peak widths. Very high laser resolution and the ability to excite single resonances are thus essential if the peak widths are to be identified with a dissociation rate. This approach is thus not advisable for dissociation rate measurements of stable molecules which have high density of states above the dissociation limit because the absorption peak widths will be determined primarily by IVR.

5.6.4 The Measurement of Product Translational Energies

5.6.4.1 Neutral Time of Flight (TOF) Analysis

When a molecule or ion dissociates, some of the excess energy invariably is released as translational energy of the two departing fragments. By conservation of momentum, $m_a u_a + m_b u_b = 0$. Combining this with the conservation of energy, $E_t = E_a + E_b = \frac{1}{2} m_a u_a^2 + \frac{1}{2} m_b u_b^2$, leads to the result that $E_t = [M/m_b]E_a$, where M is the mass of the precursor molecule. Thus, measuring the velocity or kinetic energy of one of the fragments, determines the other as well. A further constraint is imposed by energy conservation: $E_{avl} = E_t + E_r + E_v + E_e$, because the energy available to the products (E_{avl}) is partitioned among the translational, rotational, vibrational, and electronic degrees of freedom of the products.

While the determination of the internal energy distributions requires some sort of spectroscopic approach, the translational energy distribution can be obtained from time- or energy-resolved product analysis. Some of the earliest energy-selected reactions investigated involved the determination of the translational energy of the products (Riley and Wilson, 1972). This was done by timing (relative to the photodissociation laser pulse) the arrival of the neutral products at the mass spectrometer. From such experiments it was determined that the alkyl iodide molecules dissociate along a repulsive surface to produce electronically excited $^2P_{1/2}$ I atoms. Because neutral species are not deflected by electric or magnetic fields, their trajectories are undistorted. Although long flight paths are possible, the solid angle over which atoms can be collected is small. If the flight path is 5 cm and the detector has a sensitive area of 1 mm^2, the solid angle is only 4×10^{-4} steradians. Thus for isotropic product distributions, the fragment collection efficiency is only 3×10^{-5}. The ideal location for product detection is one which matches the desired resolution.

5.6.4.2 Ionized Product TOF Analysis

When the fragment species is an ion, formed either from the dissociation of a parent ion or ionized after the dissociation event, the kinetic energy release distribution (KERD) can be measured with much greater efficiency. Suppose that an electric field of ϵ V/cm is applied across the interaction region so that the ionized products are extracted toward a drift region which is terminated by the ion detector. With a sufficiently high draw-out field, all ions will be detected, regardless of their initial translational energy and angle. The total TOF will be a function of the initial velocity vector as well as the applied draw-out field. Often, the extraction is accomplished with a two-stage acceleration plus a drift region as in Figure 5.21 (p. 141). In that case, the total TOF is given by

$$\text{TOF} = T_{a1} + T_{a2} + T_d = \frac{2a_1}{u_0 + u_1} + \frac{2a_2}{u_1 + u_2} + \frac{d}{u_2}, \quad (5.25)$$

where the distances a_1, a_2, and d are defined in Figure 5.21 and the velocities are given by

$$u_0 = \sqrt{\frac{cE_0}{M}}, \quad (5.26a)$$

$$u_1 = \sqrt{\frac{c\epsilon_1 a_1}{M} + u_0^2}, \quad (5.26b)$$

$$u_2 = \sqrt{\frac{c\epsilon_2 a_2}{M} + u_1^2}, \quad (5.26c)$$

When the factor $c = 1.929$, the initial energy, E_o, can be expressed in electron-volts; the electric fields, ϵ_1 and ϵ_2, in regions 1 and 2 as volts per centimeter; the mass, M in atomic mass units; the distances in centimeters; and the velocities at the end of the first and second acceleration regions, u_1 and u_2, in centimeters per microsecond.

Suppose that product ions of a single translational energy are formed at a point source so that a_1 is fixed. Consider first the case of ions with initial velocity u_o ejected toward and away from the detector, that is, $u_o = \pm|u_o|$. The velocities and flight times in the second acceleration region and drift tube will be identical for these two ions because u_o is squared in the expressions for u_1. This is because the ion which initially headed in the wrong direction is turned around by the electric field and comes back to its initial position, where it has the same initial velocity but is now pointed toward the ion detector. Thus, the difference in the TOF of the two ions arises solely because of the "turn around time" in the acceleration region. The time for the ion to turn around is given by: $4|u_o|M/(c\,\epsilon_1)$. That is, the dispersion in ion TOF is directly related to the initial velocity and inversely related to the applied field. If the ion is ejected at some angle, relative to the extraction field, the relationship remains the same except that u_o is replaced by its projection on the experimental z-axis, $u_o\cos\theta$.

When reaction products are formed in a finite region, rather than a point source, it is best to extract the ions with a combination of acceleration and drift regions which are consistent with the Wiley-McLaren space focusing condition (Wiley and McLaren, 1955). Specific ratios of acceleration and drift distances for given electric fields are imposed so that the ion TOF is independent of the precise initial position of ionization. This is easily explained as follows. Suppose that two ions (with initially zero kinetic energy) are formed at positions $a_1 + \delta$ and $a_1 - \delta$ as measured from the second acceleration, or drift region. Clearly the ion closer to the drift region will reach this region before the other ion. On the other hand, in the ϵ_1 accelerating field, the ion that has been accelerated over $a_1 + \delta$ cm has gained a kinetic energy of $(a_1 + \delta)\epsilon_1$ eV, while the other ion has gained only $(a_1 - \delta)\epsilon_1$ eV. Thus, the higher-kinetic-energy ion can catch up with the slower ion at some point in the drift region. If a detector is located at this point the two ions will have identical flight times. This Wiley-McLaren condition is established by setting the $d(\text{TOF})/da_1 = 0$, which for the case of a single acceleration region, requires that the drift distance is twice the acceleration distance. For two acceleration regions, the second region is generally very short but adds considerable voltage to the ions. The result is that the drift distance can be lengthened, thereby increasing the mass resolution. Any combination of acceleration and drift region lengths can be operated in the space focusing condition simply by adjusting the applied voltage ratios.

Another property of the Wiley-McLaren condition is that, to first order, ions of equal initial velocity but ejected in opposite angles will arrive at times of flight that are symmetrical about the TOF of an ion with zero kinetic energy. This holds as long as the energy gained by the particles in the acceleration field is considerably greater than their initial energy. Under these conditions, the initial ion velocity is simply given by

$$u_o = [0.5c\epsilon_1/M](T - T_o), \tag{5.27}$$

where T and T_o are the flight times of an ion of initial velocity u_o and 0, respectively. As in Eqs. (5.26), the factor $c = 1.929$ again permits the use of convenient units for the electric field, mass, time, and velocity. If the Wiley-McLaren conditions are not satisfied, the width of the peak is still the same, but the center will be slightly shifted from T_o. This makes the analysis of a distribution of kinetic energy releases somewhat more difficult. Figure 5.25 shows the expected TOF distribution for three single energy releases for a mass-30 ion under the assumption that all ions had the same origin. It is apparent that even when the drift distance is far from the Wiley-McLaren condition (55 cm), the shift of the distribution from the T_o line is barely perceptible. The worst case is the 1 eV kinetic energy release distribution with $d = 15$ cm in which the center of the distribution is 9.505 μsec while the 0 kinetic energy release TOF is 9.385 μsec, a shift of 0.12 μsec.

5.6.4.3 The Doppler Shift Method

A related approach to product energy analysis is a spectroscopic one in which the Doppler shift in high-resolution absorption peaks is related to the molecule's velocity. Suppose that ν_o is the frequency of the light that is absorbed by the molecule at rest relative to the light source. When a molecule travels with speed u in the same direction as the laser light, the absorption is detected at a shifted frequency given by $\nu_{obs} = \nu_o[1 - u/c]$, where c is the speed of light. The velocity can thus be determined from the following relation:

$$u = [c/\nu_o](\nu_{obs} - \nu_o). \tag{5.28}$$

Note that the velocity is directly related to the shift in the absorption peak.

The use of a pulsed laser with a transform limited resolution of 0.001 cm^{-1} (5 nsec laser pulse width) at $\lambda = 500$ nm, permits a resolution and minimum measurable velocity of 1500 cm/sec. For a molecule such as NO with a mass of 30 amu, this

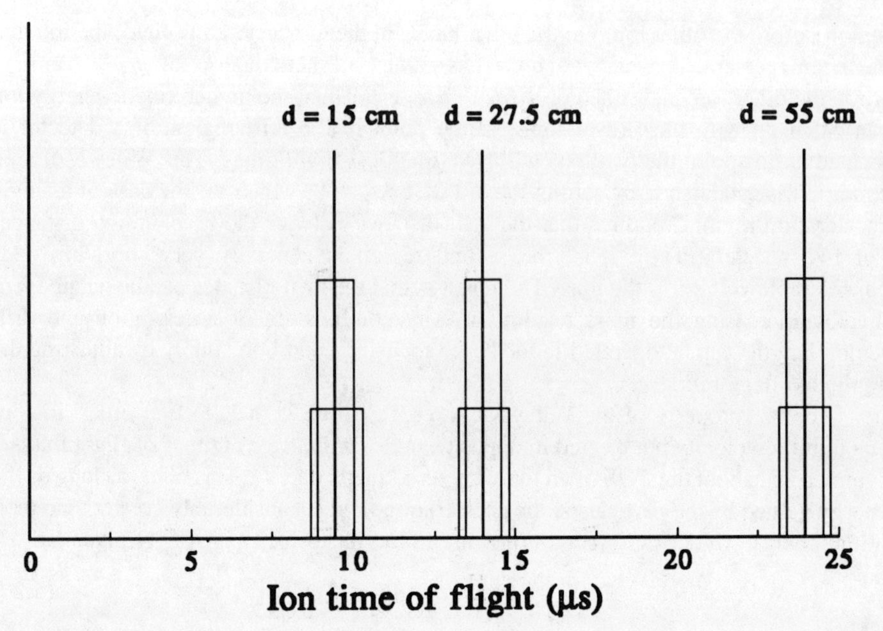

Figure 5.25 Ion TOF distributions for isotropic angular distribution of dissociation products. The mass of the fragment ion is 30 amu, the electric fields in the first two acceleration regions are 10 and 100 V/cm and the acceleration distances are 2 and 1 cm, respectively. The drift distances are indicated in the figure. The d = 55 cm corresponds to the Wiley-McLaren space focusing condition. For each drift distance, the ion TOF was calculated for three product translational energies of 0, 0.5, and 1.0 eV.

corresponds to a translational energy of 3.4 J/mol, which is 700 times less than the 300 K average thermal energy! This method, as with TOF measurements, scales *linearly* with the velocity so that the velocity resolution is constant. However, the resolution in the energy, $dE/dv = m(c/v_o)^2\Delta v$, which is normally of more interest, becomes worse as the energy of the molecule increases. At the thermal energy of about 2.5 kJ/mol, where the absorption is shifted by 0.02 cm^{-1}, the resolution is only 263 J/mol.

5.6.4.4 Kinetic Energy Analysis with an Electric Sector

A third method for measuring the product energy distribution for ionized products is with electrostatic energy analyzers, such as hemispherical sectors, or cylindrical mirrors. These measure the energy directly rather than the velocity, as is the case in the two previous methods. However, all three methods have in common the fact that they measure only the one-dimensional projection of a three-dimensional distribution. The design and operation of various energy analyzers are discussed in several sources (Moore et al., 1989; Illenberger and Momigny, 1992).

A particularly interesting and important aspect of the energy analysis is the possibility of obtaining product energies with extremely high resolution when the dissociation takes place in a fast ion beam. Suppose that the ion beam has been accelerated to an energy of E_b and that the dissociation (possibly photodissociation) releases a translational energy E_o in the center of mass reference frame. The velocity of the parent ion beam is given by: $u_b = (2E_b/M)^{1/2}$, where M is the mass of the parent ion. The daughter ion velocity in the center of mass is given by $u_o = [2E_o(M - m)/mM]^{1/2}$, where m is the daughter ion mass. The extra $(M - m)/M$ term arises because the total center of mass release energy is partitioned between the two products. Now in the laboratory reference frame, in which the energy measurements are carried out, the velocity of the product ion is given by the vector sum of $\mathbf{u}_b + \mathbf{u}_o$. Converting this velocity back into laboratory energy, we obtain

$$E_{lab} = \frac{m}{M}E_b + E_o\sqrt{\frac{4E_b m(M - m)}{E_o M^2}} \cos\theta , \tag{5.29}$$

where the condition that $u_b \gg u_o$ was used to drop the third term in the binomial expansion. The extrema in the kinetic energy of the daughter ion peak will be given by the difference between the forward ($\theta = 0$) and backward ($\theta = 180°$) scattered product ion. The maximum energy width is then

$$\Delta E_{max} = E_o\sqrt{\frac{16E_b m(M - m)}{E_o M^2}} . \tag{5.30}$$

It is evident that the maximum energy width of the product ion peak for a single release energy, E_o, as measured in the laboratory frame is given by E_o multiplied by a function which depends upon $(E_p/E_o)^{1/2}$. For a beam energy of 8 keV, a release energy of 1 meV (8 cm^{-1}) would be amplified by a factor of 3000 in the laboratory reference frame. The resulting 3 eV is an easily measured energy, especially when the ion beam is slowed down (after dissociation) to permit operation of the energy analyzer at a lower pass energy.

Experiments which take advantage of this resolution have been carried out on a routine basis in mass analyzed ion kinetic energy spectrometry (MIKES) measurements (Cooks et al., 1973). Often the ions are activated by collisions in which case the internal energy is not well specified (Boyd et al., 1984a,b). In other studies, long-lived meta-stable ions formed some several microseconds earlier in the ion source dissociate spontaneously in a drift region and then are analyzed (Griffiths et al., 1983; Cooks et al., 1974). In this case, the ion internal energy spread is rather narrow because only very low-energy and long-lived ions are measured. However, the internal energy is still not known. Some experiments in which the ion beam is photodissociated with a laser have permitted ions in relatively well selected energies to be investigated. Figure 5.26 shows the laboratory energy distribution for the dissociation of nitrogen dimer ions which have been photodissociated by a cw 488 nm laser (Jarrold et al., 1984). The original dimer ion was accelerated to 8 keV, which results in a daughter ion energy centered at 4 keV. The spread of the daughter ion energies in the center of mass is about 130 eV, which translates into a center of mass energy release of about 0.5 eV. The two curves correspond to the laser polarization parallel (0°) and perpendicular (90°) to the ion velocity axis.

5.6.4.5 Projections of the Angular Distribution onto the Experimental Axis

The TOF under Wiley-McLaren conditions and the Doppler techniques are similar in that the initial velocity produces a linear response in the measured quantity ($T - T_o$ and ν_{obs}-ν_o, respectively). Likewise the energy analysis of product ions produces a signal

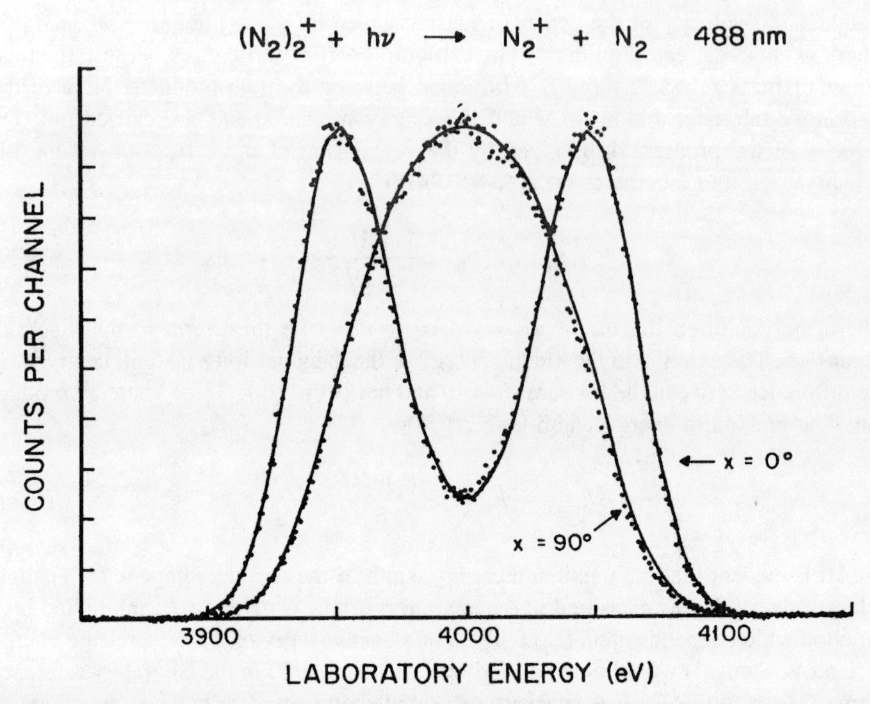

Figure 5.26 Kinetic energy of N_2^+ ions obtained from the photodissociation of nitrogen dimer ions accelerated to 8 keV. The two different spectra were obtained with the laser polarization parallel and perpendicular to the molecular beam axis. Taken with permission from Jarrold et al. (1984).

which is directly proportional to the energy of the fragment. We thus treat these experiments together.

In general, the reaction products are formed in a distribution of kinetic energies as well as in a distribution of angles with respect to the detecting laser or the TOF axis. In the experiment, these distributions in energy as well as in three-dimensional space are projected onto a single dimension, which is the shift in either the absorption or emission frequency, or the time of flight, both of which are directly related to the projection of the fragment velocity on the detection axis.

Suppose that we have a reaction in which the products are formed with a single kinetic energy, but distributed isotropically in space. According to Stanton and Monahan (1964) an isotropic distribution of equal velocity vectors will result in a rectangular projected velocity distribution, $P'(u_z)$ as shown in Figure 5.25 for the case of 0.5 and 1.0 eV kinetic energy of the products. This can be shown as follows. Suppose that the energy release is sufficiently small so that there is no angular discrimination against products that are ejected perpendicular to the direction of detection. If a fragment is ejected with velocity u_i in a small range of angles from θ to $\theta + d\theta$, the projection of this velocity on the z-axis will be $u_z = u_i\cos\theta$. The weight, or probability of this contribution will be $P(\theta)d\theta = 2\pi u_i \sin\theta d\theta$. Expressing the probability in terms of the projected velocities gives $P'(u_z)du_z = P(\theta)d\theta$, so that $P'(u_z) = P(\theta)[d\theta/du_z]$. The derivative in brackets is just $1/[u_i\sin\theta]$, so that the $P'(u_z)$ function is a constant between the upper and lower limits imposed by the maximum projected velocity on the z-axis. A similar argument holds for the projection of the three-dimensional ion kinetic energy distribution when its projection in one dimension is measured by dispersive energy analyzers.

Rectangular velocity or energy spectra are not often observed because slow statistical dissociations of polyatomics form the products in a distribution of the translational energies, or because nonstatistical and direct dissociations generally are not characterized by isotropic product velocity distributions.

Let us consider first the case of a direct dissociation in which the products are formed with a nonisotropic angular distribution but with a single kinetic energy. It is easiest to consider a rapidly dissociating diatomic in which the velocity vector of the departing atoms lies along the internuclear axis. If the molecules are prepared by linearly polarized light and the detection axis is perpendicular to the photon propagation vector, the signal intensity at the detector will be given by (Zare, 1972; Bersohn and Lin, 1992):

$$W(\theta) = \frac{1}{4\pi}[1 + \beta P_2(\cos\theta)], \tag{5.31}$$

where P_2 is the second Legendre polynomial given by $P_2(x) = \frac{1}{2}(3x^2 - 1)$, and θ is the angle between the product velocity vector ($\mathbf{u}$) and the laser polarization vector ($\mathbf{E}$). The interesting parameter in this equation is β, the anisotropy parameter, whose value ranges between -1 and 2. When $\beta = 0$, the laboratory frame distribution is isotropic which means that the signal intensity is independent of the laser polarization. If $\beta = -1$, the intensity will be zero when $\theta = 0$ and maximum when $\theta = 90°$. This would be the case when the transition moment is perpendicular to the internuclear axis. A parallel transition moment, on the other hand, would be associated with a $\beta = 2$.

As recently pointed out by several authors (Taatjes et al., 1993; Wu et al., 1994), the interpretation of the β parameter in terms of the molecule fixed frame angular distribution is fraught with difficulties. Since the alignment imposed by a linearly polarized laser only provides experimental sensitivity to the P_o and P_2 terms in the Legendre polynomial expansion of the molecule fixed frame angular distribution, a $\beta = 0$ does not ensure that the latter is isotropic, even though it gives rise to an isotropic laboratory frame distribution. To obtain sensitivity to the odd and higher-order terms in this expansion, the molecules need to be oriented with a distribution of angles that is more confined than that obtained by excitation by a linearly polarized laser.

In an experiment that is sensitive to the velocity of the fragments (e.g., Doppler shift or TOF spectrometry or energy analysis with an electric sector), a one-dimensional projection of the three-dimensional angular distribution can be obtained directly without varying the laser polarization direction. It is thus convenient to transform Eq. (5.31) into a set of variables appropriate for the experiment. The new variables are the angles between detection axis and the dissociation product velocity vector and laser polarization. The new equation is (Schmiedl et al., 1982; Jarrold et al., 1984):

$$I(\chi) = \frac{1}{4\pi}[1 + \beta P_2(\cos\theta')P_2(\cos\chi)], \tag{5.32}$$

where θ' is the angle between the laser polarization vector $\mathbf{E}$ and the detection axis, $\mathbf{z}$; and χ is the angle between the product velocity vector and the detection axis (see Fig. 5.27). Hence, $\cos(\chi)$ is the projection of the velocity onto the detection axis. Equation

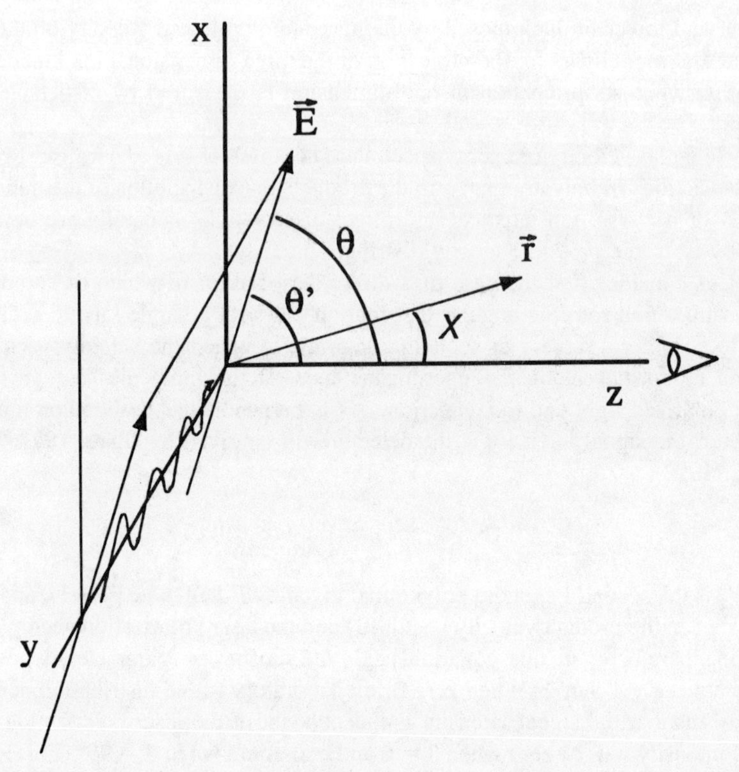

Figure 5.27 The relationship between the angles which relate the laser electric vector (**E**), the experimental detection axis, **z**, and the product velocity vector, **r**. Note that the definitions of θ′ and χ are sometimes interchanged.

(5.32) gives then the distribution of velocity vectors projected onto the detection axis for a given angle between the laser polarization and the detection axis. Some examples of Doppler shifts (or TOF distributions) for single kinetic energy releases with two laser polarizations and several β values are shown in figure 5.28 (Houston, 1989).

There is one particular angle, the so-called "magic angle" at which the $P_2(\theta)$ function vanishes. This is when $\theta' = \sqrt{1/3} = 0.955$ radians $= 54.7°$. If the electric vector of the plane polarized light is at the magic angle with respect to the experimental detection axis, then the whole second term in the brackets of Equation (5.32) vanishes and $I(\chi) =$ constant. This means that independent of the β value, the TOF distribution or the Doppler profile for a single kinetic energy release will be square. This is of great benefit when neither the translational energy distribution nor the β value is known. It permits determination of the former with no knowledge of the latter. Figure 5.29 shows the kinetic energy spectrum with the laser at the magic angle for the previously mentioned nitrogen dimer ion dissociation to N_2^+ ions. This is not a perfect rectangle because the ions were ejected with a distribution of kinetic energies. However, this distribution was strongly peaked at a kinetic energy release (KER) of 0.45 eV which means that the nitrogen neutral and ion were produced with a total of four quanta of vibrational energy (Jarrold et al., 1984).

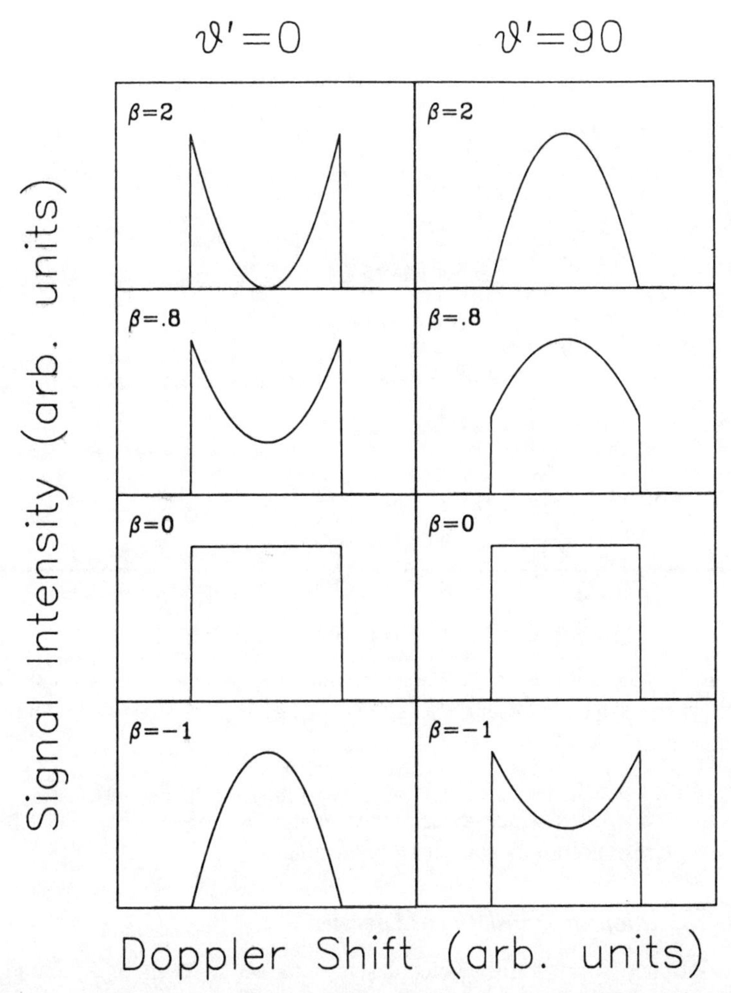

Figure 5.28 Examples of calculated Doppler profiles or TOF distributions for a dissociation with a single kinetic energy release but for various values of the anisotropy parameter, β. Taken with permission from Houston (1989).

5.6.4.6 Velocity Aligned Spectroscopy

As shown above, an isotropic distribution for a single kinetic energy release results in a rectangular Doppler profile, TOF distribution, or energy spectrum. If there are two or more velocity components, each with a rectangular distribution, information can be lost due to signal to noise problems. Xu et al. (1989) have proposed the application of a delayed analysis of the product in order to let products with large velocity components perpendicular to the analysis axis to leave the interaction region. The effect of this is to replace the rectangle with two peaks at the extrema of the Doppler profile or TOF distribution. The effect of a delayed detector pulse in a TOF analysis is shown in Figure 5.30. Dramatic enhancement in resolution was obtained by delaying the probe laser pulse for 400 nsec (Xu et al., 1989). A similar effect is achieved in a passive manner by

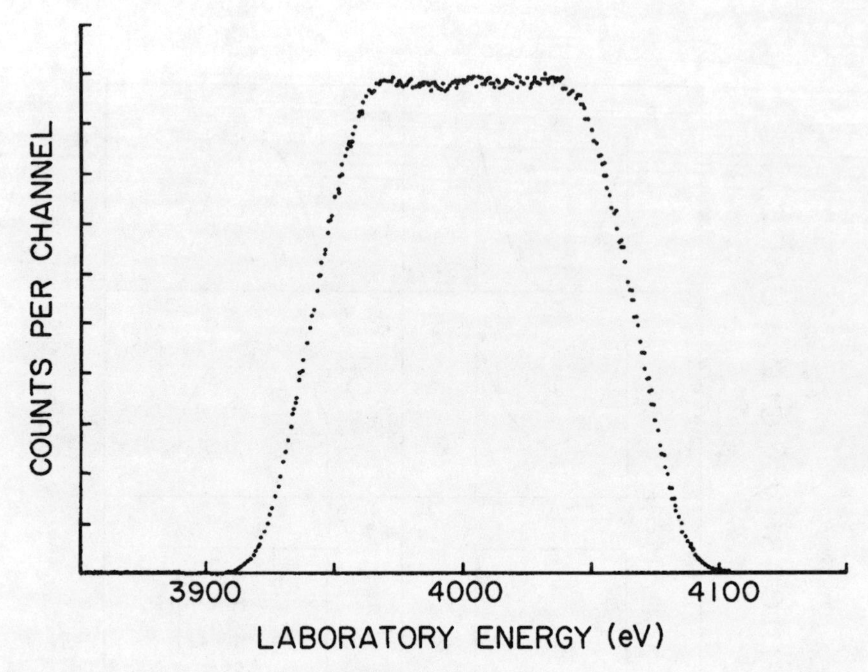

Figure 5.29 Laboratory energy distribution for the $(N_2)_2{}^+$ dissociation taken at the magic angle. This is the same dissociation as shown in figure 5.25. From M.T. Bowers (unpublished)

installing small apertures which discriminate against all off-axis ions. The latter approach, however, is not applicable to Doppler analysis.

5.6.4.7 Distribution of Translational Energies

The dissociation of polyatomic molecules or ions often result in a kinetic energy release distribution (KERD). In order to extract such a KERD from the experimental TOF distribution or Doppler profile, the experimental distribution must be deconvoluted. If the dissociation can be assumed to proceed isotropically, or if the experiment is carried out at the magic angle, and there is no experimental discrimination against high-kinetic-energy products, then the data will consist of a series of rectangles which can be deconvoluted simply by taking the derivative of the function. If there is discrimination against energetic ions, then the data can be fitted with a set of calculated single energy release "basis functions" by expressing the data in terms of a set of linear equations (Mintz and Baer, 1976).

Often it is easier to model the experiment with an assumed KERD. But before discussing this, let us consider the TOF distribution, or the equivalent Doppler profile, of a nondissociating gas at room temperature. The three-dimensional speed distribution at a given temperature, T, is given by

$$P(c) = 4\pi \left[\frac{m}{2\pi k_B T} \right]^{3/2} c^2 e^{-mc^2/2k_B T}, \tag{5.33}$$

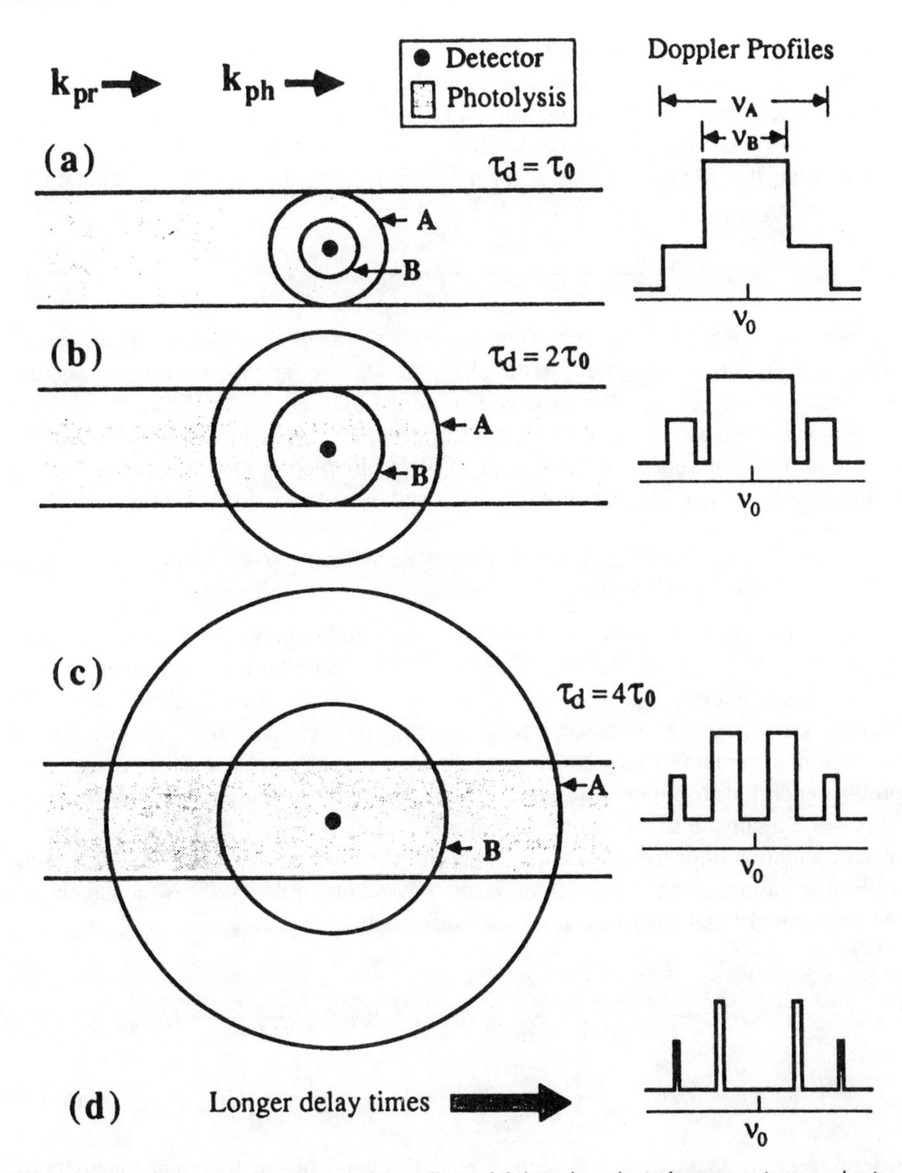

Figure 5.30 Schematic depiction of the effect of delayed analysis for a reaction producing fragments with two velocities, *A* having twice the velocity of *B*. Taken with permission from Xu et al. (1989).

where c is the scaler speed. However, the experiment is sensitive only to the distribution along the experimental axis, for which the speed (in the case if an isotropic distribution) is given by a simple Gaussian function:

$$f(u_z) = \sqrt{\frac{m}{2\pi k_B T}}\, e^{-mu_z^2/2k_B T}\,. \tag{5.34}$$

Because the TOF or the Doppler profile is linearly proportional to the velocity, the resulting experimental distribution is again a Gaussian function (Stanton and Monahan, 1964; Franklin et al., 1967; Stockbauer, 1977). A single parameter, the width, is thus sufficient to determine the whole distribution. For the case of an ion time of flight distribution, the average translational energy of the gas, $\langle E \rangle$, can be determined from the following equation (Stockbauer, 1977):

$$\langle E \rangle = \frac{3}{16 \ln(2) M} (q\epsilon)^2 (fwhm)^2 = \frac{0.261 \epsilon^2 (fwhm)^2}{M}, \tag{5.35}$$

where fwhm is the full width at half-maximum of the TOF distribution, ϵ is the electric field used to extract the ions, q is the electronic charge, and M is the ion mass. The translational temperature of the gas is determined by the relation $\langle E \rangle = 3/2\, RT$. The second part of the equation is written in terms of the electric field in V/cm, the fwhm in μsec, the mass in amu, and the energy in electron-volts. A similar equation for the fwhm of a Gaussian Doppler profile is obtained by solving Eq. (5.3) for $3/2RT$:

$$\langle E \rangle = \frac{3c^2 M}{16 \ln 2} \left(\frac{fwhm}{\nu_o} \right)^2 = 2.03 \times 10^{12} M \left(\frac{fwhm}{\nu_o} \right)^2, \tag{5.36}$$

where the fwhm and ν_o are in any convenient, but similar units. The second expression is evaluated for the energy in wave numbers (cm^{-1}) and the mass is in amu.

Of more interest than the thermal energy of the parent molecule or ion, is the average kinetic energy released ($\langle$KER$\rangle$) in the dissociation. If we ignore for the moment the problem of angular momentum conservation, a reaction with a statistical product energy distribution will release the translational energy as a three-dimensional Maxwell-Boltzmann distribution. This is the case even for a thermal sample with its own distribution of parent translational energies (Stanton and Monahan, 1964). For ion TOF distributions in an electric field, ϵ, the projection on the z-axis is then again just the one-dimensional distribution already discussed, which is given by (Stockbauer, 1977):

$$\langle \mathrm{KER} \rangle = \frac{M}{(M-m)m} \frac{3}{16 \ln(2)} (q\epsilon)^2 (fwhm)^2 - \frac{m}{(M-m)} \langle E \rangle_{\mathrm{th}} \tag{5.37a}$$

$$\langle \mathrm{KER} \rangle = 0.261 \frac{M}{(M-m)m} \epsilon^2 (fwhm)^2 - \frac{m}{(M-m)} \langle E \rangle_{\mathrm{th}}, \tag{5.37b}$$

where M is the mass of the molecule, m is the mass of the fragment that is being detected, and $\langle E \rangle_{\mathrm{th}}$ is the mean thermal energy of the parent molecule or ion. The $M/(M - m)$ mass terms appear in this equation because of the partitioning of the energy between the two fragments. The term involving the thermal energy must be subtracted because it corrects for the inherent width of the peak due to the molecule's thermal energy. The second equation is again expressed in terms of V/cm, eV, amu, and μ sec.

The corresponding equation can be set up for the Doppler profile by solving for $\langle E \rangle$ and subtracting the parent thermal kinetic energy.

5.6.4.8 Photofragment Imaging

An approach particularly helpful for direct dissociations is photofragment imaging, in which an ionized reaction product is accelerated toward a two-dimensional imaging detector (Chandler and Houston, 1987; Chandler et al., 1989; Baldwin et al., 1990). Rather than collecting only the projection of the velocities on the ion extraction axis, the product spacial dispersion in the x and y direction are measured as well. The data then consist of two-dimensional images for various total ion flight times. The traditional one-dimensional TOF experiment cannot extract simultaneously the angular distribution of the products and the kinetic energy distribution. However, the additional information provided by imaging all three dimensions, eliminates this problem.

5.6.5 Correlations in Unimolecular Dissociations

The preceding discussion applies to uncorrelated TOF distributions or Doppler profiles. These are obtained by measuring only a single parameter, namely the product velocity. This is all that can ever be derived from measuring the TOF distribution, or from a Doppler profile of, for instance, a hydrogen atom product. However, when the velocity distribution of a diatomic is measured from the Doppler profile, then the information content is considerably enhanced because the absorption is from a particular v and J state. Thus, the experiment measures the velocity of state selected products.

Figure 5.31 shows the experimental Doppler profiles for CO(J) products from the photodissociation of OCS with 222 nm laser light. The interesting feature is that horizontal ($\theta' = 0$) and vertical ($\theta' = 90°$) laser polarization does not give the profiles expected for uncorrelated distributions. For instance, the profile for the $Q(58)$ transition appears to have a β value of about 0.8 when measured with horizontal polarization, while it has a β value of slightly less than 0 when measured with vertical polarization. Furthermore, the Doppler profiles are completely different in the $P(58)$ transition even though the same initial state ($J = 58$) is being excited.

The reason for the discrepancy from the simple theory is that dynamical variables of the two particles arc correlated. Correlation can be between a number of variables. In the case of the data in Figure 5.31, the correlation is between the rotational alignment vector of the CO and the velocity vector of the fragments, that is, a **v**-**J** vector correlation. Consider the case of triatomic dissociation to a rotating diatom and an atom which necessarily takes place in a plane. Because the departing atom can induce a rotation only in the plane of dissociation, the angular momentum vector is forced to be perpendicular to the plane of dissociation so that **v** is perpendicular to **J**. Suppose that the laser is vertically polarized with respect to the experimental axis as illustrated in Figure 5.32. If the CO fragment has its velocity along the experimental **k** axis, resulting in signal at the wings of the Doppler profile, then **J** is distributed about the circle indicated in the Figure. All of these **J** vectors correspond to the excitation of the $M_J = 0$ component. On the other hand, if the **v** vector lies in the plane of the light polarization, resulting in signal in the middle of the Doppler profile, then all $\mathbf{M}_J$ values contribute. In other words, different $\mathbf{M}_J$ components contribute to the different parts of the Doppler profile. It turns out that a Q-branch in CO should have a dip in the center of the Doppler profile, while for the P and R branches, there will be peak (Houston, 1989; Hall et al., 1988).

Nonisotropic angular distributions are most obvious in direct dissociations from a repulsive potential energy surface because the molecule has no time to rotate and thereby wash out the asymmetry. In fact it has often been stated that the observation of anisotropy in angular distributions is an indication of a lifetime less than the rotational

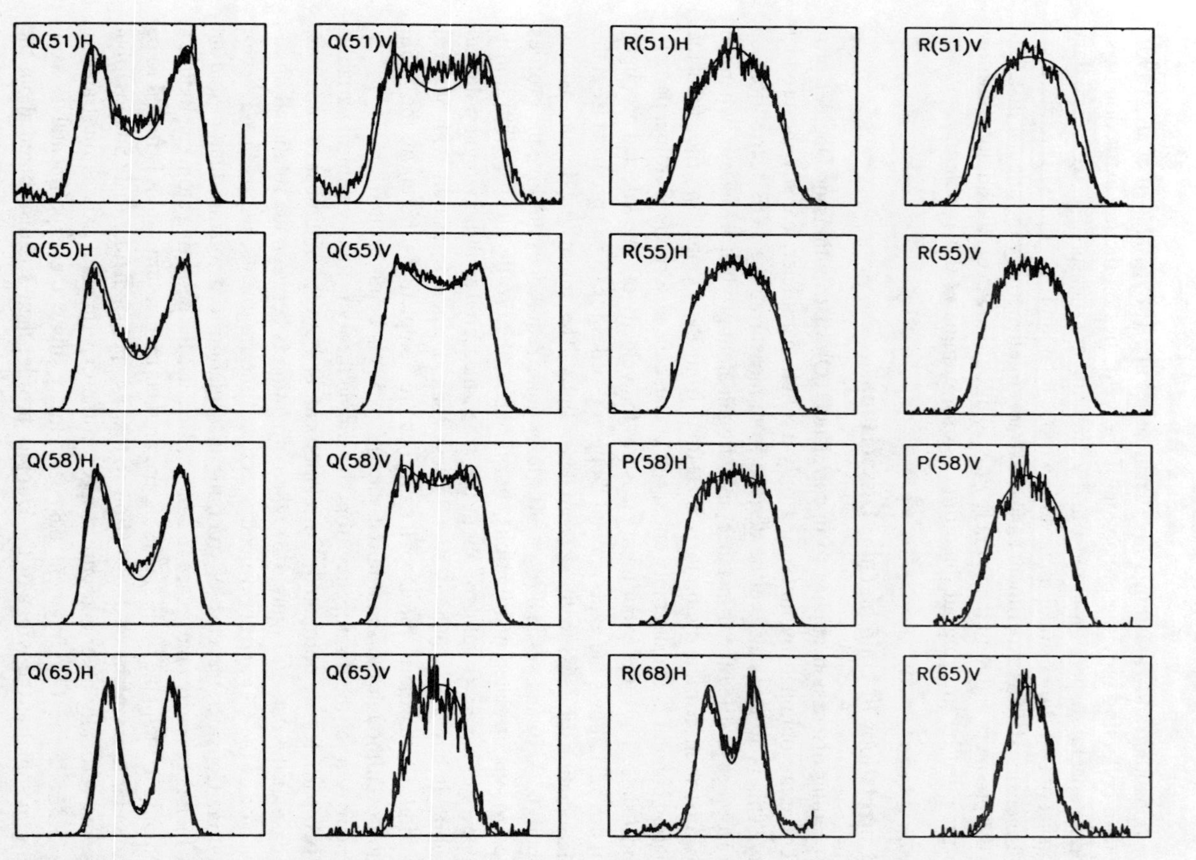

Figure 5.31 The experimental Doppler profiles for the OCS dissociation. The rotationally resolved product CO was measured by FE. Taken with permission from Sivakumar et al. (1988).

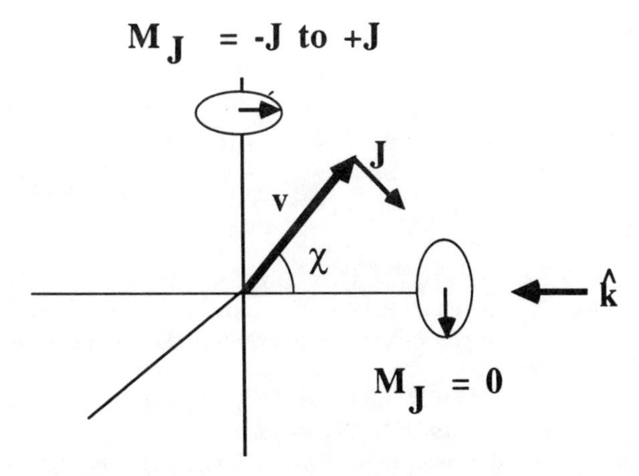

Figure 5.32 The different M_J contributions (circles) to the Doppler profile. The **v** vector represents the fragment velocity which can be directed toward the detection axis (**k**), or in the plane of the laser polarization vector (vertical). Taken with permission from Houston (1987).

period of the molecule. This is certainly true if the molecule is prepared in a distribution of rotational levels. In that case, all molecules are aligned at the time of excitation. But because they then rotate at different rates, the ensemble of molecules will dephase so that the alignment is lost after several rotational periods (Yang and Bersohn, 1974). However, the anisotropy arising from the correlated spectroscopy is independent of the parent lifetime because it originates from the moment that dissociation takes place. That is, the parent–daughter correlations disappear but the daughter–daughter correlations remain. Thus, in the case of the dissociation of long-lived glyoxal, Doppler profiles measured at the *R* and *Q* branches of the product CO show some differences (Burak et al., 1987).

If the parent molecules are state-selected, that is, in specific ro-vibrational states, and the products are also rotationally resolved, the dissociation anisotropy will never be totally lost even if the molecules dissociate slowly and with a distribution of lifetimes (Butenhoff et al., 1991; Marshall et al., 1992). This is because the phase relationship is preserved for this set of eigenstate resolved molecules.

5.7 SUMMARY

This chapter is no more than a cursory review of a few experimental methods used in chemical dynamics. It is apparent that most of these techniques have been developed only in the past decade, and that new methods continue to be invented. Most of the new methods have been made possible by the development of pulsed lasers and pulsed molecular beams. Together, they permit the state selection of molecules, the measurement of rates down to femtoseconds, and the analysis of product state distributions down to single quantum state level. Only a decade ago, the state of the art for measuring unimolecular reactions for ions and neutrals was about equal, and the amount of data for ionic systems far surpassed that for neutral reactions. However, the rapid advances in experimental techniques applicable to neutral reactions has clearly reversed the situation. Perhaps the most important consequences of these developments are the capabilities to measure the dissociation rate constants down to the reaction threshold, and to measure them over a large energy and dynamic range.

REFERENCES

Adamson, G.W., Zhao, X., and Field, R.W. (1993). *J. Mol. Spectrosc.* **160,** 11.

Anderson, J.B. (1974). In *Molecular Beams and Low Density Gasdynamics,* Wegener, P.P., Ed. Marcel Dekker, New York.

Andlauer, B., and Ottinger, Ch. (1971). *J. Chem. Phys.* **55,** 1471.

Andlauer, B., and Ottinger, Ch. (1972). *Z. Naturforsch.* **A27,** 293.

Baer, T., Peatman, W.B., and Schlag, E.W. (1969). *Chem. Phys. Lett.* **4,** 243.

Baer, T. (1979). In *Gas Phase Ion Chemistry*, Bowers, M.T., Ed. Academic Press, New York, pp. 153–196.

Baer, T., Guyon, P.M., Nenner, I., Tabche-Fouhaille, A., Botter, R., Ferreira, L.F.A., and Govers, T.R. (1979). *J. Chem. Phys.* **70,** 1585.

Baer, T. (1986). *Adv. Chem. Phys.* **64,** 111.

Baer, T. (1989). *Ann. Rev. Phys. Chem.* **40,** 637.

Baer, T., Booze, J.A., and Weitzel, K.M. (1991). In *Vacuum Ultraviolet Photoionization and Photodissociation of Molecules and Clusters*, Ng, C.Y., Ed. World Scientific, Singapore, pp. 259–298.

Baer, T., and Carney, T.E. (1982). *J. Chem. Phys.* **76,** 1304.

Baggott, J.E. (1985). *Chem. Phys. Lett.* **119,** 47.

Baldwin, D.P., Buntine, M.A., and Chandler, D.W. (1990). *J. Chem. Phys.* **93,** 6578.

Bergmann, K., Demtroder, W., and Hering, P. (1975). *Appl. Phys.* **8,** 65.

Berkowitz, J. (1979). *Photoabsorption, Photoionization, and Photoelectron Spectroscopy*. Academic Press, New York.

Berry, M.T., Loomis, R.A., Giancarlo, L.C., and Lester, M.I. (1992). *J. Chem. Phys.* **96,** 7890.

Bersohn, R., and Lin, S.H. (1969). *Adv. Chem. Phys.* **16**, 67.

Boesl, U., Weinkauf, R., and Schlag, E.W. (1992). *Int. J. Mass Spectrom. Ion. Proc.* **112,** 121.

Bohac, E.J., Marshall, M.D., and Miller, R.E. (1992). *J. Chem. Phys.* **96,** 6681.

Boyd, R.K., Kingston, E.E., Brenton, A.G., and Beynon, J.H. (1984a). *Proc. R. Soc. Lond.* **A392,** 59.

Boyd, R.K., Kingston, E.E., Brenton, A.G., and Beynon, J.H. (1984b). *Proc. R. Soc. Lond.* **A392,** 89.

Brehm, B., and Puttkamer, E.v. (1967). *Z. Naturforsch.* **A22,** 8.

Burak, I., Hepburn, J.W., Sivakumar, N., Hall, G.E., Chawla, G.K., and Houston, P.L. (1987). *J. Chem. Phys.* **86,** 1258.

Butenhoff, T.J., Carleton, K.L., van Zee, R.D., and Moore, C.B. (1991). *J. Chem. Phys.* **94,** 1947.

Butler, L.J., Ticich, T.M., Likar, M.D., and Crim, F.F. (1986). *J. Chem. Phys.* **85,** 2331.

Cannon, B.D., and Crim, F.F. (1981). *J. Chem. Phys.* **75,** 1752.

Chandler, D.W., Thoman, J.W., Janssen, M.H.M., and Parker, D.H. (1989). *Chem. Phys. Lett.* **156,** 151.

Chandler, D.W., and Houston, P.L. (1987). *J. Chem. Phys.* **87,** 1445.

Choe, J.C., and Kim, M.S. (1991). *J. Phys. Chem.* **95,** 50.

Choi, Y.S., Teal, P., and Moore, C.B. (1992). *J. Opt. Soc. Amer. B.* **7,** 1829.

Choi, Y.S., and Moore, C.B. (1989). *J. Chem. Phys.* **90,** 3875.

Choi, Y.S., and Moore, C.B. (1991). *J. Chem. Phys.* **94,** 5414.
Choi, Y.S., and Moore, C.B. (1992). *J. Chem. Phys.* **97,** 1010.
Conway, W.E., Morrison, R.J.S., and Zare, R.N. (1985). *Chem. Phys. Lett.* **113,** 429.
Cooks, R.G., Beynon, J.H., Caprioli, R.M., and Lester, G.R. (1973) *Metastable Ions*. Elsevier, Amsterdam.
Cooks, R.G., Kim, K.C., Keough, T., and Beynon, J.H. (1974). *Int. J. Mass Spectrom. Ion. Proc.* **15,** 271.
Cottrell, T.L., and McCoubrey, J.C. (1961). *Molecular Energy Transfer in Gases*. Butterworths, London.
Crim, F.F. (1990). *Science* **249,** 1387.
Cromwell, E.F., Lu, D.J., Vrakking, M.J.J., Kung, A.H., and Lee, Y.T. (1991). *J. Chem. Phys.* **95,** 297.
Dagdigian, P.J. (1988). In *Atomic and Molecular Beam Methods*, Scoles, G., Ed. Oxford University Press, New York, pp. 596–630.
Dai, H.L., Field, R.W., and Kinsey, J.L. (1985a). *J. Chem. Phys.* **82,** 2161.
Dai, H.L., Field, R.W., and Kinsey, J.L. (1985b). *J. Chem. Phys.* **82,** 1606.
Dai, H.L., Korpa, C.L., Kinsey, J.L., and Field, R.W. (1985c). *J. Chem. Phys.* **82,** 1688.
Dai, H.L., and Field, R.W.(Eds.) (1994). *Molecular Spectroscopy and Dynamics by Stimulated Emission Pumping*. World Scientific Co., Singapore.
Dannacher, J., Rosenstock, H.M., Buff, R., Parr, A.C., Stockbauer, R., Bombach, R., and Staddelmann, J.P. (1983). *Chem. Phys.* **75,** 23.
Darcy, M.G., Rogers, D.E., and Derrick, P.J. (1978). *Int. J. Mass Spectrom. Ion. Proc.* **27,** 335.
Das, P.R., Nishimura, T., and Meisels, G.G. (1985). *J. Phys. Chem.* **89,** 2808.
DePaul, S., Pullman, D., and Friedrich, B. (1993). *J. Phys. Chem.* **97,** 2167.
Derrick, P.J., and Burlingame, A.L. (1974). *Acc. Chem. Res.* **7,** 328.
Dietz, W., Neusser, H.J., Boesl, U., Schlag, E.W., and Lin, S.H. (1982). *Chem. Phys.* **66,** 105.
Dunbar, R.C., Chen, J.H., So, H.Y., and Asamoto, B. (1987). *J. Chem. Phys.* **86,** 2081.
Dunbar, R.C. (1990). *Int. J. Mass Spectrom. Ion. Proc.* **100,** 423.
Durant, J.L., Rider, D.M., Anderson, S.L., Proch, F.D., and Zare, R.N. (1984). *J. Chem. Phys.* **80,** 1817.
Dutuit, O., Baer, T., Metayer, C., and Lemaire, J. (1991). *Int. J. Mass Spectrom. Ion. Proc.* **110,** 67.
Faulk, J.D., Dunbar, R.C., and Lifshitz, C. (1990). *J. Am. Chem. Soc.* **112,** 7893.
Feinberg, T.N., Baer, T., and Duffy, L.M. (1992). *J. Phys. Chem.* **96,** 9162.
Felker, P.M., and Zewail, A.H. (1988). *Adv. Chem. Phys.* **70,** 265.
Fleming, P.R., Li, M., and Rizzo, T.R. (1991). *J. Chem. Phys.* **94,** 2425.
Fleming, P.R., and Rizzo, T.R. (1991). *J. Chem. Phys.* **95,** 1461.
Franklin, J.L., Hierl, P.M., and Whan, D.A. (1967). *J. Chem. Phys.* **47,** 3148.
Gaubatz, U., Rudecki, P., Schiemann, S., and Bergmann, K. (1990). *J. Chem. Phys.* **92,** 5363.
Gefen, S., and Lifshitz, C. (1984). *Int. J. Mass Spectrom. Ion. Proc.* **58,** 251.
Gentry, W.R. (1988). In *Atomic and Molecular Beam Methods*, Scoles, G., Ed. Oxford University Press, New York, pp. 54–82.
Gilbert, R.G., and Smith, S.C. (1990). *Theory of Unimolecular and Recombination Reactions*. Blackwell Scientific, Oxford.
Griffiths, I.W., Howe, I., March, R.E., and Beynon, J.H. (1983). *Int. J. Mass Spectrom. Ion. Proc.* **54,** 323.
Hall, G.E., Sivakumar, N., Chawla, G.K., and Houston, P.L. (1988). *J. Chem. Phys.* **88,** 3682.
Hamilton, C.E., Kinsey, J.L., and Field, R.W. (1986). *Ann. Rev. Phys. Chem.* **37,** 493.
Hefter, U., and Bergmann, K. (1988). In *Atomic and Molecular Beam Methods*, Scoles, G., Ed. Oxford University Press, New York, pp. 193–253.

Hepburn, J.W. (1991). In *Vacuum Ultraviolet Photoionization and Photodissociation of Molecules and Clusters*, Ng, C.Y., Ed. World Scientific, Singapore, pp. 435–486.
Hilbig, R., Hilber, G., Lago, A., Wolff, B., and Wallenstein, R. (1986). *Comments At. Mol. Phys.* **18,** 157.
Houston, P.L. (1987). *J. Phys. Chem.* **91,** 5388.
Houston, P.L. (1989). *Acc. Chem. Res.* **22,** 309.
Huang, Z.S., Jucks, K.W., and Miller, R.E. (1986). *J. Chem. Phys.* **85,** 3338.
Huang, Z.S., and Miller, R.E. (1989). *J. Chem. Phys.* **90,** 1478.
Illenberger, E., and Momigny, J. (1992). *Gaseous Molecular Ions: An Introduction to Elementary Processes Induced by Ionization*. Steinkopff Verlag, Darmstadt.
Jarrold, M.F., Illies, A.J., and Bowers, M.T. (1984). *J. Chem. Phys.* **81,** 214.
Jasinski, J.M., Frisoli, J.K., and Moore, C.B. (1983). *J. Chem. Phys.* **79,** 1312.
Johnson, M.A., and Lineberger, W.C. (1988). In *Techniques for the Study of Ion-Molecule Reactions*, Farrar, J.M., and Saunders, W.H.Jr., Eds. Wiley, New York, pp. 591–635.
Khundkar, L.R., Marcus, R.A., and Zewail, A.H. (1983). *J. Phys. Chem.* **87,** 2473.
Khundkar, L.R., and Zewail, A.H. (1990). *Ann. Rev. Phys. Chem.* **41,** 15.
Kluft, E., Nibbering, N.M.M., Stringer, M.B., Eichinger, P.C., and Bowie, J.H. (1988). *Int. J. Mass Spectrom. Ion. Proc.* **85,** 215.
Knee, J.L., and Zewail, A.H. (1988). *Spectroscopy* **3,** 44.
Kong, W., Rodgers, D., and Hepburn, J.W. (1993a). *Chem. Phys. Lett.* **203,** 497.
Kong, W., Rodgers, D., Hepburn, J.W., Wang, K., and McKoy, V. (1993b). *J. Chem. Phys.* **99,** 3159.
Kuhlewind, H., Kiermeier, A., and Neusser, H.J. (1986). *J. Chem. Phys.* **85,** 4427.
Landau, L., and Teller, E. (1936). *Phys. Zeit. Sowjetunion* **10,** 34.
Lemaire, J., Dimicoli, I., and Botter, R. (1987). *Chem. Phys.* **115,** 129.
Levine, R.D., and Bernstein, R.B. (1987). *Molecular Reaction Dynamics and Chemical Reactivity*. Oxford University Press, New York.
Levy, D.H. (1980). *Ann. Rev. Phys. Chem.* **31,** 197.
Likar, M.D., Baggott, J.E., Sinha, A., Ticich, T.M., Vander Wal, R.L., and Crim, F.F. (1988). *J. Chem. Soc. Faraday Trans. 2* **84,** 1483.
Lubman, D.M., Rettner, C.T., and Zare, R.N. (1982). *J. Phys. Chem.* **86,** 1129.
Luo, X., Fleming, P.R., and Rizzo, T.R. (1992). *J. Chem. Phys.* **96,** 5659.
Luo, X., and Rizzo, T.R. (1990). *J. Chem. Phys.* **93,** 8620.
Luo, X., and Rizzo, T.R. (1991). *J. Chem. Phys.* **94,** 889.
March, R.E., Londry, F.A., Alfred, R.L., Franklin, A.M., and Todd, J.F.J. (1992). *Int. J. Mass Spectrom. Ion. Proc.* **112,** 247.
Marshall, M.D., Bohac, E.J., and Miller, R.E. (1992). *J. Chem. Phys.* **97,** 3307.
McClelland, G.M., Saenger, K.L., Valentini, J.J., and Herschbach, D.R. (1979). *J. Phys. Chem.* **83,** 947.
McGinley, E.S., and Crim, F.F. (1986). *J. Chem. Phys.* **85,** 5748.
Merkt, F., Guyon, P.M., and Hepburn, J.W. (1993). *Chem. Phys.* **173,** 479.
Miller, D.R. (1988). In *Atomic and Molecular Beam Methods*, Scoles, G., Ed. Oxford University Press, New York, pp. 14–53.
Miller, J.C., and Compton, R.N. (1986). *J. Chem. Phys.* **84,** 675.
Miller, R.E. (1988). *Science* **240,** 447.
Miller, R.E. (1990). *Acc. Chem. Res.* **23,** 10.
Mintz, D.M., and Baer, T. (1976). *J. Chem. Phys.* **65,** 2407.
Moore, J.H., Davis, C.C., and Coplan, M.A. (1989). *Building Scientific Apparatus*, 2nd ed. Addison-Wesley, London.
Moss, M.G., Ensminger, M.D., and McDonald, J.D. (1981). *J. Chem. Phys.* **74,** 6631.
Muller-Dethlefs, K. and Schlag, E.W. (1991). *Ann. Rev. Phys. Chem.* **42,** 109.

Neusser, H.J. (1987). *Int. J. Mass Spectrom. Ion. Proc.* **79,** 141.
Neyer, D.W., Luo, X., Houston, P.L., and Burak, I. (1993). *J. Chem. Phys.* **98,** 5095.
Neyer, D.W., Luo, X., Burak, I., and Houston, P.L. (1995). *J. Chem. Phys.* **102,** 1645.
Nimlos, M.R., Young, M.A., Bernstein, E.R., and Kelley, D.F. (1989). *J. Chem. Phys.* **91,** 5268.
Northrup, F.J., and Sears, T.J. (1992). *Ann. Rev. Phys. Chem.* **43,** 127.
Norwood, K., Guo, J.H., and Ng, C.Y. (1989). *J. Chem. Phys.* **90,** 2995.
Orlando, T.M., Yang, B., and Anderson, S.L. (1989). *J. Chem. Phys.* **90,** 1577.
Page, R.H., Larkin, R.J., Kung, A.H., Shen, Y.R., and Lee, Y.T. (1987). *Rev. Sci. Instrum.* **58,** 1616.
Page, R.H., Shen, Y.R., and Lee, Y.T. (1988). *J. Chem. Phys.* **88,** 4621.
Polik, W.F., Guyer, D.R., and Moore, C.B. (1990). *J. Chem. Phys.* **92,** 3453.
Randeniya, L.K., and Smith, M.A. (1990). *J. Chem. Phys.* **93,** 661.
Reddy, K.V., and Berry, M.J. (1977). *Chem. Phys. Lett.* **52,** 111.
Reddy, K.V., and Berry, M.J. (1979a). *Chem. Phys. Lett.* **66,** 223.
Reddy, K.V., and Berry, M.J. (1979b). *Faraday Discuss. Chem. Soc.* **67,** 188.
Reid, S.A., Brandon, J.T., Hunter, M., and Reisler, H. (1993). *J. Chem. Phys.* **99,** 4860.
Reisner, D.E., Field, R.W., Kinsey, J.L., and Dai, H.L. (1984). *J. Chem. Phys.* **80,** 5968.
Riley, S.J., and Wilson, K.R. (1972). *Faraday Discuss. Chem. Soc.* **53,** 132.
Rizzo, T.R., Hayden, C.C., and Crim, F.F. (1983). *Faraday Discuss. Chem. Soc.* **75,** 223.
Rizzo, T.R., Hayden, C.C., and Crim, F.F. (1984). *J. Chem. Phys.* **81,** 4501.
Rizzo, T.R., and Crim, F.F. (1982). *J. Chem. Phys.* **76,** 2754.
Rosenstock, H.M., Stockbauer, R., and Parr, A.C. (1979). *J. Chem. Phys.* **71,** 3708.
Rosenstock, H.M., Stockbauer, R., and Parr, A.C. (1980). *J. Chem. Phys.* **77,** 745.
Samson, J.A.R. (1967). *Techniques of Vacuum Ultraviolet Spectroscopy*, Wiley. New York.
Scherer, N.F., and Zewail, A.H. (1987). *J. Chem. Phys.* **87,** 97.
Schinke, R. (1992). *Photodissociation Dynamics*. Cambridge University Press, Cambridge.
Schmiedl, R., Dugan, H., Meier, W., and Welge, K.H. (1982). *Z. Phys. A* **304,** 137.
Schwartz, R.N., Slawsky, Z.I., and Herzfeld, K.F. (1952). *J. Chem. Phys.* **20,** 1591.
Scoles, G.(Ed) (1988). *Atomic and Molecular Beam Methods*. Oxford University Press, New York.
Segall, J., and Zare, R.N. (1988). *J. Chem. Phys.* **89,** 5704.
Semmes, D.H., Baskin, J.S., and Zewail, A.H. (1987). *J. Am. Chem. Soc.* **109,** 4104.
Sivakumar, N., Hall, G.E., Houston, P.L., Hepburn, J.W., and Burak, I. (1988). *J. Chem. Phys.* **88,** 3692.
Smalley, R.E., Wharton, L., and Levy, D.H. (1977). *Acc. Chem. Res.* **10,** 139.
Snavely, D.L., Zare, R.N., Miller, J.A., and Chandler, D.W. (1986). *J. Phys. Chem.* **90,** 3544.
So, H.Y., and Dunbar, R.C. (1988). *J. Am. Chem. Soc.* **110,** 3080.
Spohr, R., Guyon, P.M., Chupka, W.A., and Berkowitz, J. (1971). *Rev. Sci. Instrum.* **42,** 1872.
Stanley, R.J., Cook, M., and Castleman, A.W. (1990). *J. Phys. Chem.* **94,** 3668.
Stanley, R.J., and Castleman, A.W. (1991). *J. Chem. Phys.* **94,** 7744.
Stanton, H.E., and Monahan, J.E. (1964). *J. Chem. Phys.* **41,** 3694.
Steinfeld, J.I., Francisco, J.S., and Hase, W.L. (1989). *Chemical Kinetics and Dynamics*. Prentice Hall. Englewood Cliffs.
Stevens, B. (1967). *Collisional Activation in Gases*. Pergamon Press, Oxford.
Stewart, G.M., and McDonald, J.D. (1981). *J. Chem. Phys.* **75,** 5949.
Stockbauer, R. (1973). *J. Chem. Phys.* **58,** 3800.
Stockbauer, R. (1977). *Int. J. Mass Spectrom. Ion. Phys.* **25,** 89.
Sulkes, M. (1985). *Chem. Phys. Lett.* **119,** 426.
Syage, J.A., Lambert, Wm.R., Felker, P.M., Zewail, A.H., and Hochstrasser, R.M. (1982). *Chem. Phys. Lett.* **88,** 266.

Taatjes, C.A., Janssen, M.H.M., and Stolte, S. (1993). *Chem. Phys. Lett.* **203,** 363.
Ticich, T.M., Rizzo, T.R., Dubal, H.R., and Crim, F.F. (1986). *J. Chem. Phys.* **84,** 1508.
Tobiason, J.D., Dunlop, J.R., and Rohlfing, E.A. (1995). *J. Chem. Phys.* **103,** 1448.
Tonkyn, R.G., and White, M.G. (1989). *Rev. Sci. Instrum.* **60,** 1245.
Troe, J. (1983). *J. Phys. Chem.* **87,** 1800.
Tsai, B.P., Baer, T., and Horowitz, M.L. (1974). *Rev. Sci. Instrum.* **45,** 494.
Uzer, T., and Miller, W.H. (1991). *Phys. Reports* **199,** 73.
Vidal, C.R. (1986). In *Tunable Lasers*, Mollenauer, L.F. and White, J.C., Eds. Springer-Verlag, New York.
Wannier, G.H. (1953). *Phys. Rev.* **90,** 817.
Weitzel, K.M., Booze, J.A., and Baer, T. (1991). *Chem. Phys.* **150,** 263.
Weitzel, K.M., Mahnert, J., and Baumgartel, H. (1993). *Ber. Bunsenges. Phys. Chem.* **97,** 134.
Werner, A.S., and Baer, T. (1975). *J. Chem. Phys.* **62,** 2900.
Wiley, W.C., and McLaren, I.H. (1955). *Rev. Sci. Instrum.* **26,** 1150.
Wu, M., Bemish, R.J., and Miller, R.E. (1994). *J. Chem. Phys.* **101,** 9447.
Xu, Z., Koplitz, B., and Wittig, C. (1989). *J. Chem. Phys.* **90,** 2692.
Yang, B., Chui, Y., and Anderson, S.L. (1991). *J. Chem. Phys.* **94,** 6459.
Yang, S.C., and Bersohn, R. (1974). *J. Chem. Phys.* **61,** 4400.
Zare, R.N. (1972). *Mol. Photochem.* **4,** 1.
Zare, R.N. (1984). *Ann. Rev. Phys. Chem.* **35,** 265.
Zen, M. (1988). In *Atomic and Molecular Beam Methods*, Scoles, G., Ed. Oxford University Press, New York, pp. 254–275.
Zewail, A.H. (1983). *Faraday Discuss. Chem. Soc.* **75,** 315.
Zewail, A.H. (1991). *Faraday Discuss. Chem. Soc.* **91,** 1.

6

Theory of Unimolecular Decomposition— The Statistical Approach

6.1 SOME BASIC IDEAS FROM STATISTICAL MECHANICS

The partition function and the sum or density of states are functions which are to statistical mechanics what the wave function is to quantum mechanics. Once they are known, all of the thermodynamic quantities of interest can be calculated. It is instructive to compare these two functions because they are closely related. Both provide a measure of the number of states in a system. The partition function is a quantity that is appropriate for thermal systems at a given temperature (canonical ensemble), whereas the sum and density of states are equivalent functions for systems at constant energy (microcanonical ensemble). In order to lay the groundwork for an understanding of these two functions as well as a number of other topics in the theory of unimolecular reactions, it is essential to review some basic ideas from classical and quantum statistical mechanics.

6.1.1 The Classical Hamiltonian

As discussed in chapter 2, the classical Hamiltonian, $H(\mathbf{p},\mathbf{q})$, is the total energy of the system expressed in terms of the momenta ($\mathbf{p}$) and positions ($\mathbf{q}$) of the atoms in the system. Recall that the Hamiltonian function for a molecule of N atoms may be expressed as a sum over the $3N$ Cartesian coordinates of the kinetic and potential energies of the system

$$H(\mathbf{p},\mathbf{q}) = \sum_{1}[\mathrm{p}_i^2/2m_i] + V(q_1, \ldots, q_{3n}). \tag{6.1}$$

In particular, the classical Hamiltonian for an atom in a three-dimensional box with infinite walls is given by

$$H(\mathbf{p},\mathbf{q}) = \frac{p_x^2 + p_y^2 + p_z^2}{2m}. \tag{6.2}$$

The potential energy term is absent because the potential energy inside the box is 0.

The rotational Hamiltonian also has no potential energy term, and can also be multidimensional. For instance, a one-dimensional rotational Hamiltonian describes an internal free rotation (e.g., the methyl group in toluene). The rotation of a linear molecule involves two dimensions, while the overall rotation of a nonlinear molecule takes place in three dimensions. The Hamiltonians in internal coordinates (see chapter 2) for these three situations are given by

$$H = \frac{p_\theta^2}{2I}, \qquad \text{(1-dimension)} \tag{6.3}$$

$$H = \frac{p_\theta^2}{2I} + \frac{p_\phi^2}{2I\sin^2\theta}, \qquad \text{(2-dimensions)} \tag{6.4}$$

$$H = \frac{p_\theta^2}{2I_A} + \frac{p_\psi^2}{2I_c}\ \frac{(p_\phi - p_\psi\cos\theta)^2}{2I_A\sin^2\theta}. \qquad \text{(symmetric top)} \tag{6.5}$$

The vibrational Hamiltonian, unlike the translational and rotational Hamiltonians, has a potential energy term which is often approximated by the harmonic potential:

$$H = \frac{p^2}{2\mu} + \frac{1}{2}kq^2, \tag{6.6}$$

where k is the force constant of the bond, and μ is the reduced mass of the oscillator which for a diatomic molecule with atoms of masses m_1 and m_2, is defined as $m_1m_2/(m_1 + m_2)$. The application of Hamilton's equations Eq. (2.9) leads to a differential equation for the classical harmonic oscillator which has a frequency

$$\nu = \frac{1}{2\pi}\sqrt{\frac{k}{\mu}}. \tag{6.7}$$

Thus, the force constant, k, is equal to $4\pi^2\nu\mu$.

As pointed out in chapter 2, the Hamiltonian for the vibrations of polyatomic molecules (Eq. (2.47)) can be readily constructed as sums over momenta and positions if the vibrations are separable, a condition that leads to normal modes.

6.1.2 Phase Space, Density, and Sums of States

The concept of phase space in statistical mechanics is of central importance in the statistical theory of reactions. Consider a molecule consisting of N atoms with a Hamiltonian $H(\mathbf{p},\mathbf{q})$. The momenta, $\mathbf{p}$, and position, $\mathbf{q}$, vectors will consist of $n = 3N - 6$ terms. (We exclude the three degrees of translation and three degrees of overall external rotation.) The classical phase space volume of such a system with a maximum energy E is defined by the integral

$$\text{Phase Space Volume} = \int_{H=0}^{H=E} \cdots \int dp_1 \cdots dp_n dq_1 \cdots dq_n. \tag{6.8}$$

This volume integral in $2n$-dimensional space, has units of [Joule-sec]^n.

In one dimension, the units of phase space are Joule-sec. This is often referred to as a unit of action. According to the uncertainty principle, energy and time or momentum and position are conjugate quantities which cannot be simultaneously and precisely known, that is, $\Delta p\, \Delta q \geq \hbar/2$. Hence, the smallest allowable unit in phase space must be on the order of h, so that the quantum phase space is divided up into units of h.

We can convert the phase space volume into a sum of states simply by dividing by h^n, one unit of h for each dimension. The sum of states, $N(E)$, is then

$$N(E) = \frac{1}{h^n} \int_{H=0}^{H=E} \cdots \int dp_1 \cdots dp_n dq_1 \cdots dq_n. \tag{6.9}$$

This represents the sum of all the quantum states in the system in an energy range from 0 to E. Even though the phase space has been divided into quanta of action, this is still considered the classical sum of states because the classical phase space volume is first calculated and converted into quantum states only at the end. The distinction between the classical and quantum mechanical sum of states will become evident in the discussion of the harmonic oscillator which is best treated as a quantum mechanical system.

A quantity related to the sum of states is the density of states, $\rho(E)$, which is defined as the number of states per unit energy. The number of states in the range E and $E + dE$ is denoted by $W(E)$. It is obtained by integrating $d\mathbf{p}d\mathbf{q}$ between $H = E$ and $H = E + dE$, and then dividing the resulting volume by h^n:

$$W(E) = \frac{1}{h^n} \int_{H=\mathrm{E}}^{H=E+dE} \cdots\cdots \int dp_1 \cdots dp_n dq_1 \cdots dq_n. \tag{6.10}$$

This equation can be expressed as a product of a surface integral evaluated at $H = E$ and the energy interval dE. Since $\rho(E) = W(E)/dE$, the density of states is the surface integral

$$\rho(E) = \frac{1}{h^n} \int \underset{H=\mathrm{E}}{\cdots} \int dp_1 \cdots dp_n dq_1 \cdots dq_n. \tag{6.11}$$

The density of states can also be obtained from $N(E)$ by taking the derivative with respect to the energy, E, $\rho(E) = dN(E)/dE$.

6.1.2.1 Phase Space Volume, N(E), and ρ(E) for a Particle in a Box

Consider the case of a particle in a three-dimensional box with dimensions a, b, and c and a total energy, E. The phase space volume consists of two sets of integrals, one over the dimensions, dx, dy, and dz, the other over the corresponding momenta. The triple integral over $dxdydz$ yields the volume of the box, $V = abc$, so that the phase space volume (PSV) is given by

$$PSV = \int\int\int dp_x dp_y dp_z \int_0^a \int_0^b \int_0^c dxdydz = V \int\int\int dp_x dp_y dp_z. \tag{6.12}$$

The integral over the momenta is somewhat more complicated, because the condition on the momenta is that the total energy must be less than E. However, we recognize that the Hamiltonian, rewritten as $2mE = p_x^2 + p_y^2 + p_z^2$, is the equation for a sphere with a radius of $\sqrt{2mE}$ (fig. 6.1) (Kubo, 1965; Hase, 1983). Thus, the integral over the momenta is the volume of this sphere,

$$PSV = \frac{4}{3}\pi[2mE]^{3/2}V. \tag{6.13}$$

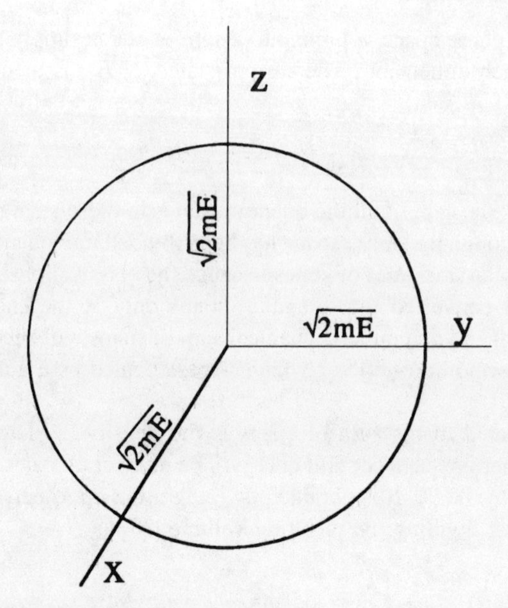

Figure 6.1 The three-dimensional translational phase space represented as the volume of the sphere with the momentum as the radius. The surface of the sphere is related to the density of states.

The sum of states is then given by

$$N(E) = \frac{4}{3h^3}\pi[2mE]^{3/2}V, \tag{6.14}$$

while the density of states is the derivative of $N(E)$ with respect to E,

$$\rho(E) = \frac{4}{h^3}\pi[m]^{3/2}(2E)^{1/2}V. \tag{6.15}$$

Note that the density of states for a particle in a three-dimensional box increases gently with the square root of the energy and linearly with the volume of the box.

A similar analysis for a particle in a one-dimensional box of length, a, results in a density of states given by:

$$\rho(E) = [2ma^2/Eh^2]^{1/2}. \tag{6.16}$$

Again, the density of states increases with the length of this box, but it decreases with energy as $E^{-1/2}$. This decrease is expected because the spacing between the non-degenerate energy levels of a particle in a one-dimensional box increase with energy since $E \propto n^2$, where n is the quantum number. In general, the density and sums of states increase with the dimensionality of the system.

6.1.2.2 Phase Space Volume, N(E), and ρ(E) for a Classical Harmonic Oscillator

The one-dimensional harmonic oscillator has two dimensions (p and q), so that its phase space volume for $H \leq E$ is given by

$$PSV = \int\int dq\, dp\,. \tag{6.17}$$

As before, the integral is subject to the condition that the total energy, kinetic plus potential, be equal to or less than E. The Hamiltonian for the one-dimensional classical oscillator can be rewritten as

$$1 = \frac{p^2}{2\mu E} + \frac{q^2}{2E/k}, \tag{6.18}$$

which identifies it as the equation for an ellipse in which the two semi-axes are $a = \sqrt{2\mu E}$ and $b = \sqrt{2E/k}$ (Hase, 1983; Steinfeld et al., 1989). The area of such an ellipse is πab so that the phase space volume for a classical harmonic oscillator is

$$PSV = 2\pi E\sqrt{\frac{\mu}{k}} = \frac{E}{\nu}. \tag{6.19}$$

The sum and density of states are then given by

$$N(E) = E/h\nu \qquad \rho(E) = 1/h\nu. \tag{6.20}$$

The constant density of states for a single harmonic oscillator is consistent with the fact that the energy levels are equally spaced by $h\nu$.

The Hamiltonian for a set of s classical, harmonic oscillators is the sum of s single oscillator Hamiltonians [Eq. 2.47)]:

$$H = \sum_i \left(\frac{p_i^2}{2\mu_i} + \frac{k_i q_i^2}{2}\right). \tag{6.21}$$

This is also the equation of an ellipse with semi axes $a_i = \sqrt{2\mu_i E}$ and $b_i = \sqrt{2E/k_i}$, but this time in $2s$-dimensional space. The volume of such an ellipse is $\pi\Pi a_i b_i/s!$, which results in a phase space volume of

$$PSV = \frac{E^s}{s!\Pi\nu_i} \tag{6.22}$$

so that the sum and density of states for a collection of classical harmonic oscillators are

$$N(E) = \frac{E^s}{s!\Pi h\nu_i} \qquad \rho(E) = \frac{E^{s-1}}{(s-1)!\Pi h\nu_i}. \tag{6.23}$$

These classical sums and densities for a single or a collection of oscillators are only accurate at very high energies in the equipartition limit. At low energies the quantum sums and densities must be used. However, the classical expressions are extremely useful because they are in the form of simple analytic functions. Equations (6.23) demonstrate, for instance, that the sums and densities are rapidly increasing functions of the energy.

6.1.2.3 Phase Space Volume, N(E), and ρ(E) for Rigid Rotors

The one-dimensional rigid rotor plays an important role in unimolecular reactions. For example, a methyl group in toluene is a free rotor which consists of the methyl group and the benzene ring rotating in opposite directions. It is thus distinguished from the overall rotation of the toluene molecule in which both the methyl group and the benzene ring rotate in the same direction. The one-dimensional rotor has a phase space volume given by

$$PSV = 2\int_0^{\sqrt{(2IE)}} dp_\theta \int_0^{2\pi} d\theta = 4\pi\sqrt{2IE}\,. \tag{6.24}$$

The factor of 2 arises from the rotor's two-fold degeneracy. The upper limit for the angular momentum ($\sqrt{2IE}$) is obtained from Eq. (6.3).

The sum and density of states are thus given by

$$N(E) = 2\sqrt{\frac{8\pi^2 IE}{h^2}} = 2\sqrt{\frac{E}{B}} \qquad \rho(E) = \sqrt{\frac{1}{BE}}\,, \tag{6.25}$$

where B is the rotational constant expressed in units of energy. As with the one-dimensional particle in a box, the one-dimensional rotor has a density of states that decreases with increasing energy.

The overall rotation of a linear molecule is given by a two-dimensional rotor, which according to Eq. (6.4) has two angle and the two conjugate momentum variables. The integration over the angles involves integrating θ from 0 to π, and ϕ from 0 to 2π. On the other hand, the two momenta must be integrated such that the total energy does not exceed E. The Hamiltonian can be rewritten as

$$1 = \frac{p_\theta^2}{2IE} + \frac{p_\phi^2}{2IE\sin^2\theta}\,, \tag{6.26}$$

which is the equation of an ellipse with semi-axes given by the square root of the two denominators. The phase space volume is then $2\pi EI\sin\theta$, which upon integration of $d\theta d\phi$ over the volume integral (4π) yields

$$PSV = 8\pi^2 EI\,, \tag{6.27}$$

so that the sum and density of states are given by

$$N(E) = \frac{8\pi^2 EI}{h^2} = \frac{E}{B} \qquad \rho(E) = \frac{1}{B}\,. \tag{6.28}$$

Evidently, the density of states for a two-dimensional rigid rotor is constant with energy, even though the quantum mechanical energy spacing of such a rotor increases with energy since $E = J(J + 1)B$. The reason for this is that the degeneracy of the quantum levels increase with the quantum number as $2J + 1$, thereby compensating for the nonconstant energy level spacings.

A good model for a three-dimensional nonlinear molecule is a symmetric top, whose Hamiltonian is given in Eq. (6.5). Two of the moments of inertia of a symmetric top are equal, that is, $I_A = I_B \neq I_C$. The derivation of this phase space volume is made more complicated because of the cross terms in the momenta. We begin by defining three new momenta:

$$p_x = \frac{p_\theta}{\sqrt{2EI_A}}\,,\quad p_y = \frac{p_\psi}{\sqrt{2EI_C}}\,,\quad p_z = \frac{p_\phi - p_\psi\cos\theta}{\sqrt{2EI_A\sin^2\theta}}\,, \tag{6.29}$$

so that $p_x^2 + p_y^2 + p_z^2 = 1$. Furthermore, note that

$$p_\phi = \sqrt{2EI_A}(\sin\theta)p_z + \sqrt{2EI_C}(\cos\theta)p_y \,. \tag{6.30}$$

The classical phase space volume is given as before by

$$PSV = \int\int\int_{H=0}^{H=E}\int\int\int dp_\theta\, dp_\phi\, dp_\psi\, d\theta\, d\phi\, d\psi\,, \tag{6.31}$$

where $0 \leq \theta \leq \pi$, $0 \leq \phi \leq 2\pi$, $0 \leq \psi \leq 2\pi$.
To solve this integral, we convert the variables

$$dp_\theta dp_\phi dp_\psi = \frac{\partial(p_\theta,p_\phi,p_\psi)}{\partial(p_x,p_y,p_z)} dp_x dp_y dp_z = J\, dp_x dp_y dp_z\,, \tag{6.32}$$

where the ratio of the partial derivatives, J, is the Jacobian which is given by the following determinant.

$$J = \begin{vmatrix} \frac{\partial p_\theta}{\partial p_x} & \frac{\partial p_\theta}{\partial p_y} & \frac{\partial p_\theta}{\partial p_z} \\ \frac{\partial p_\phi}{\partial p_x} & \frac{\partial p_\phi}{\partial p_y} & \frac{\partial p_\phi}{\partial p_z} \\ \frac{\partial p_\psi}{\partial p_x} & \frac{\partial p_\psi}{\partial p_y} & \frac{\partial p_\psi}{\partial p_z} \end{vmatrix} = \begin{vmatrix} \sqrt{2EI_A} & 0 & 0 \\ 0 & \sqrt{2EI_C}\cos\theta & \sqrt{2EI_A}\sin\theta \\ 0 & \sqrt{2EI_C} & 0 \end{vmatrix} \,. \tag{6.33}$$

With this Jacobian of $(2EI_A)(2EI_C)^{1/2}\sin\theta$, the PSV integral (Eq. (6.31)) becomes

$$PSV = \int\int\int_{H=0}^{H=E}\int\int\int dp_x dp_y dp_z (2EI_A)\sqrt{2EI_C}\sin\theta\, d\theta\, d\phi\, d\psi\,. \tag{6.34}$$

Since $p_x^2 + p_y^2 + p_z^2 = 1$, the integrals over the dp_x, dp_y, and dp_z momenta is just the volume of the unit sphere, namely $4\pi/3$, so that

$$PSV = (2EI_A)\sqrt{2EI_C}\int\int_{r=0}^{r=1}\int dp_x dp_y dp_z \int_0^\pi \int_0^{2\pi}\int_0^{2\pi} \sin\theta\, d\theta\, d\phi\, d\psi\,, \tag{6.35}$$

$$PSV = \frac{4}{3}(8\pi^2 I_A E)\sqrt{8\pi^2 I_C E}\,. \tag{6.36}$$

Division by h^3 results in the sum of states,

$$N(E) = \frac{\sqrt{\pi}}{(3/2)!}\,\frac{8\pi^2 I_A}{h^2}\sqrt{\frac{8\pi^2 I_C}{h^2}}\,E^{3/2}\,, \tag{6.37}$$

where we have replaced 4/3 by the equivalent $\sqrt{\pi}/(3/2)!$. The density of states, obtained by differentiation, is:

$$\rho(E) = \frac{\sqrt{\pi}}{(1/2)!}\,\frac{8\pi^2 I_A}{h^2}\sqrt{\frac{8\pi^2 I_C}{h^2}}\,E^{1/2}, \qquad (6.38)$$

where $\sqrt{\pi}/(1/2)!$ is just a fancy way of writing the number 2. However, it is not there simply to impress. If this density of states is used to calculate the partition function, the related Gamma function appears and is then easily canceled. The density expressed in terms of the rotational constants, B_A and B_C, is given by

$$\rho(E) = \frac{2\sqrt{E}}{B_A\sqrt{B_C}}. \qquad (6.39)$$

Equation (6.39) reduces to the correct expression for a spherical top density of states for which $B_A = B_C$.

6.1.3 The Canonical Partition Function

We began this chapter by considering the phase space volume and surface area which are related to the sum and density of states for a system at a given total energy E, that is, a microcanonical system. This is the system of major interest in this book. However, during the discussion of several topics it will be necessary to make use of the partition function, which is appropriate for constant-temperature, or canonical, systems. Because the partition functions for translations, rotations, and vibrations are derived in all undergraduate physical chemistry texts, we will not derive them here, but simply summarize the results.

The quantum mechanical canonical partition function is defined by

$$q(T) = \sum_i g_i e^{-E_i/k_B T}, \qquad (6.40)$$

where E_i is the ith quantized energy level of the system, g_i is the degeneracy of the ith energy level, k_B is the Boltzmann constant, and the sum is over all energy levels. The corresponding classical canonical partition function is given by

$$q(T) = \int \rho(E) e^{-E/k_B T}\,dE, \qquad (6.41)$$

where the degeneracy has been replaced by the density of states, $\rho(E)dE$, and the sum over discrete levels is replaced by an integral over the energy continuum.

The derived density of states for the translations, rotations, and vibrations can be used in Eq. (6.41) to obtain the corresponding classical partition functions. This will yield an accurate translational partition function at all temperatures of chemical interest because the translational energy level spacings are so dense. It will also yield accurate rotational partition functions at room temperature because molecular rotational constants are typically between 0.01 and 1 cm^{-1}. However, at the low temperatures achieved in molecular beams, the accuracy of the classical rotational partition function (especially for molecules with high rotational constants, such as formaldehyde or H_2 (B_e = 60.8 cm^{-1})) is insufficient. The energy level spacing of vibrations (ca. 2000 cm^{-1}) are considerably larger than the room temperature $k_B T$ of 207 cm^{-1} so that even at room temperature, the vibrational partition function must be evaluated by summation in Eq. (6.40).

The classical partition functions for translations and rotations and the quantum mechanical partition function for vibrations are given in Eqs. (6.42) to (6.47) (Berry et al., 1980):

$$q_t(T) = \left(\frac{2\pi^2 m k_B T}{h^2} \right)^{1/2} a \qquad \text{(1-D translation)}, \tag{6.42}$$

$$q_t(T) = \left(\frac{8\pi^2 I k_B T}{h^2} \right)^{3/2} V \qquad \text{(3-D translation)}, \tag{6.43}$$

$$q_r(T) = \frac{8\pi^2 I k_B T}{\sigma_r h^2} = \frac{k_B T}{\sigma_r B} \qquad \text{(2-D linear rotor)}, \tag{6.44}$$

$$q_r(T) = \sqrt{\pi}\, \frac{8\pi^2 I_A k_B T}{\sigma_r h^2} \left(\frac{8\pi^2 I_C k_B T}{h^2} \right)^{1/2} \qquad \text{(3-D symmetric top rotor)}, \tag{6.45}$$

$$q_r(T) = \frac{8\pi^2 (2\pi k_B T)^{3/2}}{\sigma_r h^3} (I_x I_y I_z)^{1/2} \qquad \text{(3-D asymmetric top rotor)}, \tag{6.46}$$

$$q_v(T) = \frac{1}{1 - e^{-h\nu/k_B T}} \qquad \text{(1-D harmonic oscillator)}, \tag{6.47}$$

where the a in the translational partition function is the size of the one-dimensional box, the I's in the numerator of the rotational partition functions are the moments of inertia (one for a linear molecule and three for an asymmetric top), while σ_r in the denominator of the rotational partition functions is the symmetry number (e.g., $\sigma_r = 1$ for NO, 2 for H_2, and 3 for NH_3). Although these symmetry numbers are an essential part of the partition function, when we calculate the density of states by inversion of the partition function, we normally do not include them in the rotational partition function. Rather, the symmetry numbers are explicitly taken into account in the reaction degeneracy, which is discussed under a separate heading later in this chapter.

Not only are these basic partition functions readily expressed in terms of the system parameters, mass, vibrational frequency, moment of inertia, etc., they can be readily generalized to higher dimensions. Consider, for instance, the relationship between the one- and three-dimensional translational partition functions. Since all three directions are equivalent, the same partition function applies to the other two dimensions. As a result, the three-dimensional q_t is just the product of three one-dimensional q_t. We can generalize this by saying that the partition function of a system of n degrees of freedom is given by the product of the n individual partition functions as long as the energy in each degree of freedom is independent of the energy in the other degrees of freedom (Forst, 1973). This is true for translations and vibrations, but not for rotations.

The vibrational partition function of a polyatomic molecule composed of s harmonic oscillators consists of a product of single oscillator partition functions because the energies of these oscillators are independent of each other.

$$q_v(T) = \prod_{i-1}^{s} [1 - e^{-h\nu_i/k_B T}]^{-1} . \tag{6.48}$$

It is evident that the quantum mechanical partition function of simple systems can be readily evaluated in closed form. This property turns out to be useful in evaluating the density of vibrational states of a polyatomic molecule.

6.1.4 Density of States and the Laplace Transform of the Partition Function

The determination of the density of states by the inversion of the partition function was originally suggested by Bauer (1938; 1939), who recognized that by replacing the $1/k_BT$ term in Eq. (6.41) by β, the partition function can be written as a Laplace transform:

$$Q(\beta) = \int_0^\infty \rho(E)e^{-E\beta}dE\,. \tag{6.49}$$

That is, the partition function, $Q(\beta)$, is just the Laplace transform of the density of states, which is written as

$$Q(\beta) = \mathcal{L}\,[\rho(E)] \tag{6.50}$$

so that the density of states can be expressed as an inverse Laplace transform:

$$\rho(E) = \mathcal{L}^{-1}[Q(\beta)]. \tag{6.51}$$

The Laplace transform and its inverse are well known for simple functions (Steinfeld et al., 1989; Forst, 1973). Consider, for instance, the case of a single harmonic oscillator in the classical limit, ($h\nu \ll kT$). If the exponential in Eq. (6.47) is expanded in a power series and the higher-order terms dropped, we find that $q_v(\text{classical}) = k_BT/h\nu$, or using the symbol in the Laplace transform, $q_v = (\beta h\nu)^{-1}$. The density of states of a single classical harmonic oscillator is then given as follows:

$$\rho(E) = \mathcal{L}^{-1}[(\beta h\nu)^{-1}] = (h\nu)^{-1}\mathcal{L}^{-1}(1/\beta) = (h\nu)^{-1} \tag{6.52}$$

because the inverse Laplace transform of $1/\beta$ is equal to 1. (Steinfeld et al., 1989). This is the same result as previously obtained from the classical phase space integral.

For the case of s classical harmonic oscillators, the partition function is given by:

$$Q(\beta) = \prod_i^s [\beta h\nu_i]^{-1} = \beta^{-s}\prod_i^s (h\nu_i)^{-1}\,. \tag{6.53}$$

The inverse Laplace transform of the $1/\beta^s$ function is $E^{s-1}/\Gamma(s)$, where $\Gamma(s)$ is the gamma function. For an integer s, $\Gamma(s) = (s-1)!$ so that the density of states for s classical harmonic oscillators is given (as before) by

$$\rho(E) = \prod_i^s (h\nu_i)^{-1}\frac{E^{s-1}}{(s-1)!}\,. \tag{6.54}$$

6.1.4.1 Laplace Transforms for Quantum Mechanical Oscillators

The determination of the density of states for s *classical* oscillators by the method of Laplace transforms is of limited value because this can be obtained by other methods as well. Of much greater interest is the fact that the product of the *quantum* oscillators in Eq. (6.48) can be inverted by the Laplace transform method. However, it requires solving the inverse Laplace transform integral (Forst, 1971, 1973; Hoare and Ruijgrok, 1970):

$$\rho(E) = \frac{1}{2\pi i}\int_{c-i\infty}^{c+i\infty} Q(\beta)e^{\beta E}d\beta \,. \tag{6.55}$$

This is possible via a steepest descent method (Hoare and Ruijgrok, 1970; Forst, 1971, 1973; Robinson and Holbrook, 1972; Eyring et al., 1980), which is an effective and accurate method, although not entirely trivial, for determining density and sums of vibrational and ro-vibrational states. Although it permits the use of a quantum partition function, it treats $\rho(E)$ as a smooth function, even though it is strictly speaking a series of delta functions. In the method of steepest descent, the integration in the complex plane is converted to a real integral along a line parallel to the imaginary axis by setting $c = \beta^*$. This is possible because the chosen path is such that the derivative of the integrand with respect to β at β^* is zero, thereby defining a saddle point in the two-dimensional (imaginary and real planes) space. If the curvature is very large then the integral has contributions only in the vicinity of the saddle point.

A program (in BASIC) based on the steepest descent approximation for the density of the ro-vibrational states is listed in the Appendix.

6.1.4.2 The Density of States as a Convolution

In general, the convolution integral is an extremely useful function for a variety of problems in the physical sciences. For instance, if a measured response from some instrument (e.g., a spectrometer) is a combination of the finite resolution of that instrument which is given by a function $g(x)$ and the true physical property (e.g., the natural line width of a transition), which is given by the function $h(x)$, then the observed function, $f(x)$ will be a convolution of $g(x)$ and $h(x)$, which is expressed as

$$f(x) = \int_0^x g(y)h(x-y)dy = \int_0^x h(y)g(x-y)dy \,. \tag{6.56}$$

Similarly, in timing experiments, $f(t)$ is expressed as a convolution between the true decay rate and the instrument response function.

The convolution integral also describes the manner in which two density of states are combined. Consider the case of the system consisting of rotational and vibrational energy levels each with the corresponding density of states, ρ_v and ρ_r. Now consider a molecule with energy x in vibrations and $E - x$ in rotations. The contribution to the density of states for fixed E and x, will be the product: $\rho_v(x)\,\rho_r(E-x)$. To obtain the total density of states at energy, E, we need to integrate over all possible combinations of energies in vibrations and rotations:

$$\rho_{vr}(E) = \int_0^E \rho_v(x)\rho_r(E-x)dx \quad \text{or} \quad \int_0^E \rho_v(E-x)\rho_r(x)dx \,. \tag{6.57}$$

This is just the convolution integral mentioned above. This convolution integral is also the basis of the exact, or direct count of states to be discussed later.

6.1.4.3 The Convolution Theorem for Laplace Transforms

The convolution of state densities can be derived also from the Laplace transform approach. In fact, the convolution integral plays an important role in the theory of

Laplace transforms. According to the convolution theorem, if two functions (e.g., Q's) are expressed as Laplace transforms of the functions ρ_1 and ρ_2,

$$Q_1(\beta) = \mathcal{L}[\rho_1(E)] \quad \text{and} \quad Q_2(\beta) = \mathcal{L}[\rho_2(E)], \tag{6.58}$$

then the product of Q_1Q_2 is just the Laplace transform of the convolution of ρ_1, and ρ_2. That is,

$$Q_1Q_2 = \mathcal{L}[\rho_1 * \rho_2], \tag{6.59}$$

where the symbol, *, signifies the convolution integral. We already know that the inverse transform of Q_1Q_2 is the joint density of states for the two energy sinks 1 and 2. Hence, $\rho_{12} = \mathcal{L}^{-1}[Q_1Q_2] = \rho_1 * \rho_2$, which demonstrates that the joint density of states of ρ_1 and ρ_2 is a convolution of the two density of states.

6.1.4.4 Sums of States by Laplace Transforms

A second important property of Laplace transforms is expressed in the integration theorem (Steinfeld et al., 1989; Forst, 1973). If $\rho(E)$ is the inverse Laplace transform of $Q(\beta)$, then the integral of $\rho(E)$, given by $N(E)$ is

$$N(E) = \int_0^E \rho(E) = \mathcal{L}^{-1}[Q(\beta)/\beta] \ , \tag{6.60}$$

which is the sum of states from 0 to E. Thus, the sum of states is obtained from the inverse Laplace transform of $Q(\beta)/\beta$. Equation (6.60) plus the steepest descent approximation (the integral over β is replaced by the integrand value at β^*) leads very naturally (Hoare and Ruijgrok, 1970) to an interesting relationship between the sum and density of states,

$$N(E) = \rho(E)/\beta^*, \tag{6.61}$$

which is a relation that will become useful in chapter 10. β^* corresponds to an equivalent temperature of the system with energy E, which is obtained through the molecular heat capacity, or via the well known thermodynamic relation $(\partial \ln Q(\beta)/\partial\beta)_{B=B^*} = -E$. However, the energy in this case is the microcanonical rather than the thermodynamic energy.

For the case of a single classical harmonic oscillator, the sum of states is

$$N(E) = (1/h\nu)\mathcal{L}^{-1}[\beta]^{-2} = E/h\nu, \tag{6.62}$$

because the inverse Laplace transform of β^{-2} is E. Similarly, the sum of states for s harmonic oscillators is:

$$N(E) = \Pi(h\nu_i)^{-1}\mathcal{L}^{-1}(\beta)^{-(s+1)} = \prod_i^s (h\nu_i)^{-1} \frac{E^s}{s!} \ . \tag{6.63}$$

Finally, it is interesting to combine the convolution and the integration theorem. Suppose, we have two partition functions, Q_1 and Q_2, their corresponding densities of states (ρ_1 and ρ_2), and sums of states (N_1 and N_2). The sum of states for the combined system, N_{12}, will be given by

$$N_{12}(E) = \int_0^E \rho_{12}(x)dx = \int \mathcal{L}^{-1}Q_1Q_2dx = \mathcal{L}^{-1}[Q_1Q_2/\beta] = N_1{*}\rho_2 = \rho_1{*}N_2\,. \quad (6.64)$$

It is evident that the combined sum of states, N_{12}, is given by a convolution between the sum N_1 and the density ρ_2, or by $\rho_1{*}N_2$ (not the convolution between two sum functions). This property is essential in understanding the principles underlying the direct count method for the sum of states.

6.1.5 Density and Sums of States by Direct Count

The most accurate procedure for determining the density of harmonic vibrational states is by the direct count method. A particularly clever scheme for doing this was proposed by Beyer and Swinehart (1973). As demonstrated by Gilbert and Smith (1990), this approach is based on the convolution of state densities. Suppose that the system consists of s harmonic oscillators with vibrational frequencies, $\omega_i = \nu_i/c$ (cm^{-1}). Each will have a series of equally spaced states located at $E_i = n\omega_i$ ($n = 0, 1, \ldots$). We choose the zero of energy at the molecule's zero point energy, and divide the energy into bins. The vibrational frequencies must be expressed as integral numbers of bin sizes, for example, as multiples of 10 cm^{-1} for a 10^{-1} bin size. A convenient bin size is 1 cm^{-1} so that the s frequencies can be simply rounded off to the nearest wavenumber.

We begin with a density of states vector, $\boldsymbol{\rho}(E)$, in which a state is assigned to the 0th energy level (we know that there must be a state there), and in which 0's are placed in all the other bins [1,0,0, . . .]. It is most convenient to use the zero index in the density of states vector, $\boldsymbol{\rho}(I)$, so that the index I can be directly identified with the energy. After this initialization step, the algorithm (in BASIC) for generating the density of states for all s oscillators is as follows:

Beyer-Swinehart density of vibrational states count

```
  ρ(I) = [1,0,0,0, . . . ]         (initialize the ρ vector)
1 FOR J = 1 TO s                   (s = the number of oscillators)
2    FOR I = ω(J) TO M             (M = the maximum energy bin of interest)
3    ρ(I) = ρ(I) + ρ(I − ω(J)
4    NEXT I
5 NEXT J
```

The logic of this calculation can be understood by considering the first two J loops. We begin with the initialized $\boldsymbol{\rho}(I)$ vector in which all elements except the 0th one are equal to 0. Now the first oscillator ($J = 1$) is folded in. In step 3, the $\rho(I)$'s will be zero until, $I = \omega(1)$ at which point $I - \omega(1) = 0$ and $\rho(I - \omega(1)) = \rho(0) = 1$. Thus, $\rho(\omega(1)) = 1$. Similarly, when $I = 2\omega(1)$, $\rho(I)$ will also be set equal to 1, and so on. After this first passage through the I loop we have a $\boldsymbol{\rho}(I)$ vector which consists of mostly an empty vector having one's (or more precisely, the degeneracy of the level) at the harmonics of the ground-state frequency. The second set of levels is now convoluted in by running through the $J = 2$ loop. Whenever there is a level at $I - \omega(2)$, its density is added to the one that was directly below it by the vibrational frequency in question.

It is evident that the inner loop consisting of steps 2, 3, and 4 is just a convolution integral, $\int \rho(I)\,\rho(I - J)\mathrm{d}I$. Its repeated application folds in all of the vibrational normal modes in turn so that the complete density of states calculation is obtained after s iterations.

The sum of states, $N(E)$, can be obtained either by direct numerical integration of the density of states, or it can be computed directly using the BS scheme (Stein and Rabinovitch, 1973; Gilbert and Smith, 1990). Recall, that the sum of states for two degrees of freedom is given by the convolution of a sum with a density (Eq. (6.64)). It is thus sufficient to define the initial vector $\mathbf{N}(I) = [1,1,1,1,1. \ldots,]$ which is the integral of the initialized $\boldsymbol{\rho}(I)$ vector, that is, a sum of states. All following convolutions will then automatically be sums, rather than densities. The sum is thus computed by executing steps 6–10.

Beyer-Swinehart sum of vibrational states count

```
     N(I) = [1,1;1,1 . . . . 1]       (initialize the N vector)
  6  FOR J = 1 TO s                   (s = the number of oscillators)
  7     FOR I = ω(J) TO M             (M = the maximum energy bin of interest)
  8     N(I) = N(I) + N(I − ω(J))
  9     NEXT I
 10  NEXT J
```

The BS method can also be used for calculating the ro-vibrational density of states. In the BS scheme, the initialized $\boldsymbol{\rho}$ vector, $\boldsymbol{\rho}(I) = \boldsymbol{\delta}(I)$, is the initial density of states vector. Thus, the first pass through the BS scheme is the convolution of the (1,0,0,0, . . .) vector with the state density of oscillator one. If the initialized $\boldsymbol{\rho}(I)$ vector is replaced with the rotational density of states, all of the vibrations are convoluted into the rotational density of states, thereby producing a ro-vibrational density of states. Because the rotational density of states can be readily calculated from classical mechanics [Eq. (6.25), (6.28), or (6.39)], it is easy to generate the initial $\boldsymbol{\rho}_r(I)$ vector. For instance, in the case of the HCN molecule with a rotational constant of 1.4878 cm^{-1}. It has a constant density of rotational states of $1/B = 0.672$ states/cm^{-1} so that the initial density vector would be $\boldsymbol{\rho}(I) = [0.672, 0.672, \ldots]$.

Beyer-Swinehart density of ro-vibrational states

```
     ρ(I)                             (initialize as rotational density of states
                                      vector)
 11  FOR J = 1 TO s                   (s = the number of oscillators)
 12     FOR I = ω(J) TO M             (M = the maximum energy bin of interest)
 13     ρ(I) = ρ(I) + ρ(I − ω(J))
 14     NEXT I
 15  NEXT J
```

To obtain the ro-vibrational sum of states, the $\mathbf{N}(I)$ vector must be initialized as a rotational sum vector. The procedure is then given by steps 11–15.

6.1.5.4 The Beyer-Swinehart-Stein-Rabinovitch (BSSR) Anharmonic Density Count Method

The BS method described so far, utilizes the equally spaced energy levels of harmonic oscillators to avoid solving an additional integral. If only one of the vibrational modes is to be treated as an anharmonic oscillator, it can be assigned to the initial $\boldsymbol{\rho}(I)$ vector. In this manner, the simple BS scheme can be employed by summing over the remaining $s - 1$ harmonic frequencies. However, if more than one of the oscillators is to be treated anharmonically, than the BS scheme will not work. Stein and Rabinovitch (1973) extended the BS direct count method so that any energy level scheme can be accommodated, as long as the different vibrational modes are not coupled. This is done at the expense of an additional integration so that the calculation takes more time.

The Morse oscillator energy levels are expressed in terms of a single parameter, a, the anharmonicity constant as

$$E(v) = \sigma v \left[1 - \frac{a}{1+v}\right], \text{ where } v = 0, 1, 2, \ldots, 1/(a-1). \quad (6.65)$$

This oscillator has a fixed number of states given by $1/(a - 1)$. Suppose that the molecule has s_1 independent anharmonic oscillators, and s_2 harmonic oscillators. We first generate a two-dimensional array, $\mathbf{K}(I, J)$ for the anharmonic oscillators, in which I identifies the particular oscillator in question, and J represents the energy levels of the Ith oscillator. The index I will thus run from 1 to s_1, while J will run from 0 to $1/(a_I - 1)$. The anharmonic density of states is generated with program steps 16 through 23, the output of which is then used as the initial vector for the harmonic part.

The BSSR Direct Count of Anharmonic Density of States

```
    K(I,J)                            (Initialize anharmonic energy levels)
    ρ(I) = δ(I)                       (Initialize density of states)
    ρ'(I) = 0                         (Initialize this intermediate array)
16  FOR I = 1 TO s1                   (Loop over anharmonic oscillators)
17    FOR J = 0 TO JMAX               (Loop over levels in Ith oscillator)
18      FOR L = K(I,J) TO M           (Loop over M energy levels)
19      ρ'(L) = ρ'(L) + ρ(L − K(I,J))
20      NEXT L
21    NEXT J
22  FOR J = 0 TO M: ρ(J) = ρ(J) + ρ'(J): NEXT J
23  NEXT I
```

The final $\boldsymbol{\rho}(I)$ vector contains the anharmonic density of states for the first s_1 oscillators. To include the next s_2 harmonic oscillators, we simply add (after line 23) the program lines 1–5, making certain that the $\boldsymbol{\rho}$ vector is not reinitialized.

It is evident that lines 16–23 are not limited to anharmonic oscillators. Any set of energy levels can be used. Thus, if the energy levels of a hindered rotor are available, they can be included in the $\mathbf{K}(I, J)$ vector. This BSSR extension is thus of great interest for intermediate size molecules in which a great deal of spectroscopic information, or information from *ab initio* calculations, is available. However, it is limited to independent oscillators. Realistic systems have coupled oscillators so that the energy spacing of the ith oscillator is dependent upon the energy of the jth oscillator. The treatment of such systems is deferred until the next chapter.

6.1.6 Comparison of Direct Count and Laplace Transform Methods

It is interesting to compare the direct count method with the inversion of the partition function approach. At high energies, the methods give identical results. However, at low energies, the direct count method yields densities which consist of delta functions, while the inverse Laplace transform approach smoothes over these spikes. This is shown in table 6.1, in which the two methods are compared. A major difference between the two methods is that with the Steepest Descent program it is possible to calculate the density or sum at any single energy. With the direct count one needs to calculate all energies with the same resolution. In addition, all points up to the maximum (16,000 points in this case) must be calculated!

Thus, at very low energies, the direct count method is recommended. At higher energies, the faster steepest descent approach is preferred, especially when the densities or sums of states are required at a high resolution. It must be recalled that both of these approaches are only approximate because they do not include anharmonicities.

Table 6.1. Comparison of Direct and Steepest Descent Calculations.

Energy	Density			Sum of States		
cm^{-1}	Direct	SD	Direct/SD	Direct	SD	Direct/SD
0	1			1		
100	0	.01	—	2	1.70	1.77
200	0	.02	—	3	3.37	0.89
300	0	.04	—	6	6.14	0.98
400	0	.05	—	10	10.35	0.96
500	1	.08	—	15	16.50	0.91
600	1	.11	—	24	25.23	0.95
800	0	.19	—	51	53.86	0.95
1000	1	.34	—	102	105.4	0.97
2000	6	3.3	1.82	1,421	1,466	0.97
3000	21	19.2	1.09	10,455	10,704	0.98
4000	85	83.6	1.02	53,977	55,098	0.98
5000	297	300	.99	221,815	226,143	0.98
6000	922	937	.98	775,407	789,568	0.98
8000	6,651	6,763	.98	6.72E(6)	6.83E(6)	0.98
10000	36,065	36,600	.99	4.22E(7)	4.28E(7)	0.99
12000	158,912	161,100	.99	2.10E(8)	2.13E(8)	0.99
14000	5.98E(5)	605,900	.99	8.81E(8)	8.91E(8)	0.99
16000	1.99E(6)	2.01E(6)	.99	3.22E(9)	3.25E(9)	0.99

Frequencies are: 3030 3020 3010 3005 1503 1401 1383 999
951 877 652 333 278 250 89

Running Time with a 286 PC (16 MHz) using uncompiled "QUICK BASIC":

About 2 min for the Direct Count program

About 15 sec for the Steepest Descent program

For small molecules at high energies, the true anharmonic densities can be several times greater than the calculated harmonic densities (see chapter 7). Although the errors are much smaller for large molecules, they remain significant. Thus, the "exact" count method is exact only for a very approximate harmonic model.

6.1.7 The Hindered Rotor

The hindered rotor is a hybrid between a free rotor and a vibration. An example of this important mode is the methyl group rotation in such molecules as ethane, acetaldehyde, and toluene. The potential for such a mode in lowest order is given by (Pitzer, 1953):

$$V(\theta) = \tfrac{1}{2}V_o(1 - \cos n\theta) \tag{6.66}$$

in which θ is the rotation angle, V_o is the classical barrier height, and n is the symmetry number of the rotor ($n = 3$ in the case of a methyl group rotation). Some typical barrier heights are ca 400 cm^{-1} in acetaldehyde (Herschbach, 1959; Moule et al., 1992), 1012 cm^{-1} in ethane (Moazzen-Ahmadi et al., 1988), 4.88 cm^{-1} in toluene (Rudolph et al., 1967), and 16 cm^{-1} in the toluene ion (Lu et al., 1992). The range of barriers is thus large and whether a particular mode is treated as an oscillator, or as a free rotor (or a combination of the two) depends upon the energy range of interest and the molecule. The hindered rotor is important even in molecules without internal rotors because the conversion of a bending vibration in the molecule into free product rotation can be treated as a hindered rotor problem (Jordan et al., 1991). In this application, the barrier height will be a function of the reaction coordinate.

The potential functions and the resulting energy levels of a one-dimensional hindered rotor, a free rotor, and a harmonic oscillator are compared in figure 6.2. The hindered rotor energy levels can be solved numerically as solutions to the Mathieu differential equation (Abramonitz and Stegun, 1972; Wilson, 1940) and lists of values

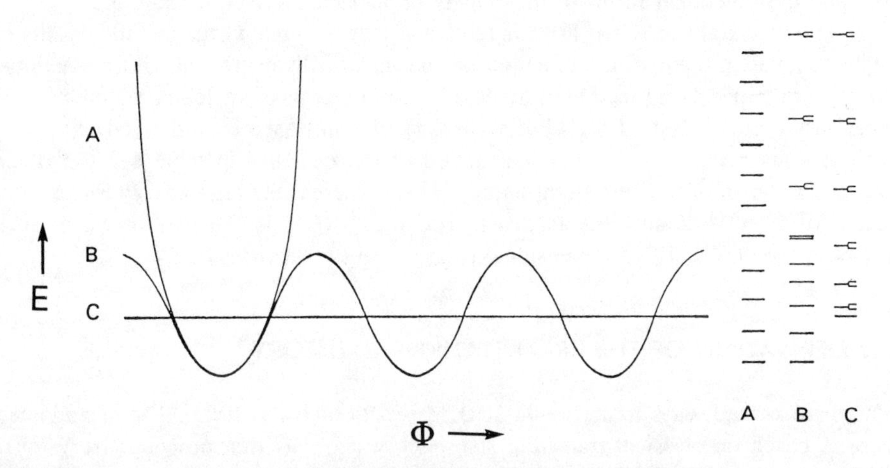

Figure 6.2 Potential energy surface for a hindered rotor such as $—CH_3$. The harmonic potential is also shown, as are the energy levels for the harmonic potential (A), the hindered rotor (B), and the free rotor (C). Adapted from Pitzer (1953).

for some systems have been tabulated (Wilson, 1940; Eyring et al., 1944). However, no simple closed form solution exits. At low energies, the motion is similar to a vibration with equally spaced, but triply degenerate, energy levels, while at energies in excess of the barrier height the motion turns into a free rotation. The latter has unequally spaced energy levels which increase with the rotational quantum number as $E_r = J^2B$ ($j = n, n + 1, n + 2, \ldots$), and which are doubly degenerate.

Pitzer (1953) pointed out that the energy levels well above the barrier are those of a free rotor but with an energy that is shifted by $V_o/2$. At high energy this "Pitzer" rotor is approximated by $\rho_{hr}(E) \approx \rho_r(E - V_o/2)$. On the other hand, at low energies it is treated as a vibrator. Various approximations for calculating hindered rotor density of states have been compared by Stein and Rabinovitch (1974). More recently Chesnavich (1986) has applied a simple approximate scheme for treating the low energy $C - X$ bending modes as hindered rotors in a $C - X$ bond dissociation reaction (see chapter 7). These methods involve explicit evaluation of the hindered rotor energy level which are then convoluted into the densities or sums of the remaining states by use of the BSSR direct count algorithm. This approach is necessary at low energies, while at high energies, the vibrator modes can be ignored and the rotor treated simply as a "Pitzer" rotor.

Because hindered rotors involve densely spaced energy levels, it is possible to treat the problem classically. This has been done for the case of a two-dimensional hindered rotor (Jordan et al., 1991) which is particularly important in the dissociation of "loose" transition states, a topic to be discussed in the following chapter. The classical phase space integral is solved using the Hamiltonian:

$$H = \frac{p_\theta^2}{2I} + \frac{p_\phi^2}{2I\sin^2\theta} + V(\theta) \ . \tag{6.67}$$

This is just Eq. (6.4) with an added hinderance potential. Jordan et al. (1991) obtained the phase space volumes, the partition functions, and density of states for various assumed forms of the potential, $V(\theta)$. The potential functions are illustrated in figure 6.3 and the functional forms of the density of states are shown in table 6.2.

The potential functions, $V(\theta)$, in table 6.2 have the great virtue that the density of states, as well as the partition functions can be analytically expressed. Other functional forms for the partition function of hindered rotors in terms of switching functions have been proposed (Truhlar, 1991), but no density of states have been derived.

A consequence of the classical nature of the densities in table 6.2 is that no account is taken of the zero-point energy. Thus, whereas the high-energy Pitzer rotor has an offset of $V_o/2$, such that its density becomes $[B(E - V_o/2)]^{-1/2}$, there is no such offset in the classical two-dimensional rotor density.

6.2 DERIVATION OF THE RRKM STATISTICAL THEORY

There are several ways to derive the RRKM equation (Forst, 1973). The one adopted here is based on classical transition state theory and was first proposed by Wigner (Wigner, 1937; Hirschfelder and Wigner, 1939). Although there are several other statistical formulations of the unimolecular rate [phase space theory (Pechukas and Light, 1965), statistical adiabatic channel model (SACM) (Quack and Troe, 1974),

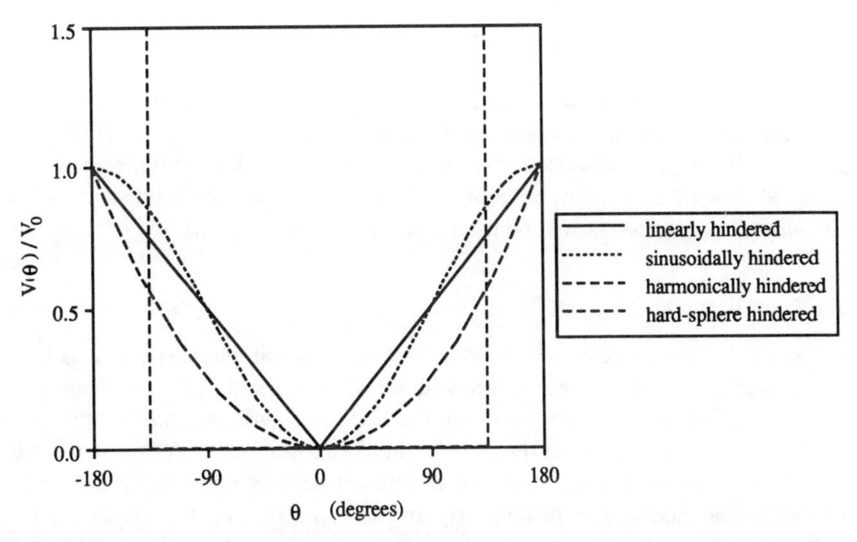

Figure 6.3 Comparison of different potentials, $V(\theta)$ for a hindered two-dimensional rotor. Taken with permission from Jordan et al. (1991).

etc.] we concentrate on RRKM because it is the most easily applied and with various modifications includes most other theories as special cases. It is worth pointing out here that what we call RRKM has also been termed quasi-equilibrium theory (QET) (Rosenstock et al., 1952) and was developed independently at about the same time as RRKM (Marcus and Rice, 1951). All formulations of statistical rate theories, except perhaps SACM, share the same basic assumptions and differ only in the detail in which rotations or the transition state are treated. The basic assumptions of statistical theories is that the rate constant $k(E, J)$ depends only on the total energy E and the total angular momentum J. It is assumed that the rate does not depend upon where the energy is initially located and that a microcanonical ensemble is maintained as the molecule dissociates. This is equivalent to assuming that IVR is rapid compared to the lifetime with respect to dissociation. That is, vibrations are assumed to be strongly coupled by higher order terms (anharmonicities, Fermi resonances, etc.) in the expansion of the potential energy function (see chapter 2). In addition, some rotational degrees of

Table 6.2. Density of States for Various Two-Dimensional Hindered Rotor Potentials.

Potential	$V(\theta)$	$\rho(E)$	
No barrier	0	$1/B$	
Linear	$V_o\,\theta$	$(1/2B)[1\text{-}\cos(\pi E/V_o)]$	$E \le V_o$
	$V_o\,\theta$	$1/B$	$E \ge V_o$
Sinusoidal	$\frac{1}{2}V_o(1\text{-}\cos\theta)$	$E/(BV_o)$	$E \le V_o$
		$1/B$	$E \ge V_o$
Hard sphere[a]	0 if $\theta \le \theta_{max}$ ∞ if $\theta > \theta_{max}$	$(1 - \cos\theta_{max})/B$	

[a]θ_{max} is defined in figure 6.3 by the vertical dashed lines.

freedom are also coupled to vibrations through coriolis interactions (see chapters 2 and 4).

This section begins with a simple derivation of the microcanonical rate constant $k(E)$ in which rotations are ignored and in which the location of the transition state is assumed to be fixed at a saddle point and is thus independent of the energy in the system. Methods for including rotations and for treating the transition state for reactions with no saddle points will be discussed in the following chapter.

6.2.1 The Dissociation as a Flux in Phase Space

A unimolecular reaction can be viewed as a reaction flux in phase space. It is best to have in mind a potential energy surface with a real barrier in the product channel, that is, a saddle point. Figure 6.4 shows both the reaction coordinate and a picture of the phase space associated with the molecule and the transition state. Recall, that a molecule of several atoms having a total of m internal degrees of freedom can be fully described by the motion of m positions (**q**) and m momenta (**p**). At any instant in time, the system is thus fully described by $2m$ coordinates. A constant energy molecule (a microcanonical system) has its phase space limited to a surface in which the Hamiltonian $H = E$. Thus, the dimensionality of this hypersurface is reduced to $2m - 1$.

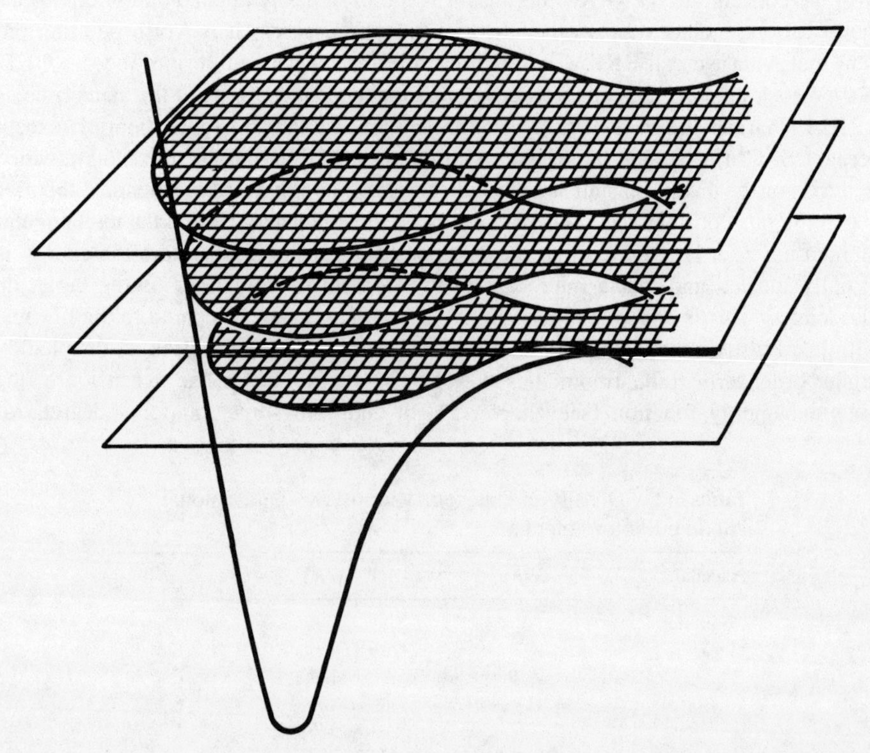

Figure 6.4 Reaction coordinate with a saddle point. Above it is a diagram of the phase space which varies with the reaction coordinate and with the total energy. Taken with permission from Baer (1986).

Whether this is called a volume or a surface is immaterial because in any case its dimensionality is far greater than 2 or 3. We simply assign the words "volume" and "surface" to indicate the relationship between them, the surface having one less dimension than the volume.

If the energy, E, is greater than the dissociation energy, E_o, the molecule has a chance to dissociate and thus to reach a part of the hypersurface which is associated with the critical surface, or the transition state. The critical surface is defined as the $2m$-2-dimensional surface which divides reactants and products and which is so located that a trajectory, once having passed through it, will proceed on to reaction products without returning. For reactions with substantial saddle points, the critical surface is located at the saddle point. For systems with small or no saddle points, that is, very small or no reverse activation energy, the location of the critical surface depends upon the total energy.

The rate of reaction is then related to the total flux of reactants which passes through this critical surface. At the critical surface, the molecule is in the process of dissociating along a one-dimensional reaction coordinate, which is the minimum energy reaction path. It is generally assumed that at the saddle point, the reaction path is perpendicular to all other coordinates, that is, the system is separable. We assign the spatial and the conjugate momentum $q^\ddagger$ and $p^\ddagger$ (without subscripts) to these two special coordinates. For the dissociation of $CH_3 \cdots H$, $q^\ddagger$ and $p^\ddagger$ correspond to the C—H bond distance and the momentum of the separating fragments. The other $n - 1$ (8 for the case of CH_4) C—H stretch and bending modes remain as vibrations perpendicular to, and are assumed to be decoupled from, the reaction coordinate.

The assumption that the total phase space is populated statistically means that the population density over the whole surface of the phase space is uniform. This permits the ratio of molecules near the critical region to the total molecules [$d\mathbb{N}(q^\ddagger, p^\ddagger)/\mathbb{N}$] to be expressed as the ratio of the phase space at the dividing surface to the total phase space. Thus, at any instant in time, the ratio of molecules whose special coordinates have values that range from $q^\ddagger$ to $q^\ddagger + dq^\ddagger$ and from $p^\ddagger$ to $p^\ddagger + dp^\ddagger$ to the total phase space is given by

$$\frac{d\mathbb{N}(q^\ddagger, p^\ddagger)}{\mathbb{N}} = \frac{dq^\ddagger dp^\ddagger \int \cdots \int_{H=E-\epsilon_t-E_o} \cdots \int dq_1^\ddagger \cdots dq_{n-1}^\ddagger dp_1^\ddagger \cdots dp_{n-1}^\ddagger}{\int\int_{H=E} dq_1 \cdots dq_n dp_1 \cdots dp_n} . \tag{6.68}$$

In this equation, E_o is the activation energy and ϵ_t is the translational energy associated with the momentum $p^\ddagger$ in the reaction coordinate. Both of these energies must be subtracted from the total energy at the saddle point because these energies are not available for the $n - 1$ momenta, $p_i^\ddagger$, and $n - 1$ coordinates, $q_i^\ddagger$. We can think of this expression as an equilibrium constant for a microcanonical system.

The rate of reaction is obtained from the time derivative of the molecules near the critical surface. That is, the reaction flux or reaction rate is given by

$$\text{Flux} = \text{reaction rate} = \frac{d\mathbb{N}(q^\ddagger, p^\ddagger)}{dt} . \tag{6.69}$$

This represents the flux of molecules passing through the critical region. Because of the assumption that the reaction coordinate is separable from and perpendicular to all other coordinates, the time derivative involves only the $dq^\ddagger\, dp^\ddagger$ term. The $dq^\ddagger\, dp^\ddagger/dt$ can be rearranged by noting that $dq^\ddagger/dt = p^\ddagger/\mu^\ddagger$, where $\mu^\ddagger$ is the reduced mass of the two separating fragments. When this is substituted in the $d\mathbb{N}(q^\ddagger, q^\ddagger)/dt$ expression, we obtain

$$\frac{d\mathbb{N}(q^\ddagger, p^\ddagger)}{dt} = \frac{\mathbb{N}\dfrac{p^\ddagger dp^\ddagger}{\mu^\ddagger}\displaystyle\int \cdots\cdots \int_{H=E-\epsilon_t-E_o} dq_1^\ddagger \cdots dq_{n-1}^\ddagger dp_1^\ddagger \cdots dp_{n-1}^\ddagger}{\displaystyle\int \cdots \int_{H=E} dq_1 \cdots dq_n dp_1 \cdots dp_n}. \tag{6.70}$$

Since the energy in the reaction coordinate is $\epsilon_t = p^{\ddagger 2}/2\,\mu^\ddagger$, its derivative is $d\epsilon_t = p^\ddagger dp^\ddagger/\mu^\ddagger$ so that Eq.(6.70) can be converted into

$$\frac{d\mathbb{N}(q^\ddagger, p^\ddagger)}{dt} = \frac{\mathbb{N} d\epsilon_t^\ddagger \displaystyle\int \cdots\cdots \int_{H=E-\epsilon_t-E_o} dq_1^\ddagger \cdots dq_{n-1}^\ddagger dp_1^\ddagger \cdots dp_{n-1}^\ddagger}{\displaystyle\int \cdots \int_{H=E} dq_1 \cdots dq_n dp_1 \cdots dp_n}. \tag{6.71}$$

This equation expresses the reaction rate (molecules per unit time) in terms of $\mathbb{N}$, the number of molecules, multiplied by the rate constant $k(E, \epsilon_t)$. The latter is expressed in terms of a ratio of the phase space areas (note that the integral is on the constant energy surface). These phase space areas can be converted into densities of states. In fact, the denominator of Eq. (6.71) is just the density of states multiplied by the factor h^n. The numerator is an integral over one less dimension, so that it is a density multiplied by h^{n-1}. Thus the rate constant, $k(E, \epsilon_t)$ becomes:

$$k(E, \epsilon_t) = \frac{\rho(E - E_o - \epsilon_t)}{h\rho(E)} \tag{6.72}$$

Equation (6.72) has a very simple and important interpretation. The rate constant is expressed in terms of the total energy, E, and ϵ_t the translational energy of the departing fragments at the transition state. As shown in figure 6.5, there are many ways for the reaction to pass through the transition state region, and these ways differ in how the available energy, $E - E_o$, is partitioned between the internal energy of the transition state and the translational energy of the fragments. Equation (6.72) is a state-to-state reaction rate constant. However, if we want to know the total dissociation rate, we must integrate over all the different translational energies in the transition state, which yields

$$k(E) = \frac{\displaystyle\int_0^{E-E_o} \rho^\ddagger(E - E_o - \epsilon_t) d\epsilon_t}{h\rho(E)} = \frac{N^\ddagger(E - E_o)}{h\rho(E)}, \tag{6.73}$$

where $N^\ddagger(E - E_o)$ is the sum of states at the transition state from 0 to $E - E_o$. This is almost the standard RRKM theory expression for the rate constant (neglecting the rotations). In the usual way of calculating densities and sums of states, the reactant and transition state symmetries are generally ignored. In order to make up for this error in $N^\ddagger(E - E_o)$ and $\rho(E)$, the rate constant in Eq. (6.73) must be multiplied by the reaction

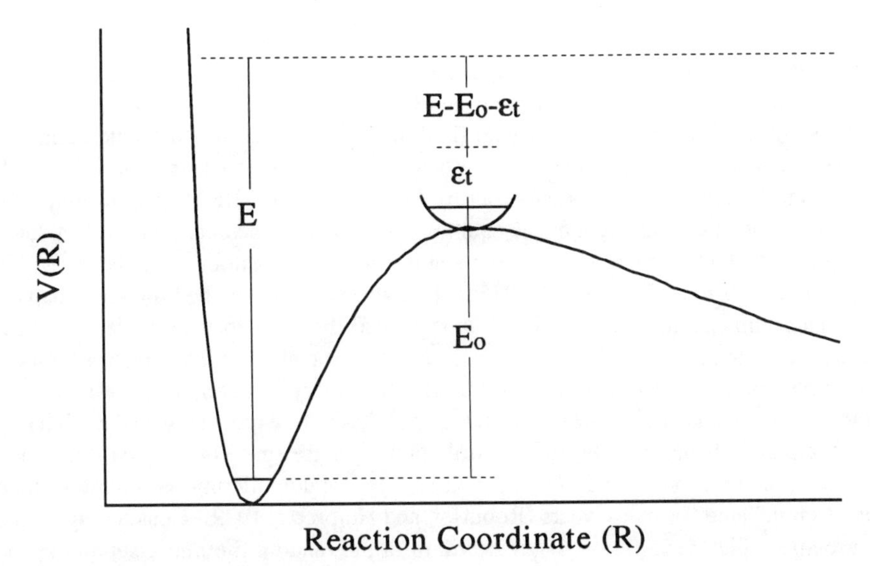

Figure 6.5 The reaction coordinate and the partitioning of the energy in the transition state ($E - E_o$) between the translational energy, ϵ_t, and the vibrational energy of the modes normal to the reaction coordinate.

symmetry. This is discussed in a later section. Until we do so, we will assume that the sums and densities are correctly calculated.

The minimum rate for a unimolecular reaction at $E = E_o$ is given by $1/h\ \rho(E_o)$. It represents the reaction rate constant when there is just one path leading over the transition state region. At the higher energy, E, the total rate is expressed as a sum of rates over all the paths, each path contributing $k' = 1/h\ \rho(E)$ to the total. Thus, the total rate is given by $k(E) = k' N^{\ddagger}(E - E_o)$.

The rate k' leading through just one path can be written in terms of its reciprocal, the mean lifetime as $\tau/\rho(E) = h$. However, the reciprocal of the density of states is just the average spacing between energy levels, so that this equation can be expressed as

$$\tau \delta E = h, \tag{6.74}$$

which has the form of the uncertainty relation. It indicates that the mean rate is such that the molecular states are always overlapped. As pointed out by Forst (1973) it means that there will always be an available molecular state for the reverse association reaction, $A + B \rightarrow AB$. Reisler and Miller have treated the unimolecular reaction in terms of overlapping resonances which become important in small molecules (Reid and Reisler, 1994; Peskin et al., 1994).

6.2.2 The Assumptions Implicit in the RRKM Theory

The RRKM equation has been derived in terms of a classical mechanical flux in phase space. It is converted into a quantum mechanical theory by simply dividing the phase space volume and surfaces by h^{n-1}, thereby converting these quantities into sums and density of states. In addtion, the molecule's zero point energy must be taken into account. This is done by referencing all energies at the zero point energy. Hence, E_o is defined as the dissociation energy at 0 K. The final RRKM equation derives its simplicity from the neglect of all dynamical aspects and from the treatment of the system in terms of only the molecular and transition state properties. The transition state is not a real state in that it has no finite lifetime. Rather, it is a saddle point and corresponds to a dividing surface in phase space, a surface which separates the products from the reactants. As pointed out by Miller (1976a), RRKM or transition state theory is not based on a chemical equilibrium between reactants and an "activated complex." In fact, if there were such a stable complex, transition state theory would not work unless modified (Miller, 1976b). This will be treated in the following chapter.

The fundamental assumption of the statistical theory is that the molecule populates all of phase space statistically throughout its dissociation so that a microcanonical ensemble is maintained. This will be true if IVR is very fast compared to the rate of reaction. It is this assumption that guarantees a single exponential decay, $\mathbb{N}(t) = \mathbb{N}(0)\exp[-k(E)t]$, because the same distribution of phase space is sampled during the course of the reaction. Rapid IVR is the basis of the random lifetime assumption which has been debated for many years (Robinson and Holbrook, 1972). Consider again the phase space picture in figure 6.4. According to the random lifetime assumption, the lifetime is independent of where in phase space the molecule happens to be located. That is, there are no preferred locations for dissociation in this phase space, locations which might shorten the lifetime.

The random lifetime assumption is perhaps most easily tested by classical trajectory calculations (Bunker, 1962; 1964; Bunker and Hase, 1973). Initial momenta and coordinates for the Hamiltonian of an excited molecule can be selected randomly, so that a microcanonical ensemble of states is selected. Solving Hamilton's equations of motion, Eq. (2.9), for an initial condition gives the time required for the system to reach the transition state. If the unimolecular dynamics of the molecule are in accord with RRKM theory, the decomposition probability of the molecule versus time, determined on the basis of many initial conditions, will be exponential with the RRKM rate constant. That is, the decay is proportional to $\exp[-k(E)t]$. The observation of such an exponential distribution of lifetimes has been identified as intrinsic RRKM behavior. If a microcanonical ensemble is not maintained during the unimolecular decomposition (i.e., IVR is slower than decomposition), the decomposition probability will be nonexponential, or exponential with a rate constant that differs from that predicted by RRKM theory. The implication of such trajectory studies to experiments and their relationship to quantum dynamics is discussed in detail in chapter 8.

The second assumption in RRKM theory is that all molecules which find themselves in the region of phase space bounded by $q^\ddagger$ and $q^\ddagger + dq^\ddagger$ and $p^\ddagger$ and $p^\ddagger + dp^\ddagger$ lead to products. Again classical trajectories provide a simple test. According to this assumption, no trajectories once having crossed the dividing surface which defines the transition state, can recross back to products. These recrossings would reduce the rate so that RRKM theory would overestimate the rate constant. Thus, in TST we assume that the trajectories pass only once through the dividing surface. Figure 6.6 shows why the transition state is a unique part of the phase space. It can be viewed as a device for determining the number of molecules that pass over to the product side. It is located at a position where the molecule, once having arrived there, has the greatest chance of continuing on to products. It is clear from figure 6.6 that if the dividing surface were located too early, for example, at TS_a, there would be many trajectories that return to

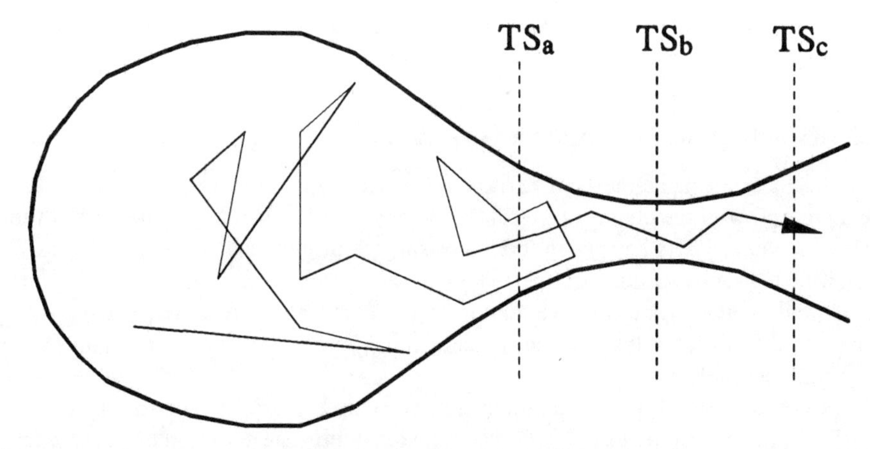

Figure 6.6 Sample trajectories on a potential energy surface. The transition state or dividing surface is located at the position of minimum flux, TS_b. It is assumed, or hoped that this represents the position where the trajectories are least likely to recross.

the left and would thus be wrongly counted. Because of microscopic reversibility, the reaction must have the same transition state in the forward and reverse reactions. For this reason transition state TS_c can be rejected because it would be subject to recrossings for the reverse recombination reaction. Operationally, the TS is not found by finding the point of minimum recrossings. Rather, the TS is located by the minimum flux criterion which in figure 6.6 is indicated by the narrowest part of the phase space, at TS_b. This should also correspond to the region where recrossings are minimized.

How good is the no recrossing assumption? Trajectory calculations on potential energy surfaces (chapter 8) have tested this effect. For instance, Garrett and Truhlar (1980) have found that the trajectories for the reaction between H and Cl_2 recross several times even at the best dividing surface for the reaction. Most such calculations have been carried out on small systems. It is expected that as the molecule increases in size, the chance for a reversal becomes rapidly smaller (Miller, 1976a). This is because the phase space volume increases away from the transition state much more rapidly in larger systems than in small molecules. In addition, recrossing will be least likely at energies close to the dissociation threshold because the phase space is very constricted and because couplings between the reaction coordinate and modes perpendicular to it are weak. Finally, reactions with large reverse activation energies will suffer less from recrossings. An interesting reaction in which recrossing of trajectories is significant is in the $Cl^-\cdots CH_3Cl$ complex (Cho et al., 1992). The interaction between the closed shell Cl^- ion and the CH_3Cl is weak so that the intermolecular modes do not couple well. As a result, the displacement of one Cl^- unit by the other is hindered.

A third assumption in RRKM theory is that the special coordinates, $dq^\ddagger$ and $dp^\ddagger$, are perpendicular to all other coordinates, and that they can therefore be separated from the other coordinates. This separability is most valid at the saddle point and at very low energies where the vibrational modes of the transition state can be treated as normal modes. However, as the energy is increased, there will be coupling between the reaction coordinate and the rest of the modes, especially for the case of a complicated reaction, for example, HCl loss from $C_2H_5Cl^+$. The separability assumption is one that is amenable to direct experimental verification, as will be seen in the following discussion.

6.2.3 Experimental Tests of the RRKM Assumptions

Most of the experimental tests of RRKM theory have supported its assumptions. At the same time, few of the claims of nonstatistical behavior have withstood the test of time. Experimental artifacts have been the major sources of the apparent nonRRKM behavior. In fact, it appears that as experiments have become more controlled and refined, the more dramatic has been the validation of the statistical assumptions. The theory has been tested from long times (msec) to short times (psec), from large molecules to the very smallest molecules.

Rigorous tests of the assumptions are not easily performed. The first assumption of RRKM theory can be tested only by single quantum state selection of the reactants. Because state selection is possible in only very small molecules, most efforts have been directed at energy selection which produces a distribution of initial states. While not ideal, this does permit testing one of the consequences of the random lifetime assumptions which is that $k(E)$ increases monotonically with increasing E.

The following case studies provide excellent examples of systems which support the statistical assumptions. These will be followed by examples in which IVR is clearly not sufficiently rapid to compete with dissociation.

6.2.3.1 The Dissociation of NO_2 and H_2CCO

Recent experiments, in which the rates were measured in the threshold region for the dissociation of $NO_2 \rightarrow NO + O$ (Brucker et al., 1992; Ionov et al., 1993) and H_2CCO (ketene) $\rightarrow H_2C + CO$ (Lovejoy et al., 1992; Chen and Moore, 1990a, b; Green et al., 1992) provide strong support for the separability assumption as well as for the legitimacy of the transition state as a real entity. Figures 6.7 and 6.8 show the $k(E)$ curves for the two reactions in the vicinity of the threshold. Clearly evident in the case of the NO_2 reaction and slightly less so in the ketene reaction are steps beginning with $k_{min} = 1/h\ \rho(E_o)$. The steps correspond to the opening up of new reaction channels as the energy of the system is increased. The new channels are ones that arise from the excitation of modes perpendicular to the reaction coordinate (fig. 6.9). At low energies, the transition state density of states in these small molecules is so sparse that each new vibrational state in the TS shows up in the $k(E)$ curve. However, after about 300–400 cm^{-1}, the density of ro-vibrational states increases so that the structure fades out.

The NO_2 dissociation rate was measured by a two-color picosecond pump-probe method in which the product NO was monitored by LIF. Of particular significance in this study is that the NO_2 density of states at the dissociation limit of 25,130.6 cm^{-1} is relatively well established from an extrapolation of experimentally determined densities at an energy of 18,500 cm^{-1}. This density (for cold samples where the rotations do not contribute significant densities) is 0.3 states per cm^{-1}, (Miyawaki et al., 1993) which leads to a minimum rate constant $1/h\ \rho(E_o) = 1 \times 10^{11}$ sec^{-1}. The experimentally measured rate increases from 0 to 1.6×10^{11} sec^{-1} at the dissociation limit. It is interesting that the subpicosecond laser pulses with their transform limited resolution of about 20 cm^{-1} do not excite individual NO_2 resonance states (see section 8.3, p. 284) but, instead, prepare a superposition of those states that are optically accessible within the laser bandwidth. It is thought that all resonance states in this bandwidth are

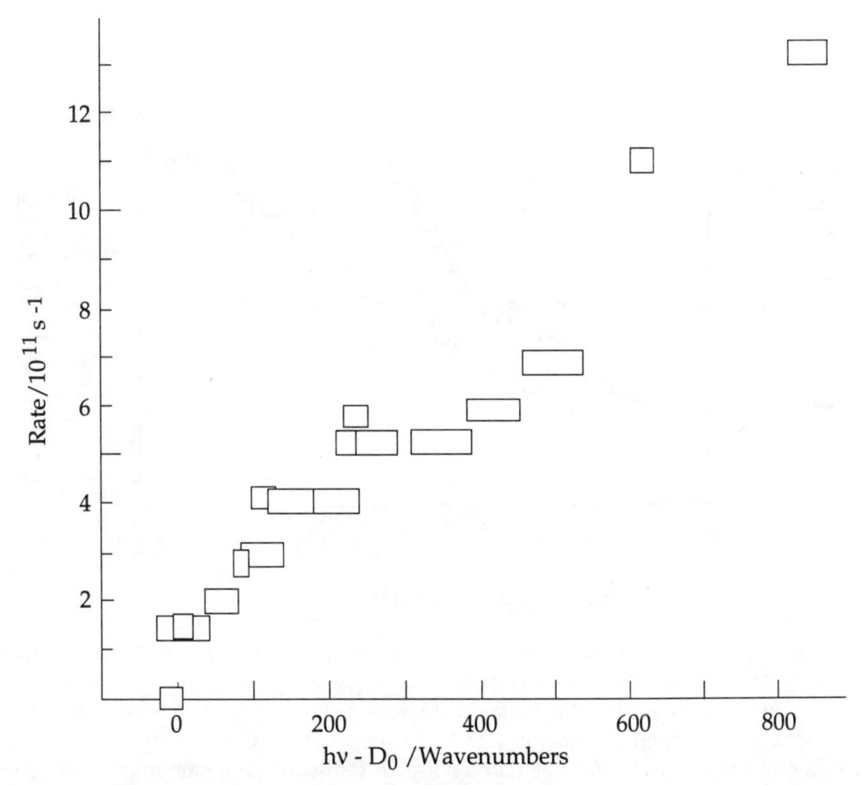

Figure 6.7 The $k(E)$ versus E curve for the NO_2 reaction at the dissociation threshold. The steps show the onset of new dissociation channels via transition state vibrational levels as the energy is increased. The rectangle lengths represent the spectral bandwidths of the pump pulse, while the heights represent the statistical uncertainties of the rate constants. Taken with permission from Ionov et al. (1993).

excited so that the number of states excited is $\rho(E)\delta E$. Coherence does not seem to play a role in the dissociation. The transition state density of states near the reaction threshold appears to be small so that structure in the form of steps is observed. These energy selected rate constants for the NO_2 dissociation are consistent with state specific studies which are discussed in section 8.3.7 (p. 297).

The NO_2 reaction has no barrier so that its TS might be expected to be very loose with free rotors. Two independent considerations suggest otherwise. First a variational transition state theory (see chapter 7) calculation (Klippenstein and Radivoyevitch, 1993) has shown that just 200 cm^{-1} above the dissociation threshold, the TS has shifted toward shorter internuclear separation so that the TS no longer has free rotors. Secondly, the rate at which $k(E)$ increases with energy indicates that the TS is moderately "tight." Hence, the structure observed in figure 6.7 is rather well resolved. This is fortunate because a truly loose TS would not have a resolvable transition state structure. While the observed structure has not been assigned, some candidates are the O atom fine structure levels at 0, 158 and 227 cm^{-1} as well as O—N—O bending nodes in the TS (Ionov et al., 1993).

The ketene dissociation to triplet CH_2 + CO was followed by a pulse probe

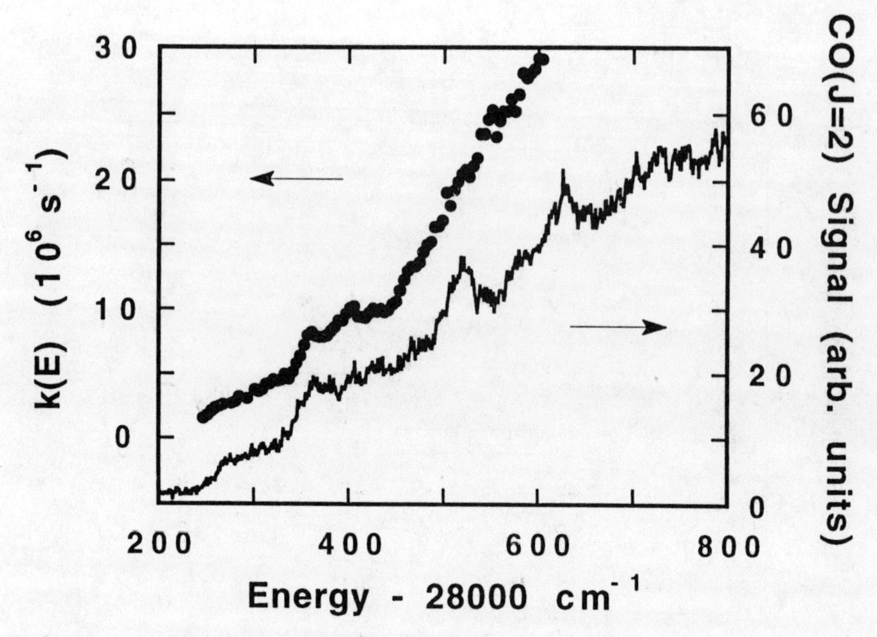

Figure 6.8 The $k(E)$ curve for the production of triplet CH_2 + CO (v = 0,J) from ketene [CH_2CO] near the dissociation threshold. The steps show the onset of new dissociation channels (via transition state vibrational levels) as the energy is increased. The lower curve is a photofragment excitation (PHOFEX) spectrum for CO (v = 0,J = 2) product spectra collected 50 nsec after the pump pulse. Taken with permission from Green et al. (1992).

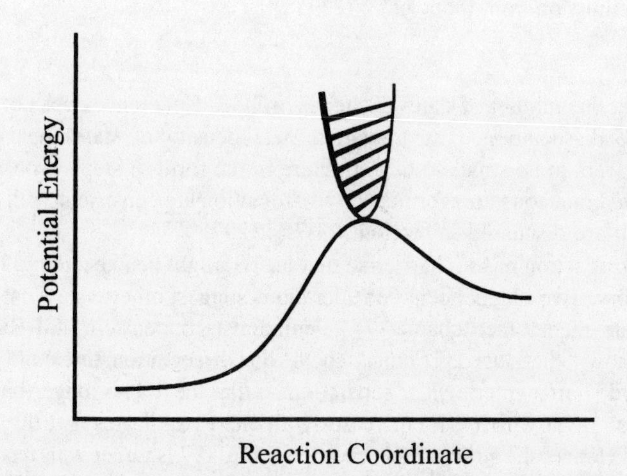

Figure 6.9 Schematic diagram of the one-dimensional reaction coordinate and the surface perpendicular to it in the region of the transition state. As the molecule's energy is increased, the number of states perpendicular to the reaction coordinate increases, thereby increasing the rate of reaction.

experiment in which the product $CO(v = 0,J)$ was monitored by LIF as a function of time after the nanosecond excitation pulse. In addition, the photofragment excitation (PHOFEX) spectra for various product channels were collected 50 nsec after the excitation pulse. The PHOFEX spectrum in Figure 6.8 is for CO in $J = 2$. The PHOFEX spectra show some of the same structure as the $k(E)$ curves. As new levels in the TS become accessible, the rate increases so that the product yield measured 50 nsec after the excitation pulse also increases. However, the $k(E)$ and PHOFEX spectra are not identical because not all transition state energy levels will populate all product states with equal probability. The $k(E)$ curve is identical only to the *sum* of all the PHOFEX spectra to the various final product states. The steps in the $k(E)$ curves figures 6.7 and 6.8 demonstrate that the reaction rate is "controlled by a flux through quantized transition state thresholds" (Lovejoy et al., 1992).

Unlike the NO_2 dissociation, the lowest dissociation limit of ketene involves a barrier of 1330 cm^{-1} and hence a real saddle point at 28,250 cm^{-1} (Lovejoy et al., 1992). According to Allen and Schaefer (1988) the ketene molecule is bent at the TS as shown in figure 6.10. The structure in figure 6.8 is ascribed to the excitation of the CH_2 wag and these bending modes of ca. 250 cm^{-1} (Green et al., 1992). Because of the saddle point, observation of structure in the ketene $k(E)$ curve is somewhat less surprising than in the NO_2 reaction. The five atoms in ketene (compared to just three in NO_2) increases the density of states which is found to be 12,000 vibrational states/cm^{-1} (Chen and Moore, 1990a). This leads to a minimum dissociation rate constant of 2.5×10^6 sec^{-1}, which is very close to the measured value of about 2×10^6 sec^{-1}. The much lower dissociation rate constant of ketene versus NO_2 is due almost entirely to the increased number of vibrational modes in ketene because the activation energies of the two reactions are very similar.

6.2.3.2 The State-Specific Dissociation of D_2CO

A frequently asked question is "how small can a molecule be and still decay statistically?" The decay rate of deuterated formaldehyde which fragments to D_2 and CO appears to answer this question. The potential energy diagram for D_2CO, based on *ab initio* molecular orbital calculations (Scuseria and Schaefer, 1989; Frisch et al., 1981; Dupuis et al., 1983) is shown in figure 6.11. The molecule is energized by laser excitation to the S_1 state, which then decays either by fluorescence back to the S_0 state or by nonradiative transition to high vibrational levels of the S_0 states. The latter dissociates via tunneling through the barrier. An important feature is the location of the S_1 state which lies just below the classical barrier for dissociation of the ground S_0 state. In the case of deuterated formaldehyde, a combination of the lower zero-point energy and the lower tunneling rates causes these rates to be small enough to permit extracting the lifetime from the broadened S_0 states. (Polik et al., 1990b). In the case of H_2CO, the dissociation rates are so fast that overlapping of S_0 states prevents measurement of individual excited state lifetimes.

Because formaldehyde is such a small molecule, the density of states in the S_1 and the high levels of the S_0 states are sufficiently small to be resolvable. In fact, the S_0 density at 28,000 cm^{-1} is so sparse (ca. 110 vibrational states per cm^{-1} of a given symmetry and 400 states/cm^{-1} of all symmetries) that most S_1 states do not overlap with vibrationally excited S_0 states. Therefore, most of the excited S_1 states decay by

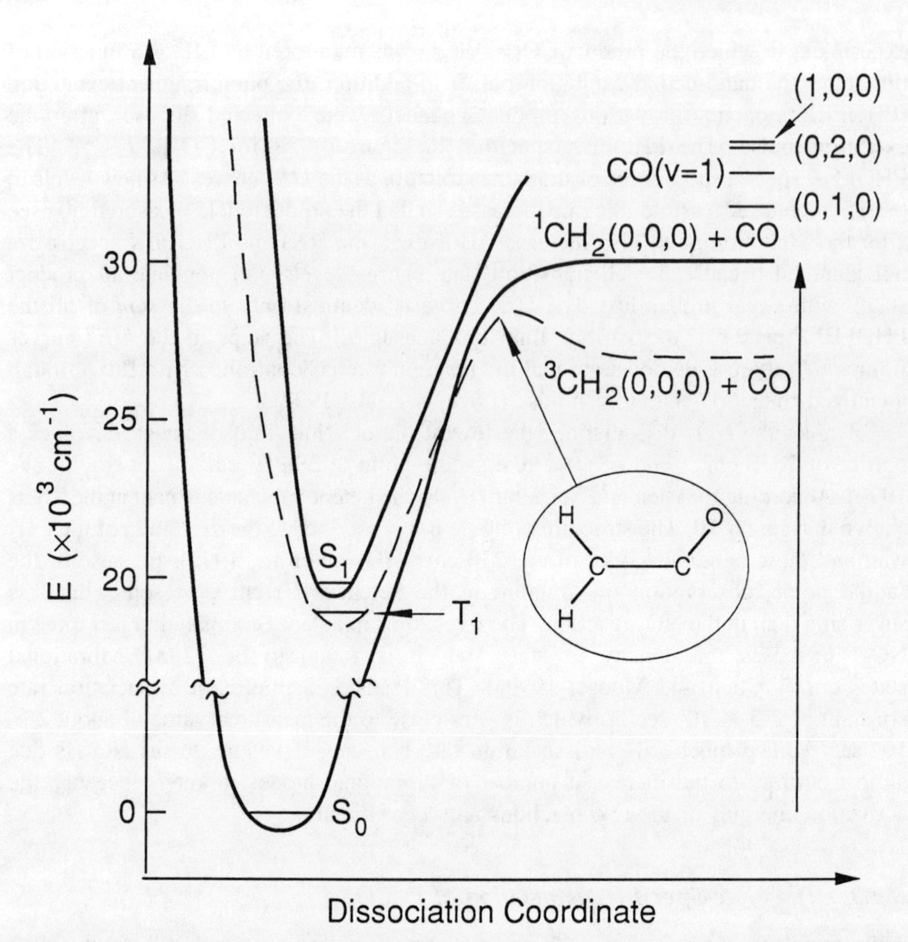

Figure 6.10 The potential energy surfaces for the ketene dissociation to both singlet and triplet products. The triplet surface contains a barrier of 1330 cm^{-1} and a bent transition state. Taken with permission from Green et al. (1992).

fluorescence. However, through the application of strong electric fields and noting that the S_1 and S_0 levels Stark shift to different degrees, it is possible to bring the levels into resonance. Polik et al. (1990b) noted that whenever the levels are in resonance, the decay rate, as measured via the fluorescence channel, increased. Furthermore, the Stark level crossing spectra showed peaks of varying widths which corresponded to the different fluorescence lifetimes.

Three rates are involved in the overall reaction: the S_1 fluorescence rate, which is constant over the small energy range investigated; the $S_1 \rightarrow S_0$ coupling, which can exceed the fluorescence rate when the levels are in resonance; and finally, the S_0 dissociation rate, which is considerably faster than the $S_1 \rightarrow S_0$ coupling rate. Through a detailed analysis of the various couplings among these rates including phase angles, it was possible to extract the S_0 dissociation rate (Dai et al., 1985; Polik et al., 1988; Miller et al., 1990; Polik et al., 1990a). This was done for hundreds of well resolved (v, J, K, and M) states, that is, individual eigenstates. The resulting decay rates in the form of line widths are shown in figure 6.12. The remarkable finding is that these

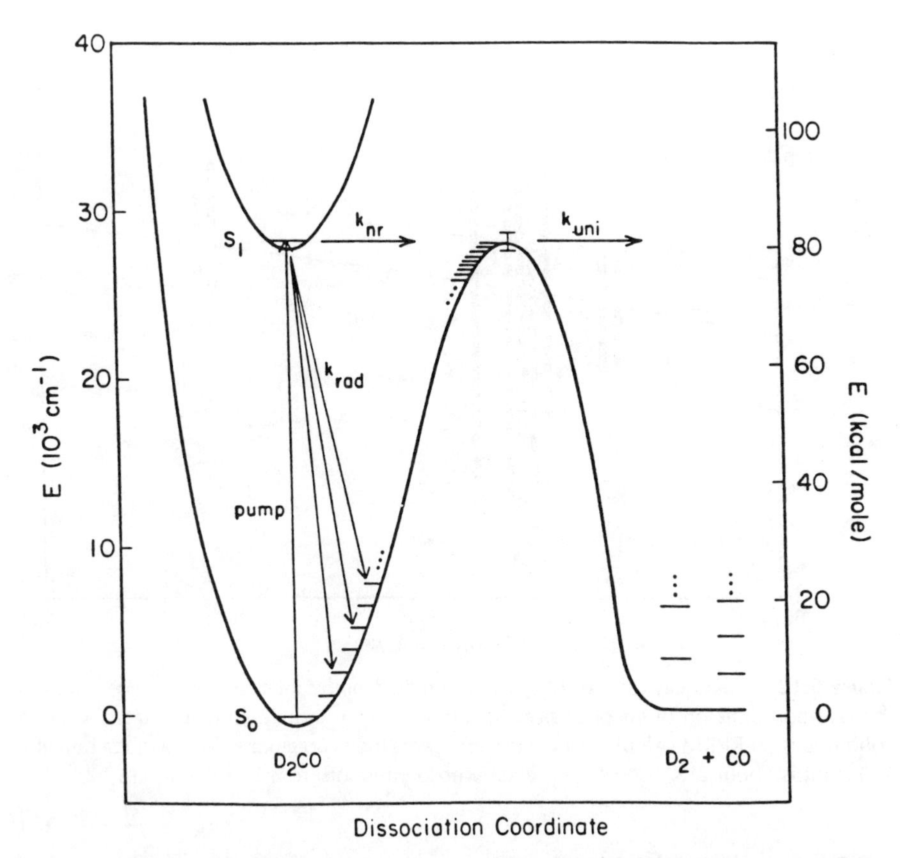

Figure 6.11 Reaction coordinate for D_2CO. After excitation of the S_1 state, D_2CO can decay either by fluorescence (k_{rad}) or by internal conversion to high vibrational levels of the S_o state (k_{nr}), followed by tunneling through the barrier to D_2 + CO products. Taken with permission from Polik et al. (1990b).

decay rates vary by at least two orders of magnitude depending upon which eigenstate is excited. This indicates that formaldehyde dissociates in a very state specific fashion. Is this evidence for non-RRKM behavior? Detailed analysis indicates precisely the opposite. These oscillations in the rate are attributed to quantum statistical fluctuations which are a necessary consequence of a complete mixing of vibrational quantum states. The source of the fluctuations is the projection of the completely and randomly mixed eigenstates onto the reaction coordinate. The explanation is based on ideas from the field of statistical spectroscopy (Porter, 1965; Zimmermann et al., 1988). As a consequence of this statistical mixing, the calculated RRKM average rates, with the inclusion of tunneling through an Eckart barrier (see section 7.6, p. 264) pass through the middle of the experimental points (solid line in fig. 6.12).

We see here then the beginnings of statistical behavior at the level of the individual quantum eigenstates. Similar results have been reported for the HO_2 dissociation (Dobbyn et al., 1995). It is only in molecules where such state-specific experiments can be carried out. In most molecules, the density of states is so great that individual quantum state excitation is not possible because they are overlapped.

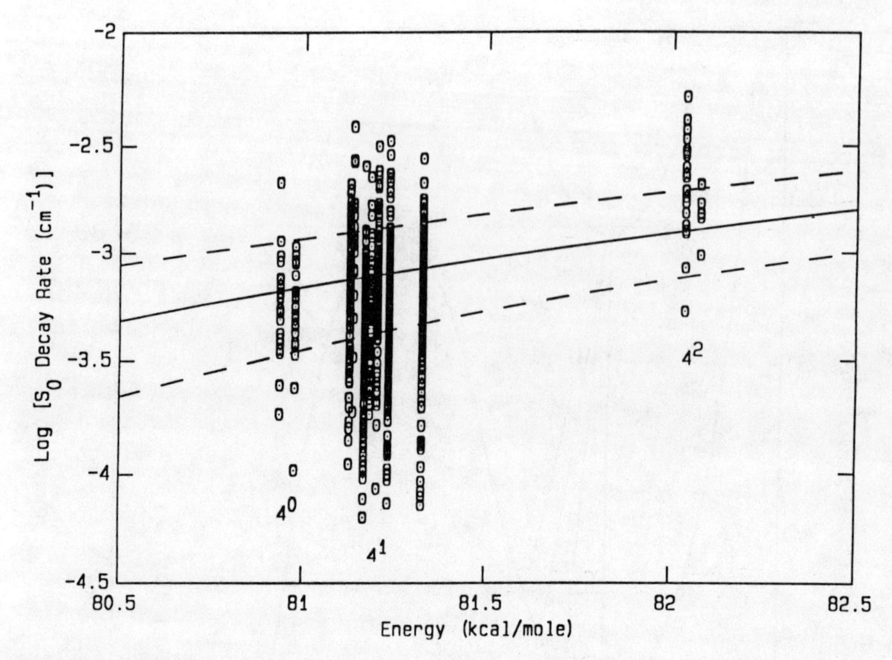

Figure 6.12 The decay rates of D_2CO as a function of the internal energy. The points are the rate constants (in terms of peak widths in cm^{-1}) for individual eigen states while the solid line is an RRKM calculation which represents the average rate. The average rate at 81 kcal/mol is about 2×10^7 sec^{-1}. Taken with permission from Polik et al. (1990b).

6.2.3.3 Experimental Evidence for Nonstatistical Behavior

The historically first example of non-RRKM behavior was one reported by Rynbrandt and Rabinovitch (1970, 1971). Even today, it represents virtually the only example of non-RRKM behavior in a normal molecule. As described in chapter 1, a molecule consisting of a cyclopropane ring attached to a CF$=$CF$_2$ group [*c*-CF_2CH_2CF-CF$=$CF$_2$] was chemically activated by the addition of a CD_2 group across the double bond, thereby producing a molecule consisting of two partially fluorinated cyclopropane rings, whose energy is above that required to open one of the three-membered rings. The initially activated ring contained a CD_2 group while the other ring contained the CH_2 group. The activated molecule could either be stabilized by collisions, or it could isomerize at either end by converting the ring into an alkene group. It was found that at short times (nanoseconds) the ring initially activated reacted more rapidly than the original ring. This shows that reaction was more rapid than IVR. At the high pressures needed to investigate these fast rates, the vast majority of the molecules were stabilized with both rings intact. It is probable that many reactions of large molecules exhibit this behavior at very short times. However, the small number of molecules that dissociate within this short time prohibits observation of these nonstatistical events. Thus the regime in which this nonstatistical behavior is observed is rather restricted.

The most dramatic evidence for nonstatistical behavior has been reported in the dissociation of loosely bound van der Waals dimers (see chapter 10) where the coupling between the high-frequency modes of the monomers and the low-frequency intermolecular modes is weak. In a dimer such as H—F···H—F, infrared radiation can be used to excite the two HF bonds which differ slightly in their vibrational frequency. The excitation energy of about 4000 cm^{-1} is sufficient to break the van der Waals bond, which is bound by just 1017 cm^{-1}. However, when the "free" HF bond is excited, the dimer lifetime was found to be 24 nsec, while the excitation of the "bonded" HF bond resulted in a lifetime of just 1 nsec (Huang et al., 1986). Many other examples of nonstatistical dissociations have been found in the infrared photodissociation of dimers such as HF-C_2H_2 (Huang and Miller, 1989); HCN-C_2H_2 (Block et al., 1989); and $(HCCCN)_2$ (Kerstel et al., 1993). In addition to the dramatic differences in the lifetimes when different parts of the molecule are excited, the dissociation rates also are orders of magnitude slower than predicted by RRKM. Nonstatistical decays of electronically excited dimers have also been reported (Semmes et al., 1987). It is apparent that dimers as a class of molecules represent the clearest evidence for non-RRKM behavior. The origin of this nonstatistical behavior lies in the large difference in vibrational frequencies between the chromophore (e.g., the H—F stretch) mode and the van der Waals modes which results in weak intermolecular couplings. Under these conditions, IVR is slow and the molecule fragments by transferring the chromophore energy directly to the van der Waals bond.

Another example of non-RRKM dissociation is that of HCO (Adamson et al., 1993; Tobiason et al., 1995), in which the sparse density of states permits excitation of individual resonances above the dissociation limit to H + CO. However, unlike HO_2 in which the resonances are intrinsically unassignable (Dobbyn et al., 1995), the HCO resonances are assignable. Furthermore, the decay rates, as determined from the absorption peak widths, are highly mode specific. The measured variations between the narrow resonances, involving mostly the C-O stretch mode, and those containing a significant C-H component, were as high as two orders of magnitude. The theoretical treatment of such resonances are discussed in chapter 8.

6.3 THE RRKM EQUATION IN THE CLASSICAL LIMIT, THE RRK THEORY

In the classical limit, the vibrational density of states for s oscillators and the sum of states for the $s - 1$ transition state oscillators are given by

$$\rho(E) = \frac{E^{s-1}}{(s-1)!\prod_{i=1}^{s} h\nu_i} \qquad N^{\ddagger}(E - E_o) = \frac{(E - E_o)^{s-1}}{(s-1)!\prod_{i=1}^{s-1} h\nu_i^{\ddagger}}, \tag{6.75}$$

where the ‡ indicates vibrational frequencies of the transition state. When these expressions are substituted into the RRKM Eq. (6.73), we obtain the classical RRKM rate constant

$$k_{cl}(E) = \left(\frac{E - E_o}{E}\right)^{s-1} \frac{\prod_{i=1}^{s} \nu_i}{\prod_{i=1}^{s-1} \nu_i^{\ddagger}}. \tag{6.76}$$

In the products over the frequencies, the numerator has one additional frequency which can be identified with the reaction coordinate. The ratio of these frequencies is a frequency itself, so that the classical RRKM equation can be written in a form that is identical to the usual RRK result:

$$k(E) = \nu \left(\frac{E - E_o}{E} \right)^{s-1} . \tag{6.77}$$

Whereas in the old RRK theory the ν was simply an adjustable parameter (Rice and Ramsperger, 1927, 1928), it can here be calculated from the vibrational frequencies of the TS and the molecule. The classical rate constant in Eq. (6.77) cannot be compared to experimentally measured rate constants because the vibrational density of states is dominated by quantum effects. On the other hand, classical RRKM rate theory is highly useful for comparing with rate constants obtained from classical trajectory calculations.

6.4 THE CANONICAL $k(T)$ RATE FROM THE MICROCANONICAL $k(E)$ RATE

The conversion of the microcanonical rate constant to a rate expression appropriate for a constant temperature system requires an averaging over the distribution of internal energies at the temperature T. Suppose that this distribution function is given by $P(E, T)$:

$$P(E, T) = \frac{\rho(E) e^{-E/k_B T}}{\int_0^\infty \rho(E) e^{-E/k_B T} dE} = \frac{\rho(E) e^{-E/k_B T}}{Q(T)} , \tag{6.78}$$

where Q is the partition function for the system. The rate constant at some temperature T is then given by

$$k(T) = \int_0^\infty k(E) P(E, T) dE = \int_0^\infty \frac{N^\ddagger(E - E_o)\rho(E)}{h\, \rho(E)\, Q(T)} e^{-E/k_B T} dE . \tag{6.79}$$

We now note that $N^\ddagger$ is nonzero only when $E > E_o$. Thus,

$$k(T) = \frac{1}{hQ(\mathrm{T})} \int_{E_o}^\infty N^\ddagger(E - E_o) e^{-E/k_B T} dE = \frac{e^{-E_o/k_B T}}{hQ(\mathrm{T})} \int_0^\infty N^\ddagger(E) e^{-E/k_B T} dE . \tag{6.80}$$

The last integral, which is the Laplace transform of the sum of states ($\mathcal{L}[N^\ddagger]$), can be converted by the integration theorem of Laplace transforms (or by integration by parts) to

$$\mathcal{L}[N^\ddagger(E)] = \mathcal{L}\left[\int_0^E \rho^\ddagger(E) dE \right] = \frac{\mathcal{L}[\rho^\ddagger(E)]}{\beta} = Q^\ddagger(T) \mathrm{k_B} T . \tag{6.81}$$

Hence, it follows that

$$k(T) = \frac{k_B T Q^\ddagger(T)}{h\, Q(T)} e^{-\mathrm{E}_o/k_B T} , \tag{6.82}$$

which is the high-pressure canonical unimolecular rate constant obtained from transition state theory. This demonstrates that RRKM theory is identical to canonical TST when averaged over the Boltzmann energy distribution. In fact, RRKM theory is simply the microcanonical version of the canonical rate constant.

Forst (1990; 1982) showed that the microcanonical RRKM equation can also be derived from canonical TST by the use of an inverse Laplace transform. If the first portion of Eq. (6.79) is written as

$$Q(T)k(T) = Q(\beta)k(\beta) = \int_0^\infty k(E)\,\rho(E)\,e^{-E/k_BT}\,dE \tag{6.83}$$

we recognize this integral as a Laplace transform of $k(E)\,\rho(E)$ whose inverse is

$$k(\mathrm{E})\,\rho(E) = \mathcal{L}^{-1}\,[Q(\beta)k(\beta)]\,. \tag{6.84}$$

The high-pressure unimolecular rate constant, $k(T)$ [or $k(\beta)$] is usually written as an Arrhenius equation:

$$k(T) = A_\infty \exp[-E_\infty/k_BT] = A_\infty \exp[-E_\infty\beta] = k(\beta), \tag{6.85}$$

where A_∞ and E_∞ are the preexponential constant and high-pressure activation energy, respectively. The latter is related to, but not identical to the 0 K activation energy, E_o. The relationship among the various energies [see Robinson and Holbrook (1972, p. 152)] is given by

$$E_\infty = E_o + k_BT + \langle E^\ddagger(T)\rangle - \langle E(T)\rangle, \tag{6.86}$$

where the terms in brackets refer to the average thermal energies of the transition state and molecule. Substitution of Eq. (6.85) into (6.84) yields

$$k(E)\,\rho(E) = A_\infty\mathcal{L}^{-1}\,[Q(\beta)\exp(-E_\infty\beta)]. \tag{6.87}$$

The exponential function in the inverse Laplace transform has the effect of simply shifting the energy so that this expression reduces to (Forst, 1972):

$$k(E) = \frac{A_\infty\rho(E - E_\infty)}{\rho(E)}\,. \tag{6.88}$$

This equation is not precisely the RRKM equation, although part of the difference is deceptive. The density of states in the numerator is based on the inversion of the partition function, $Q(\beta)$, which refers to the molecule, not the transition state. Thus, it is equivalent to the sum of states of the transition state since the latter has one less dimension.

6.5 THE REACTION DEGENERACY, σ

The RRKM equation is usually written as

$$k(E) = \frac{\sigma N^\ddagger(E - E_o)}{h\,\rho(E)}, \tag{6.89}$$

where σ is the reaction degeneracy. The σ term is explicitly included in Eq. (6.89) because we have chosen to ignore this degeneracy in calculating the sums and densities of states. The evaluation of the reaction degeneracy is one of the more confusing aspects of statistical theory calculations. Two approaches have been proposed for its evaluation. One is based on the number of equivalent paths leading from reactants to the transition state (Laidler, 1987). The other approach involves evaluating only the symmetry numbers of reactant and transition state (Pollak and Pechukas, 1978). The reaction degeneracy is then given by their ratio. With sufficient care, both methods will lead to the correct answer. However, we prefer the symmetry number approach because it addresses directly the origin of the problem, which is failure to include symmetry in the partition functions. Yet, it also must be used with considerable care. According to this method, the reaction degeneracy is given by:

$$\text{Reaction Degeneracy} = \sigma = \sigma_r/\sigma^\ddagger, \tag{6.90}$$

where σ_r and $\sigma^\ddagger$ are the symmetry numbers of the reactant and the transition state structures, respectively.

The symmetry number approach for evaluating the reaction degeneracy is based on the obvious requirement that the ratio of the reaction degeneracies for the forward and backward reactions must lead to the correct equilibrium constant. Consider first the equilibrium: $A \leftrightarrows B$. The canonical equilibrium constant at some temperature, T, is given by

$$K_{eq} = \frac{[B]}{[A]} = \frac{Q_B}{Q_A} e^{-E/k_B T}, \tag{6.91}$$

where Q_A and Q_B are the total partition functions for molecules A and B. Each is a product of the rotational and vibrational partition functions, $Q_r Q_v$. Because the rotational partition function contains the molecular symmetry in its denominator, the equilibrium constant includes the term σ_a/σ_b.

Now consider the reaction proceeding via a $TS^\ddagger$. The pair of forward and backward reactions will be

$$A \rightarrow T^\ddagger \rightarrow B \tag{6.92}$$

$$B \rightarrow T^\ddagger \rightarrow A \tag{6.93}$$

for which A, $T^\ddagger$, and B have the molecular symmetries given by: σ_a, $\sigma^\ddagger$, and σ_b, respectively. The reaction degeneracy for the forward and reverse reactions will then be given by

$$\text{forward reaction degeneracy} = \sigma_a/\sigma^\ddagger, \tag{6.94}$$

$$\text{reverse reaction degeneracy} = \sigma_b/\sigma^\ddagger. \tag{6.95}$$

It follows that the symmetry factor for the equilibrium reaction, $A \leftrightarrows B$, is given by the ratio of the forward and backward reaction degeneracy as: σ_a/σ_b, which is correct. We see then that the reaction degeneracy has its roots in the molecular and transition-state symmetry.

It is interesting to consider the relationship between the reaction degeneracy and the molecular symmetries in canonical transition state theories. In the latter, the rate constants are expressed in terms of the partition functions, including the rotational partition functions, so that the molecular symmetries are automatically included. On the other hand, in the microcanonical TST, the rotational density of states is often not part of the rate constant expression (see discussion of rotational effects in the following chapter). Thus, the reaction degeneracy must be included separately.

6.5.1 Optically Active Reactants or Transition States

The symmetry number method must be used with some care when dealing with optically active reactants or transition states (Pollak and Pechukas, 1978). Suppose that the transition state is optically active. The reaction can then be written as

$$A \rightarrow \ell\text{-}TS \quad \text{and} \quad A \rightarrow d\text{-}TS$$

where the ℓ and d prefixes indicate ℓ and d optical isomers. Since the reactants can produce either one of the optical isomers, the reaction degeneracy must be multiplied by two. This is because the reaction has two different, but energetically equivalent, transition states. This problem is also discussed by Laidler (1987) in terms of the statistical factor approach.

Another possibility is that the reactants are optically active, while the transition state is not. In this case, the relevant reactions are

$$\ell\text{-}A \rightarrow TS \quad \text{and} \quad d\text{-}A \rightarrow TS.$$

There is no problem with the forward reaction. However, it is evident that the transition state can break up into either ℓ or d enantiomers. Thus, the symmetry factor for the reaction must be multiplied by 1/2. That is, the backward reaction has a multiplicity of 2 which the forward reaction does not have.

If both the reactants and the transition state are optically active, and if the optical activity is maintained during the course of the reaction, then no extra correction is required. The reaction simply consists of two separate reactions which are treated independently.

6.5.2 Some Examples

The definition of molecular symmetry numbers can be found in most physical chemistry texts. Briefly, the symmetry number of a molecule (or the transition state) is defined as the "number of indistinguishable framework orientations obtainable by proper rotations of the nuclear framework" (Berry et al., 1980). For example, the symmetry numbers for HD, H_2, and CH_3Cl are 1, 2, and 3, respectively. Benzene has a symmetry number of 12 because it has six indistinguishable rotations about the axis perpendicular to the plane and a single C_2 rotation about an axis through two of the carbon atoms. All other rotations are linear combinations of these two. Similarly, CH_4 has a symmetry number of 12 (four C_3 rotation axes).

We begin with some simple problems, such as the H-atom loss from benzene, in which the transition state is simply benzene with an elongated C—H bond. While the benzene symmetry number is 12, the TS symmetry number is now just 2 since it has only a single C_2 rotation. Thus the symmetry factor for the H loss from benzene is 12/2 = 6; exactly as expected. However, the benzene ion is slightly different. Its ground electronic state ($^2E_{1g}$) can be considered either doubly degenerate, or Jahn Teller distorted, which produces two slightly nonsymmetric molecular ions with a very low barrier separating the two structures. If both benzene states were to correlate with the transition state, then the reaction symmetry would remain 6. However, the two electronic states correlate to two different products, namely the doublet H atom with either the singlet or the triplet phenyl ion. If only one of the benzene ion electronic states correlates to the products, as suggested by Klippenstein et al. (1993), the reaction degeneracy is reduced to 3. Determining to what extent mixing of the states takes place is not a simple task.

A similar problem arises in H atom loss from CH_4. The TS structure consists of the methane molecule with an elongated C—H bond, which gives the TS C_{3v} symmetry. Thus, the symmetry factor for the neutral reaction is 12/3 = 4, which is as expected. On the other hand, the methane ion is electronically degenerate, which must be taken into account if the methane ion states do not all correlate with the lowest energy H loss transition state.

Now consider the reaction of H_2 loss from benzene, and suppose that the TS structure involves the loss of adjacent H atoms. If the TS structure is a planar benzene ring with two adjacent H atoms leaving in the benzene plane, the TS symmetry number will be 2. Thus the reaction degeneracy for the reaction

$$C_6H_6 \rightarrow TS^{\ddagger} \rightarrow C_6H_4 + H_2 \tag{6.96}$$

$$\sigma_r = 12; \quad \sigma^{\ddagger} = 2$$

will be 6. This is true even if the experimental evidence indicates that there is extensive scrambling of the H atoms prior to H_2 loss. However, if the TS is distorted and has σ = 1, then the symmetry factor will be 12. Why should this be so? Suppose that one of the H atoms moves over toward its neighbor, thereby breaking the C_2 symmetry. There are now two energetically equivalent TS structures with the H leaning either to one side or the other so that $\sigma = 12$.

Some care must be applied in deciding whether a *TS* is symmetric or slightly distorted. It makes little physical sense to claim that the reaction degeneracy changes by a factor of two with a slight distortion of the transition state from symmetric to asymmetric. As pointed out by Gilbert and Smith (1990), the barriers to distortion must be considered. If an asymmetric TS can vibrate or internally rotate between its two equivalent structures, thereby passing through a symmetric structure, the overall symmetry number of the TS should be determined by its average symmetric structure. It is ultimately a question of the activation energy separating the species. Thus, if the H_2 leaving group from benzene can wag back and forth to either side of the plane of the benzene ring, $\sigma^{\ddagger}$ remains 2, and $\sigma = 6$.

The benzene ion also loses the linear C_3H_3 moiety to form the cyclopropenium ion, $C_3H_3^+$. The transition state clearly has no symmetry. Hence, the reaction degeneracy for this path is either 12 or 6 depending upon whether the two electronic states mix, or do not mix.

Another subtle example is the H_2 loss from cyclopentene (Pollak and Pechukas, 1978). The symmetry number for cyclopentene is 2. The most symmetric of transition states, in which the H_2 leaves in the plane of the ring, has a symmetry of 1. Thus, the symmetry factor for this reaction is 2. This is in spite of the fact that four combinations of adjacent H atoms (*cis* to each other) can be lost. However, if the H_2 leaving group is out of the plane (i.e., the departing hydrogens are *cis*) the transition state (TS) will be optically active. In that case, the symmetry factor must be multiplied by 2, thereby making it 4. As in the benzene H_2 loss reaction, the barriers separating the planar and nonplanar transition state must be considered. The correct symmetry factor may depend upon the energy of the reacting molecule.

The final example illustrates the difficulty involved in applying these rules without careful thought. Consider the two H atom loss reactions:

$$\underset{\sigma_r = 2}{Cl_2C{=}C{=}C{=}C{=}CH_2} \longrightarrow \underset{\sigma^{\ddagger} = 1}{Cl_2C{=}C{=}C{=}C{=}C{\cdots}H_2} \tag{6.97}$$

$$\underset{\sigma^{\ddagger} = 1}{ClBrC{=}C{=}C{=}C{=}CH_2} \longleftarrow \underset{\sigma_r = 1}{ClBrC{=}C{=}C{=}C{=}CH_2} \longrightarrow \underset{\sigma^{\ddagger} = 1}{ClBrC{=}C{=}C{=}C{=}CH_2} \tag{6.98}$$

The application of the rules outlined above indicate that the reaction degeneracies for these two reactions are 2 and 1 respectively. Yet, there is clearly something wrong with this conclusion. The replacement of a Cl by a bromine atom 6 Å from the reaction center can hardly change the rate by a factor of 2. Furthermore, if this reaction were treated by canonical TST including the rotational partition functions with their symmetries properly included, a similarly wrong answer would emerge. This illustrates a classic chemical dilemma in the application of symmetry arguments. The resolution is clear once it is recognized that reaction (6.98) involves the loss of hydrogens which lead to transition states (and products) that are chemically distinct, but energetically (nearly) equivalent. Thus, the total rate for reaction (6.98) is the sum of the two rates to the two different products. If the transition states are really of equal energy, we recuperate the factor 2 which was lost in the blind application of the rules. This is a case in which the reaction path degeneracy approach would not have lead us astray.

REFERENCES

Abramonitz, M., and Stegun, I.A. (1972). *Handbook of Mathematical Functions*. Dover, New York.

Adamson, G.W., Zhao, Z., and Field, R.W. (1993). *J. Mol. Spectrosc.* **160,** 11.

Allen, W.D., and Schaefer, H.F. (1988). *J. Chem. Phys.* **89,** 329.

Baer, T. (1986). *Adv. Chem. Phys.* **64,** 111.

Bauer, S.H. (1938). *J. Chem. Phys.* **6,** 403.

Bauer, S.H. (1939). *J. Chem. Phys.* **7,** 1097.

Berry, R.S., Rice, S.A., and Ross, J. (1980). *Physical Chemistry*. Wiley, New York.

Beyer, T., and Swinehart, D.R. (1973). *ACM Commun.* **16,** 379.

Block, P.A., Jucks, K.W., Pedersen, L.G., and Miller, R.E. (1989). *Chem. Phys.* **139,** 15.

Brucker, G.A., Ionov, S.I., Chen, Y., and Wittig, C. (1992). *Chem. Phys. Lett.* **194,** 301.

Bunker, D.L. (1962). *J. Chem. Phys.* **37,** 393.

Bunker, D.L. (1964). *J. Chem. Phys.* **40,** 1946.

Bunker, D.L., and Hase, W.L. (1973). *J. Chem. Phys.* **54,** 4621.

Chen, I.C., and Moore, C.B. (1990a). *J. Phys. Chem.* **94,** 263.

Chen, I.C., and Moore, C.B. (1990b). *J. Phys. Chem.* **94,** 269.

Chesnavich, W.J. (1986). *J. Chem. Phys.* **84,** 2615.

Cho, Y.J., Vande Linde, S.R., Zhu, L., and Hase, W.L. (1992). *J. Chem. Phys.* **96,** 8275.

Dai, H.L., Field, R.W., and Kinsey, J.L. (1985). *J. Chem. Phys.* **82,** 1606.

Dobbyn, A.J., Stumpf, M., Keller, H.-M., Hase, W.L., and Schinke, R. (1995). *J. Chem. Phys.* **102,** 5867.

Dupuis, M., Lester, W.A., Lengsfield, B.H., and Liu, B. (1983). *J. Chem. Phys.* **79,** 6167.

Eyring, H., Walter, J., and Kimball, E.G. (1944). *Quantum Chemistry*. Wiley, New York.

Eyring, H., Lin, S.H., and Lin, S.M. (1980). *Basic Chemical Kinetics*. Wiley, New York.

Forst, W. (1971). *Chem. Rev.* **71,** 339.

Forst, W. (1972). *J. Phys. Chem.* **76,** 342.

Forst, W. (1973). *Theory of Unimolecular Reactions*. Academic Press, New York.

Forst, W. (1982). *J. Phys. Chem.* **86,** 1771.

Forst, W. (1990). *J. Chim. Phys.* **87,** 715.

Frisch, M.J., Kirshnan, R., and Pople, J.A. (1981). *J. Phys. Chem.* **85,** 1467.

Garrett, B.C., and Truhlar, D.G. (1980). *J. Phys. Chem.* **84,** 805.

Gilbert, R.G., and Smith, S.C. (1990) *Theory of Unimolecular and Recombination Reactions*. Blackwell Scientific, Oxford.

Green, W.H., Moore, C.B., and Polik, W.F. (1992). *Ann. Rev. Phys. Chem.* **43,** 591.

Hase, W.L. (1983). *J. Chem. Ed.* **60,** 379.

Herschbach, D.R. (1959). *J. Chem. Phys.* **31,** 91.

Hirschfelder, J.O., and Wigner, E. (1939). *J. Chem. Phys.* **7,** 616.

Hoare, M.R., and Ruijgrok, Th.W. (1970). *J. Chem. Phys.* **52,** 113.

Huang, Z.S., Jucks, K.W., and Miller, R.E. (1986). *J. Chem. Phys.* **85,** 3338.

Huang, Z.S., and Miller, R.E. (1989). *J. Chem. Phys.* **90,** 1478.

Ionov, S.I., Brucker, G.A., Jaques, C., Chen, Y., and Wittig, C. (1993). *J. Chem. Phys.* **99,** 3420.

Jordan, M.J.T., Smith, S.C., and Gilbert, R.G. (1991). *J. Phys. Chem.* **95,** 8685.

Kerstel, E.R.Th., Scoles, G., and Yang, X. (1993). *J. Chem. Phys.* **99,** 876.

Klippenstein, S.J., Faulk, J.D., and Dunbar, R.C. (1993). *J. Chem. Phys.* **98,** 243.

Klippenstein, S.J., and Radivoyevitch, T. (1993). *J. Chem. Phys.* **99,** 3644.

Kubo, R. (1965) *Statistical Mechanics*. North-Holland, Amsterdam.

Laidler, K.J. (1987). *Chemical Kinetics*, 3rd ed. Harper & Row, New York.

Lovejoy, E.R., Kim, S.K., and Moore, C.B. (1992). *Science* **256,** 1541.

Lu, K.T., Eiden, G.C., and Weisshaar, J.C. (1992). *J. Phys. Chem.* **96,** 9742.

Marcus, R.A., and Rice, O.K. (1951). *J. Phys. Colloid Chem.* **55,** 894.

Miller, W.H. (1976a). *Acc. Chem. Res.* **9,** 306.

Miller, W.H. (1976b). *J. Chem. Phys.* **65,** 2216.

Miller, W.H., Hernandez, R., Moore, C.B., and Polik, W.F. (1990). *J. Chem. Phys.* **93,** 5657.

Miyawaki, J., Yamanouchi, K., and Tsuchiya, S. (1993). *J. Chem. Phys.* **99,** 254.

Moazzen-Ahmadi, N., Gush, H.P., Halpern, M., Jagannath, H., Leung, A., and Ozier, I. (1988). *J. Chem. Phys.* **88,** 563.

Moule, D.C., Bascal, H.A., Smeyers, Y.G., Clouthier, D.J., Karolcyak, J., and Nino, A. (1992). *J. Chem. Phys.* **97,** 3964.

Pechukas, P., and Light, J.C. (1965). *J. Chem. Phys.* **42,** 3281.

Peskin, U., Reisler, H., and Miller, W.H. (1994) *J. Chem. Phys.* **101,** 9672.

Pitzer, K.S. (1953). *Quantum Chemistry*. Prentice-Hall, New York.

Polik, W.F., Moore, C.B., and Miller, W.H. (1988). *J. Chem. Phys.* **89,** 3584.

Polik, W.F., Guyer, D.R., Miller, W.H., and Moore, C.B. (1990a). *J. Chem. Phys.* **92,** 3471.

Polik, W.F., Guyer, D.R., and Moore, C.B. (1990b). *J. Chem. Phys.* **92,** 3453.

Pollak, E., and Pechukas, P. (1978). *J. Am. Chem. Soc.* **100,** 2984.

Porter, C.E. (1965). *Statistical Theories of Spectra: Fluctuations*. Academic Press, New York.

Quack, M., and Troe, J. (1974). *Ber. Bunsenges. Phys. Chem.* **78,** 240.

Reid, S.A., and Reisler, H., (1994). *J. Chem. Phys.* **101,** 5683.

Rice, O.K., and Ramsperger, H.C. (1927). *J. Am. Chem. Soc.* **49,** 1617.

Rice, O.K., and Ramsperger, H.C. (1928). *J. Am. Chem. Soc.* **50,** 617.
Robinson, P.J., and Holbrook, K.A. (1972). *Unimolecular Reactions.* Wiley-Interscience, London.
Rosenstock, H.M., Wallenstein, M.B., Wahrhaftig, A.L., and Eyring, H. (1952). *Proc. Nat. Acad. Sci.* **38,** 667.
Rudolph, H., Dreizler, H., Jaeschke, A., and Wendling, P. (1967). *Z. Naturforsch.* **22A,** 940.
Rynbrandt, J.D., and Rabinovitch, B.S. (1970). *J. Phys. Chem.* **74,** 4175.
Rynbrandt, J.D., and Rabinovitch, B.S. (1971). *J. Phys. Chem.* **75,** 2164.
Scuseria, G.E., and Schaefer, H.F. (1989). *J. Chem. Phys.* **90,** 3629.
Semmes, D.H., Baskin, J.S., and Zewail, A.H. (1987). *J. Am. Chem. Soc.* **109,** 4104.
Stein, S.E., and Rabinovitch, B.S. (1973). *J. Chem. Phys.* **58,** 2438.
Stein, S.E., and Rabinovitch, B.S. (1974). *J. Chem. Phys.* **60,** 908.
Steinfeld, J.I., Francisco, J.S., and Hase, W.L. (1989). *Chemical Kinetics and Dynamics.* Prentice-Hall, Englewood Cliffs, NJ.
Tobiason, J.D., Dunlop, J.R., and Rohlfing, E.A. (1995). *J. Chem. Phys.* **104,** 1448.
Truhlar, D.G. (1991). *J. Comp. Chem.* **12,** 266.
Wigner, E. (1937). *J. Chem. Phys.* **5,** 720.
Wilson, E.B. (1940). *Chem. Rev.* **27,** 31.
Zimmermann, T., Köppel, H., Cederbaum, L.S., Persch, G., and Demtroeder, W. (1988). *Phys. Rev. Lett.* **61,** 3.

7 Applications and Extensions of Statistical Theories

7.1 APPLICATION OF THE RRKM EQUATION

7.1.1 Anharmonicity Effects

The RRKM rate constant as given by equation (6.73) in the previous chapter is expressed as a ratio of the sum of states in the transition state and the density of states in the reactant molecule. An accurate calculation of this rate constant requires that all vibrational anharmonicity and vibrational/rotational coupling be included in calculating the sum and density. The vibrational energy levels in units of wavenumbers can be represented by a power series:

$$E(v_1, v_2, \ldots) = \sum \omega_i(v_i + \tfrac{1}{2}) + \sum\sum x_{ij}(v_i + \tfrac{1}{2})(v_j + \tfrac{1}{2}) + \text{higher-order terms.} \tag{7.1}$$

The first-order anharmonicity includes the diagonal terms ($i = j$) as well as the off-diagonal terms ($i \neq j$). These can be taken into account by the Beyer-Swinehart-Stein-Rabinovitch (BSSR) direct count method (Stein and Rabinovitch, 1973). The first-order bond stretch anharmonicity which leads to dissociation is often modeled by the Morse oscillator [Eq. (6.65)]. The effect of these modes on the density of states has been handled semiclassically by Haarhoff (1963) with moderate success (Forst, 1973). In a further simplification, Troe (1977, 1979) treated the molecule in terms of m classical Morse oscillators with identical dissociation energies, and $s - m$ harmonic oscillators. The anharmonic correction for this system is $[(s - 1)/(s - 3/2)]^m$. These approximations are expected to be best when the average energy per oscillator is low. However, at higher energies, the off-diagonal, and higher-order terms in equation (7.1) become dominant. For instance, cross-terms such as bend–stretch coupling which originates from the attenuation of a bending force constant as one of the bonds, which defines the bond angle, is stretched have been found to be important at high energies (Bhuiyan and Hase, 1983; Swamy and Hase, 1986).

A convenient way to represent the RRKM rate constant with anharmonicity is as a product:

$$k(E, J) = f_{\text{anh}}(E, J)\, k_h(E, J), \tag{7.2}$$

where $k_h(E, J)$ is the rate constant calculated by assuming harmonic frequencies and separable degrees of freedom, while $f_{\text{anh}}(E, J)$ is a correction factor for anharmonicity and vibration/rotation coupling. A qualitative plot of $f_{\text{anh}}(E)$ vs. E/E_o is shown in figure 7.1. The trend with energy is easily understood in terms of the effects of

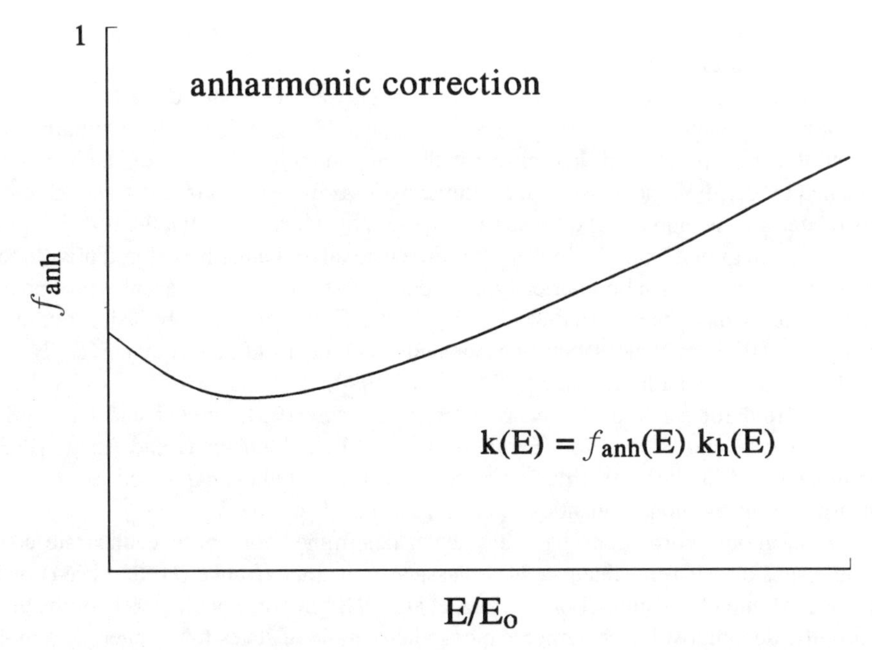

Figure 7.1 Schematic plot of $f_{anh}(E)$ variation with the energy. The rate constant $k(E) = f_{anh}\ k_h(E)$, where k_h is the rate constant determined with density and sums of states using the harmonic oscillator approximation.

anharmonicity on the numerator and denominator of the RRKM equation. At low energies, near the reaction threshold, the transition-state modes are well represented by the harmonic model, while the molecular density of states is not. Because the x_{ij} terms in equation (7.1) are generally negative (the higher energy levels are more closely spaced), anharmonicity increases the density of states in the denominator, making the correction factor much less than 1. As the energy increases, the effect becomes still greater because the anharmonicity will affect the density more than the sum of states since the latter is evaluated at a low energy. However, eventually, as the energy is increased further, the anharmonicity will reduce the sum of states as well, so that the two effects will tend to cancel. At very high energies, the correction factor may approach 1.

Only a few experimental studies have investigated the effect of anharmonicity on the density of states. To do so requires spectroscopic information up to very high energies. Such information is available only for a few small molecules such as H_2CO (McCoy et al., 1991; Reisner et al., 1984). Polik et al. (1990b) were able to measure experimentally the H_2CO density of ro-vibrational states at an excitation energy of approximately 80 kcal/mol. They found that the experimental density (ca 400 states/cm^{-1}) exceeded the harmonic density by a factor of 11. Interestingly, they found that when the density of H_2CO at 80 kcal/mol was calculated by using first-order anharmonic correction with diagonal and cross-terms from spectroscopic data at 26 kcal/mol (Reisner et al., 1984), the derived density was too low by a factor of 6. Apparently the potential energy surface must have higher-order anharmonicities which are dominant at higher energies. Another well understood molecule is acetylene, for which Abramson et al. (1985) found the ratio of anharmonic to harmonic density of states at a total energy of 27,000 cm^{-1} [ρ_a/ρ_h] to be about 6.

The bulk of the information about anharmonicity comes from classical mechanical calculations in which the phase space for a classical Hamiltonian with an anharmonic potential energy function is determined with equation (6.8). The sum of states is then obtained by dividing the phase space volume by h^n as in equation (6.9), while the density of states is the numerical derivative of the sum of states. Although the PSV integral cannot be integrated analytically for multidimensional and anharmonic potential functions, it can be evaluated numerically by Monte Carlo methods. In recent work, novel enhancements have been introduced to the Monte Carlo procedure by using multiple histograms (Bichara et al., 1988; Labastie and Whetten, 1990), Nose dynamics (Nose, 1984a,b), and adiabatic switching (Reinhardt, 1990). Calculations of this type have been reported for a number of Hamiltonians (Bunker, 1964; Bunker and Pattengill, 1968; Bhuiyan and Hase, 1983; Farantos et al., 1982; Weerasinghe and Amar, 1993; Berblinger and Schlier, 1992b). Some results of these studies, expressed as ratios of anharmonic to harmonic densities of states, are listed in table 7.1.

Anharmonic corrections have also been determined for unimolecular rate constants using classical mechanics. In a classical trajectory (Bunker, 1962, 1964) or a classical Monte Carlo simulation (Nyman et al., 1990; Schranz et al., 1991) of the unimolecular decomposition of a microcanonical ensemble of states for an energized molecule, the initial decomposition rate constant is that of RRKM theory, regardless of the molecule's intramolecular dynamics (Bunker, 1962; Bunker, 1964). This is because a

Table 7.1. Anharmonic Corrections in Small Molecules.

Molecule	E (kcal)	D_e(kcal)	$\rho_{anh}(E)/\rho_h(E)$	Method[a]	Reference
H—C—C	90	90	3–5	Phase space	(Bhuiyan & Hase, 1983)
Al_3	40	40	2.5	Phase space	(Peslherbe & Hase, 1984)
Ar_{14}	E/D_e = 10		130	Nose Dyn.	(Weerasinghe & Amar, 1993)
HCN	124	124	8	Quantum	(Wagner et al. 1992)
H_2CO	80	80	11	Experim.	(Polik et al. 1990b)
H—C≡C—H	77	127	6	Experim.	(Abramson et al. 1985)
Reaction	**E (kcal)**	**D_e(kcal)**	**$k_{anh}(E)/k_h(E)$**	**Method**	**Reference**
$CH_4 \rightarrow H + CH_3$	131.9	110.6	0.42	Traject.	(Hu & Hase, 1991)
$C_2H_5 \rightarrow H + C_2H_4$	100.0	43.5	0.20	Traject.	(Hase & Buckowski, 1982)
HCC → H + CC	46.0–60.0	43.5	0.43–0.20	Traject.	(Hase & Wolf, 1981)
$M_5 \rightarrow M_4 + M$ (1-D)	E/D_e=1.1		0.4 ± 0.2	Traject.	(Schranz et al. 1991)
$M_5 \rightarrow M_4 + M$ (1-D)	E/D_e=1.5		0.26 ± 0.05	Traject.	(Schranz et al. 1991)

[a]Phase space = density of states from classical phase space volume; Nose dynamics see Nose (1984a, 1984b); quantum = quantum mechanical calculation; experim. = spectroscopic method; traject. = classical trajectory calculations.

microcanonical ensemble exists at $t = 0$. The classical anharmonic RRKM rate constants determined from such calculations can be compared with classical harmonic values [Eq. (6.76)]. Comparisons of this type are also listed in table 7.1.

Ideally one would like quantum anharmonic corrections for the densities of states and for the unimolecular rate constants. This requires knowing anharmonic energy levels of both the molecule and the transition state, so that a direct count can be performed for the anharmonic density and sums of states. However, such calculations represent a formidable task, since as outlined in section 3 of chapter 2, a large-scale variational calculations is required. At this time such calculations are only feasible for relatively small molecules with low density of states. A relatively well studied molecule is HCN for which the eigenvalues have been enumerated in the vicinity of the isomerization barrier (Bentley et al., 1993) as well as at the dissociation limit for H loss (Wagner et al., 1992) For instance, Wagner et al. (1992) found that large angle bending motions near the dissociation limit increased the density of states by a factor of 8 relative to the harmonic density in which the rotations and vibrations were assumed to be uncoupled. In order to avoid the difficulties in determining exact anharmonic quantum densities, a semiclassical determination of the quantum density of states has been proposed (Berblinger and Schlier, 1992a,b). It is based on correcting the sum of states obtained from the classical phase space volume for zero-point energy effects in analogy with the Whitten-Rabinovitch procedure (Whitten and Rabinovitch, 1963). Comparison of the exact quantum mechanical and the semiclassical results for the H_3^+ ion suggests that this may be a viable approach to calculating accurate energy level densities for medium to large molecules (Berblinger and Schlier, 1992a).

In most molecules of moderate to large size, the anharmonicities and vibration/rotation coupling terms are not known. Thus a common approximation is the neglect of all anharmonicity and coupling terms. Such approximations become progressively more justifiable as the molecular size increases. Consider the benzene molecule with its 30 normal modes. At a total energy of 4.5 eV (36,000 cm^{-1}), the average energy content per oscillator is only 1210 cm^{-1}. Thus, on the average, only one quantum of energy is located in each oscillator. This is consistent with a low f_{anh} = 1.4 factor found by Klippenstein et al. (1993) for the benzene ion. As molecular size increases, the contribution to the total state density of configurations in which the vibrational states are more or less uniformly populated is far greater than the configurations in which energy is concentrated in a few degrees of freedom. As a result, anharmonicity tends to become less important in larger molecules, a rather convenient trend since our spectroscopic information decreases with the size of the molecule.

7.1.2 The RRKM Equation without Anharmonic Corrections and $J = 0$

Often the amount of information concerning a molecule or ion is insufficient to justify the inclusion of anharmonic terms, or even rotational effects. The RRKM equation can nevertheless be successfully employed, and it can yield relatively accurate rate constants. We begin this section with the simplest use of the RRKM equation, in which we assume harmonic frequencies and assume that $J = 0$.

This section is divided into reactions that are treated in terms of vibrator transition states and those that are treated with flexible, or rotator type transition states. Reactions with reverse activation energies have saddle points in the potential energy surface [fig.

7.2(a)]. The transition state, which is located at or near this saddle point, has a well defined set of $n - 1$ vibrational frequencies because the bending modes have not yet been converted into rotations of the products. If, as is often the case, the vibrational frequencies of such a vibrator transition state are greater than the frequencies of the molecule, the *TS* is called "tight." However, even reactions with no reverse barriers [figure 7.2(b)] can be treated with similar vibrator transition states simply by reducing some or all of the transition state vibrational frequencies. Such a *TS* is called "loose."

For some reactions with no reverse barriers it is convenient or necessary to treat the transition state as flexible. This is the case when the vibrational degrees of freedom have already evolved to free rotations. The transition state is then treated in terms of a combination of vibrational and rotational degrees of freedom, the exact number of which depends on the number of rotational degrees of freedom in the products. The flexible transition state model is equivalent to phase space theory (Pechukas and Light, 1965) when the transition state properties are the same as those of the products. Phase space theory can be viewed as a low-energy limiting form of variational transition state theory because the transition state evolves from a vibrator type at high energies to flexible at low energies.

7.1.2.1 Vibrator Transition States

7.1.2.1.a Reactions with Saddle Points and Tight Transition States. The evaluation of the RRKM equation requires a knowledge of the activation energy, E_o, the n vibrational frequencies of the molecule, and the $n - 1$ vibrational frequencies of the transition state. All, or part of these parameters can be treated as adjustable parameters when fitting experimentally determined rate constants. The molecular frequencies are often known from experiment. On the other hand, the transition-state frequencies are, in general, not known. Furthermore, they are extremely difficult to guess because they may change dramatically during the course of the reaction (Waite et al., 1983; Hase and Duchovic, 1985). Examples are shown in figures 3.6 and 3.7. In a few cases

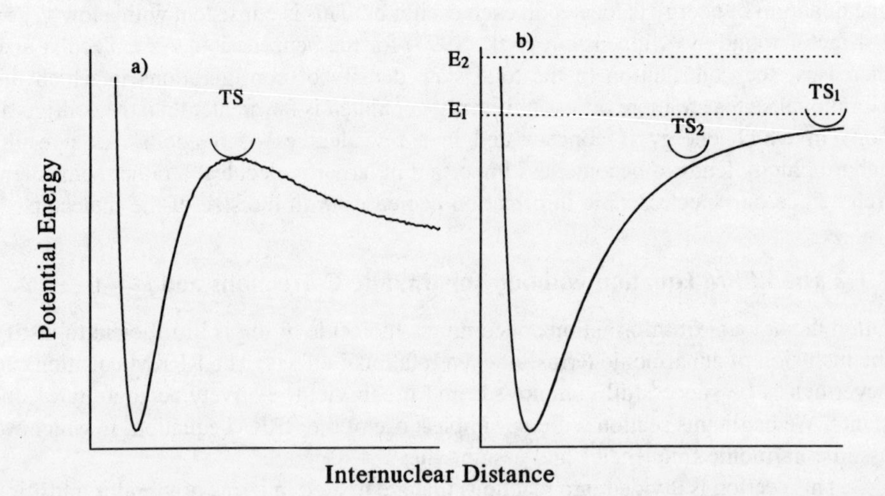

Figure 7.2 Schematic potential energy surfaces for reactions with and without a saddle point.

involving triatomic molecules, transition-state frequencies have been measured by electron detachment photoelectron spectroscopy (Waller et al., 1990; Neumark, 1992). Although this is practical for certain small molecules which have stable negative ions, such a IHI, it is not a general spectroscopic tool for most transition states.

The transition-state frequencies can be obtained by three means. They can be used simply as adjustable parameters in order to fit the data. They can be estimated by schemes such as one developed by Benson (1976) and reviewed by Gilbert and Smith (1990). Finally, they can be calculated by *ab initio* molecular orbital techniques (see chapter 3).

By far the most satisfying approach to obtaining the frequencies for reactions with saddle points is by *ab initio* molecular orbital calculations. Not only do these calculations provide the vibrational frequencies, but the structure of the *TS* and therefore its moments of inertia are also acquired. As will be discussed in the following section, the rotational constants are needed to evaluate the role of angular momentum on the reaction rates. *Ab initio* calculations are particularly valuable in the case of ions or free radicals for which experimental vibrational frequencies are usually lacking. The accuracy of the vibrational frequencies for calculations with even modest efforts are generally sufficient for RRKM calculations. If, in addition to the frequencies, the activation energy is also calculated, then it is possible to calculate from first principles the dissociation rate constants with the RRKM equation with no adjustable parameters. However, several important approximations are made including the use of harmonic frequencies.

In practice, activation energies of sufficient accuracy to be of use in RRKM calculations are extremely difficult to calculate for moderate to large molecules. This is because the reaction rate constant is highly sensitive to the assumed activation energy so that even small errors in E_o will result in major errors in the calculated rate. Even with the most sophisticated *ab initio* calculations, it is difficult to obtain energies to better than 2 kcal/mol. Hence, accurate RRKM rate calculations with no adjustable parameters are rarely possible.

The $2n - 1$ vibrational frequencies of the molecule and the transition state are not completely independent parameters. This is more evident in the canonical, thermodynamic version of transition state theory in which the rate constant is expressed in terms of an activation entropy and an activation enthalpy.

$$k(T) = \frac{k_B T}{h} e^{\Delta S^{\ddagger}/R} e^{-\Delta H_0^{\ddagger}/RT}. \tag{7.3}$$

Although the entropy is evaluated in terms of the vibrational frequencies of the reactant and the transition state, it is a single parameter. Thus, in spite of the large number of frequencies, the RRKM equation, is in first order a low-parameter theory.

In order to take advantage of the inherently small number of parameters in the RRKM equation, it is often convenient to determine the activation entropy from the vibrational frequencies. In this way, we can also obtain a quantitative measure of the degree of "tightness" or "looseness" of the transition state. The activation entropy, $\Delta S^{\ddagger}$, as determined from the canonical partition function is

$$\Delta S^{\ddagger} = k_B \ln \frac{Q^{\ddagger}}{Q} + \frac{U^{\ddagger} - U}{T} = k_B \ln \frac{\prod q_i^{\ddagger}}{\prod q_i} + \frac{U^{\ddagger} - U}{T}, \tag{7.4}$$

where the Q's are the total partition functions, the q_i's are the molecular vibrational and rotational partition functions, the U's are the thermodynamic internal energies at the temperature, T, at which the entropy is to be calculated. All of these terms can be calculated from the assumed vibrational frequencies and the moments of inertia. Tight and loose TS can then be simply defined as follows:

$$\Delta S^{\ddagger} > 0 \qquad \text{Loose TS}$$
$$\Delta S^{\ddagger} < 0 \qquad \text{Tight TS}$$

The effect on the slopes of the $k(E)$ curves for the dissociation of the bromobenzene ion with various assumed entropies, based on the frequencies in table 7.2, and energies of activation are shown in figure 7.3 (Baer et al., 1991). While the bromobenzene ion has no barrier in the dissociation channel, it is here treated with a vibrator TS. It is evident that two parameters can generate a whole family of $k(E)$ curves. Thus, if neither the activation energy nor the transition state structure is known, any set of data can be fit with RRKM theory. The lower the TS frequencies, the steeper the slope. However, if either the activation energy is known from other information, or if the frequencies are known from calculations, then the RRKM equation reduces to a one parameter model in which either the magnitude of the rate or the slope can be adjusted, but not both.

The frequencies along with their corresponding activation energies, were arbitrarily chosen so that the curves would pass through a common point on the graph. The experimentally obtained rates are consistent with an E_o of 2.76 eV and a $\Delta S^{\ddagger}$ of 8.3 cal/mol-K (Baer and Kury, 1982; Rosenstock et al., 1980). Clearly from a given data set in which $k(E)$ has been measured over some energy range, it is possible to extract both the $\Delta S^{\ddagger}$ and the activation energy. Not only does this provide a simple and quantitative means for distinguishing loose and tight transition states, it also provides a means for comparing the microcanonical rate data with canonical data in which the $\Delta S^{\ddagger}$ is extracted directly from the TST analysis of the data. The usual temperatures at which $\Delta S^{\ddagger}$ are reported are 600 and 1000 K because these are in the range of the usual thermal data.

The results of figure 7.3 might suggest that the $2n - 1$ vibrational frequencies are completely equivalent to $\Delta S^{\ddagger}$. This has indeed been suggested by Troe (1988) who claimed that for the case of the cycloheptatriene isomerization to toluene, it makes little difference whether a single frequency is adjusted (as in the above calculation), or whether all frequencies are simply multiplied by a common factor. The $k(E)$ curve is the same over ten orders of magnitude in the calculated rates. However, Troe's calculation is for a case in which the molecule and TS frequencies are rather similar and in which the different TS frequencies are also similar. We show in figure 7.4 what happens for a dissociation in which the transition state is loose and in which the

Table 7.2. The Vibrational Frequencies (and Degeneracies) Used for $k(E)$ Curves of Figure 7.3.

Molecular Ion:	3054(5) 1518(4) 1292(2) 1053(8) 821(3) 654(3) 460(1) 409(1) 315(1) 254(1) 181(1)
TS	3054(5) 1518(4) 1292(2) 1053(8) 821(3) 654(3) Y(5)

where Y is 650, 450, 332, 245, 160, 115, 92, and 40 cm^{-1} for curves 1 through 8.

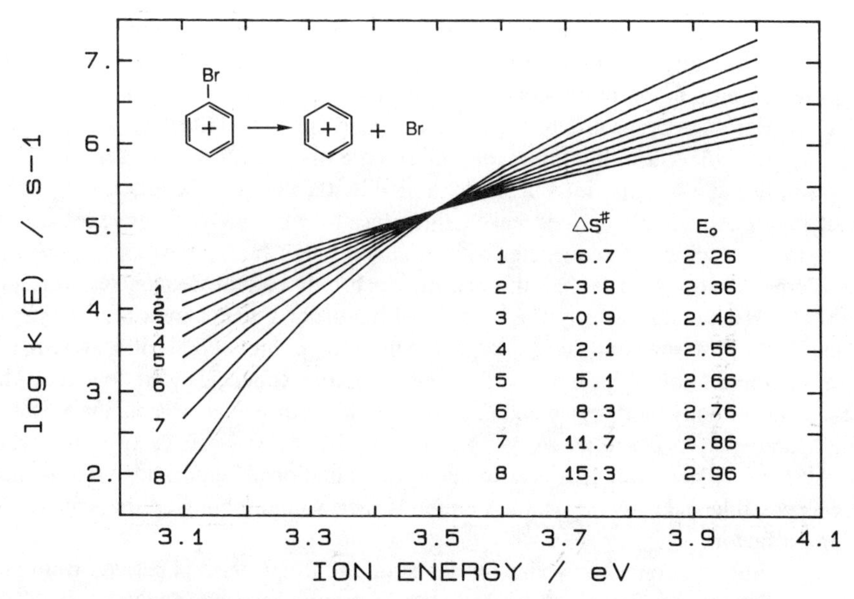

Figure 7.3 Calculated $k(E)$ curves for the bromobenzene ion dissociation in which the vibrational frequencies and the dissociation energies are varied, thereby providing $k(E)$ curves with different slopes. Taken with permission from Baer et al. (1991).

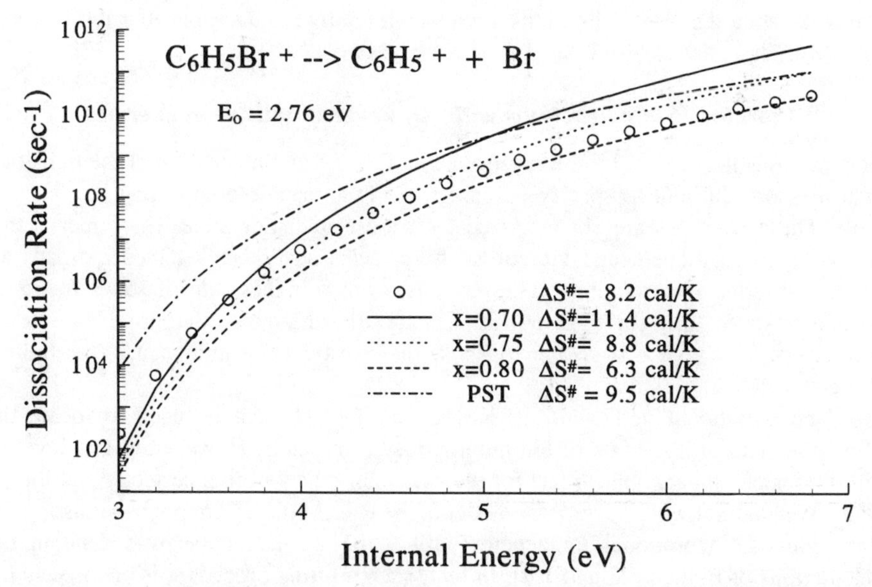

Figure 7.4 Calculated $k(E)$ curves for the bromobenzene ion. The open circles are obtained by lowering the last five vibrational frequencies to 115 cm^{-1} in the transition state. The PST (phase space theory) rate constant had the lowest two frequencies replaced by free rotors, while in the other three lines all transition state frequencies were multiplied by the indicated factor. The E_0 and the parent ion vibrational frequencies were the same for all calculations. Note the different slopes.

molecular ion and the TS frequencies are very different. We compare here the effect of lowering drastically the lowest five frequencies, and alternatively, lowering all frequencies by a common multiplicative factor. The molecular ion frequencies listed in table 7.2 were used. The circles are obtained when the lowest five frequencies are set equal to 115 cm^{-1}. The line, dots, and dashes are obtained when the TS frequencies are those of the molecular ion but all multiplied by the common factor, x, as indicated in the graph. Apparently, it does make a difference how the TS frequencies are chosen. Arbitrarily reducing all frequencies by a common factor makes the slope of the $k(E)$ curve considerably steeper than the one in which only certain frequencies were lowered. This is a quantum effect. The high C—H frequencies do not contribute very much to the density of states at these energies. Thus, the circles represent a system with fewer effective frequencies than the lines in which all frequencies were lowered. Thus the entropy of activation, evaluated at some specific intermediate temperature, is not a valid parameter to characterize completely the RRKM $k(E)$ curve over the whole energy range. Therefore, it is best to chose the vibrational frequencies with as much care as possible and to avoid arbitrary methods such as multiplying all frequencies by a common factor.

A useful additional constraint for adjusting the frequencies is derived from rates obtained for several isotopically substituted molecules or ions (Dutuit et al., 1991; Baer and Kury, 1982; Kuhlewind et al., 1987). With this approach it is possible to determine directly from the experimental rates the values of certain classes of transition state frequencies. In the fitting of the $k(E)$ data for various isotopically substituted benzene ions, it was determined that the transition state C—H frequencies are increased, while the C—C vibrations are lowered relative to the molecular ion frequencies (Kuhlewind et al., 1987).

7.1.2.2 Vibrator TS for Reactions with No Reverse Activation Energies

For reactions that do not have a reverse activation energy the location of the transition state is more difficult to specify because there is no saddle point along the reaction path. These reactions are characterized by a loose transition state which means that bonds have lengthened, and vibrational frequencies have been reduced, perhaps already converted into rotations. The result is a much higher sum of states for loose transition states than for tight transition states. Furthermore, the slope, or rate of increase of $k(E)$ with E is greater for loose than for tight transition states (see figures 7.3 and 7.4).

The variational version of RRKM theory (VTST) can be used to locate the transition state on the basis of the minimum sum of states. However, if this level of effort does not appear appropriate for the particular reaction, it is perfectly possible to fit a given data set with the vibrator model of the RRKM theory simply by adjusting the transition-state vibrational frequencies until a fit is obtained (as was done in the calculations of figures 7.3 and 7.4). In fact, such a fitting procedure is one means for determining whether the reaction is characterized by a loose or a tight transition state.

Although the vibrational frequencies can be varied arbitrarily, there is un upper limit to the $\Delta S^{\ddagger}$. The loosest of all transition states (highest $\Delta S^{\ddagger}$) is one in which the TS has all of the properties of the dissociated products. If a molecule such as ethane dissociates to two methyl radicals, a total of six vibrations are converted to five rotations and one translation (for a detailed accounting of all the ethane modes see section 7.4.3). Thus, the loosest of all transition states for the ethane dissociation is one which has five free rotors and the 12 remaining vibrations of the two CH_3 units. The maximum entropy of activation is then calculated from these frequencies and moments of inertia. This limiting RRKM rate constant for forming products with a total translational/rotational energy in the range $E_{tr} \rightarrow E_{tr} + dE_{tr}$ is given by

$$k(E, E_{tr}) = \rho_v^{\ddagger}(E - E_o - E_{tr})\, \rho_{tr}(E_{tr})/h\rho(E), \tag{7.5}$$

where ρ_v and ρ_{tr} are the product vibrational and translational/rotational density of states, respectively. $\rho(E)$ remains the density of states of the reactant molecule. Thus the total RRKM unimolecular rate constant becomes

$$k(E) = \frac{1}{h\rho(E)} \int_0^{E-E_0} \rho_v^{\ddagger}(E - E_0 - E_{tr})\, \rho_{tr}(E_{tr})\, dE_{tr}$$

$$= \frac{1}{h\rho(E)} \int_0^{E-E_0} \rho_{vr}^{\ddagger}(E - E_0 - E_t)\, dE_t. \tag{7.6}$$

Both the rotational and translational density of states can be treated classically. In the second expression, the rotational and vibrational density of states have been combined through the usual convolution of these functions. The rate constant described in equation (7.6) is similar to a phase space theory rate constant (see section 7.3) except that angular momentum is not conserved.

An example of the limiting RRKM $k(E)$ curve is shown in figure 7.4 for the bromobenzene ion dissociation. In this case, only three vibrational frequencies are lost, two going to rotations and one to translation. The two lowest frequencies were replaced by two rotors with moments of inertia equal to 80 amu-Å^2. Although the rates are about a factor of 10 too high, the slope is much closer to that of the calculation with the five lowest frequencies reduced to 115 cm^{-1}.

7.1.3 Examples

7.1.3.1 The Dissociation of the Butylbenzene Ion

An ionic reaction that exhibits both loose and tight transition states is the dissociation of *n*-butylbenzene ion:

$$C_6H_5C_4H_9^+ \rightarrow \begin{matrix} C_6H_6CH_2^+ + C_3H_6 & E_0 = 0.99 \text{ eV}, & (7.7a) \\ C_6H_5CH_2^+ + C_3H_7 & E_0 = 1.61 \text{ eV}. & (7.7b) \end{matrix}$$

These rates were measured at low energies from asymmetric PEPICO TOF distributions (Baer et al., 1988). At these energies, only the lower-energy channel [(reaction 7.7(a)] is important. At energies sufficiently high to observe the loss of C_3H_7, the dissociation rate is already too fast to be measurable by PEPICO. Thus only the *relative* rates of reactions 7.7(a) and (b) were determined. However, by extending the RRKM calculation for reaction 7.7(a) to higher energies, and using the measured branching ratios for the two rates, the rates for reaction 7.7(b) could be determined (see figure 7.5).

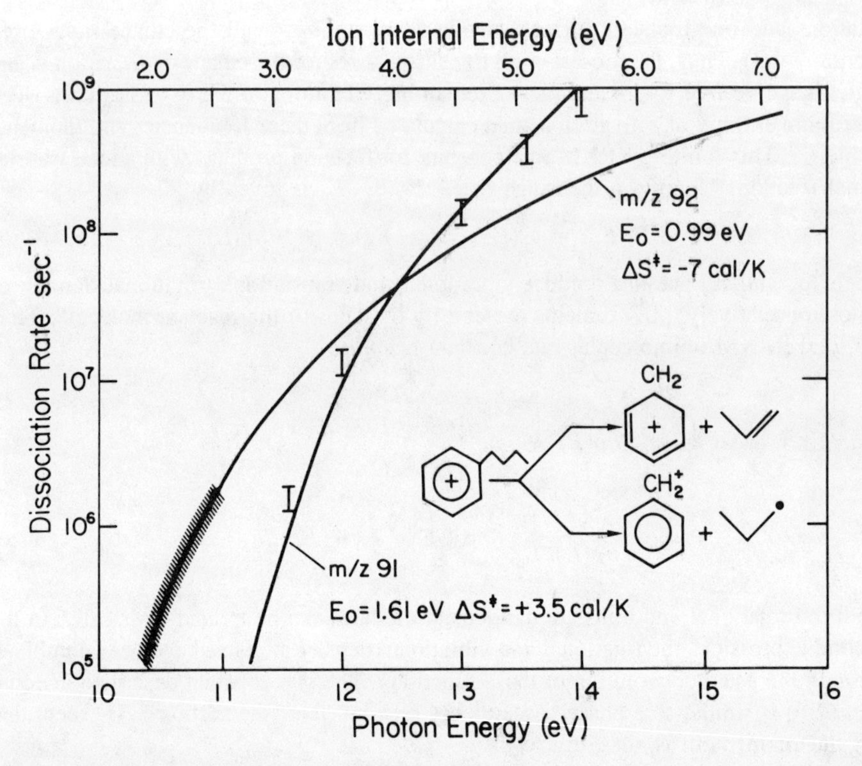

Figure 7.5 The $k(E)$ curves, experimental (shaded region and vertical lines) and RRKM calculations (lines), for the dissociation of *n*-butyl benzene ions. The two reactions proceed via "tight" and "loose" transition states. Taken with permission from Baer et al. (1988).

This is a classic reaction in which a low-energy channel involving a rearrangement, in this case an H-atom transfer from the leaving group to the benzene ring, competes with a high-energy, direct cleavage reaction. As the rates in figure 7.5 indicate, at low energies, the low-energy channel dominates. But its slope is small, so that once the direct cleavage reaction becomes energetically possible, it rapidly dominates the reaction dynamics. The reaction entropies shown in figure 7.5 are consistent with these ideas.

7.1.3.2 c-C_4H_6 (Cyclobutene) → C_4H_6 (Butadiene)

This is a prototype of a concerted ring opening reaction. The reaction simply requires the breaking of the C—C bond between the two CH_2 groups. This reaction is highly stereospecific in that 3-methylcyclobutene isomerization yields only *trans*-penta-1,3-diene (Frey et al., 1965). The reaction is exothermic by 11.2 kcal/mol (298 K values) and has an activation energy determined from the thermal reactions (Elliot and Frey, 1966) of 32.7 kcal/mol. This yields a substantial reverse barrier of 43.9 kcal/mol. The transition state, which we expect to be "tight," has a $\Delta S^‡$ at 450 K which ranges from $-.75$ to -2.3 cal/mol-K (Elliot and Frey, 1966), which translates at 1000 K to a $\Delta S^‡$ ranging from -1.8 to -4.54 cal/mol-K.

Jasinski et al. (1983) have measured the microcanonical rate of the isomerization by $v = 5$ and 6 overtone excitation of the CH modes in cyclobutene. These span a range from 13,342 to 16,602 cm^{-1}, which is well above the assumed activation energy of 11,500 cm^{-1}. Several energies in this range can be excited because of the two different CH bonds (methylenic and olefinic). The rates were derived from the slope of the (quantum yield)$^{-1}$ versus the inert gas pressure Stern-Volmer plots (see chapter 5). The isomerization rates obtained are plotted in figure 7.6.

The Jasinski et al. data were compared with RRKM calculations using two different transition-state models neither of which considered the role of rotations. The dashed line is one taken from Frey's calculation with the looser of the two transition

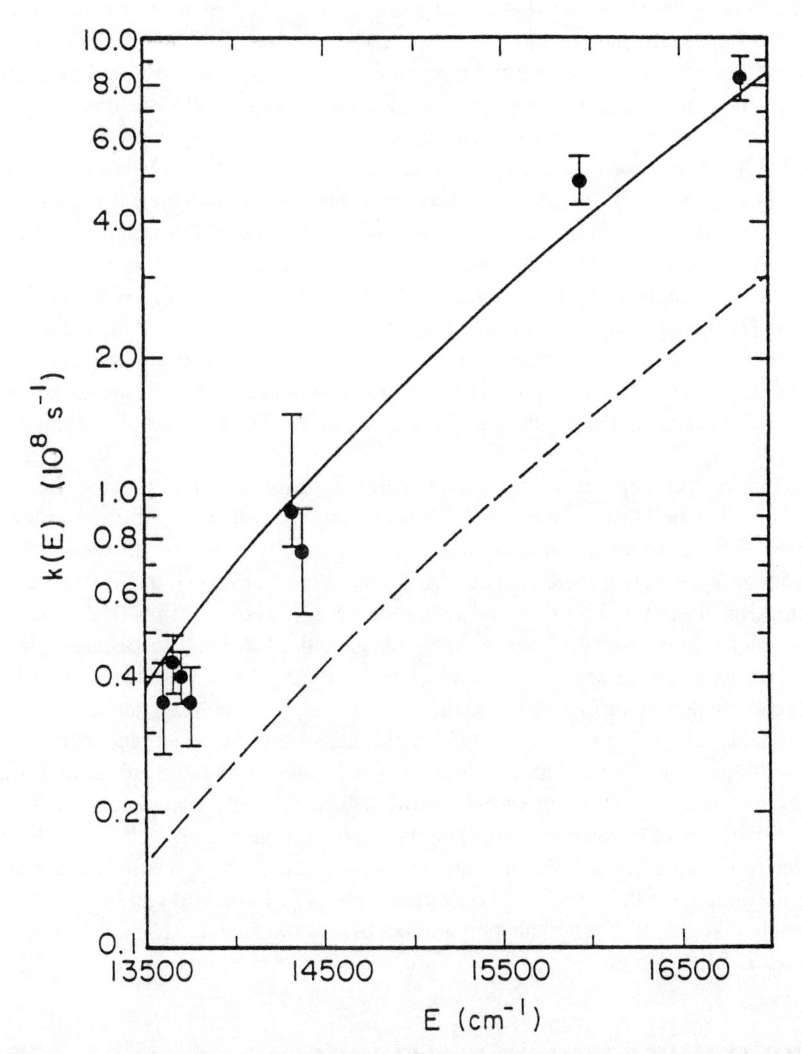

Figure 7.6 Plot of experimental and RRKM calculated $k(E)$ vs. E for the isomerization of cyclobutene to butadiene. The dashed line is an RRKM calculation using the transition state parameters of Frey et al. (1966) while the solid line is an RRKM calculation of Jasinski et al. (1983). Taken with permission from Jasinski et al. (1983).

states (Elliot and Frey, 1966). When the assumed activation energy is 11,500 cm^{-1}, the rates are too low by about a factor of 3. However, by a modest change in the TS vibrational frequencies (five TS frequencies were lowered from 660 to 450 cm^{-1}), it was possible to increase the rates sufficiently to achieve the fit shown by the solid line in the figure 7.6. The calculated entropy of activation at 1000 K for the Jasinski et al. calculation is −0.94 cal/mol-K, which is slightly looser than the Elliott-Frey model.

The assumed vibrational frequencies depend on the assumed reaction degeneracy. In the Elliott and Frey paper, this reaction degeneracy was taken as 2. However, the application of the rules outlined in chapter 6, suggests that $\sigma = 1$. The symmetry of cyclobutene is 2. According to Elliott and Frey, the reaction producing butadiene is highly stereospecific, a fact that is not evident in the cyclobutene reaction. But if a methyl group is attached at the 3 position, only *trans*-penta-1,3-diene is formed as a product. They thus propose that the transition state is a nonplanar twisted cyclobutene in which the single bond opposite the double bond is in the process of breaking. In the case of cyclobutene, such a twisted structure still has a symmetry factor of 2. Hence, $\sigma = 2/2 = 1$. In the case of 3-methylcyclobutene, neither the molecule nor the transition state has any symmetry, so that the reaction degeneracy in this case is also 1.

The solid line in figure 7.6 can be obtained with the use of a reaction degeneracy of 1, the assumption of adiabatic rotations, an activation energy of 11,050 cm^{-1}, and the following molecular frequencies: 2989(6), 1477(3), 1238(4), 1067(4), 842(4), 637(2), 325(1); and transition-state frequencies: 3300(6), 1320(12), 400(5). The degeneracy of these grouped frequencies are given in parenthesis. The TS frequencies were taken from previous studies with only the lowest one adjusted to fit the experimental results. The calculated reaction rate at the onset of 11,050 cm^{-1}, is determined to be 3.7×10^5 sec^{-1}.

This reaction appears to be a good candidate for further study because the system is sufficiently small that good *ab initio* MO calculations could be performed on both the reactant molecule and the transition state. This would yield the 23 transition-state vibrational frequencies, thereby reducing the number of unknown parameters required for the RRKM calculation to just the activation energy. The experimental data could be improved by more precise energy selection gained from the use of supersonically cooled samples. In the analysis of Jasinski et al. (1983), the average thermal energy of the cyclobutene was simply added to the photon energy. However, such a procedure is questionable in the case of bound-to-bound transitions in which the narrow laser bandwidth excites specific initial to specific final states. Another need is for a greater variety in the cyclobutene energies. Unfortunately, the only one-photon method for state preparation of cyclobutene appears to be via overtone excitation because there are no electronic states in the vicinity of the dissociation limit. Because the density of overtone states is rather small, only a limited number of energies can be selected. An approach that might permit the preparation of cyclobutene in a greater number of energies is SEP (see chapter 5).

7.2 THE ROLE OF ROTATIONS IN RRKM THEORY

The role of rotations, which has so far been ignored, affects the unimolecular rate in two ways. First, an energy barrier resulting from the conservation of angular momentum appears in the exit channel of the dissociation. This rotational energy is not constant during the course of reaction because it depends on the moments of inertia and therefore the evolving geometry. For reactions with "tight" transition states and reverse activation energies, the rotational energy in the transition state may be larger or smaller than the rotational energy in the molecule. For reactions with no reverse activation energy in the $J = 0$ potential energy surface, rotational energy gives rise to a centrifugal barrier along the reaction path. This barrier tends to be much smaller than the molecule's rotational energy because it is determined by moments of inertia at a large internuclear separation.

A second rotational effect comes into play when rotations are strongly coupled to the vibrations, via, for instance, coriolis interactions. In that case, the projection of the principle rotational quantum number, the K quantum number in symmetric top molecules, is no longer conserved. The energy associated with this quantum number then gets mixed in with the molecule's vibrational energy, thereby increasing the density and sums of states. When this happens we say that the K-rotor is "active." If the K-rotor does not couple with the vibrations, it is "inactive." We first discuss what happens when a diatom dissociates and follow that with the dissociation of polyatomic molecules.

7.2.1 Rotational Effects in the Dissociation of a Diatom

The dissociation of a diatom differs from that of polyatomic molecules in two ways. First, the dissociation has no real barrier and is given by a simple one-dimensional interaction potential which is often expressed as a Lennard-Jones or a Morse potential. Secondly, the product atoms have no angular momentum. Thus, the initial angular momentum of the diatom is converted exclusively into orbital angular momentum (which is relative translational energy) of the products. For this reason, the diatom dissociation is not an appropriate model for the much more complex dissociation of polyatomic molecules. Nevertheless, there are certain features that carry over.

The long range attractive potential for a diatomic dissociation is given by $V(r) = r^{-n}$, where n is the interaction parameter. Typically $n = 6$ for a neutral molecule (recall the Lennard-Jones 6-12 potential). On the other hand, for an ionic dissociation into an ion and a polarizable neutral, $n = 4$. The Hamiltonian for such a two-body central force system can be expressed in the center of mass as a one-body problem with reduced mass μ and relative velocity v. In polar coordinates, this takes the form [(Eq. 2.15)]

$$H = \frac{1}{2\mu}\left(p_r^2 + \frac{p_\theta^2}{r^2}\right) + V(r), \tag{7.8}$$

where p_r and p_θ are the linear and angular momenta, respectively. We can obtain the equations of motion from the four Hamilton's equations [(Eq. (2.9)] involving the four variables p_r, p_θ, r, and θ, which yields

$$dr/dt = p_r/\mu, \tag{7.9a}$$

$$d\theta/dt = p_\theta/\mu r^2, \tag{7.9b}$$

$$dp_r/dt = -dV(r)/dr + p_\theta^2/\mu r^3, \tag{7.9c}$$

$$dp_\theta/dt = 0. \tag{7.9d}$$

The first equation is just a redundancy stating that the velocity is the velocity. The second equation gives the angular frequency in terms of the angular momentum. The third equation is the radial equation of motion which contains the physically interesting part of the dynamics, namely the interaction potential. Finally, the fourth equation states that the angular momentum, p_θ, does not vary with time, that is, it is a conserved quantity. When integrated, it yields $p_\theta = C$, a constant.

The angular momentum can be expressed (for our diatom) in terms of the rotational quantum number, J, as $p_\theta = \sqrt{J(J + J)}\hbar$. As the atoms separate, this angular momentum is converted into orbital angular momentum which is given by $\mu v b$ (see figure 7.7), where b is the impact parameter for the collision between the two atoms (Johnston, 1966). The angular momentum is thus converted from a highly quantized into a nearly continuous quantity. Because p_θ is a constant of the motion, the second term in equation (7.8) depends only upon r so that it can be incorporated into the $V(r)$ term. This effective potential, given by

$$V_{\text{eff}} = \frac{p_\theta^2}{2\mu r^2} + V(r) \tag{7.10}$$

is the sum of the centrifugal potential (often called the fictitious potential) and the real potential. The centrifugal potential is always positive and monotonically vanishes at large r.

The manner in which the rotational energy is converted is readily apparent when the effective potential $V_{\text{eff}}(r, p_\theta)$ is plotted for several Lennard-Jones interaction potentials. For this purpose, we plot the function

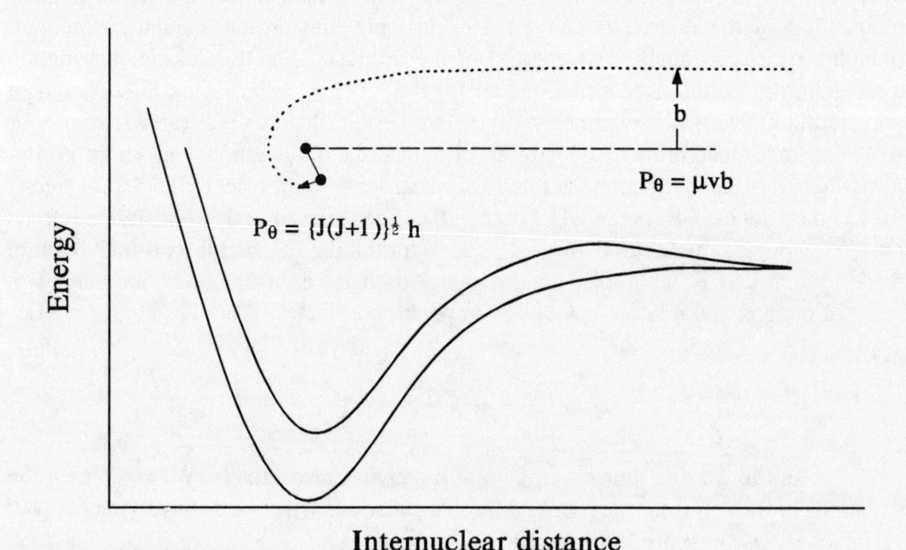

Figure 7.7 The effect of rotational energy on the effective potential for a diatomic molecule. The centrifugal barrier is related to the impact parameter for the reverse association reaction. The upper part of the figure shows the dissociation of a rotating diatom (in terms of its reduced mass, μ) and the conversion of its angular momentum into orbital angular momentum.

$$V_{\text{eff}} = 4\epsilon\left[\left(\frac{\sigma}{r}\right)^{12} - \left(\frac{\sigma}{r}\right)^{n}\right] + \frac{p_\theta^2}{2\mu r^2} \tag{7.11}$$

in figure 7.8 for three values of the interaction potential, n in $V(r) = r^{-n}$. The σ term is the internuclear separation when the potential (in the absence of rotations) crosses the $V(r) = 0$ axis. The minimum in this rotationless potential energy function is at $-\epsilon$, and the value of r at this minimum energy is $2^{1/6}\sigma$. The angular momentum introduces a barrier in the exit channel, the magnitude and position of which is a strong function of n. It is used in phase space theory for locating the position of the transition state for polyatomic reactions that have no inherent barrier for dissociation. The barrier can also be used to determine the collision cross section for the reverse recombination reaction (Steinfeld et al., 1989).

The position and magnitude of the centrifugal barrier can be determined by setting the derivative of V_{eff} with respect to r equal to 0 and solving for r. Substituting this value of r into the equation for V_{eff}, yields the barrier height. Because the repulsive part of the potential is of importance only at short r, this term is usually ignored when determining the centrifugal barrier. Thus, for the potential of the form

$$V_{\text{eff}} = \frac{p_\theta^2}{2\mu r^2} - \frac{a}{r^n} \tag{7.12}$$

the barrier location and size $V_{\max}$ are given in terms of the angular momentum by

$$r_c = \left(\frac{n\mu a}{p_\theta^2}\right)^{1/(n-2)} \tag{7.13a}$$

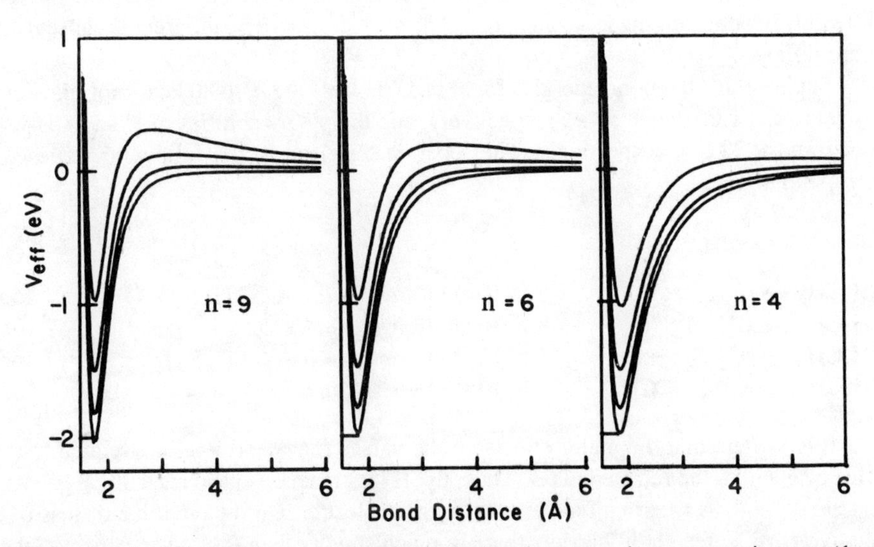

Figure 7.8 The effect of the interaction potential and rotational energy on the centrifugal potential for a diatomic dissociation. The four effective potentials are for $E_r = 0$, 0.2, 0.5, and 1.0 eV and the assumed reduced mass is 14 amu. Taken with permission from Baer (1986).

and

$$V_{max} = \left(\frac{p_\theta^2}{\mu}\right)^{n/(n-2)} \frac{1}{(na)^{2/(n-2)}} \left(\frac{1}{2} - \frac{1}{n}\right). \tag{7.13b}$$

Two important cases are the dissociation of an ion into a fragment ion and a polarizable neutral, and the dissociation of a neutral molecule into two polarizable fragments. The parameters for these two situations are

Reaction	n	a	
$AB^+ \rightarrow A^+ + B$	4	$\alpha q^2/2$,	(7.14a)
$AB \rightarrow A + B$	6	$2\,\epsilon\, r_e^6$,	(7.14b)

where α is the polarizability of fragment B in units of cm^3, q is the charge of 4.0803×10^{-10} esu [g$^{1/2}$cm$^{3/2}$/s], ϵ is the bond energy in any convenient energy units, and r_e is the equilibrium bond distance in same units as r.

It is interesting to compare the centrifugal barrier for the loss of H and CH_3 from ethane neutrals and ethane ions. For this purpose, we treat ethane as a diatomic molecule with a reduced mass of 7.5 amu. The ethane neutral and ion can be assumed to have the same moment of inertia which is approximately $I = \mu r^2 = 3 \times 10^{-46}$ kg $\cdot$ m^2. Thus at room temperature the assumed two-dimensional rotor with its average rotational energy of RT = 207 cm^{-1} will have an average J of 15 and an angular momentum of 1.5×10^{-33} J $\cdot$ sec. As the molecule or ion dissociates to 2 CH_3 or C_2H_5 + H, the reduced mass associated with methyl loss remains the same at 7.5 amu, whereas it will reduce to about 1 amu for the H-atom loss. Because the reduced mass plays a role in the centrifugal barrier, the barriers for H and CH_3 loss will be very different. In addition, the ionic and neutral dissociations will be different because n and a are different.

Suppose that the bond energies for H and CH_3 loss are both 80 kcal/mol, the bond distances are 1.08 and 1.54 Å, respectively, and the polarizabilities of H and CH_3 are 0.667 and 2.22 Å^3, respectively. These parameters lead to the following values for a, r_c, and V_{max}:

Reaction	a	r_c	V_{max}	
$CH_3CH_3 \rightarrow C_2H_5 + H$	1.76×10^{-78} J $\cdot$ m^6	2.9 Å	271.2 cm^{-1},	(7.15a)
$CH_3CH_3 \rightarrow CH_3 + CH_3$	1.48×10^{-77} J $\cdot$ m^6	8.4 Å	4.3 cm^{-1},	(7.15b)
$CH_3CH_3^+ \rightarrow C_2H_5^+ + H$	5.55×10^{-59} J $\cdot$ m^4	4.0 Å	111.5 cm^{-1},	(7.15c)
$CH_3CH_3^+ \rightarrow CH_3^+ + CH_3$	1.86×10^{-58} J $\cdot$ m^4	20.2 Å	0.56 cm^{-1}.	(7.15d)

It is evident that the major effects on the values for r_c and V_{max} are related to the value of n and to the reduced mass. Thus, the H loss reactions have centrifugal barriers that are as high as the rotational energy in the molecule. On the other hand, loss of a massive particle results in barriers that are much smaller than the molecular rotational energy. While this model drastically simplifies the reaction by ignoring the rotations of the products, the uniqueness of the H loss reaction demonstrated by these results is accurate.

7.2.2 Angular Momentum in Polyatomic Systems

7.2.2.1 Angular Momentum in Vibrator Transition States—Energy Effects

Polyatomic molecules differ from diatoms in that the departing fragments can themselves rotate so that the angular momentum can be conserved in many different ways. Secondly, the dissociation of polyatomic species may involve a complex series of rearrangements in which the transition state may have a structure that is very different from the molecule so that its rotational constants may differ as well. If the transition state has a real barrier and is described solely in terms of vibrational oscillators (plus perhaps one or two internal rotors), that is, a vibrator transition state, angular momentum conservation results in a much larger rotational barrier than the previously discussed centrifugal barrier in the diatom dissociation.

Many nonlinear molecules can be treated as symmetric top rotors in which two of the moments of inertia are equal. The moment of inertia about the symmetry axis is I_z, while the two other moments of inertia are $I_x = I_y$. A symmetric top can be visualized as a rotating cylinder. For a given J, the cylinder can rotate in a total of $2J + 1$ orientations, each with a different K quantum number which determines its projection along the symmetry axis. Figure 7.9 shows the case of prolate and oblate tops rotating with $K \approx J$ and $K = 0$.

If we define $A = \hbar^2/2I_z$, and $B = \hbar^2/2I_x$, the rotational energy, given in terms of the J and K quantum numbers, is given by

$$E_r(J, K) = B\,J(J+1) + (A - B)K^2 \qquad J = 0, 1, 2, \ldots \quad K = 0, \pm 1, \pm 2, \ldots, \pm J \tag{7.16}$$

If $I_z < I_x$ the molecule is a prolate top, while if $I_z > I_x$, the molecule is an oblate top. It is clear that for prolate tops, such as ethane or formaldehyde for which $A > B$, that the rotational energy increases as K increases. On the other hand, in oblate tops, such as benzene for which $A < B$ the rotational energy decreases with K. (Note that the above definitions of A and B are not those in most spectroscopy texts.)

If the molecule is a near symmetric top, it can be converted to the above form by replacing I_x by the average of I_x and I_y (Townes and Schalow, 1955). The treatment of asymmetric rotors is considerably more complex and will not be discussed in this book.

The effect of rotations on the activation energy of the reaction can be determined only if the geometry of the transition state is known. As previously pointed out, this is possible if the TS is located at a saddle point so that *ab initio*, or semi-empirical molecular orbital, or molecular mechanics calculations can be used to generate transition state geometries (Hehre et al., 1986; Minkin et al., 1990). Of these three, the *ab initio* MO calculations are much more reliable for transition state structures because semi-empirical and molecular mechanics programs are parametrized by equilibrium geometries. In the absence of such calculations, it may be possible to use chemical intuition to deduce the TS geometry. However, the results will be far less reliable. If the molecular and TS moments of inertia are known, then the RRKM rate constant can be expressed in terms of the total energy, $E = E_v + E_r(J, K)$, and the rotational quantum numbers as

$$k(E, J, K) = \frac{N^{\ddagger}[E - E_0 - E_r^{\ddagger}(J, K)]}{h\rho[E - E_r(J, K)]}, \tag{7.17}$$

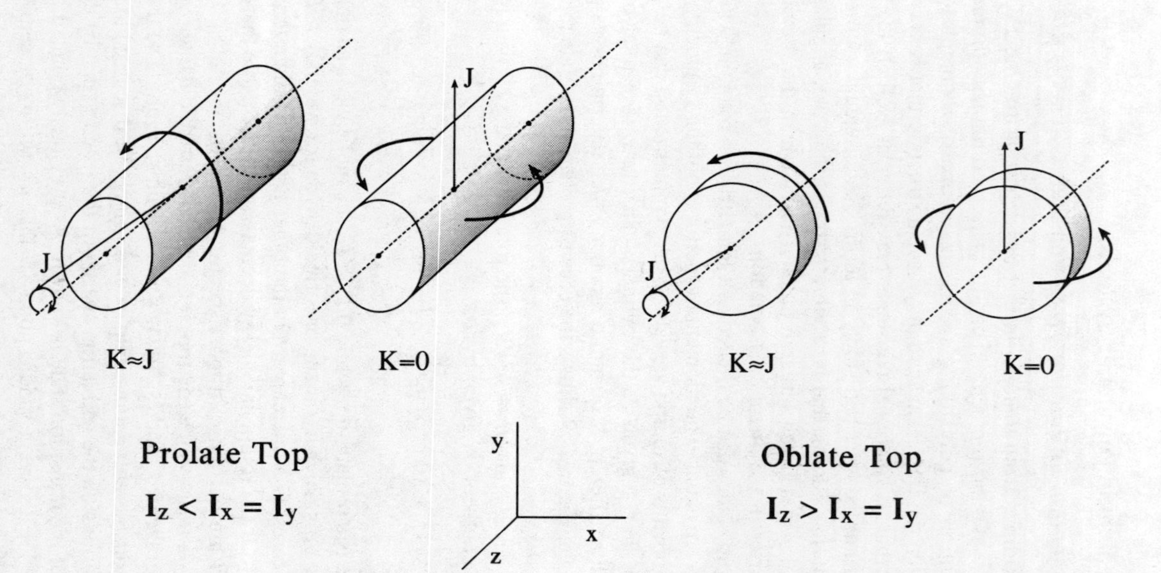

Figure 7.9 Prolate and oblate symmetric top molecules represented as cylinders. The rotational motion due to *J* and *K* are indicated.

where the E_r and $E_r^\ddagger$ are obtained from equation (7.16). Equation (7.17) takes into account only the effect of angular momentum on the energetics. It assumes that the rotations are adiabatic or nonactive. Figure 7.10 shows the effect of rotational energy on the activation energy. Whether angular momentum raises or lowers the dissociation rate constant depends upon the rotational constants of the molecule and the transition state as well as the total energy. If the rotational constants are approximately the same, the activation energy remains independent of *J*, a case characteristic of tight transition states. For a system at constant energy and near the dissociation threshold, an increase in *J* lowers the dissociation rate because it affects the bottleneck at the transition state more than it does the density of states in the molecule. However, if the transition state has significantly lower rotational constants than the molecule (loose transition states) an increase in *J* will not affect the transition state sum of states as much as it does the density of states. The latter will decrease, thereby raising the dissociation rate constant.

Frequently, ionic dissociations with no reverse activation energy have very loose transition states with a large separation (r_c) of the products [see Eq. (7.15)]. The large moment of inertia results in a rotational energy in the transition state that is very small or nearly negligible. This has been tested in dissociative photoionization experiments carried out with variable temperature samples. In the case of CH_4 (McCulloh and Dibeler, 1976) and C_2H_2 (Dibeler and Walker, 1973), the photoionization onsets were shifted to higher energy as the sample was cooled from room to liquid nitrogen temperatures. The magnitude of the shift could only be explained if the rotational energy associated with at least two rotational degrees of freedom became available for dissociation. That is, there is no rotational barrier in the transition state. More recently,

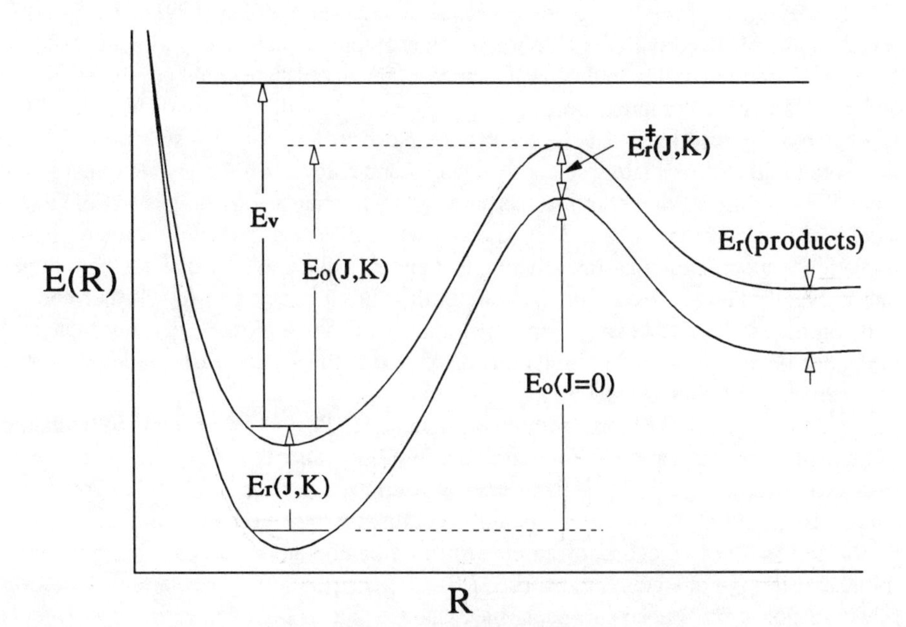

Figure 7.10 The potential energy diagram for a reaction with a saddle point and a "tight" transition state. Note that the rotational energy in the transition state is not equal to that in the molecule.

PEPICO experiments have tested this with room temperature and supersonically cooled isobutane. Again, the shift of 0.123 eV in the dissociation onset is equal to the sum of the vibrational and total rotational energy (Weitzel et al., 1991). This shift in the onset upon cooling has led Weitzel (1991) to suggest its use as a simple test to determine if the transition state is loose or tight. If it is loose, the shift should include the vibrational as well as the rotational energy. If it is tight, the shift should include only the vibrational energy.

Equation (7.17) is appropriate for chemical systems in which the J and K quantum numbers have been resolved. This is possible only in the case of small molecules such as formaldehyde (Green et al., 1992). If the sample is cooled in a supersonic jet, the initial J and K levels will be very low. In such experiments the sample is prepared vibrationally excited but rotationally cold because only low ΔJ transitions are allowed by selection rules so that equation 7.17 reduces to the RRKM equation (6.73) without rotations. In order to prepare the molecule in higher J and K levels, it is necessary to use a room-temperature sample which has a rich distribution of rotationaı states. However, the experiment and the data analysis become much more complicated. Consider the case of neutral systems in which the sample is excited with a single photon to a predissociative state. In most cases, the congested states will not be resolvable so that at any given photon energy, neither the initial, nor the final J and K states will be known. Hence, the total energy in the molecule is also uncertain. Such situations have been documented in the dissociation study of allyl isocyanide (Reddy and Berry, 1979; Segall and Zare, 1988) and 1,3-cyclohexadiene (Chuang and Zare, 1985).

At energies sufficiently elevated to cause dissociation, the spectrum of most molecules is congested so that resolving the J and K quantum numbers will be impossible. However, in IR/vis two-color experiments (Luo and Rizzo, 1991), the J and K levels can be resolved with the IR photon. An example of such an experiment is shown for HOOH in figure 5.11. In this particular case, the upper state could also be resolved. However, in more complex molecules the upper state will be too congested, or the levels too broadened by IVR to be resolvable. Nevertheless, J and K selection with the IR photon will approximately select the upper J and K levels. If the visible transition is a parallel one, the second photon absorption step will change J by only ± 1 while the K level remains constant. Thus, K is known, while J is approximately known. If the transition is a perpendicular one, then both J and K will change by one, so that neither one is exactly known. However, they are both approximately known. Furthermore, J will continue to be fixed even if the transition has been to a higher electronic state and the molecule undergoes a rapid internal conversion to the ground-state potential energy surface prior to dissociation.

If an energy-selected ionic sample is prepared via PEPICO by directly ionizing the sample into the ionization continuum, the resulting distribution of states becomes manageable as long as the electron energy resolution is broader than the rotational energy distribution. Because it is a bound→continuum transition, every initial state can be ionized (with the ejection of an electron in a bandwidth of ΔE) so that the whole initial distribution of states is transposed to the ion distribution. The rotational selection rules are not strict, but experiments have shown that in photoionization ΔJ remains small (Bryant et al., 1992; Muller-Dethlefs and Schlag, 1991). Thus, it can be assumed that the initial thermal rotational and vibrational distribution in the neutral sample is similar to the distribution in the final ionic sample. With a known distribution, it is possible then to express the total rate in terms of a distribution of rates given by equation (7.17). If the precision of the data do not warrant treating the data in terms of a distribution, they can be analyzed by using simply the average E_v, J, and K for the ensemble.

In a formal sense, we can average the rate constant over the K levels for a constant total energy, E. This results in (Zhu and Hase, 1990; Aubanel et al., 1991b):

$$k(E, J) = \sum_{-J}^{+J} P(K)\, k(E, J, K), \tag{7.18}$$

where:

$$P(K) = \frac{\rho(E, J, K)}{\sum_{-J}^{+J} \rho(E, J, K)} \tag{7.19}$$

so that

$$k(E, J) = \frac{\sum_{-J}^{+J} N^{\ddagger}[E - E_0 - E_r^{\ddagger}(J, K)]}{h \sum_{-J}^{+J} \rho(E, J, K)}, \tag{7.20}$$

where $\rho(E, J, K) = \rho[E - E_r(J, K)]$. Although this $k(E, J)$ represents an average rate constant over the K ensemble, the ensemble of molecules with a given J will in fact decay with a distribution of rate constants, each K giving its own rate constant. However, model system studies have shown that the sum of exponential functions will in general not deviate significantly from an average single exponential function (Zhu and Hase, 1990). Thus, from an experimental point of view, the two may be indistinguishable.

How does this relate to experimental data? In parallel transitions ($\Delta J = \pm 1$; $\Delta K = 0$), figure 7.11 shows that a sample may be prepared in a given J level, but in a distribution of K levels. Although this appears to be the situation implied by equations (7.18) to (7.20), problems arise because the total energy, E, is not fixed. That is, a single photon energy will excite molecules from a distribution of initial K levels, whose energies must be added to the photon energy so that the total energy is given by: $E_T = h\nu + E_r(J, K)$, where the rotational energy refers to the ground state. Thus, the experimental distribution of states will result in a distribution of total energy states, and therefore also a distribution of decay constants.

In a thermal sample at a temperature T, the canonical distribution of K levels in the ground state will be

$$P(K) = \exp\left(-\frac{(A - B)K^2}{k_B T}\right) \qquad K = 0, \pm 1, \ldots, \pm J \tag{7.21}$$

in which the definitions of A and B are those from equation (7.16). The distribution of exponential decay curves is given by

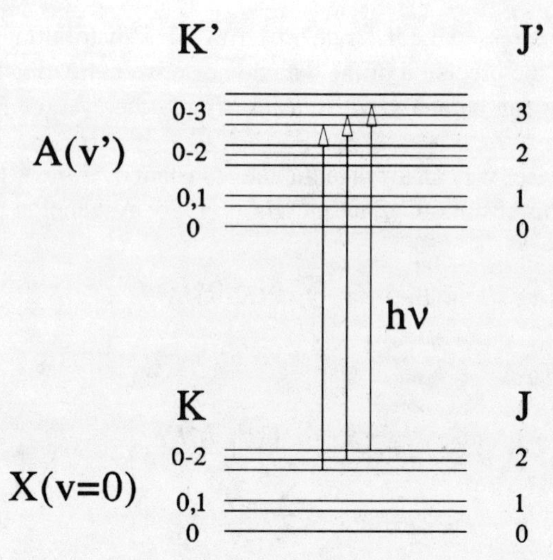

Figure 7.11 Parallel transitions in a J-resolved absorption. Note that the excited state is produced in a distribution of K levels and therefore in a distribution of internal energies. The example is for a prolate top in which the energy increases with K.

$$[AB](t) = [AB]_0 \sum_K P(K) \exp[-k(h\nu, J, K)\, t], \tag{7.22}$$

where $[AB](t)$ is the concentration of AB molecules at time t, and where $k(h\nu, J, K)$ is the rate constant at the total energy, $E_T = h\nu + E_r(J, K)$. The conversion of this ensemble to one implied by equation (7.17) is not straightforward. In an experiment on benzene ions, it was possible to select J with relatively little dispersion in the K values because it was found that the dominant transition was to $K = J$ levels (Kiermeier et al., 1988). However, this is not a general case.

7.2.3 The Active *K*-Rotor

Although J is always conserved during the course of the reaction, K need not be. In fact, in most reactions, it is likely that the K-rotor quantum number, due to coriolis and other interactions, is not conserved so that this rotor should be treated as active (Forst, 1973; Schneider and Rabinovitch, 1962; Berblinger and Schlier, 1992b; Fahrer and Schlier, 1992). The energy associated with an active K-rotor can pass to and from the vibrational energy sink during the course of the reaction. Hence, this one-dimensional rotor should be treated as part of the vibrational degrees of freedom when calculating density and sums of states. Hard experimental evidence in support of active K-rotors is difficult to obtain and thus largely lacking. In the case of the D_2CO dissociation (the most carefully and most state resolved unimolecular reaction so far investigated), the conclusion is not entirely clear, but tends to support an active K-rotor (Polik et al., 1990a; Hernandez et al., 1993). However, the extent to which this mixing is a result of the 20 kV/cm Stark field used to derive the experimental results remains unresolved (Hernandez et al., 1993). In another well studied molecule, HOOH, Luo and Rizzo (1991) found partial mixing of the K states in their high resolution spectra of the fifth overtone bands.

The K-rotor can be active or inactive in both the molecule or the transition state. Thus four situations can, in principle, be envisioned (Zhu et al., 1993).

Molecule K-rotor	TS K-rotor
1. inactive	inactive
2. active	active
3. active	inactive
4. inactive	active

At energies close to the dissociation threshold, the calculated rates are sensitive to the disposition of the K-rotor. However, at higher energies, the active K-rotor effects tend to cancel if they are applied to both the molecule and the transition state so that the rate constants change only a few percent when going from inactive/inactive to active/active. But, if one is treated as active, and the other as inactive, then the rates can vary by one or more orders of magnitude because making the K-rotor active is equivalent to adding a very low frequency vibration to the set of vibrational modes. In the following discussions we consider only the active/active or inactive/inactive cases because they appear to be the most reasonable situations.

One approach to convoluting the K-rotor into the vibrational density and sums of states (Quack and Troe, 1974; Miller, 1979; Troe, 1983; Aubanel et al., 1991b) is as follows:

$$N^{\ddagger}(E, J) = \sum_{-J}^{+J} N^{\ddagger}[E - E_0 - E_r^{\ddagger}(J, K)], \tag{7.23a}$$

$$\rho(E, J) = \sum_{-J}^{+J} \rho[E, E_r(J, K)]. \tag{7.23b}$$

These sums are equivalent to convolutions with K treated as a continuous variable. It is interesting that this treatment of the active K-rotor [Eq. (7.23)] is formally identical to the rate expressed as an average over the K quantum number but treated as an inactive rotor [Eq. (7.20)]. However, the fundamental difference is that in this case, the ensemble does decay with a single rate constant given by $k(E, J) = N^{\ddagger}(E, J)/h\rho(E, J)$. That is, it makes no difference which K level is initially accessed. They all decay with the same rate constant. Because the energy associated with the K-rotor is not great, the sum and density of states in equations (7.23) increase approximately by $2J + 1$.

Another approach to including an active K-rotor is to convolute the density of the one dimensional K-rotor with the sum and density of vibrational levels (Schneider and Rabinovitch, 1962; Forst 1973, p. 79):

$$N(E') = \int_0^{E'} N_V(E)\, \rho_r(E' - E)dE, \tag{7.24a}$$

$$\rho(E_V) = \int_0^{E_V} \rho_V(E)\, \rho_r(E_V - E)dE, \tag{7.24b}$$

where E' is the total active energy in the transition state and is given by: $E' = E_v + E_r(J) - E_r^{\ddagger}(J) - E_o$, where the two rotational energies include only the conserved and inactive J part. E_v is the active (vibrational) energy in the molecule so that its total energy is given by: $E_T = E_v + E_r(J)$. Although this formulation does not insure that $|K| \leq J$, it places indirect limits on the value of K. This is especially true when the molecule consists of many vibrational oscillators, because the N_v and ρ_v terms in the convolution will then dominate by keeping the energy associated with the ρ_r term small. The formulation in equations (7.24) is particularly easy to implement when the densities and sums are obtained by the inversion of the partition function with the steepest descent program. The classical one-dimensional rotational density of states is given by $(BE)^{-1/2}$ [Eq. (6.25)]. Although equations (7.24) are reasonably accurate for systems with room-temperature J distributions, they are not appropriate when J is small (i.e., low-temperature systems) because the limitation on K imposed by $|K| \leq J$ would be grossly violated.

7.2.4 Examples: $C_6H_6^+$ and $C_4H_8^+$

Although the role of rotations in unimolecular decay has received considerable attention, clear experimental evidence for it is largely lacking. The difficulty in preparing a molecule in selected J levels has prohibited extensive studies. One of the few investigations in which J level selection has been achieved has been in the dissociation of the benzene ion (Neusser, 1989; Kiermeier et al., 1988). By adjusting the molecular beam conditions, it was possible to "warm" the benzene molecules to a rotational temperature of 25 K so that higher rotational levels could be excited. The first photon accessed J resolved levels of the S_1 state. The absorption of a second photon (same color) resulted in the formation of a ground state ion with J approximately that of the S_1 state. The excitation function showing the partially J-resolved S_1 structure is illustrated in figure 7.12. Levels up to about $J = 15$ were resolved, while J levels up to 60 were partially resolved. Absorption of a third photon (color #2) above the dissociation limit resulted in dissociation. Because K was found to be approximately equal to J, the energy of the S_1 state was fixed. In this fashion it was possible to prepare ions in selected J states while keeping the total energy constant at 5.3 eV. The general trend of the decay rate shown in figure 7.13 indicates that the rates decrease gently with J. Solid lines B and A were calculated by assuming inactive and active K-rotors, respectively. From the better fit of curve A, it was concluded that the K-rotor is active in this dissociation. However, this conclusion should be viewed in light of several assumptions in the calculation. The major one is that the H loss transition state moments of inertia were assumed to be the same as those of the benzene ion. As previously mentioned, without accurate knowledge of the TS structure, it is virtually impossible to determine the effect of angular momentum on the dissociation rate. Since H loss transition states often have higher rotational constants, the data in figure 7.13 could probably be accommodated by a small change in the TS structure.

The effect of angular momentum on the dissociation rates has been observed in another ionic dissociation. The butene ion dissociates at low energies to four products (Booze et al., 1993) which are:

$$C_4H_8^+ \rightarrow \begin{cases} C_4H_7^+ + H & E_0 = 2.01 \text{ eV}, & (7.25a) \\ C_3H_5^+ + CH_3 & E_0 = 2.28 \text{ eV}, & (7.25b) \\ C_3H_4^+ + CH_4 & E_0 = 2.09 \text{ eV}, & (7.25c) \\ C_2H_4^+ + C_2H_4 & E_0 = 2.57 \text{ eV}. & (7.25d) \end{cases}$$

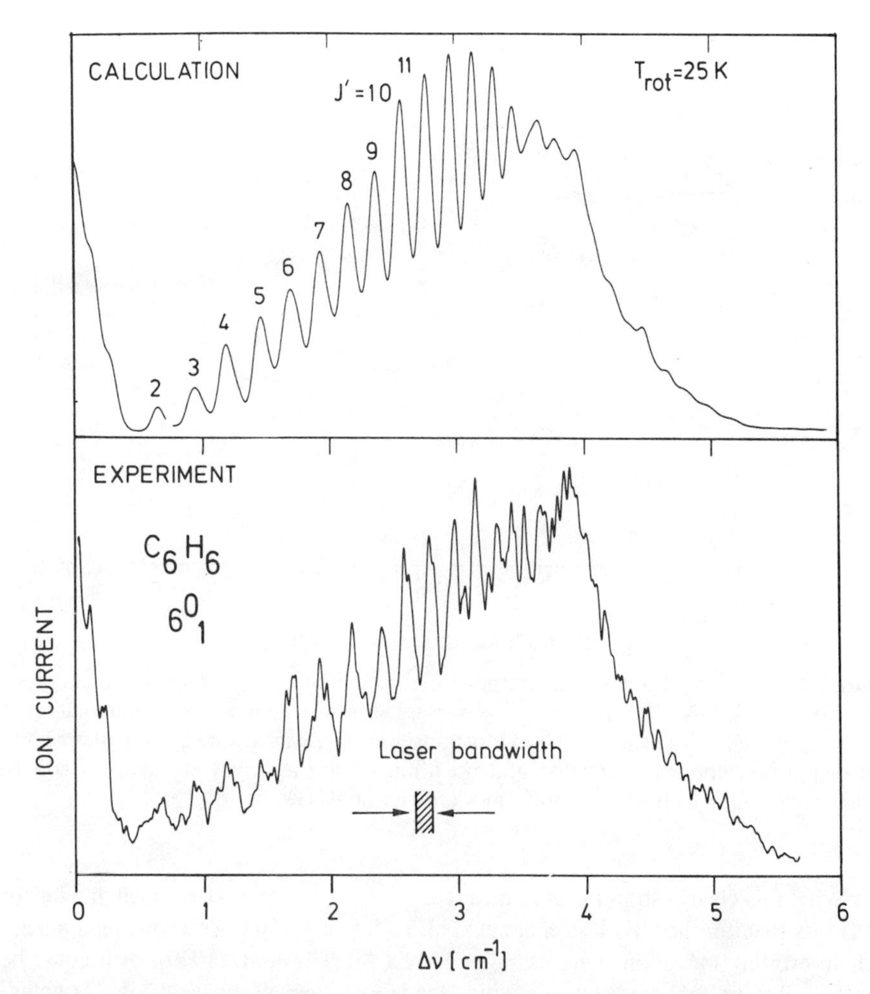

Figure 7.12 Ultraviolet spectrum of the S_1 state of benzene with partially resolved rotational structure. Although the K levels are not resolved, the transition probability is peaked for $K \approx J$. The absorption of one additional photon produced the benzene ion. Taken with permission from Kiermeier et al. (1988).

The loss of CH_4 occurs via a tight transition state and is a rather minor product, while the last reaction is of importance only at higher energies. On the other hand, the H loss and the CH_3 loss reactions proceed with nearly equal rates near threshold, with the CH_3 loss channel overtaking the H loss channel at higher energies. The ratio of H to CH_3 loss has been investigated by a number of different methods which vary in the amount of angular momentum in the sample. The butene ion can be produced (a) in a cold molecular beam with very little angular momentum (Booze et al., 1993), (b) in a room-temperature sample (Meisels et al., 1979), and (c) via an ion–molecule complex from the low-energy collision between $C_2H_4^+$ and C_2H_4 (Franklin et al., 1956). The latter will impart a significant amount of angular momentum, μvb, because of the large impact parameter of the reaction. The ratio of the H to CH_3 channels at the same total energy of 2.6 eV shows that the H loss channel is greatly suppressed by angular momentum. The ratio of k_H/k_{CH_3} for the three conditions are listed in table 7.3.

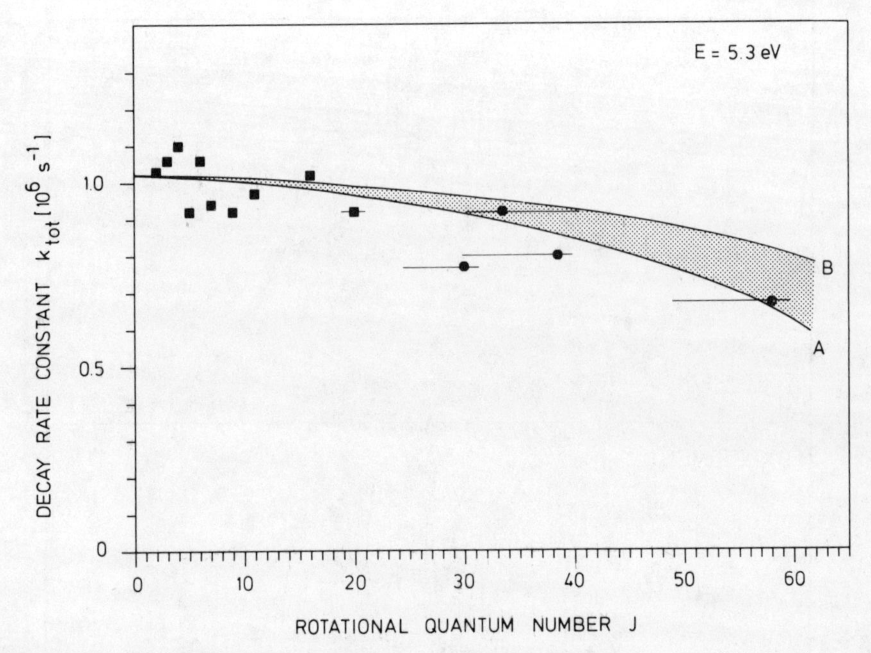

Figure 7.13 The effect of J on the benzene ion dissociation rate at a constant total energy ($E_{vib} + E_{rot}$) of 5.3 eV. The solid lines are numerical results from an RRKM calculation in which the rotational constant of the H loss transition state was assumed to be identical to that of the benzene ion. K-mixing and no K-mixing are assumed in curves A and B, respectively. Taken with permission from Kiermeier et al. (1988).

What this clearly shows is that the angular momentum barrier is much greater for the H loss than for the CH_3 loss channel. This could be a result of a centrifugal barrier with an orbiting transition state, as suggested by Meisels et al. (1979), or it could be due to a vibrator transition states in which the H loss channel proceeds via a "tighter" transition state (Booze et al., 1993; Bowers et al., 1983).

The available data, including *ab initio* molecular orbital calculations of the transition-state structures, indicate that a vibrator model for the transition state accounts for the $k(E, J)$ curves. Both the *ab initio* calculations and the experimental rate constants indicate that the H loss transition state has a slightly "tight" transition state. Figure 7.14 shows how the rates for reactions (7.25(a))–(7.25(c)) vary with J. Note that the rate constants for H loss channel decrease much more rapidly with J because

Table 7.3. Effect of Angular Momentum on Butene Ion Branching Ratio.

Method of $C_4H_8^+$ preparation	$\langle J \rangle$	k_H/k_{CH_3}
"Cold" $C_4H_8^+$ dissociation	5	0.61
"Warm" $C_4H_8^+$ dissociation	35	0.39
$C_2H_4^+/C_2H_4$ collision experiment	80	0.11

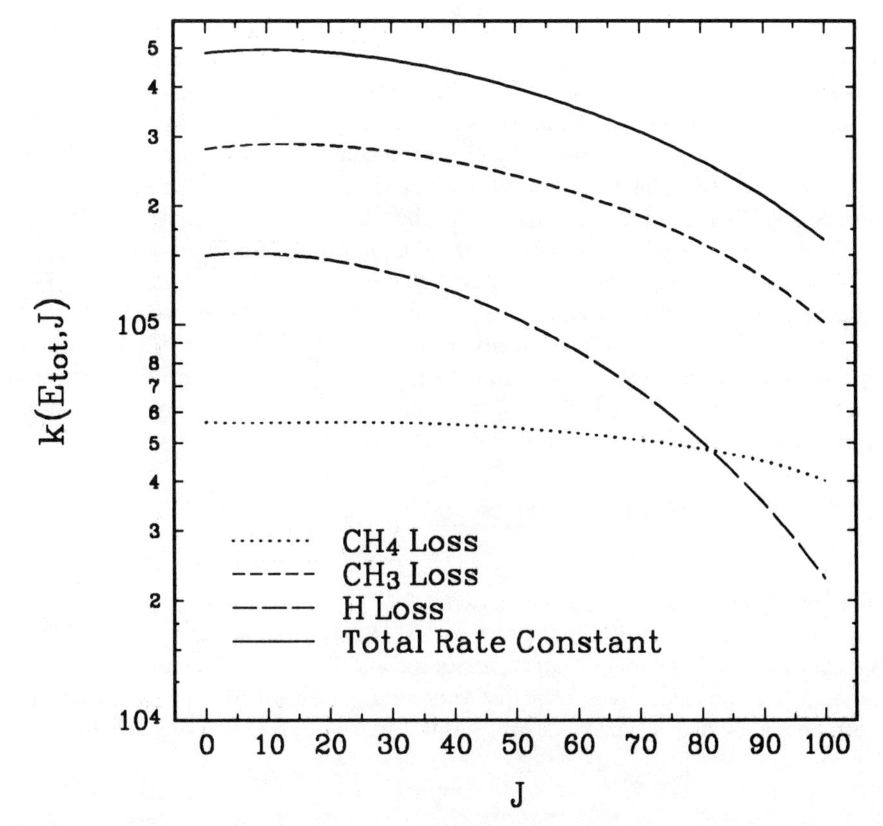

Figure 7.14 The variation of the $k(E, J)$ with J for the three reactions of the butene ion. The variation in the rates is due entirely to the different rotational constants of the molecular ion and the corresponding transition states. Taken with permission from Booze et al. (1993).

the activation energy of this channel actually increases with J. The decrease in the butene ion $k(E, J)$ with J can be largely accounted for by differences in the transition-state moments of inertia. The data appear rather insensitive to the question of active versus inactive K-rotors.

7.3 VARIATIONAL TRANSITION STATE THEORY

In order to deal properly with reactions that have no saddle point, it is necessary to go back to the notion that a unimolecular reaction is represented by a flux in phase space. Recall that the TS is defined as the surface in phase space which divides reactants from products, and at which the phase space is a minimum. For reactions with a substantial energy barrier, the dividing surface will be located at the saddle point because energy is such a dominating factor in determining the transition-state sum of states. However, for loose transition states it is necessary to search directly for the minimum flux configuration. The existence of such a minimum flux configuration is due to the interplay between the potential energy, which is constantly rising as the reaction proceeds, and the decrease in the vibrational frequencies for transitional modes which are evolving into product rotations and translations. As the reaction proceeds, the reduction in the available energy tends to reduce the density (or sum) of states while the lowering of the transitional vibrational frequencies increases the density of states. As shown in figure 7.15, these two opposing forces result in a minimum in the density or sum of states at some $R^{\ddagger}$. The TS, located at $R^{\ddagger}$, has been called an entropic bottleneck. (In canonical VTST, the bottle neck corresponds to the minimum in the free energy.)

The modification to the RRKM theory that makes possible accurate modeling of loose transition states is variational transition state theory (Pechukas, 1981; Miller, 1983; Forst, 1991; Wardlaw and Marcus, 1984, 1985, 1988; Hase, 1983, 1987). In this approach the rate constant $k(E, J)$ is calculated as a function of the reaction coordinate, R. The location of the minimum flux is found by setting the derivative of the sum of states equal to zero and solving for $R^{\ddagger}$. Thus, we evaluate

$$\frac{dN^{\ddagger}(E,\ J,\ R)}{dR} = 0, \tag{7.26}$$

solve for $R^{\ddagger}$, and obtain the variational $k(E, J, R^{\ddagger}) = N^{\ddagger}(E, J, R^{\ddagger})/h\rho(E, J)$. This was first suggested by Keck (1967) and further developed by Bunker and Pattengill (1968) as well as others (Truhlar and Garrett, 1980; Rai and Truhlar, 1983; Hase, 1972, 1976). Although the procedure is in principle very straightforward, it can become involved depending upon the accuracy desired. Many names have been attached to models

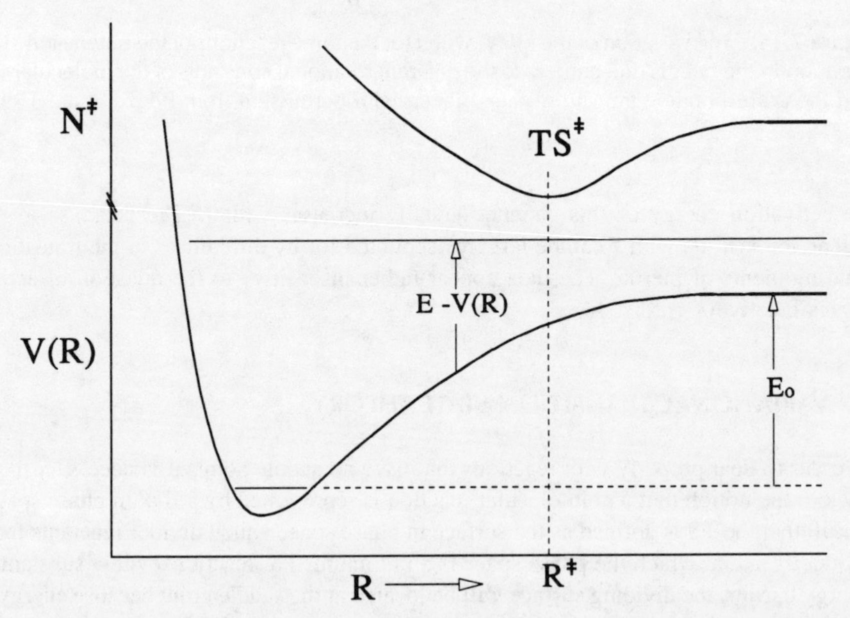

Figure 7.15 The variation of the sum of states with R for a reaction with no exit barrier. The position of the transition state at a given total energy is shown at $R = R^{\ddagger}$.

describing the same general idea: microcanonical variational TST, canonical VTST, variational RRKM theory, transition-state switching model, and flexible transition-state theory. Even though they may differ in detail, we call them all VTST.

7.3.1 A (Nearly) Analytical Approach to VTST

We illustrate the VTST with a particularly simple analytical approach which makes a number of approximations (Hase, 1987). Its main virtue is that it provides physical insight, while its major fault is its quantitative inadequacy. Suppose that the one-dimensional interaction potential for the dissociation reaction is given by $V(R)$ in figure 7.15. Further assume that the sum of states can be expressed classically as

$$N(E^{\ddagger}, R) = \frac{[E^{\ddagger}]^{s}}{s! \prod_{i=1}^{s} h\nu_i(R)} = \frac{[E - V(R)]^{s}}{s! \prod_{i=1}^{s} h\nu_i(R)}, \tag{7.27}$$

where the vibrational frequencies have been expressed as a function of the reaction coordinate, R. As before, s is the number of vibrations orthogonal to the reaction coordinate. When equation (7.27) is differentiated with respect to the reaction coordinate, we obtain (after factoring out $N(E^{\ddagger}, R)$):

$$0 = \frac{s \dfrac{dV(R)}{dR}}{E - V(R)} + \sum_{i=1}^{s} \frac{d\ln\nu_i(R)}{dR} \tag{7.28}$$

It is now convenient to make the approximation that most of the frequencies in the molecule remain constant and that it is only the transitional frequencies which change with R. Thus, in the case of H loss from CH_4, it is only the two C—H bends that are transitional. These are the ones that are converted into orbital angular momentum and are thus lost in the course of the reaction. The C—H stretch is the reaction coordinate, while the others remain intact. Hase (1972) and Quack and Troe (1974) introduced the idea of modeling the frequencies in a smooth manner from molecule to products by the use of switching functions. In the spirit of the bond energy bond order (BEBO) model (Johnston, 1966), the transitional frequencies are assumed to vary exponentially according to

$$\nu_i(R) = \nu_i(R_e) \exp[-a_i R] \tag{7.29}$$

so that the sum of derivatives in equation (7.29) reduces to a sum over a_i's. These a_i parameters vary with each type of bond and must be determined either from some other method, for example, *ab initio* MO calculation, or they must be estimated. (For C—H in CH_4, a good value appears to be $a = 1.65$ Å^{-1}.) The condition for the minimum flux is then given by

$$A = \sum_{i=1}^{s} \frac{a_i}{s} = \frac{\dfrac{dV(R)}{dR}}{E - V(R)}, \tag{7.30}$$

where the sum over the a_i's is a constant for a given reaction, independent of the total energy in the system. While the sum is over all the oscillators, the conserved vibrational modes have $a_i = 0$, so that the sum is only over the transitional modes.

All that is now needed is an expression for the interaction potential. The simplest of the possible functional forms is $V(R) = -c/R^n$ in which only the attractive, long-range interaction is included. The energy in the transition state is given by $E^{\ddagger} = E^* - V(R)$, where E^* is the energy above the dissociation limit. The internuclear distance at the transition state, $R^{\ddagger}$, can then be determined from the equation

$$E^* = \frac{cn}{A(R^{\ddagger})^{(n+1)}} - \frac{c}{(R^{\ddagger})^n}. \tag{7.31}$$

Unfortunately, it is not possible to solve this equation analytically for $R^{\ddagger}$, so that further progress can only be achieved by resorting to numerical or graphical procedures. Equation (7.31) has the form of an interaction potential. However, it has meaning only for $E^* > 0$. In this domain, the function is single-valued which means that there is only one value of $R^{\ddagger}$ which leads to a bottleneck. The only exception is at the dissociation threshold, where $E^* = 0$ for which $R^{\ddagger} = n/A$ or ∞. Hase (1987) has calculated values of $R^{\ddagger}$ for various reactions, values of a's, and energies. Not unexpectedly, as the attenuation of the transitional modes is increased [Σa_i increases], $R^{\ddagger}$ becomes smaller. Furthermore, as the attractive exponent, n, becomes smaller and the potential widens, the TS moves toward longer $R^{\ddagger}$. The values of $R^{\ddagger}$ for the various values of a_i, internal energies, and values of n range from 2 to 17 Å. It is therefore evident that the modification to RRKM which is embodied in the VTST is important.

It is helpful here to recapitulate the approximations made in this approach to VTST. First, the one-dimensional reaction coordinate is approximated by a simple $-cR^{-n}$ function. Second, it is assumed that the vibrational frequencies of the transitional modes decrease exponentially with R, thereby vanishing at large R. Finally, the density and sums of states are assumed to be given by their classical expressions.

Some of these assumptions can be softened at the expense of added complexity. For instance, the Whitten-Rabinovitch approximation (Forst, 1971; Whitten and Rabinovitch, 1963, 1964) for $N(E^{\ddagger})$:

$$N(E^{\ddagger}, R^{\ddagger}) = \frac{[E^{\ddagger} - V(R^{\ddagger}) - (1 - b)\Sigma\frac{1}{2}h\nu_i(R^{\ddagger})]^s}{s!\,\Pi h\nu_i(R^{\ddagger})} \tag{7.32}$$

can be used to improve on the classical approximation. The parameter b, which varies with the energy, takes account of the neglect of zero-point energy in the classical theory. A similar correction holds for the density of states.

The assumption that the vibrational frequencies of the transitional modes decrease with R in an exponential manner is common to a number of the more sophisticated RRKM formulations. The exponential function is one of several switching functions that have been proposed to take into account the fact that several frequencies change dramatically in going from the molecule to products. However, these switching functions add new variables which ultimately become empirical fitting parameters. The a_i's in the exponential function differ for different types of bonds and they also differ for neutral and ionic reactions. For instance, theoretical studies have shown that for neutral covalent bonds, the a's typically range between 1.0 and 2.0, while for ionic bonds they are closer to 0.1 (Hase, 1987; Mondro et al., 1986). Few accurate comparisons between theory and experiments have been carried out to determine optimum values for a variety of molecules and internal energies.

7.3.2 A Simple Quantum Version of VTST

Two problems with the simple analytical approach to VTST are the assumption of classical oscillators and the use of the exponential switching function. The problem with this switching function is that the frequencies go to zero as $R \rightarrow \infty$. Therefore $N^{\ddagger}(E^{\ddagger}, R)$ goes to infinity as $R \rightarrow \infty$. Since the TS is never located at $R = \infty$, this divergence is not a serious problem. Nevertheless, an interpolation function that switches a vibration into a rotation would be much more realistic. This is not easily accomplished since the energy level spacings of vibrations and rotations differ in their functional form.

Forst (1991) introduced a particularly simple approach in which the partition function, rather than the vibrational frequencies are switched. Furthermore, the switching for the transitional modes is between the vibrational and the rotational partition functions. If Q_{m} is the partition function of a transitional mode in the reactant molecule and Q_{rot} is the partition function of the product rotation which is correlated to the molecular vibration, and $S(R)$ is the switching function, then the logarithm of the partition function as a function of R is given by:

$$\ln Q^{\ddagger}(R) = [\ln Q_{\mathrm{m}} - \ln Q_{\mathrm{rot}}]S(R) - \ln Q_{\mathrm{rot}}. \tag{7.33}$$

The switching is done in terms of the logarithm because of its particularly convenient form when determining sums and density of states by the method of the inverse Laplace transform (see chapter 6). It should also be noted that this function interpolates between a vibration and the final rotational partition function. Because the interpolation is made with the logarithm of the partition function, the recommended interpolation function is

$$S(R) = 1 - \tanh[a(R - R_{\mathrm{e}})^{b}], \tag{7.34}$$

where a and b are constants. The b exponent, which is related to the curvature of the potential energy surface, typically has a value near 2. A larger b is appropriate for a surface in which the transition state "looks like" products at small R, that is, an early TS. The constant a, which typically has values ranging from 0.05 to 0.15 Å^{-2}, determines the slope at the mid range of R for a given b.

In this version of VTST, there is no need to distinguish between transitional modes and conserved modes. All modes of the molecule can be treated as transitional, some changing toward product vibrations, others toward rotations.

Because this is a quantum formulation, care must be taken to account for the change in the zero-point energies as the vibrations evolve with R. The energy available for redistribution, $E^{\ddagger}(R)$, is given by

$$E^{\ddagger}(R) = E_{\mathrm{v}} - V(R) + ZPE - ZPE^{\ddagger}(R) + E_{\mathrm{r}}(J) - E_{\mathrm{r}}^{\ddagger}(J, R), \tag{7.35}$$

where ZPE and $ZPE^{\ddagger}$ are the zero-point energies of the molecule and the transition state, the E_r's are the energies in rotation which are assumed to be inactive. In order to account for the rotations, a knowledge of the moments of inertia as a function of R is required.

In common with the previously discussed analytical formulation, the Forst treatment neglects coupling among the various degrees of freedom. That is, each mode is treated as an independent harmonic oscillator or rotor. Second, the use of switching functions again introduces new parameters which are not well characterized. So far, insufficient rate data near the reaction threshold are available to help provide much guidance for determining the appropriate values for a and b in equation (7.34).

7.3.3 VTST from *Ab Initio* Potential Energy Functions

Analytic potential energy functions for unimolecular reactions without reverse activation energies can be obtained by semi-empirical methods or by *ab initio* calculations, and enhanced by experimental information such as vibrational frequencies, bond energies, etc. To determine microcanonical VTST rate constants from such a potential function, the minimum in the sum of states along the reaction path must be determined. Two approaches have been used to calculate this sum of states.

7.3.3.1 Vibrator Transition States

The vibrator variational transition state model treats all modes orthogonal to the reaction, except torsions, as quantum harmonic oscillators (Vande Linde et al., 1987; Aubanel et al., 1991a; Booze et al., 1993), and is similar to the loose vibrator TS model described above. However, the potential energy, vibrational frequencies, and moments of inertia for the transition state are determined from the potential energy function and are not arbitrarily chosen to fit experimental rates. They are determined from reaction path following calculations as described in chapter 3. Calculations of this type, using accurate potential energy surfaces, have been reported for $CH_4 \rightarrow H + CH_3$ (Aubanel et al., 1991b), $Cl^-[CH_3Br] \rightarrow Cl^- + CH_3Br$ (Wang et al., 1994). A more approximate approach was used for $C_4H_8^+ \rightarrow C_3H_5^+ + CH_3$ (Booze et al., 1993).

7.3.3.2 Flexible Transition States

Wardlaw and Marcus (1984, 1985, 1988) have developed a flexible variational TS model for calculating the transition-state sum of states. This method treats the molecule's conserved vibrations in the normal quantized manner, while treating the transitional modes by classical mechanics. Thus for the bent NO_2 molecule which dissociates to NO + O, three vibrations are converted into one vibration and two rotations of the NO fragment. The variables that describe the potential energy surface of the transitional modes are two bond distances, N—O and the distance between the center of mass of the NO and the departing O, as well as two angles.

In the more complex dissociation of ethane into two methyl radicals, there are 12 conserved and 6 transitional modes. The conserved modes are associated with the CH stretch and bending motions of the two methyl fragments. The fate of the conserved and disappearing vibrational modes is shown in table 7.4. The 6 transitional modes for ethane dissociation is the maximum number of disappearing modes and arises whenever the dissociation produces two nonlinear products.

The Hamiltonian for the flexible variational transition state model is based on the separable long-range product Hamiltonian:

$$H = H_{v1} + H_{v2} + T_{r1} + T_{r2} + \frac{\ell^2}{2\mu R^2} + E_t + V, \tag{7.36}$$

where the H_v's and the T_r's are the vibrational Hamiltonians and rotational kinetic energies of the reaction fragments, ℓ is the orbital angular momentum, μ is the reduced mass of the fragments, R their center-of-mass separation, E_t is the relative translational energy projected on the fragments' center-of-mass separation, and V is the interaction potential. The unimolecular system's total angular momentum is the vector sum of the rotational angular momenta $\mathbf{j}_1$ and $\mathbf{j}_2$ for the two fragments and the orbital angular momentum $\boldsymbol{\ell}$. That is,

$$\mathbf{J} = \mathbf{j}_1 + \mathbf{j}_2 + \boldsymbol{\ell}. \tag{7.37}$$

The interaction potential V is anisotropic and depends on the relative orientation of the two fragments as well as their separation, R. The frequencies for the fragments, that is, the conserved modes, versus R have been determined by either interpolation between reactant and products frequencies (Wardlaw and Marcus, 1984, 1985, 1988), or by reaction path following (Aubanel and Wardlaw, 1989).

Table 7.4. Correlation of Reactant and Product Modes $CH_3CH_3 \longrightarrow 2\ CH_3$.

			Conserved Modes				
CH_3CH_3 Modes			Attr.	CH_3 Modes			
ν_1	A_{1g}	2954	CH_3 str	ν_1	A_1'	3044	str
ν_5	A_{2u}	2896	CH_3 str				
ν_2	A_{1g}	1388	CH_3 def	ν_2	A_2''	580	out of pl.bend
ν_6	A_{2u}	1379	CH_3 def				
ν_7	E_u	2985	CH_3 str	ν_3	E'	3162	str
ν_{10}	E_g	2969	CH_3 str				
ν_8	E_u	1472	CH_3 def	ν_4	E'	1396	in pl. bend
ν_{11}	E_g	1469	CH_3 def				

			Disappearing Modes	
CH_3CH_3 Modes			Attr.	Product Modes
ν_3	A_{1g}	995	C—C str	Translation
ν_4	A_{1u}	289	Torsion	Rotation
ν_9	E_u	821	CH_3 rock	Rotation (2)
ν_{12}	E_g	1206	CH_3 rock	Rotation (2)
J		Rotation		Orb.ang. mom. translation (2)
K		K-rotor		Rotation

Taken from Quack and Troe (1974); Wardlaw and Marcus (1986).

The transitional mode Hamiltonian is given by the last four terms in equation (7.36). In the work of Wardlaw and Marcus, the phase space volume for the transitional modes was calculated versus the center of mass separation R, so that R is assumed to be the reaction coordinate. In recent work, Klippenstein (1990, 1991) has considered a more complex reaction coordinate. The multidimensional phase space volume for the transitional modes can not be determined analytically, but must be evaluated numerically, for example, by a Monte Carlo method of integration (Wardlaw and Marcus, 1984). The density of states is then obtained by dividing the phase space volume by h^n, where n is the dimensionality of the integral, and differentiating with respect to the energy. The total sum of states of the transition state is obtained by convoluting the density of the transitional modes with the sum of the conserved modes, $N(E,J)^{\ddagger}$ so that

$$N(E^{\ddagger}, J) = \int_0^{E^{\ddagger}} N(E^{\ddagger} - \epsilon)\rho_J(\epsilon, R^{\ddagger})d\,\epsilon, \qquad (7.38)$$

where $E^{\ddagger} = E - V(R)$ is the energy that is available for redistribution as shown in figure 7.15.

In this flexible transition-state model all of the interactions among the various transitional normal modes are taken into account, including anharmonic effects. Furthermore, the need for transitional mode switching functions for the vibrational frequencies is circumvented because the density of states at various points in the reaction are directly calculated from the multidimensional Hamiltonian. However, because the phase space integral is a classical quantity, the derived density of states is also classical. This is a rather good approximation for very loose transition states in which the transitional modes are mostly rotations. Thus it is most accurate near the dissociation threshold at which the TS state is located at large R. The other approximation is the neglect of coupling between the conserved and transitional modes. If reaction path following is used, the conserved vibrational frequencies will be correct. However, the transitional mode Hamiltonian alone will not give the correct transitional or conserved mode frequencies for small R.

To date, there have been few comparisons between vibrator and flexible VTST rate constants for the same analytic potential. For the CH_4 and $Cl^-[CH_3Br]$ reactions the two methods are in excellent agreement (Aubanel et al., 1991b; Wang et al., 1994). However, the agreement is poor for $CH_2CO \rightarrow CH_2 + CO$ (Yu and Klippenstein, 1991).

7.3.4 Examples: NCNO, $C_6H_6^+$, and $C_6H_5Br^+$

The VTST model is most important when analyzing data that spans over a large energy range, and reaches down to the reaction threshold. One such example is the dissociation of NCNO which has been studied with picosecond resolution by transient absorption of the parent NCNO (probe 2 in figure 7.16) and by laser-induced fluorescence of the product CN (probe 1 in figure 7.16) (Khundkar et al., 1987). The sample was cooled in a free-jet expansion in order to eliminate hot bands and rotationally excited states. Examples of the NCNO decay are shown in figure 7.17.

The four-dimensional potential energy surface was generated by splicing together

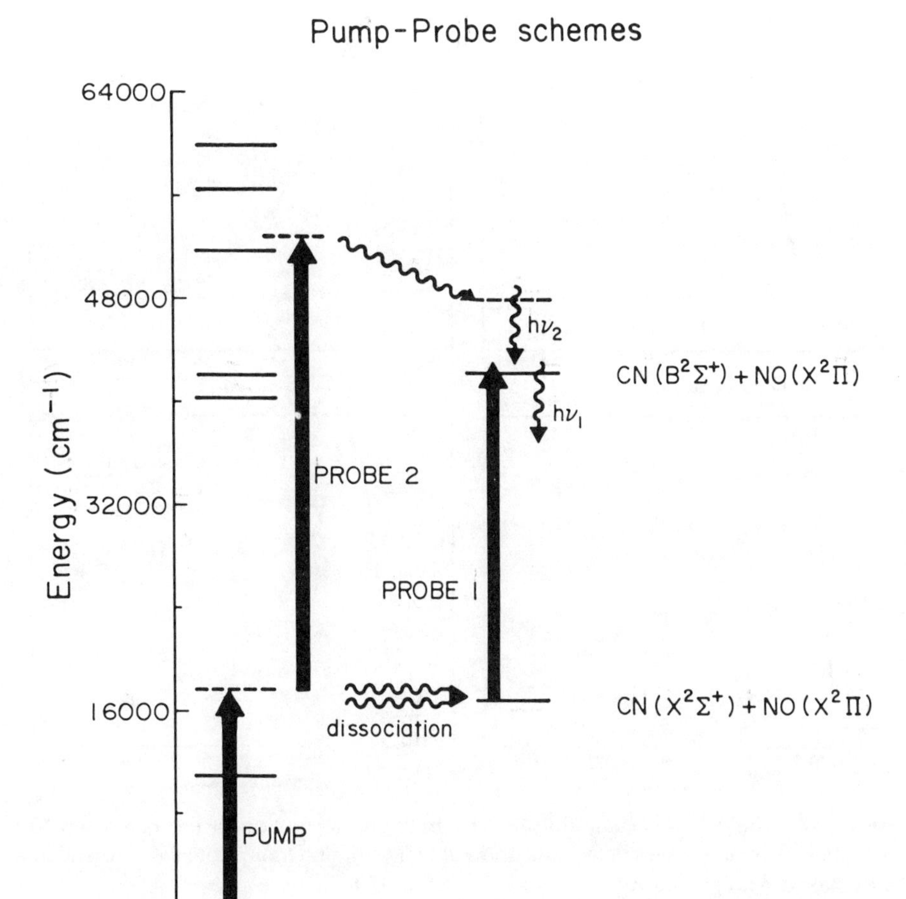

Figure 7.16 Energy level diagram for the NCNO → NC + NO dissociation including the scheme for monitoring the NCNO decay rate and the product NO appearance rate (which were shown to be equal). Taken with permission from Khundkar et al. (1987).

various interaction potentials including a Varshni potential (Varshni, 1957) for the C—N bonding potential and Lennard-Jones 6-12 potentials for nonbonding interfragment potentials (Klippenstein et al., 1988). The coordinates for the disappearing modes are shown in figure 7.18, while the fate of the conserved and disappearing modes are shown in table 7.5. It is evident that "conserved" modes are not really conserved. To account for their significant change, an interpolation function was used to map them from reactant to product values.

The transition-state sum of states was evaluated as a function of R for several excess energies. Figure 7.19 shows that at low energies ($E - E_0 = 50$ cm^{-1}) the sum of states is a minimum at large internuclear separations. That is, the transition state is located at R in excess of 6 Å. This corresponds to a very loose transition state in which

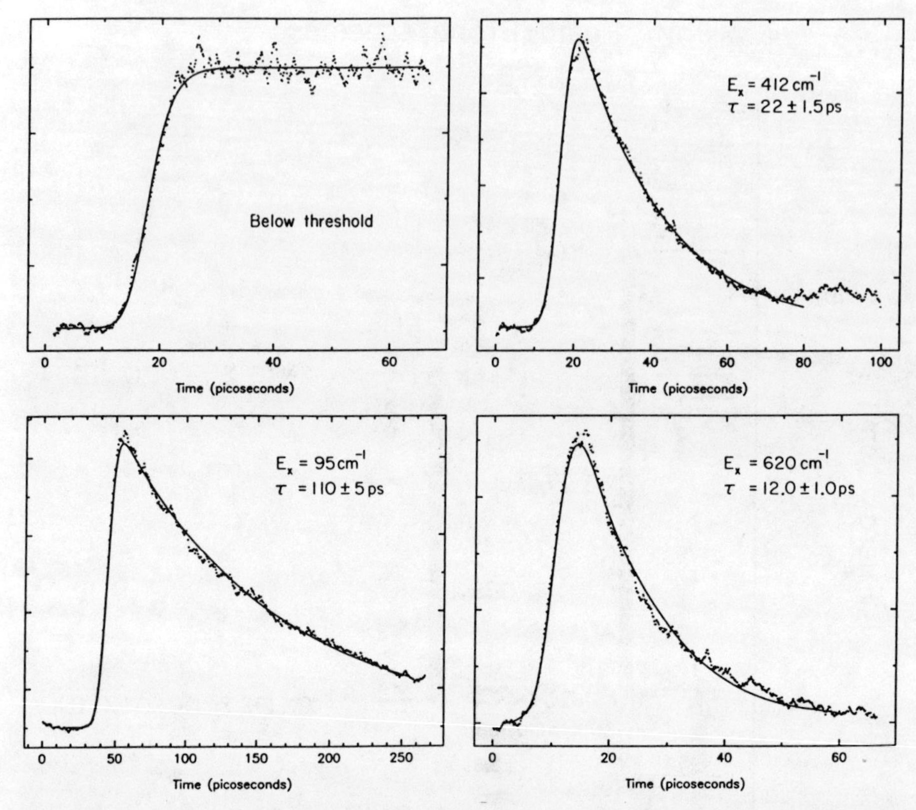

Figure 7.17 The NCNO decay followed by transient absorption of the parent NCNO molecule at several energies below and above the dissociation limit. Note the different time scales. Taken with permission from Khundkar et al. (1987).

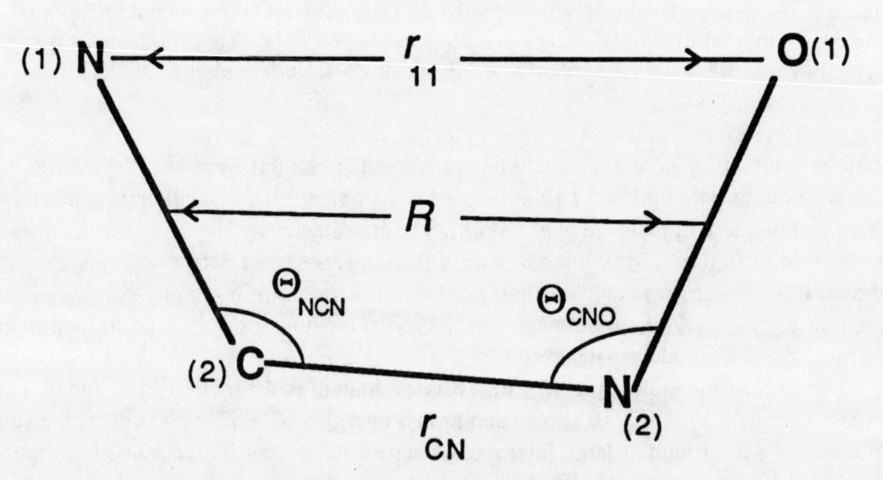

Figure 7.18 The NCNO coordinates for the VTST treatment of the production of CN + NO. Taken with permission from Klippenstein et al. (1988).

Table 7.5. Correlation of Reactant and Product Modes NCNO ⟶ CN + NO

		Conserved Modes		
NCNO modes		Attr.	NC	NO
ν_1	2170	NC str	2068.7	
ν_2	1501	NO str		1904
		Disappearing Modes		
NCNO modes		Attr.	Product Modes	
ν_3	820	CN str	Translation	
ν_5	216	CNO bend	Rotation	
ν_4	588	NCN bend	Rotation	
ν_6	270	NCNO wag	Rotation	
J		Rotation	Orb.ang.mom. translation(2)	
K		K-rotor	Rotation	

Taken from Klippenstein et al. (1988); Nadler et al. (1984).

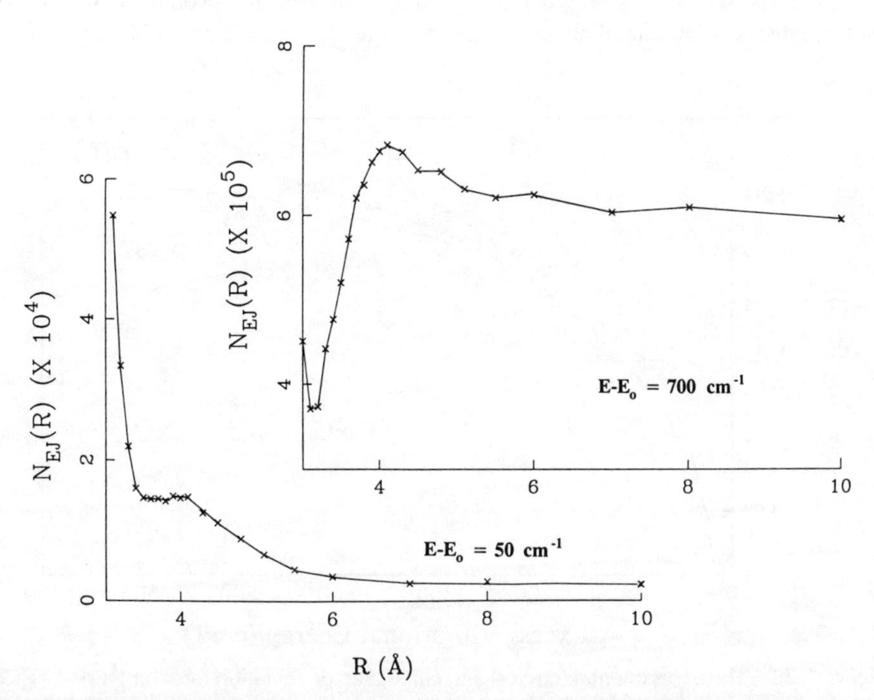

Figure 7.19 The calculated sums of states as a function of the reaction coordinate in the NCNO dissociation at two energies above the dissociation limit. Taken with permission from Klippenstein et al. (1988).

the CN and NO units are nearly freely rotating. On the other hand, at an excess energy of 700 cm^{-1}, the sum of states exhibits a sharp minimum at about 3.3 Å, which moves even closer in at 2000 cm^{-1} (Klippenstein et al., 1988).

It is most revealing to compare the best RRKM calculation with a fixed transition state with the VTST result. This is shown in figure 7.20. The improvement offered by the variationally located TS is remarkable. To be sure, this is a somewhat extreme example because the transitional modes outnumber the conserved modes and the latter hardly contribute to the density of states since their frequencies are so high.

Another example of the VTST application is to the benzene ion dissociation. The benzene ion fragments via four channels by loss of H, H_2, C_2H_2, and C_3H_3. The lowest energy channel involves the production of the phenyl ion, $C_6H_5^+$ via H loss. The phenyl ion heat of formation is not known very accurately so that questions remain about the reaction energetics. The major methods for measuring the phenyl ion heat of formation have been through RRKM fitting of $k(E)$ vs. E curves for the X loss from $C_6H_5X^+$, where X = H, Cl, Br, and I. These rates have been investigated by a number of techniques including charge exchange mass spectrometry (Andlauer and Ottinger, 1971, 1972), photoionization (Pratt and Chupka, 1981), PEPICO (Eland and Schulte, 1975; Baer et al., 1979) multiphoton ionization with TOF mass spectrometry (Kuhlewind et al., 1984, 1986, 1987; Neusser, 1987), and decay in an ICR cell (Klippenstein et al., 1993). Unlike the NCNO reaction, the large density of states in the benzene ion causes the rate of dissociation at the reaction threshold to be immeasurably slow. Extrapolation of the RRKM curve to the reaction threshold predicts lifetimes in excess of 100 sec. As has been demonstrated by Dunbar and co-workers (1987, 1990), below a dissociation rate of about 10^3 sec^{-1}, infrared fluorescence cools the parent ion

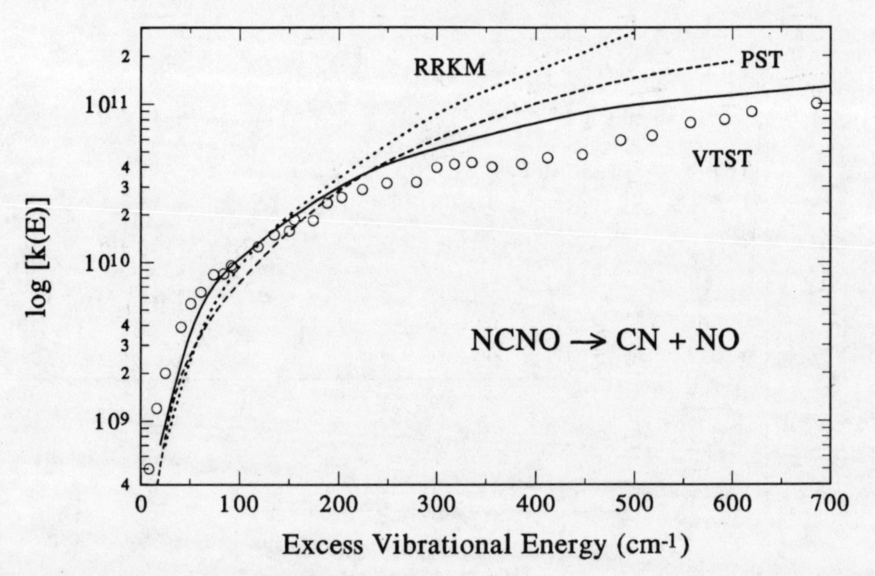

Figure 7.20 The experimental (circles) and calculated dissociation rates for NCNO → CN + NO. The curves labeled as RRKM and PST are the best of several calculated models taken from Khundkar et al. (1987), while the solid line labeled VTST is the $k(E)$ curve calculated with flexible VTST (adapted from Klippenstein et al., 1988).

so that decay by dissociation effectively stops. For these reasons, the X loss from substituted benzene ions cannot be measured down to the reaction threshold. Hence, extrapolation by RRKM is the only method for determining the E_0. Because these extrapolations often exceed 1 eV (23 kcal/mol), the reliability of the derived onset is less than desired.

Klippenstein et al. (1993) used the flexible variational transition state model to analyze the benzene ion dissociation rates. The comparison between normal RRKM and VTST is shown in figure 7.21. The normal RRKM theory with a fixed transition state assumed a threshold energy of 3.65 eV, while the VTST was fitted with an assumed activation energy of 3.88 eV. The difference of 0.23 eV (5 kcal/mol) is a result of the location of the transition state at high excess energy. It is situated at short internuclear distance where the interaction potential is below the dissociation limit so that the effective E_0 is below the true dissociation limit. It is interesting that in the benzene ion, both normal RRKM and VTST can fit the data over its limited energy range. This is because the energy range of the measured rates are well above threshold where the vibrator transition state is already fully developed. Because the dissociation rates cannot be measured any closer to threshold, it appears that in the benzene ion dissociation, it is impossible to verify experimentally that VTST is better than normal RRKM. Nevertheless, in view of the VTST analysis, it seems appropriate to adjust the phenyl ion heat of formation to the higher value.

A final example is that of the bromobenzene ion dissociation. Several studies have measured the dissociation rates (Baer et al., 1976; Baer and Kury, 1982; Rosenstock et

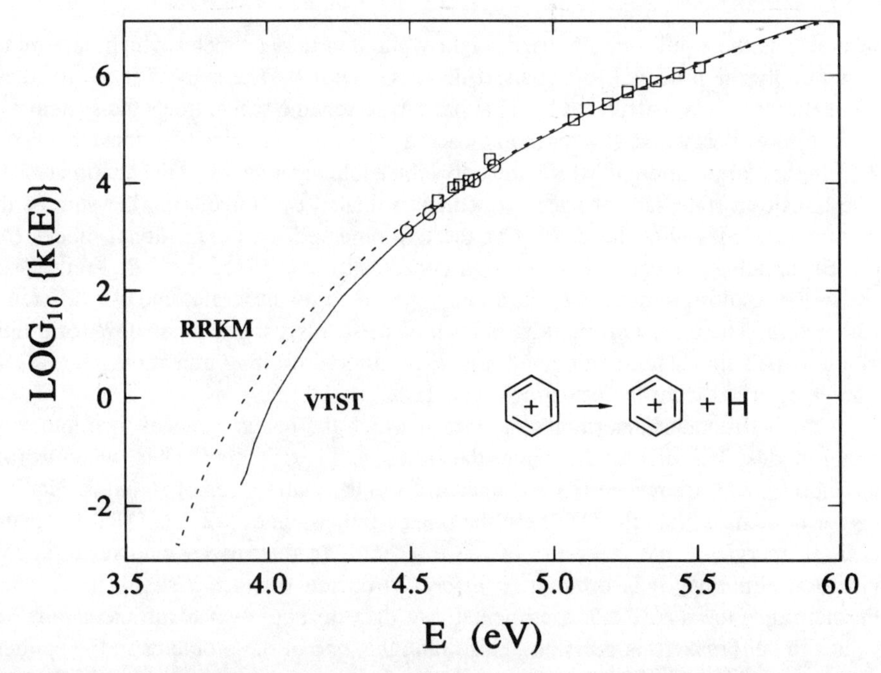

Figure 7.21 The decay rate constants for the benzene ion obtained by MPI-time of flight mass spectrometry and ICR. The solid line is a VTST theory calculation. Taken with permission from Klippenstein et al. (1993).

al., 1980; Pratt and Chupka, 1981; Gefen and Lifshitz, 1984). Lifshitz et al. (1991) have applied a modified vibrator transition state model of the VTST due to Chesnavich (1986). In this treatment, coupling between rotations and vibrations are ignored and the three disappearing modes treated independently. The C—Br stretch was modeled with the following expression which includes both short range R^{-6} and the R^{-4} ion-induced dipole attractions, as well as an exponential repulsion at short distances:

$$V_{\mathrm{C-Br}}(R) = [D_e/(c_1 - 6)][2(3 - c_2)\exp[c_1(1 - x)] - (4c_2 - c_1c_2 + c_1)x^{-6} - (c_1 - 6)c_2x^{-4}], \quad (7.39)$$

where $x = R/R_e$, R_e is the C—Br equilibrium bond distance, D_e is the C—Br dissociation energy measured from the "bottom of the well." The difference between D_e and E_o, the critical energy, is the sum of the zero-point energies of the three transitional modes. The constants c_1 and c_2 are chosen in order to give the correct ion-induced dipole interaction at large R and to give the correct harmonic C—Br vibrational stretch frequency of 670 cm^{-1}.

The two C—Br bending transitional frequencies of 254 and 181 cm^{-1} were treated as hindered rotors with interaction potentials given by

$$V_b = \tfrac{1}{2}V_o(R)[1 - \cos 2\theta]. \quad (7.40)$$

The rotor barrier height was assumed to be a smoothly decreasing function of the reaction coordinate, R and given by

$$V_o(R) = V_e\exp[-a(R - R_e)^2], \quad (7.41)$$

where V_e is the equilibrium barrier height while a is the parameter which determines how rapidly the hindered rotor barrier decreases with R. The sum of states for these hindered rotors was calculated by an approximate scheme which treats the system as a rotor at low energy and as a harmonic oscillator at high energies (Chesnavich, 1986).

In this formulation of VTST, the only adjustable parameter is a in equation (7.41). The transition state sum of states was then evaluated by convoluting the sum of the conserved modes with the density of the two hindered rotor transitional modes (the C—Br stretch is of course the reaction coordinate) as a function of R. Figure 7.22 shows the resulting sum of states as a function of R for two energies and two different a parameters. There are two minima in each of these curves: one at small R for a tight vibrator transition state, and one at large R for a loose orbiting transition state (OTS). The latter arises from the centrifugal barrier.

The a parameter determines the rate at which the bending modes turn into rotational modes. The smaller the value, the slower is this transition. Thus, when $a = 1$, the vibrator transition state has a chance to form at small values of R and results in a bigger bottleneck than the OTS. On the other hand, when $a = 2$, the OTS dominates even at energies 1 eV in excess of the threshold. In this model with no rotational vibration coupling, it is difficult, *a priori*, to decide whether a should be 1 or 2. Furthermore, the available experimental rate data do not cover a sufficient range of values to help make this decision. This highlights one of the problems in the application of approximate VTST to reactions in which only a limited range of rate constants is available. Both normal RRKM and VTST fit the data. However, in the VTST treatment, the derived activation energies depend upon the assumed values for the

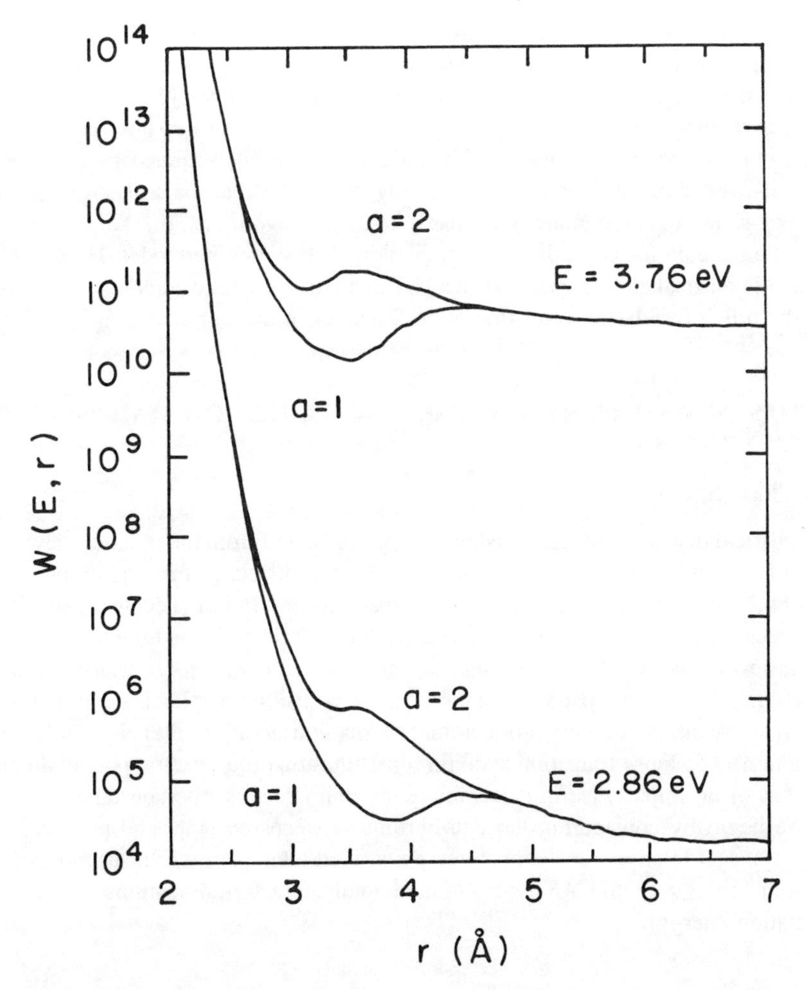

Figure 7.22 The sum of states calculated by a one dimensional VTST calculation of the bromobenzene ion dissociation. The "a" is a switching function parameter which determines how fast the vibrations are converted to rotations. Taken with permission from Lifshitz et al. (1991).

adjustable variables. More systems need to be analyzed in order to place tighter limits on the adjustable parameters.

7.3.5 The Transition State Switching Model

The transition state switching model (Chesnavich et al., 1981; Chesnavich and Bowers, 1982; Chesnavich, 1986) is a special case of the more general VTST. A program is available from the quantum chemistry program exchange (Chesnavich et al., 1988). Rather than assuming that the TS position changes slowly with R as the excess energy increases, the switching model assumes that the $N^{\ddagger}(E, R)$ vs. R curve has two minima, one at short R and one at long R, as indeed suggested by the calculations on bromobenzene in figure 7.22. Furthermore, it assumes that the lower of the two minima is the rate-determining step and that the switch between the two is sudden. The transition state at long R is an orbiting TS, while at short R it is the vibrator TS. Sufficient tests to determine whether there are always two minima in the $N^{\ddagger}(R)$ function have not been carried out. As shown in figure 7.22, small variations in assumed parameters can change the shape of the function considerably. It appears that ionic dissociations with their long-range interaction are more likely to exhibit two transition states and thus to be more amenable for analysis by the transition state switching model (Hu and Hase, 1989). It is most likely the case that the TST switching model cannot be distinguished from the full VTST because insufficient differences in their predicted $k(E)$ curves.

7.4 PHASE SPACE THEORY AND ORBITING TRANSITION STATE PHASE SPACE THEORY

7.4.1 Phase Space Theory

The statistical dissociation rate constant can be calculated from the point of view of the reverse reaction, namely the recombination of the products to form a complex. This approach, commonly referred to as phase space theory (PST) (Pechukas and Light, 1965; Pechukas et al., 1966; Nikitin, 1965; Klots, 1971, 1972) is limited to reactions with no reverse activation energy, that is, reactions with very loose transition states. PST assumes the decomposition of a molecule or collision complex is governed by the phase space available to each product under strict conservation of energy and angular momentum. The loose transition state limit assumes that the reaction potential energy surface is of no importance in determining the unimolecular rate constant.

We begin by considering the equilibrium between reactants and products, at a constant $A + B$ relative translational energy, a fixed total energy E measured from the zero-point energy of the AB reactant, and total angular momentum, J (E_o is the dissociation energy).

$$AB \underset{k_b}{\overset{k_u}{\rightleftharpoons}} A + B. \tag{7.42}$$

By microscopic reversibility, the forward and backward rates are equal at equilibrium so that

$$k_u = k_b \frac{[A][B]}{[AB]} = \sigma u \frac{\rho_A(E - E_o, j_A)^{*}\rho_B(E - E_o, j_B)}{\rho_{AB}(E, J)}, \tag{7.43}$$

where we have expressed the equilibrium constant, $[A][B]/[AB]$ in terms of densities of states (a convolution thereof in the numerator), and have replaced the bimolecular rate constant with the collision cross section times the relative velocity (u) of A and B. Because energy can be shared between the two products, the two densities in the numerator are convoluted. The combined product densities of state can be written as:

$$\rho_A(E - E_o, j_A)^{*}\rho_B(E - E_o, j_B) = \rho_t(E_t)\, \rho_{A,B}(E - E_o - E_t, j_A, j_B) \tag{7.44}$$

in which $\rho_{A,B}$ represents the total product vibrational/rotational density of states and ρ_t is the translational density of states at a relative energy, E_t. Use of the three-dimensional translational density of states [Eq. (6.15)], and some rearrangement of terms, yields the unimolecular rate constant:

$$k_u = \frac{\sigma[8\pi/h^2]\mu E_t\, \rho_{A,B}(E - E_o - E_t, j_A, j_B)}{h\, \rho_{AB}(E, J)}. \quad (7.45)$$

This equation bears some similarity to the RRKM equation. However, noticeably absent is the concept of the transition state. The densities of states in equation (7.45) refer to the products and to the molecule, but not the transition state. This equation is sometimes written in terms of the de Broglie wavelength of the products, $\lambda = h/p = h/(2\mu E_t)^{1/2}$, where μ is the reduced mass of the colliding A and B units. Thus,

$$k_u = \frac{\sigma\, \rho_{A,B}(E - E_0 - E_t, j_A, j_B)}{\pi ƛ^2 h\, \rho_{AB}(E, J)}, \quad (7.46)$$

where $ƛ$ is the de Broglie wavelength, $\hbar/p$ expressed in terms of $\hbar$ rather than h.

So far, the derivation has been general. However, to proceed further it is convenient to consider the dissociation for a spherical top to a spherical top product plus an atom. In that case, the conservation of angular momentum involves just the three quantum numbers, J, j, and ℓ. Angular momentum conservation is expressed as

$$\mathbf{J} = \mathbf{j} + \boldsymbol{\ell} \quad (7.47)$$

or

$$\mathbf{J} = |\mathbf{j} + \boldsymbol{\ell}| \cdots |\mathbf{j} - \boldsymbol{\ell}|. \quad (7.48)$$

Equation (7.47) or (7.48) can be illustrated graphically as shown in figure 7.23. The allowed combination of j and ℓ states for the dissociation of a molecule with an initial J is shown as the shaded region. For the case of a rotationally cold sample, the ℓ–j plot is just a line with unit slope so that $\ell = j$.

To obtain a final expression for k_u, recall that the total collision cross section is defined in terms of the impact parameter, b, as πb^2, and that the orbital angular momentum of the system, ℓ, is given by $\mu ub = \sqrt{\ell(\ell + 1)}\hbar$, where u is the relative velocity of the two fragments. Thus, the total cross section is

$$\sigma_T = \frac{\pi\ell(\ell + 1)\hbar^2}{2\mu E_t} = \pi\ell(\ell + 1)ƛ^2 \quad (7.49)$$

and the partial collision cross section for a particular ℓ, $\sigma(\ell)$, is given by $\pi(2\ell + 1)ƛ^2$. However, not all collisions will lead to the production of AB in the angular momentum state J. The fraction of collisions that do is given by $(2J + 1)[(2\ell + 1)(2j + 1)]^{-1}$, where the denominator represents all the different ways of combining ℓ and j. Hence the cross section for combining j and ℓ into J is

$$\frac{\sigma(\ell + j \rightarrow J)}{\pi\lambda} = (2\ell + 1)\frac{(2J + 1)}{(2\ell + 1)(2j + 1)} = \frac{(2J + 1)}{(2j + 1)}. \quad (7.50)$$

Thus, the PST rate constant for forming products with specific becomes E_v, E_t, j, and ℓ becomes

$$k_u(E, J \rightarrow E_v, E_t, j, \ell) = \frac{(2J + 1)n(\ell)\rho_R(j)\rho_A(E_v)}{h(2j + 1)\rho_{AB}(E, J)}. \quad (7.51)$$

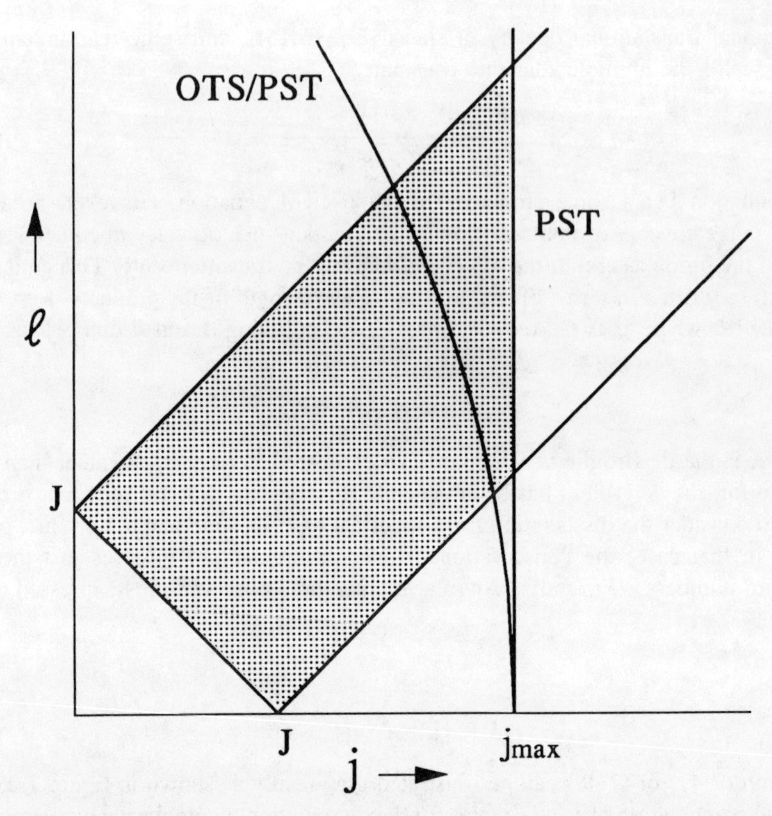

Figure 7.23 Angular momentum conservation in dissociation reactions. ℓ vs. j plot showing the region of allowed ℓ and j product combinations for an initial angular momentum of J. The vertical, PST line corresponds to the maximum product j_{max} which is limited by energy conservation. The curved vertical line (Eq. 7.61) shows the additional constraint imposed by the interaction potential and the resulting centrifugal barrier.

In this equation, the product vibrational energy is given by $E_v = E - E_o - E_{tr}$, where $E_{tr} = E_t + E_r$. In addition, the product density of states in the numerator has been separated into vibrational and rotational parts.

The total rate is obtained by summing over all the allowed values of ℓ, given by $n(\ell)$. The rotational density of states (or the rotational degeneracy) for a spherical top product is $(2j + 1)^2$. The value of $n(\ell)$ can be deduced from figure 7.23. For $j < J$, the number of allowed ℓ values increases as $2j + 1$. However, beyond $j = J$, the number of allowed ℓ values is constant at $2J + 1$. The PST dissociation rate summed over all the allowed ℓ values is thus given by

$$k_u(E, J \rightarrow E_v, E_t, j) = \frac{(2J + 1)(2j + 1)^2 \rho_A(E_v)}{h\rho_{AB}(E, J)} \qquad \text{for } j < J \qquad (7.52)$$

and

$$k_u(E, J \rightarrow E_v, E_t, j) = \frac{(2J + 1)^2(2j + 1)\rho_A(E_v)}{h\rho_{AB}(E, J)} \qquad \text{for } j > J. \qquad (7.53)$$

As before, $E_v = E - E_o - E_{tr}$. Thus, in order to obtain $k_u(E, J)$ the rates must be integrated over the product translational and rotational energies. When these equations are summed over all allowed j values up to $j_{max} = [(E - E_v)/B]^{1/2}$, we obtain [Klots (1971, 1972)]:

$$k(E,J) = \frac{\dfrac{1}{2J+1}\displaystyle\int_0^{BJ^2} \frac{4}{3}\,\rho_A(E - E_o - E_{tr})\left(\frac{E_{tr}}{B}\right)^{3/2} dE_{tr} + \int_{BJ^2}^{E-E_o} \rho_A(E - E_o - E_{tr})\left(\frac{E_{tr}}{B} - \frac{J^2}{3}\right) dE_{tr}}{h\rho(E,J)}. \tag{7.54}$$

The integration limits, BJ^2 are the energies associated with $j = J$ (see fig. 7.23), where B is the rotational constant of the spherical product. The final integration limit, $E - E_o$ is the maximum available energy (vertical PST line in fig. 7.23). With the assumption that the molecule is initially cold, so that only second integral is important, a useful PST rate constant for a molecule breaking apart into a spherical top plus an atom is

$$k(E,J) = \frac{1}{h\rho(E,J)} \int_0^{E-E_o} \rho_A(E - E_o - E_{tr})\,\frac{E_{tr}}{B}\,dE_{tr}. \tag{7.55}$$

This equation deserves some comment. The integral is the convolution of the vibrational density of states with the translational/rotational density. The difference between it and the usual RRKM rate constant is that the transition state is assumed to be product like, consisting of the product vibrations plus the spherical product rotations. The term E_{tr}/B represents the vibrational/translational density of states (two degrees of freedom from each).

It is interesting to note that the flexible variational RRKM model approaches PST as the energy is lowered and the separation between the product fragments becomes very large at the variational transition state. This is because the $\ell^2/2\mu R^2$ and V terms in equation (7.36) become negligible so that the flexible Hamiltonian becomes

$$H = H_{v1} + H_{v2} + T_{r1} + T_{r2} + E_t, \tag{7.56}$$

which is the Hamiltonian for PST. PST and the flexible variational RRKM model use different sets of conjugate coordinates and momenta to determine phase space volumes for the degrees of freedom that are treated classically. However, both models use the same conservation laws, namely total energy and that of angular momentum. When the flexible Hamiltonian approaches that of PST, the flexible variational RRKM rate constant approaches that of PST. However, as the energy increases, PST continues to assume the loosest of all transition states, whereas it tightens up in VTST. Hence, PST always represents an upper limit to the rate which is generally considerably higher than experimentally measured rates.

7.4.2 Orbiting Transition State Phase Space Theory

Chesnavich and Bowers (1977a,b; 1979) modified the phase space theory model by assuming (a) an orbiting transition state located at the centrifugal barrier, and (b) that orbital rotational energy at this transition state is converted into relative translational energy of the products. The Hamiltonian used for this orbiting transition state/phase space theory (OTS/PST) is identical to that in equation (7.36) for the flexible transition state model, except that there is no anisotropy in the intermolecular potential.

The differential unimolecular rate constant [similar to Eqs. (7.52) and (7.53)] can be written in a general way as

$$k(E, J, E_{tr}) = \frac{\rho_{A,B}(E - E_o - E_{tr})\Gamma_{ro}(E_{tr}, J)}{h\rho(E, J)} \tag{7.57}$$

so that the integrated rate constant becomes

$$k(E, J) = \frac{1}{h\rho(E, J)} \int_{E_{tr}^{*}}^{E-E_o} \rho_{A,B}(E - E_o - E_{tr})\Gamma_{ro}(E_{tr}, J)dE_{tr}, \tag{7.58}$$

where $\rho_{A,B}$ is the vibrational density of states of the two products, both of which may be polyatomics, and Γ_{ro} is the product sum of rotational-orbital states with rotational-orbital energy less than or equal to E_{tr}. The statistical factor has not been included in equation (7.58).

The essence of PST or OTS/PST lies in the evaluation of the Γ_{ro} function. It is obtained by evaluating the following integral in the j–ℓ plane:

$$\Gamma_{ro}(E_{tr}, J) = \int\int \Gamma(E_r, j)djd\ell, \tag{7.59}$$

where $\Gamma(E_r,j)$ is the product rotational sum of states. The functional forms for various combinations of product rotational states have been tabulated (Chesnavich and Bowers, 1979), and are listed in table 9.2. The limits of j and ℓ in equation (7.59), which are determined by conservation of energy and angular momentum, are shown as the shaded area in figure 7.23, but this time limited on the right by the OTS/PST line. The difference between PST and OTS/PST is that in the former the integration limits are determined with the TS at infinity, where as in the latter, they are determined with the TS at the centrifugal barrier. This means that the integration in the $j - \ell$ plane of figure 7.23 is limited by the OTS/PST line rather than in the vertical PST line.

The OTS/PST line in figure 7.23 is determined by noting that E_t is greater than the centrifugal barrier. This barrier for an interaction potential of the form $-a/r^n$ is given in equation 7.13b. Thus, the minimum translational energy $E_t^{\ddagger}$ is given by

$$E_t^{\ddagger} = \frac{\ell^{2n/(n-2)}}{\left(\frac{n\mu}{\hbar^2}\right)^{n/(n-2)} a^{2/(n-2)} \left(\frac{2}{n-2}\right)} = \frac{\ell^{2n/(n-2)}}{\Lambda}, \tag{7.60}$$

where Λ is the expression in the denominator. In converting equation (7.13b) into (7.60), P_θ was replaced by $\hbar[\ell(\ell + 1)]^{1/2} \approx \hbar\ell$. Noting that $E_{tr} = E_t + j^2B_r$, where B_r is the effective rotational constant of the products, we can solve equation (7.60) for the maximum ℓ:

$$\ell_{max} = \Lambda^{(n-2)/2n} (E_{tr} - j^2B_r)^{(n-2)/2n}, \tag{7.61}$$

which is the line labeled OTS/PST in figure 7.23. In general the integrals over j and ℓ in equation (7.59) must be evaluated numerically. However, Olzmann and Troe (1992) have proposed an approximate method for evaluation of the integrals in equations

(7.58) and (7.59). It is based on an empirical interpolation between analytical expressions for these integrals in the low and high J limits. Additional information is given in chapter 9.

7.5 THE STATISTICAL ADIABATIC CHANNEL MODEL (SACM)

In the adiabatic theory of chemical reaction rates, the reactive system is assumed to remain in the same vibrational/rotational quantum level as it moves along the reaction path (Eliason and Hirschfelder, 1959; Hofacker, 1963; Mies, 1969; Marcus, 1966; Quack and Troe, 1974, 1981; Troe, 1983). This gives rise to adiabatic potential energy curves, a schematic diagram of which is shown in figure 7.24 for the case of a unimolecular reaction. For the same reason that electronic potentials of the same symmetry cannot cross (non-crossing rule), vibronic potential curves of the same symmetry in figure 7.24 do not cross (Gilbert and Smith, 1990). Since transitions are not allowed between the adiabatic curves, it may seem incongruous that a statistical theory can be derived from such a model. However, this is done in SACM by assuming that the energized molecules have a statistically distributed set of quantum numbers (Quack and Troe, 1974, 1981; Troe, 1979). Instead of determining the actual adiabatic potential energy curves, diabatic potential energy curves are constructed by assuming

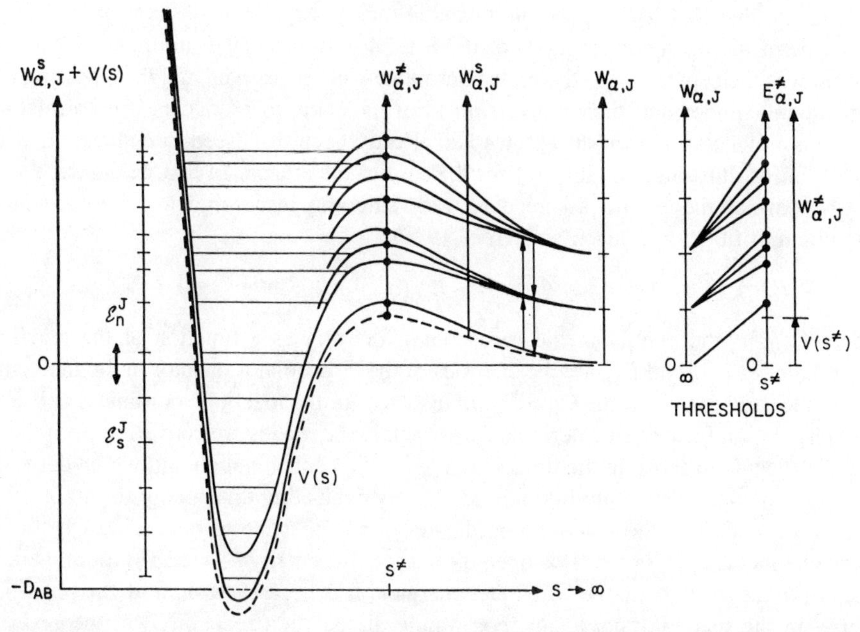

Figure 7.24 Adiabatic vibronic reaction channels in the statistical adiabatic channel model (SACM) as a function of the reaction coordinate, s. ϵ^J_n and ϵ^J_n are, respectively, the continuum and bound state energy levels of the molecule. $W_{\alpha,J}$ are the final product energies when $s = \infty$. These degenerate final product states are adiabatically related to the various barriers at $s^{\ddagger}$. This figure is taken with permission in slightly modified form from Mies (1969).

that quantum numbers for a potential curve are conserved in going from reactants to products. If the diabatic curve rises above the total energy, the rate for that molecular state (i.e., the diabatic state) is zero. Otherwise, the rate is given by

$$k(E, v, J) = \frac{1}{h\,\rho(E, J)}, \tag{7.62}$$

where the v refers to the vibrational quantum numbers for a diabatic curve, and $\rho(E, J)$ is the molecular density of states. The total rate constant is the sum of all the initial molecular states which have sufficient energy to surmount all barriers along their diabatic curves. An example of a molecular state in the dissociation of $H_3C—X \rightarrow H_3C + X$ that does not surmount the barrier would be one in which the energy residing in the C—H vibrations exceeds the excess energy, $E - E_o$.

The total decay rate is then given by

$$k(E, J) = \sum_i \frac{\mathrm{h}(E - E_i^{max})}{h\,\rho(E, J)} = \frac{N_p(E - E_o)}{h\,\rho(E, J)}, \tag{7.63}$$

where the sum extends over all states that lead to products. In the numerator of this equation, $\mathrm{h}(E - E_i^{max})$ is the step (or heavyside) function, which is either 0 if the total energy $E < E_i^{max}$, or is 1 if $E \geq E_i^{max}$. E_i^{max} is the maximum energy in each diabatic channel. This model conserves angular momentum by including centrifugal barriers along the reaction path. Since SACM uses diabatic instead of adiabatic curves, it is unfortunate that SACM is called an adiabatic theory.

One of the main contributions of the SACM to the theory of unimolecular dissociation was the separation of the molecular modes into conserved and disappearing (or transitional) modes and their conversion from molecule to products. The correlation between molecular and product states has already been discussed in connection with VTST and is illustrated in table 7.4 for the case of the ethane. In SACM the energy of all the normal modes are assumed to vary smoothly according to an exponential interpolation function (Quack and Troe, 1974):

$$E_i(q) = E_{i,\mathrm{m}}(q_\mathrm{e}, v_i)\, e^{-\alpha(q-q_\mathrm{e})} + E_{i,\mathrm{p}}(q_\infty, v_i)(1 - e^{-\alpha(q-q_\mathrm{e})}) + V_{\mathrm{cent}}, \tag{7.64}$$

where $E_i(q)$ is the energy in the ith oscillator or rotor as a function of the reaction coordinate, q; $E_{i,\mathrm{m}}$ and $E_{i,\mathrm{p}}$ are the energies of the ith oscillator or rotor in the molecule and in the products, q_e is the equilibrium distance for the reaction coordinate, and α is an interpolation factor that determines how rapidly the modes are converted from those in the molecule to those in the products. V_{cent} is the centrifugal energy which depends upon the angular momentum in the products. By evaluating this energy for all oscillators as a function of the reaction coordinate q, it is possible to determine whether a given channel is open or not. It is open if the available energy $E > \Sigma E_i(q)$ for all values of q. In order to evaluate these channel energies, it is necessary to know the potential energy in the reaction coordinate. For simple dissociation reactions with no reverse activation barriers, it is generally assumed to be a Morse potential. The major adjustable parameters are then α in equation (7.64) and β which enters into the Morse potential. It is worthwhile noting that some dissociation reactions with no reverse barrier have radial potentials that are not well described by a Morse function.

This interpolation function [Eq. (7.64)] is a reasonable approximation for a reaction that has no barrier. However, for reactions with barriers, the SACM needs to consider the structure of the transition state and it thus becomes very similar to standard RRKM theory. In fact, SACM can be considered as a theory which includes standard RRKM and PST as special cases (Troe, 1983). In this respect it is similar to variational transition state theory.

The SACM has been applied to the NO_2 dissociation (Quack and Troe, 1974; Troe, 1983). However, no $k(E)$ curves over a sufficiently broad energy range have been measured to test the SACM derived rate curves. On the other hand, the NO_2 dissociation has been extensively studied with the VTST (Wardlaw and Marcus, 1985). The comparison of $N(E, J)$ vs. E (where it is understood that the $N(E, J)$ is summed differently in the two theories) shows that they agree up to the investigated excess energy of 40 kcal/mol. This is rather significant because over this energy range, the location of the transition state in the VTST model ranges from about 2.3 to 3 Å.

Another system which has been treated in a rather complete manner is the dissociation of HOOH (Brouwer et al., 1987). The rates as well as the product energy distributions were calculated. As with the NO_2 reaction, the interaction potential was assumed to have no barriers so that E_i^{max} for each HOOH reaction channel is assumed to be associated with the centrifugal barrier. In order to calculate this barrier, the reaction is treated as a triatomic dissociation, $ABC \rightarrow AB + C$. The effective rotational constant, B_{cent}, at the centrifugal barrier is calculated according to formulas derived by Troe (1983). In addition, the model was simplified by replacing two adjustable parameters, α (from the interpolation function) and β (from the Morse potential), by their ratio, α/β. A value of 0.44 was found to adequately account for the data. Figure 7.25 shows the comparison of the SACM $k(E)$ curves with those obtained from experiments or trajectory calculations.

The full implementation of SACM is a large task because of the many vibronic channels that must be individually followed along the reaction path. In order to reduce the problem to manageable proportions, simplified forms of SACM have been proposed (Troe, 1983). In addition to the α parameter in the interpolation function, a β value must be chosen for the Morse function. These are related and Troe has suggested that a single parameter, the previously mentioned ratio of α/β, is sufficient to characterize most reactions. Additional parameters that enter into the calculation of the energy maxima have been adjusted to yield "reasonable" rate vs. energy curves (Troe, 1981, 1983). However, even with these simplifications, there remain difficulties in applying the SACM to large molecules.

The large number of adjustable constants that have been used in most applications of SACM to date undoubtedly accounts to some extent for the ability of SACM to fit data. It is thus difficult to evaluate the validity of the SACM. There are some fundamental differences between the assumptions in SACM and RRKM. In the latter, energy is statistically distributed among all modes as the molecule evolves toward the transition-state structure. Hence a molecule which is initially prepared in a single energy state is assumed to distribute its energy among all the available states during the course of reaction. On the other hand, in SACM, the rather artificial assumption must be made that a molecule which is initially prepared in a single state distributes its energy by IVR (contrary to the adiabatic hypothesis) in order to establish a microcanonical distribution of states. Once the energy is dispersed and the molecule begins to dissociate, the molecule remains in the same diabatic state. This has been justified

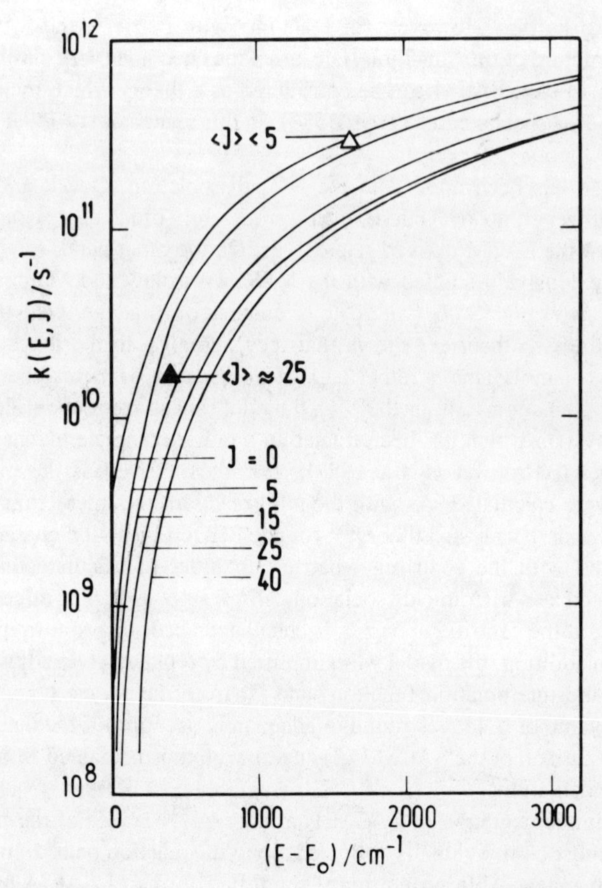

Figure 7.25 Calculated and measured dissociation rates for the HOOH reaction. The open triangle is the upper limit of the rate inferred from a line widths measurement at the fifth overtone transition. The solid triangle is from a time resolved measurement (Scherer and Zewail, 1987). The solid lines are SACM calculated $k(E, J)$ curves for various values of J. Taken with permission from Brouwer et al. (1987).

by proposing a time lag prior to dissociation during which time IVR can mix the modes (Troe, 1991). However, during the much shorter final separation of products, the states remain diabatic.

Recent experiments on the ketene dissociation appear to have shed some light on the validity of the diabatic assumption of SACM (Green et al., 1991). As the potential energy surface in figure 6.10 illustrates, the production of singlet CH_2 + CO from ketene involves no barriers since the dissociation products appear at the thermochemical threshold. The reaction was followed by measuring the product fluorescence excitation spectra (PHOFEX) signal for CH_2 in various rotational and vibrational states. As the exciting laser light was scanned to higher energies, the PHOFEX signal increased in steps which are related to the opening up of new product states, including the CO rotational levels. An example is shown in figure 7.26 where the $[0,2,0]1_{01}$ level of CH_2

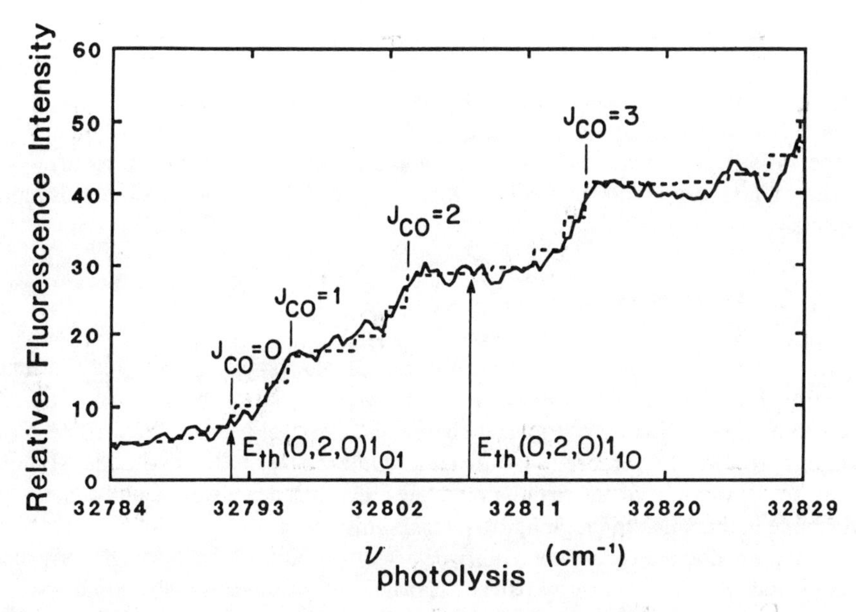

Figure 7.26 The dissociation of H_2CCO (ketene) produces singlet CH_2 + CO without a barrier. The spectrum above is an excitation function (PHOFEX spectrum) in which the singlet methylene is monitored as a function of the photolysis energy. The signal is integrated over the whole dissociation time so that the steps in the PHOFEX spectrum correspond to the opening up of new CO(J) channels. These appear at their thermochemical onsets which indicates the absence of dynamical barriers in this reaction. Taken with permission from Green et al. (1991).

is monitored. At the onset near 32,793 cm^{-1} only the $J = 0$ level of CO is energetically possible. However, as the excitation energy is increased, new CO rotational levels are observed. Of particular importance is the fact that these steps appear precisely where expected on the basis of product energies.

The rate of the ketene reaction is considerably less than that predicted by PST (Potter et al., 1989; Kim et al., 1991). While the VTST is capable of fitting these measured rates with no adjustable parameters (Klippenstein et al., 1994), the SACM accounts for rates less than the PST limit by imposing barriers thereby preventing certain diabatic channels from reacting. In particular it predicts rotational barriers which increase in size with product rotational energy. These predictions appear to be contrary to the experimental findings which exhibit no dynamical barriers in the ketene reaction to within at least 0.5 cm^{-1} and up to at least $J = 3$ (Green et al., 1991). A similar situation exists for the NCNO dissociation in which the rates are significantly slower than the PST limit, yet all final product channels are open.

It appears that a strictly diabatic (or adiabatic) theory cannot account for the above data. The SACM has been modified by including some nondiabaticity (Dashevskaya et al., 1990, 1992), and it may be that with an appropriate mix of diabatic and nondiabatic curves, the ketene and NCNO decay rates could be accommodated. In this regard, it is interesting that some adiabaticity has recently been introduced into variational RRKM theory (Wardlaw and Marcus, 1988). However, this does not affect the predicted dissociation rate because it imposes adiabaticity only to the conserved modes from the transition state to products. It is introduced in order to permit RRKM theory to account for product energy distributions (see section 9.6.1, page 359). In summary it is perhaps fair to state that the statistical adiabatic channel model, while containing a number of attractive features, is cumbersome and complex to apply to ordinary size molecules, and that its fundamental assumption of adiabaticity (or diabaticity) is highly suspect.

7.6 TUNNELING AND RRKM THEORY

Reactions with saddle points are subject to tunneling, and energy-selected molecules are particularly good systems in which to observe this effect. This is because a molecule can be prepared in an energy state below the classical barrier where 100% of the reactions are forced to proceed via tunneling. On the other hand, in thermal systems, the broad distribution of internal energy states, combined with the rapidly increasing $k(E)$ curve in the tunneling region, often masks this effect. Yet, even in energy-selected systems, tunneling effects are not readily detected without either a thorough knowledge of the classical potential energy barrier, or data with isotopically labeled atoms.

One of the first studies that incorporated exact tunneling calculations with a calculated reaction coordinate into transition state theory was that of Truhlar and Kuppermann (1971). Similar studies have been carried out more recently (Truhlar et al., 1982; Skodje et al., 1984; Isaacson et al., 1987; Miller et al., 1990; Gray et al., 1980, 1981; Moiseyev et al., 1989). However, these methods have been applied mostly to tri- or tetra-atomic molecules because a sufficient knowledge of the reaction path Hamiltonian for larger systems is difficult to obtain. We thus outline below an approximate scheme for implementing tunneling in RRKM calculations. It assumes that the reaction coordinate is fully separable from the other modes so that tunneling proceeds only through the barrier in the one-dimensional reaction path.

The RRKM rate constant with tunneling is given by (Miller, 1979, 1987; Troe, 1986):

$$k(E) = \int_{-E_o}^{E-E_o} \kappa(\epsilon_t) k(E, \epsilon_t) d\epsilon_t = \frac{1}{h\,\rho(E)} \int_{-E_o}^{E-E_o} \kappa(\epsilon_t) \rho^{\ddagger}(E - E_o - \epsilon_t) d\epsilon_t, \qquad (7.65)$$

where ϵ_t is the translational energy in the reaction coordinate and $\kappa(\epsilon_t)$ is the tunneling probability. This expression is very similar to the normal RRKM rate constant except that the integral now extends to energies below the barrier, that is, negative energies. Below the barrier ($\epsilon_t < 0$), the value of κ is less than 1 and it approaches 1 as $\epsilon_t \rightarrow \infty$. At the barrier, the probability is just 0.5 because half of the trajectories are reflected and the other half transmitted. Figures 7.27(a) and b show the two situations in which the total energies are below or above the classical barrier. Because both the density of states function, $\rho^{\ddagger}$, and $\kappa(\epsilon_t)$ increase with the energy, the two functions in the integrand go in opposite directions.

7.6.1 The Tunneling Probability

In order to obtain the tunneling probability, a functional form of the one-dimensional potential barrier, $V(R)$, is required. This is then substituted into the one-dimensional

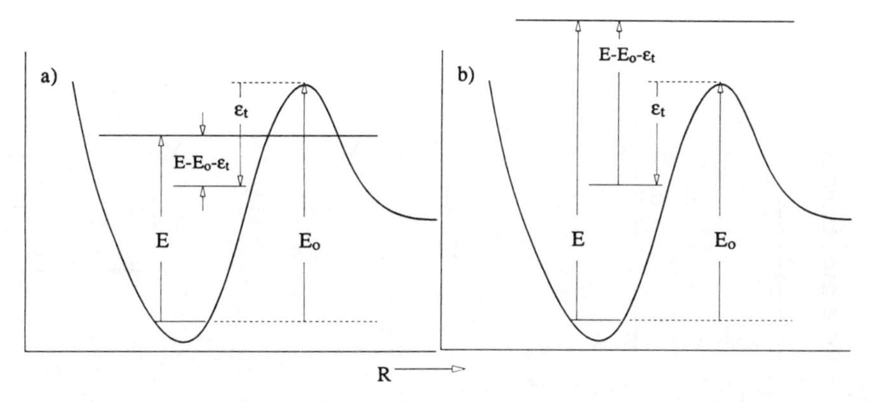

Figure 7.27 Potential energy diagram for a reaction that includes tunneling. The two examples are for energies below and above the barrier.

Schrödinger equation, for which solutions can be obtained to the left of the barrier, inside the barrier where the kinetic energy is less than zero, and again to the right of the barrier where the kinetic energy is positive. By making the wave functions continuous at the interfaces, the transmission probability for any potential function can be solved, although not necessarily in a closed analytical form (Steinfeld et al., 1989; Nikitin, 1974).

There are several functional forms for $V(R)$, such as the inverted parabola, that yield analytical solutions for the transmission probability. Of these, the most realistic and the most commonly used model for determining the $\kappa(\epsilon_t)$ function is the Eckart barrier (Eckart, 1930; Johnston and Heicklen, 1962). This three-parameter function with unsymmetric forward and backward reaction coordinates is given by

$$V = \frac{Ay}{1+y} + \frac{By}{(1+y)^2}, \tag{7.66a}$$

$$y = e^{2\pi R/L}, \tag{7.66b}$$

$$A = V_0 - V_1, \tag{7.66c}$$

$$B = [V_0^{1/2} + V_1^{1/2}]^2, \tag{7.66d}$$

$$L/2\,\pi = [2/k^*)^{1/2}[V_0^{-1/2} + V_1^{-1/2}]^{-1}, \tag{7.66e}$$

$$k^* = 4\pi^2 \nu_c^2 \mu, \tag{7.66f}$$

where V_0 and V_1 are the barrier heights in the forward and backward directions, and ν_c is the absolute value of the critical frequency, which is related to the curvature at the top of the barrier, k^*. The latter is the force constant for this imaginary frequency, while μ is the reduced mass of the critical oscillator. L is a characteristic width of the barrier. The Eckart potential function is plotted in figure 7.28.

The transmission coefficient can be expressed in various forms, one of which is (Miller, 1979):

$$\kappa(\epsilon_t) = \frac{\sinh(a)\sinh(b)}{\sinh^2[(a+b)/2] + \cosh^2(c)}, \tag{7.67}$$

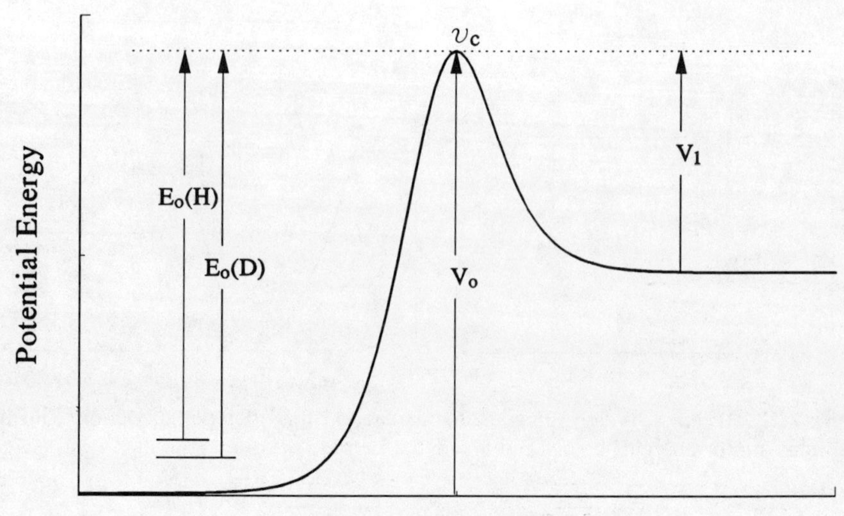

Figure 7.28 The Eckart barrier. The three parameters are V_0, V_1, and the curvature expressed as the imaginary (critical) vibrational frequency, ν_c. The zero-point energy in the reaction coordinate must be taken into account. $E_o(H)$ and $E_o(D)$ are the activation energies for normal and deuterated samples for the case of an H-atom transfer reaction.

where

$$a = \frac{4\pi}{h\nu_c}\sqrt{\epsilon_t + V_0}\,\frac{1}{V_0^{-1/2} + V_1^{1/2}}, \tag{7.68}$$

$$b = \frac{4\pi}{h\nu_c}\sqrt{\epsilon_t + V_1}\,\frac{1}{V_0^{-1/2} + V_1^{1/2}}, \tag{7.69}$$

$$c = 2\pi\sqrt{\frac{V_0 V_1}{(h\nu_c)^2} - \frac{1}{16}}\,. \tag{7.70}$$

As before, the energy in the reaction coordinate, ϵ_t, is measured from the top of the barrier. Solutions to equation (7.67) for the barrier of figure 7.28, are plotted in figure 7.29 for three values of $h\nu_c$, 3000, 2122, and 1000 cm^{-1}, but for the same curvature at the top of the barrier. These are chosen to correspond approximately to a reaction in which the critical coordinate involves an H atom transfer, a D atom transfer, and perhaps a C—C bond break. The marked differences show the well known effect of the reduced mass on the transmission probabilities. Note that even at the top of the barrier, the transmission coefficient equals 0.5 exactly. Note also that while the Eckart equation is a function of the curvature, k^*, which depends upon the critical frequency and the reduced mass, the transmission coefficient depends only upon the critical frequency.

7.6.2 Treatment of Zero-Point Energy

So far, the zero-point energy (ZPE) in both the reaction coordinate and in the rest of the normal modes has been ignored. We normally measure the energy from the ZPE level.

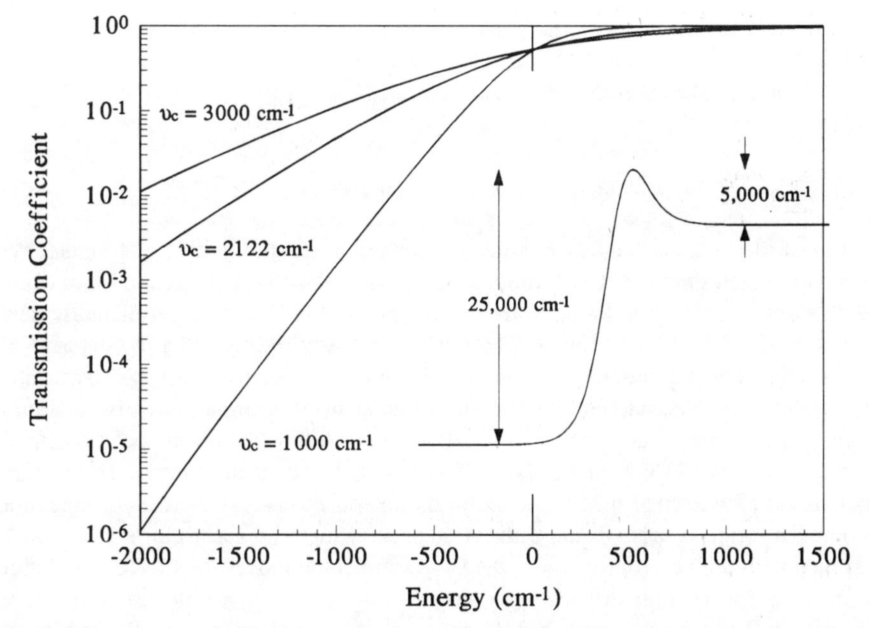

Figure 7.29 Tunneling probability for the indicated Eckart barrier with three different critical vibrational frequencies. The latter are related to the curvature at the top of the barrier. Note that the transmission coefficient is equal to 0.5 at the saddle point.

In the molecule, each normal mode, including the one for the reaction coordinate, contains ZPE. However, the Eckart potential is unbound on the reactant and product side so that there is no ZPE energy in the reaction coordinate. In reconciling these two energy references, we need consider only the energy in the one-dimensional reaction coordinate. The energy in the other modes, which are perpendicular to the reaction coordinate, can continue to be referenced to the ZPE level.

One of the advantages of referencing the reaction coordinate energy, ϵ_t, at the top of the barrier is that there is no ZPE in the reaction coordinate at this point. Some care is required, however, when reaction rates for an H and D atom transfer are calculated. Because the normal energy reference is at the ZPE of the molecule, the energies at the top of the barrier will not coincide (fig. 7.28). In other words, the deuterated molecule has a higher activation energy (but the same V_o) so that the same excitation energy places less energy in the deuterated molecule, and thus results in a slower reaction. This is a ZPE effect which is in addition to the dynamical isotope effect shown in figure 7.29.

7.6.3 Effect of Reaction Path Curvature

The above treatment ignores any effects resulting from the coupling of the reaction coordinate with the other degrees of freedom. The main feature in the potential energy surface that causes coupling is curvature in the reaction path (Miller et al., 1980). These effects cannot be corrected in the above approach in which an Eckart potential is substituted for the true reaction path since it is *a priori* a one-dimensional function. In order to include the effects of reaction path curvature it is necessary to treat the reaction path in far greater detail as has been done by a number workers (Marcus and Coltrin, 1977; Skodje et al., 1984; Kreevoy and Truhlar, 1986; Miller, 1987; Waite and Miller, 1981).

7.6.4 Examples

The most thoroughly studied unimolecular tunneling reaction is that of formaldehyde which dissociates to H_2 + CO (Troe, 1984; Miller et al., 1990; Dai et al., 1985; Miller, 1979; Gray et al., 1981; Polik et al., 1990a; Forst, 1983). The experimentally determined rates of deuterated formaldehyde (Polik et al., 1988, 1990a,b; Guyer et al., 1986) have been supported by high-level calculations. Details about this dissociation are discussed in the chapters 6 and 8. It is interesting to compare the dissociation of formaldehyde with that of the formaldehyde ion. Although the dissociation on the ground state of CD_2O^+ proceeds with no barrier and is thus rapid, D loss after excitation to the excited $\tilde{A}\ ^2A''$ state slows the dissociation rate by orders of magnitude (Bombach et al., 1981). Because the A state photoelectron spectrum is vibrationally resolved with a long progression of the C=O stretching mode, the PEPICO technique permitted measuring the dissociation rates as a function of the A state vibrational level. These data have been interpreted as tunneling through the barrier created by the intersection of the upper and lower electronic state potential energy surfaces (Lorquet and Takeuchi, 1990). The model for nonadiabatic processes is discussed in chapter 8. In both of the above reactions, only the rate constant for the deuterated species could be measured experimentally.

An example of tunneling in which both the normal and the deuterated species were measured is in HCl or DCl loss from the ethyl chloride ion (Booze et al., 1991). *Ab initio* molecular orbital calculations (Morrow and Baer, 1988) demonstrated that the HCl loss channel proceeds via a substantial barrier of 1 eV (23 kcal/mol). Yet, the onset for HCl loss occurs at about 0.3 eV. Furthermore, the reaction is very slow ($k \approx 10^5$ sec^{-1}). These facts all point to a reaction that proceeds via tunneling. The *ab initio* calculations provide the vibrational frequencies for both the ion and the transition state, which is located at a saddle point. In addition, they furnish the activation energy as well as the curvature of the barrier (the imaginary frequency). Thus, all the information required in the utilization of the Eckart barrier and equation (7.65) is given.

In order to fit the data, it was however necessary to adjust the barrier height as well as the barrier curvature. This was done in order to fit the data for normal ethylchloride ($C_2H_5Cl^+$). The *ab initio* and the experimentally determined barrier heights and curvatures are shown in table 7.6.

Table 7.6. Comparison of *ab initio* and Experimental Tunneling Parameters[a]

	V_0 (kJ/m)	f (kg/sec^2)	$\nu_H^\ddagger$(cm^{-1})	$\nu_D^\ddagger$(cm^{-1})
Ab initio	104	−144.5	1499	1096
Experiment	72 ± 3	−109 ± 3	1300 ± 15	903 ± 12

[a]Taken from Booze et al. (1991).

With the adjusted parameters, but using the calculated vibrational frequencies, it was possible to fit the rate data for the normal ethyl chloride ion. However, with no additional adjustable parameters, these same values (adjusted only for the mass change) permitted the fitting of the deuterated data. The resulting rates for the normal and deuterated ions are shown in figure 7.30. These calculated curves passed through the data in the region between 10^5 and 10^6 sec^{-1}. In this region the isotope effect is about three orders of magnitude which is the combined effect of the ZPE shift and the different tunneling probabilities for HCl and DCl loss.

The integral in equation (7.65) has two terms: the transmission coefficient which increases as ϵ_t increases (becomes more positive) and the transition state density of states which goes in the opposite direction. We can determine the major contribution to tunneling by plotting the integrand as a function of ϵ_t. As shown in figure 7.31 the major contribution to the total reaction of ethyl chloride ions is from tunneling at the highest possible energies. Larger molecules with larger densities of states will have proportionately greater contribution from lower energies. It depends simply on the rate of increase of the density of states and of the tunneling probability with energy. In the case of ethyl chloride, the tunneling probability varies more rapidly with energy than does the density of states.

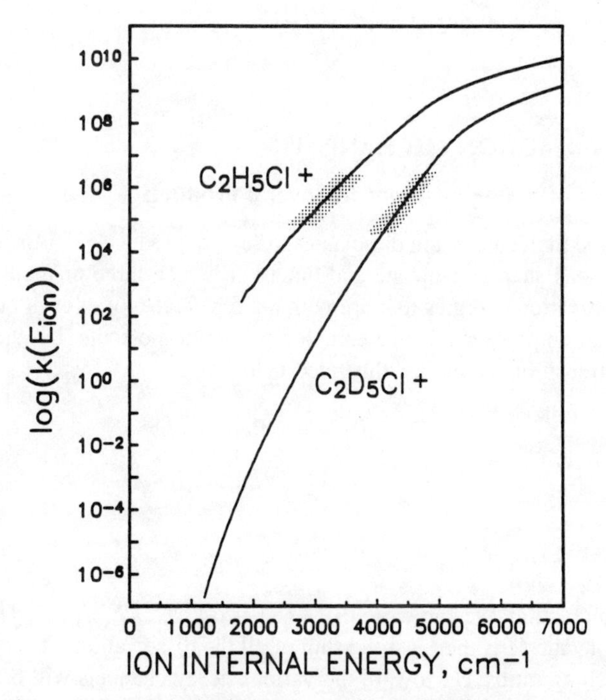

Figure 7.30 The experimental (shaded areas) and calculated decay rates via tunneling for normal and deuterated ethyl chloride. The reaction coordinate for HCl or DCl loss is primarily the transfer of the H (D)-atom from the terminal carbon to the Cl atom. Taken with permission from Booze et al. (1991).

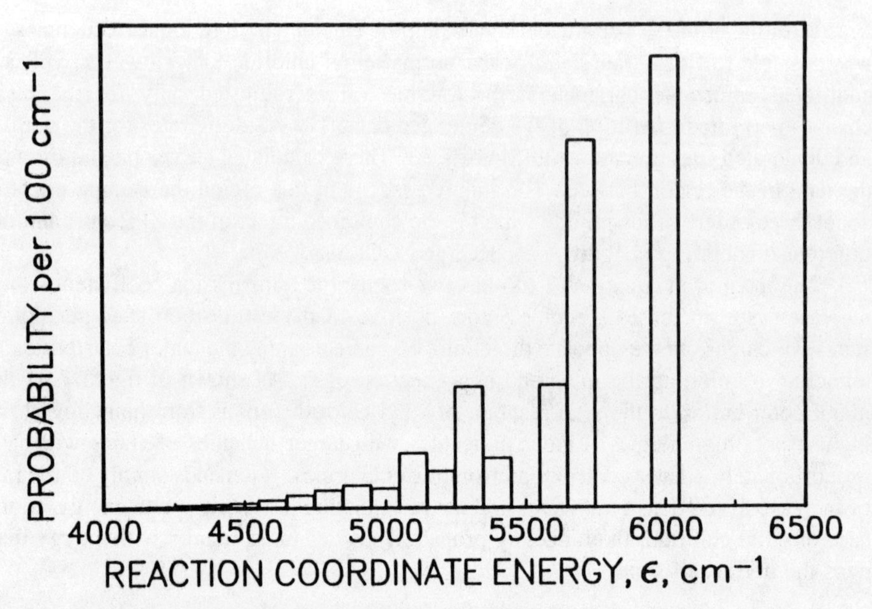

Figure 7.31 The distribution of tunneling rates for HCl loss from the ethylchloride ion as function of the energy in the reaction coordinate. The maximum energy is 6000 cm^{-1}. Taken with permission from Booze et al. (1991).

7.7 COMPLEX REACTION MECHANISMS

7.7.1 Reactions of a Single Isomer to Several Products

Often, an excited molecule or ion dissociates to several products. This is well known in such common ions such as benzene and butene in which three or more products are formed with activation energies that are within a few kcal/mol of each other. Suppose that the dissociation proceeds from a common precursor molecule, but the dissociation is via several transition states, as illustrated below:

$$\begin{array}{ccc} P_1 & & P_2 \\ \nwarrow & & \nearrow \\ & M & \\ \swarrow & & \searrow \\ P_3 & & P_4 \end{array} \tag{7.71}$$

Then, for each reaction, we can write that $k_i(E) = \sigma N_i^\ddagger(E - E_o)/\rho(E)$. The total decay rate for the molecule M is then just the sum of all the dissociation rate constants, $k_T = \Sigma k_i$. The branching ratios, B_i, toward the various decay channels will be given by the ratio of rate constants, k_i/k_T. The concentration of M as a function of time continues to be given by a single exponential decay with the form, $\exp(-k_T t)$. Thus, from the rate measurement of any one channel and the branching ratios to the various dissociation paths, it is possible to extract the absolute rates for all dissociation paths.

7.7.2 Reaction from an Equilibrated Mixture of Isomers

A more complicated situation arises when a reacting species is in equilibrium, or partial equilibrium, with other isomers. This happens in both large neutral molecules (Orchard and Thrush, 1974; Cromwell et al., 1991) as well as in ions (Rosenstock et al., 1982; Baer, 1986; Bunn and Baer, 1986). An example of a two-state problem is shown in figure 7.32 with isomers A and B. Four rates are involved in the formation of products 1 and 2 by the following reaction mechanism:

$$P_1 \overset{k_1}{\leftarrow} A \underset{k_3}{\overset{k_2}{\rightleftharpoons}} B \overset{k_4}{\rightarrow} P_2. \tag{7.72}$$

Consider first the case in which $k_1 = 0$, so that only a single product, P_2, is formed. The total rate of decay is then given by

$$\text{Decay rate} = k_4[\mathbb{N}_b] = \frac{N_4^{\ddagger}(E - E_0)}{\mathrm{h}\,\rho(E_b)}\,[\mathbb{N}_b], \tag{7.73}$$

where $[\mathbb{N}_b]$ is the concentration of molecules which can be identified as having structure B, $\rho_b(E_b)$ is the density of states of molecules having the structure B, while $N_4^{\ddagger}(E - E_o)$ is the sum of states involved in the dissociation step. The energy, E_b here is referenced to the bottom of the B isomer well.

If A and B are in equilibrium, the condition $k_2, k_3 >> k_4$ is fulfilled and a microcanonical distribution of molecules is maintained at all times. The microcanonical equilibrium condition is:

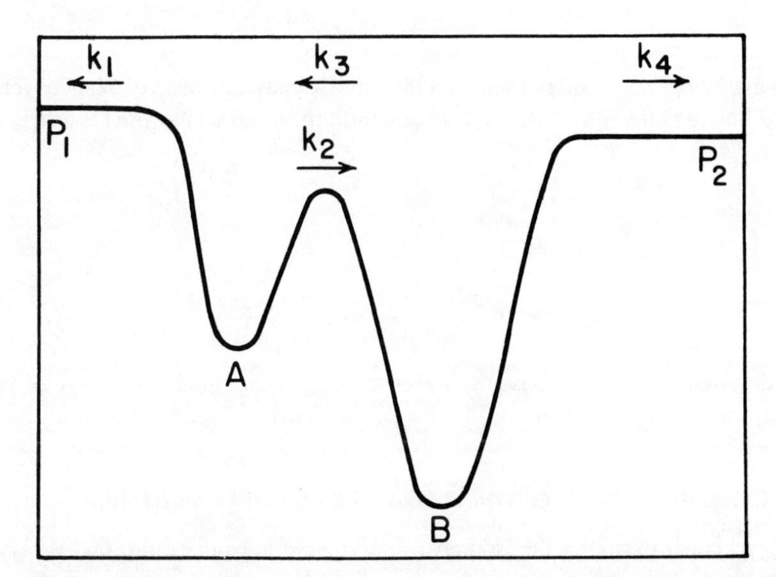

Figure 7.32 The potential energy diagram for a reaction in which isomers A and B are in partial equilibrium with two dissociation channels. Taken with permission from Baer et al. (1983).

$$k_2\mathbb{N}_a = k_3\mathbb{N}_b \quad \text{or} \quad k_2/k_3 = \mathbb{N}_b/\mathbb{N}_a. \tag{7.74}$$

The equilibration steps with rate constants k_2 and k_3 can be expressed in terms of the density of states and the sum of states for isomerization, N_{iso}, through the RRKM equation so that $k_2 = N^{\ddagger}_{iso}(E - E_o)/\rho_a(E_a)$ and $k_3 = \sigma N^{\ddagger}_{iso}(E - E_o)/\rho_b(E_b)$. Hence, the concentrations can be related to the density of states as:

$$[\mathbb{N}_a]/\rho_a(E_a) = [\mathbb{N}_b]/\rho_b(E_b) \quad \text{or} \quad [\mathbb{N}_a]/[\mathbb{N}_b] = \rho_a(E_a)/\rho_b(E_b). \tag{7.75}$$

This is the microcanonical equilibrium constant. The canonical counterpart is written in terms of the partition functions.

In many measurements it is not possible to distinguish isomers A and B so that the concentration of B is not known. However, the total concentration, $\mathbb{N} = \mathbb{N}_a + \mathbb{N}_b$ is known. Of more interest is therefore the total decay rate constant, expressed as:

$$\textit{Decay rate} = k_{obs}[\mathbb{N}] = k_{obs}[\mathbb{N}_a + \mathbb{N}_b]. \tag{7.76}$$

We can express the concentration, $[\mathbb{N}_b]$, in terms of the total concentration, $[\mathbb{N}]$, and the density of states and thus obtain the total, or observed, decay rate constant as

$$k_{obs} = \frac{N^{\ddagger}_4(E - E_0)}{h[\rho_a(E_a) + \rho_b(E_b)]}. \tag{7.77}$$

Equation (7.77) shows that the additional well adds to the density of states and thus slows down the reaction.

However, what happens when reaction can proceed from both isomers to P_1 and P_2? The two products might be the same or different. The decay rate will now be given by

$$\textit{Decay rate} = k_1[\mathbb{N}_a] + k_4[\mathbb{N}_b], \tag{7.78}$$

where the two dissociation rates are not the same because the density of states for A and B are different. Still, $[\mathbb{N}_a]$ and $[\mathbb{N}_a]$ are at equilibrium and $[\mathbb{N}] = [\mathbb{N}_a] + [\mathbb{N}_b]$, so that

$$[\mathbb{N}_a] = \frac{\rho_a(E_a)}{\rho_a(E_a) + \rho_b(E_b)} \quad [\mathbb{N}_b] = \frac{\rho_b(E_b)}{\rho_a(E_a) + \rho_b(E_b)}. \tag{7.79}$$

Thus, the total decay rate will be

$$\textit{Decay rate} = \frac{N^{\ddagger}_1(E - E_0) + N^{\ddagger}_4(E - E_0)}{h[\rho_a(E_a) + \rho_b(E_b)]}[\mathbb{N}]. \tag{7.80}$$

Thus the total decay rate constant is proportional to the number of ways of passing through the two transition states, divided by the density of states of both of the wells.

7.7.3 Competition between Direct Dissociation and Isomerization

In the preceding examples the isomers A and B were in equilibrium, that is, k_2 and k_3 were assumed to be much greater than k_1 and k_4. However, it is also possible that the various rates are more nearly equal. In that case the coupled differential rate equations must be solved (Baer et al., 1975, 1983). The formation rate of P_1 and P_2 is

$$\frac{dP_1}{dt} = k_1[A] \quad \text{and} \quad \frac{dP_2}{dt} = k_4[B] \,. \tag{7.81}$$

The rate is thus dependent upon the concentrations of A and B which change with time. The time dependencies of $[A]$ and $[B]$ are given by a set of coupled, linear, homogeneous differential equations:

$$\frac{dA}{dt} = (-k_1 - k_2)A + k_3B = a_1A + b_1B \,, \tag{7.82a}$$

$$\frac{dB}{dt} = k_2A + (-k_3 - k_4)B = a_2A + b_2B \,, \tag{7.82b}$$

which has solutions of the form

$$A = \alpha e^{-\Gamma t} \quad \text{and} \quad B = \beta e^{-\Gamma t}. \tag{7.83}$$

Substituting these solutions into the coupled differential equations yields the matrix equation

$$\begin{bmatrix} a_1 + \Gamma & b_1 \\ a_2 & b_2 + \Gamma \end{bmatrix} \begin{bmatrix} \alpha \\ \beta \end{bmatrix} = \begin{bmatrix} 0 \\ 0 \end{bmatrix} . \tag{7.84}$$

This equation only has solutions when the determinant of the 2×2 matrix vanishes. This condition yields the following values of Γ:

$$\Gamma_\pm = -\tfrac{1}{2}(a_1 + b_2) \pm \tfrac{1}{2}[(a_1 + b_2)^2 + 4a_2b_1 - 4a_1b_2]^{1/2}. \tag{7.85}$$

Because the general solutions for A and B are given by linear combinations of the plus and minus solutions, the time dependent solutions of A and B are

$$A(t) = c_1\alpha_+e^{-\Gamma_+t} + c_2\alpha_-e^{-\Gamma_-t}, \tag{7.86a}$$

$$B(t) = c_1\beta_+e^{-\Gamma_+t} + c_2\beta_-e^{-\Gamma_-t}. \tag{7.86b}$$

The α and β coefficients are related by the conditions set by equation (7.86). However, only the ratio of α and β is determined, so that we can arbitrarily set one of them (for instance β) equal to 1. The other constraints which determine the coefficients c_1 and c_2 are the initial conditions. Suppose that A is initially formed at an energy above the dissociation limit of both P_1 and P_2. As a result, $B(0) = 0$ so that $A(t)$ and $B(t)$ reduce to

$$A(t) = \alpha_-e^{-\Gamma_-t} + \alpha_+e^{-\Gamma_+t}, \tag{7.87a}$$

$$B(t) = e^{-\Gamma_-t} - e^{-\Gamma_+t}, \tag{7.87b}$$

where we have written the slow rate, Γ_- first. The rates of product formation can now be expressed in terms of the time evolution of $A(t)$ and $B(t)$. However, because the Γ's are rather complicated, it is useful to consider several limiting cases. One such case is the previously treated rapid isomerization in which $k_2 \approx k_3 >> k_1$, k_4. The net result is that both products are formed with only a slow rate constant which is a weighted sum of the two slow rates:

$$\Gamma_- = k_1\left(\frac{k_3}{k_2 + k_3}\right) + k_4\left(\frac{k_2}{k_2 + k_3}\right). \tag{7.88}$$

When the sums and densities of states are introduced for each of the rates, we obtain the k_{obs} of equation (7.80).

7.7.3.1 Initial Preparation of Isomer A

Consider now a system in which the isomerization of A is in competition with direct dissociation. In that case $k_1 \approx k_2 >> k_3 \approx k_4$. The derived fast and slow rates are now

$$\Gamma_+ \cong k_1 + k_2 \quad \text{and} \quad \Gamma_- \cong k_4 + k_3\frac{k_1}{k_1 + k_2} \tag{7.89}$$

and the resulting decay rates for A and B are given by

$$\frac{dP_1}{dt} = k_1[A] = k_1\left(\frac{k_3}{k_1 + k_2}\, e^{-\Gamma_- t} + \frac{k_1 + k_2}{k_2}\, e^{-\Gamma_+ t}\right) \tag{7.90a}$$

$$\frac{dP_2}{dt} = k_4 B = k_4(e^{-\Gamma_- t} - e^{-\Gamma_+ t}). \tag{7.90b}$$

Unlike the previously considered case of fast isomerization, products P_1 and P_2 are produced with different rates. Product P_2 will be formed only with a slow rate constant because the second term in equation (7.90b) becomes negligible when $t > 0$. On the other hand, P_1 will be formed with a two-component exponential function because the coefficient for the second fast rate term in equation (7.90a) is much larger than the coefficient for the first term. Thus, their integrated areas will be comparable.

When the slow and fast rates in equation (7.89) are expressed in terms of the density and sums of states, we obtain

$$k_{slow} = \frac{1}{h\rho_b(E)}\left(N_4 + \frac{N_{iso}N_1}{N_1 + N_{iso}}\right) \qquad k_{fast} = \frac{N_1 + N_{iso}}{h\rho_a(E)}. \tag{7.91}$$

These dissociation rates are expressed in terms of the four rate constants. The experimental data could contain as many as four observables. These are the fast and slow dissociation rates, the branching ratio between P_1 and P_2, and finally, the branching ratio between fast and slow components in the P_1 production. Thus, all rate constants can be measured. If only the slow rate constant is measurable, then a single RRKM calculation for one of the rates, permits all the others to be determined experimentally.

7.7.3.2 Initial Preparation of Isomer B

Suppose that the rate constants are as they were in the previous case, but now the more stable B isomer (see fig. 7.32) is initially prepared. The Γ_- and Γ_+ will remain the same, but now $A(0) = 0$ and the resulting rates of product formation are given by

$$\frac{dP_1}{dt} = k_1[A] = k_1(e^{-\Gamma_- t} - e^{-\Gamma_+ t}), \tag{7.92a}$$

$$\frac{dP_2}{dt} = k_4[B] = k_4\left(\frac{k_1 + k_2}{k_3}\, e^{-\Gamma_- t} + \frac{k_3}{k_1 + k_2}\, e^{-\Gamma_+ t}\right). \tag{7.92b}$$

It is evident that both products will now be formed with just a slow rate constant because the coefficients for the exponential functions are all of the order of one so that the fast component will be negligibly small. Furthermore, the slow rate constant is the same as the slow rate constant when A is initially prepared.

An example of this situation has been found in the dissociation of the cyclopentane ion and five of its isomers (Brand and Baer, 1984; Duffy et al., 1995). Figure 7.33 shows the product TOF distributions for two of the ions at approximately the same total energy. The cyclopentane ion has both a fast and a slow path for the loss of C_2H_4, but only a slow path for the loss of CH_3. On the other hand, the lowest energy $C_5H_{10}^+$ isomer, 2-methyl-2-butene ion, has only a slow path for the loss of both ions. The slow rate is the same in both cases.

7.7.3.3 Product Formation via a High-Energy Intermediate

The last case to be considered is one which may be more commonly encountered than realized. This is the case in which the dissociation proceeds to a single product via an isomerization to a relatively long-lived, but high-energy, complex. The potential energy diagram, in which the isomerization barrier is below the dissociation limit, is shown in figure 7.34. We use the same analysis as before. But now, $k_1 = 0$ and $k_3 \approx k_4 >> k_2$. Again, we have two rates. But only the slow one is of importance as long as the lower energy isomer A is initially prepared. This slow rate constant is given by

$$\Gamma_- = k_2 \frac{k_4}{k_3 + k_4} . \tag{7.93}$$

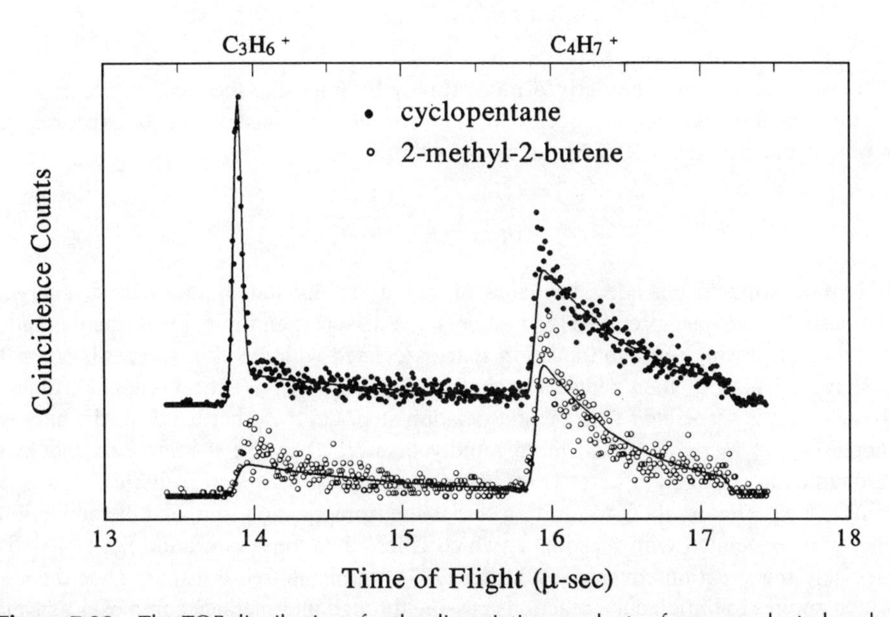

Figure 7.33 The TOF distributions for the dissociation products of energy-selected cyclopentane and 2-methyl-2-butene ions at similar total energies. The former looses C_2H_4 via a biexponential decay. The slow rates for the two isomers are the same because some of the cyclopentane ions isomerize to the lower energy 2-methyl-2-butene ions prior to dissociation.

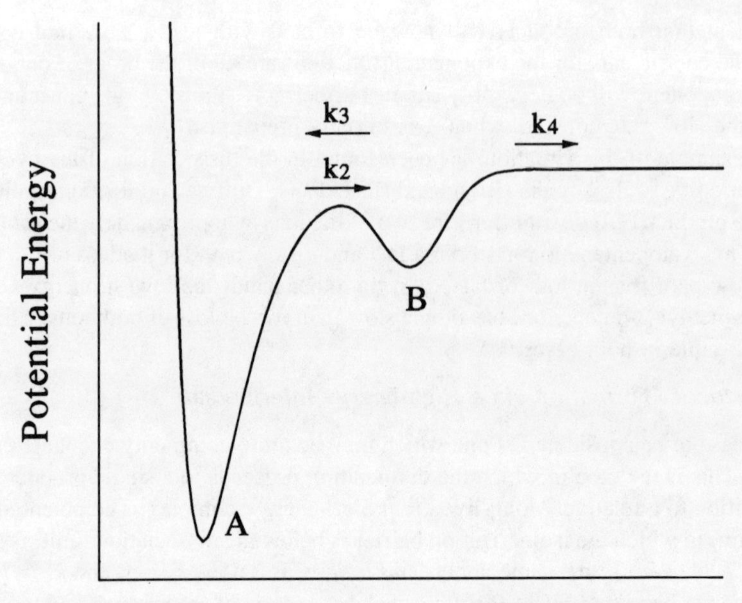

Figure 7.34 The potential energy diagram for a reaction which passes through a well near the dissociation limit. At low energies, the rate-limiting step is k_4, while at high energies, it shifts to k_2.

This equation can also be derived more simply by imposing the steady-state approximation for the intermediate species. If the observed rate constant is now expressed in terms of the individual RRKM rates, we obtain

$$k_{\mathrm{obs}} = \frac{N^{\ddagger}_{\mathrm{iso}}}{h\rho_{\mathrm{a}}}\left(\frac{N^{\ddagger}_4}{N^{\ddagger}_{\mathrm{iso}} + N^{\ddagger}_4}\right). \tag{7.94}$$

It is interesting to consider the value of k_{obs} as a function of the internal energy. Suppose that we have prepared the system at the dissociation limit. Then, there is only one way of passing through transition state associated with reaction 4. That is $N^{\ddagger}_4 = 1 << N^{\ddagger}_{\mathrm{iso}}$. The k_{obs} then reduces to $k_{\mathrm{obs}} = N^{\ddagger}_4/h\rho_{\mathrm{a}}$. That is, the bottleneck in the dissociation is associated with the dissociation step. On the other hand, as the energy increases, $N^{\ddagger}_4$ increases much more rapidly than $N^{\ddagger}_{\mathrm{iso}}$ because it represents the flux through a loose transition state. Thus, at higher energies, $N^{\ddagger}_4 >> N^{\ddagger}_{\mathrm{iso}}$. In that case, $k_{\mathrm{obs}} = N^{\ddagger}_{\mathrm{iso}}/h\,\rho_{\mathrm{a}}$. That is, the rate-limiting step is the isomerization step, and the activation energy is associated with reaction 2, which is less than the dissociation limit. This is precisely the situation consider by Miller (1976) in his unified transition state theory, which applies to bimolecular reactions passing through intermediate complexes as well as to unimolecular reactions such as shown in figure 7.34. In addition, this idea was employed by Chesnavich and Bowers (1977b) in their attempt to fit several ionic dissociation reactions. It is also related to the transition-state switching model (Chesnavich et al., 1981) in that the transition state switches from a loose one at low energies to the tighter one at higher energies. However, now the two transition states are separated by a real well in the potential energy surface.

In concluding this section it is worthwhile considering the possible nonstatistical aspect of these complex reactions. In the previous chapter it was pointed out that a single exponential decay is one of the hallmarks of a statistical dissociation. Just as fundamental is the assumption that the decay rate, $k(E)$, depends only upon the energy and angular momentum and not on how the system is prepared. Both of these assumptions appear to be violated for systems such as shown in figures 7.32 and 7.34. Consider the dissociation in which the system is prepared in either the A or the B well. The two rates are different and in at least one case, the rate is bi-exponential. In one sense these are failures of the statistical theory. That is, there are bottlenecks in the phase space which prevent a full microcanonical equilibration of all the states. Or put another way, IVR is slow compared to dissociation. However, we have analyzed these reacting systems in terms of the RRKM theory by considering the phase space of each isomer separately and by considering explicitly the rates across these bottlenecks. In this manner, each individual reacting system is still statistical. The reason, we choose to call these reactions statistical thereby distinguishing them from truly nonstatistical reactions, is that we can associate each bottleneck in the phase space with a particular molecular structure. Furthermore, there are real energy barriers that separate these structures so that we can treat the flux across them in terms of a series of transition states. This is in sharp contrast to an inherently nonstatistical system in which there are dynamical constraints to energy flow without real energy barriers. Such hidden bottlenecks cannot be treated by the usual RRKM theory in terms of a flux in phase space through a transition state (see chapter 8).

REFERENCES

Abramson, E., Field, R.W., Imre, D., Innes, K.K., and Kinsey, J.L. (1985). *J. Chem. Phys.* **83,** 453.

Andlauer, B., and Ottinger, Ch. (1971). *J. Chem. Phys.* **55,** 1471.

Andlauer, B., and Ottinger, Ch. (1972). *Z. Naturforsch.* **A27,** 293.

Aubanel, E.E., Robertson, S.H., and Wardlaw, D.M. (1991a). *J. Chem. Soc. Faraday Trans.* **87,** 2291.

Aubanel, E.E., Wardlaw, D.M., Zhu, L., and Hase, W.L. (1991b). *Int. Rev. Phys. Chem.* **10,** 249.

Aubanel, E.E., and Wardlaw, D.M. (1989). *J. Phys. Chem.* **93,** 3117.

Baer, T., Werner, A.S., and Tsai, B.P. (1975). *J. Chem. Phys.* **62,** 2497.

Baer, T., Tsai, B.P., Smith, D., and Murray, P.T. (1976). *J. Chem. Phys.* **64,** 2460.

Baer, T., Willett, G.D., Smith, D., and Phillips, J.S. (1979). *J. Chem. Phys.* **70,** 4076.

Baer, T., Brand, W.A., Bunn, T.L., and Butler, J.J. (1983). *Faraday Discuss.* **75,** 45.

Baer, T. (1986). *Adv. Chem. Phys.* **64,** 111.

Baer, T., Dutuit, O., Mestdagh, H., and Rolando, C. (1988). *J. Phys. Chem.* **92,** 5674.

Baer, T., Booze, J.A., and Weitzel, K.M. (1991). In *Vacuum Ultraviolet Photoionization and Photodissociation of Molecules and Clusters,* Ng, C.Y., Ed. World Scientific, Singapore, pp. 259–298.

Baer, T., and Kury, R. (1982). *Chem. Phys. Lett.* **92,** 659.

Benson, S.W. (1976). *Thermochemical Kinetics,* 2nd ed. Wiley, New York.

Bentley, J.A., Huang, C.M., and Wyatt, R.E. (1993). *J. Chem. Phys.* **98,** 5207.

Berblinger, M., and Schlier, C. (1992a). *J. Chem. Phys.* **96,** 6842.

Berblinger, M., and Schlier, C. (1992b). *J. Chem. Phys.* **96,** 6834.

Bhuiyan, L.B., and Hase, W.L. (1983). *J. Chem. Phys.* **78,** 5052.

Bichara, C., Gaspard, J.P., and Mathieu, J.C. (1988). *J. Chem. Phys.* **89,** 4339.

Bombach, R., Dannacher, J., Stadelmann, J.P., and Vogt, J. (1981). *Chem. Phys. Lett.* **77,** 399.

Booze, J.A., Weitzel, K.M., and Baer, T. (1991). *J. Chem. Phys.* **94,** 3649.

Booze, J.A., Schweinsberg, M., and Baer, T. (1993). *J. Chem. Phys.* **99,** 4441.

Bowers, M.T., Jarrold, M.F., Wagner-Redeker, W., Kemper, P.R., and Bass, L.M. (1983). *Faraday Discuss. Chem. Soc.* **75,** 57.

Brand, W.A., and Baer, T. (1984). *J. Am. Chem. Soc.* **106,** 3154.

Brouwer, L., Cobos, C.J., Troe, J., Dubal, H.R., and Crim, F.F. (1987). *J. Chem. Phys.* **86,** 6171.

Bryant, G.P., Jiang, Y., Martin, M., and Grant, E.R. (1992). *J. Phys. Chem.* **96,** 6875.

Bunker, D.L. (1962). *J. Chem. Phys.* **37,** 393.

Bunker, D.L. (1964). *J. Chem. Phys.* **40,** 1946.

Bunker, D.L., and Pattengill, M. (1968). *J. Chem. Phys.* **48,** 772.

Bunn, T.L., and Baer, T. (1986). *J. Chem. Phys.* **85,** 6361.

Chesnavich, W.J., Bass, L., Su, T., and Bowers, M.T. (1981). *J. Chem. Phys.* **74,** 2228.

Chesnavich, W.J. (1986). *J. Chem. Phys.* **84,** 2615.

Chesnavich, W.J., Bass, L., Grice, M.E., Song, K., and Webb, D.A. (1988). *QCPE* **8,** 557.

Chesnavich, W.J., and Bowers, M.T. (1977a). *J. Chem. Phys.* **66,** 2306.

Chesnavich, W.J., and Bowers, M.T. (1977b). *J. Am. Chem. Soc.* **99,** 1705.

Chesnavich, W.J., and Bowers, M.T. (1979). In *Gas Phase Ion Chemistry, Vol. 1,* Bowers, M.T., Ed. Academic Press, New York, pp. 119–151.

Chesnavich, W.J., and Bowers, M.T. (1982). *Prog. Reaction Kinetics* **11,** 137.

Chuang, M.C., and Zare, R.N. (1985). *J. Chem. Phys.* **82,** 4791.

Cromwell, E.F., Liu, D.J., Vrakking, M.J.J., Kung, A.H., and Lee, Y.T. (1991). *J. Chem. Phys.* **95,** 297.

Dai, H.L., Field, R.W., and Kinsey, J.L. (1985). *J. Chem. Phys.* **82,** 1606.

Dashevskaya, E.I., Nikitin, E.E., and Troe, J. (1990). *J. Chem. Phys.* **93,** 7803.

Dashevskaya, E.I., Nikitin, E.E., and Troe, J. (1992). *J. Chem. Phys.* **97,** 3318.

Dibeler, V.H., and Walker, H.A. (1973). *Int. J. Mass Spectrom. Ion. Proc.* **11,** 49.

Duffy, L.M., Keister, J.W., and Baer, T. (1995). *J. Phys. Chem.* **99,** 17862.

Dunbar, R.C., Chen, J.H., So, H.Y., and Asamoto, B. (1987). *J. Chem. Phys.* **86,** 2081.

Dunbar, R.C. (1990). *Int. J. Mass Spectrom. Ion. Proc.* **100,** 423.

Dutuit, O., Baer, T., Metayer, C., and Lemaire, J. (1991). *Int. J. Mass Spectrom. Ion. Proc.* **110,** 67.

Eckart, C. (1930). *Phys. Rev.* **35,** 1303.

Eland, J.H.D., and Schulte, H. (1975). *J. Chem. Phys.* **62,** 3835.

Eliason, M.A., and Hirschfelder, J.O. (1959). *J. Chem. Phys.* **30,** 1426.

Elliot, C.S., and Frey, H.M. (1966). *Trans. Faraday Soc.* **62,** 895.

Fahrer, N., and Schlier, C. (1992). *J. Chem. Phys.* **97,** 7008.

Farantos, S.C., Murrell, J.N., and Hajduk, J.C. (1982). *Chem. Phys.* **68,** 109.

Forst, W. (1971). *Chem. Rev.* **71,** 339.

Forst, W. (1973). *Theory of Unimolecular Reactions*. Academic Press, New York.

Forst, W. (1983). *J. Phys. Chem.* **87,** 4489.

Forst, W. (1991). *J. Phys. Chem.* **95,** 3612.

Franklin, J.L., Field, F.H., and Lampe, F.W. (1956). *J. Am. Chem. Soc.* **78,** 5697.

Frey, H.M., Marshall, D.C., and Skinner, R.F. (1965). *Trans. Faraday Soc.* **61,** 861.

Gefen, S., and Lifshitz, C. (1984). *Int. J. Mass Spectrom. Ion. Proc.* **58,** 251.

Gilbert, R.G., and Smith, S.C. (1990). *Theory of Unimolecular and Recombination Reactions*. Blackwell Scientific, Oxford.
Gray, S.K., Miller, W.H., Yamaguchi, Y., and Schaefer, H.F. (1980). *J. Chem. Phys.* **73,** 2733.
Gray, S.K., Miller, W.H., Yamaguchi, Y., and Schaefer, H.F. (1981). *J. Am. Chem. Soc.* **103,** 1900.
Green, W.H., Mahoney, A.J., Zheng, Q.K., and Moore, C.B. (1991). *J. Chem. Phys.* **94,** 1961.
Green, W.H., Moore, C.B., and Polik, W.F. (1992). *Ann. Rev. Phys. Chem.* **43,** 591.
Guyer, D.R., Polik, W.F., and Moore, C.B. (1986). *J. Chem. Phys.* **84,** 6519.
Haarhoff, P.C. (1963). *Mol. Phys.* **7,** 101.
Hase, W.L. (1972). *J. Chem. Phys.* **57,** 730.
Hase, W.L. (1976). *J. Chem. Phys.* **64,** 2442.
Hase, W.L. (1983). *Acc. Chem. Res.* **16,** 258.
Hase, W.L. (1987). *Chem. Phys. Lett.* **139,** 389.
Hase, W.L., and Buckowski, D.G. (1982). *J. Comp. Chem.* **3,** 335.
Hase, W.L., and Duchovic, R.J. (1985). *J. Chem. Phys.* **83,** 3448.
Hase, W.L., and Wolf, R.J. (1981). *J. Chem. Phys.* **75,** 3809.
Hehre, W.J., Radom, L., Schleyer, P.v.R., and Pople, J.A. (1986). *Ab Initio Molecular Orbital Theory*. Wiley-Interscience, New York.
Hernandez, R., Miller, W.H., Moore, C.B., and Polik, W.F. (1993). *J. Chem. Phys.* **99,** 950.
Hofacker, L. (1963). *Z. Naturforsch.* **18A,** 607.
Hu, X., and Hase, W.L. (1989). *J. Phys. Chem.* **93,** 6029.
Hu, X., and Hase, W.L. (1991). *J. Chem. Phys.* **95,** 8073.
Isaacson, A.D., Truhlar, D.G., Rai, S.N., Steckler, R., and Hancock, G.C. (1987). *Comp. Phys. Comm.* **47,** 91.
Jasinski, J.M., Frisoli, J.K., and Moore, C.B. (1983). *J. Chem. Phys.* **79,** 1312.
Johnston, H.S. (1966). *Gas Phase Reaction Rate Theory*. Ronald Press, New York.
Johnston, H.S., and Heicklen, J. (1962). *J. Phys. Chem.* **66,** 532.
Keck, J.C. (1967). *Adv. Chem. Phys.* **13,** 85.
Khundkar, L.R., Knee, J.L., and Zewail, A.H. (1987). *J. Chem. Phys.* **87,** 77.
Kiermeier, A., Kuhlewind, H., Neusser, H.J., Schlag, E.W., and Lin, S.H. (1988). *J. Chem. Phys.* **88,** 6182.
Kim, S.K., Choi, Y.S., Pibel, C.D., Zheng, Q.K., and Moore, C.B. (1991). *J. Chem. Phys.* **94,** 1954.
Klippenstein, S.J., Khundkar, L.R., Zewail, A.H., and Marcus, R.A. (1988). *J. Chem. Phys.* **89,** 4761.
Klippenstein, S.J. (1990). *Chem. Phys. Lett.* **170,** 71.
Klippenstein, S.J. (1991). *J. Chem. Phys.* **94,** 6469.
Klippenstein, S.J., Faulk, J.D., and Dunbar, R.C. (1993). *J. Chem. Phys.* **98,** 243.
Klippenstein, S.J., East, A.L.L., Allen, W.D. (1994). *J. Chem. Phys.* **101,** 9198
Klots, C.E. (1971). *J. Phys. Chem.* **75,** 1526.
Klots, C.E. (1972). *Z. Naturforsch.* **27A,** 553.
Kreevoy, M.M., and Truhlar, D.G. (1986). in *Investigation of Rates and Mechanisms of Reactions,* Bernasconi, C.R., Ed. Wiley, New York, pp. 13–95.
Kuhlewind, H., Neusser, H.J., and Schlag, E.W. (1984). *J. Phys. Chem.* **88,** 6104.
Kuhlewind, H., Kiermeier, A., and Neusser, H.J. (1986). *J. Chem. Phys.* **85,** 4427.
Kuhlewind, H., Kiermeier, A., Neusser, H.J., and Schlag, E.W. (1987). *J. Chem. Phys.* **87,** 6488.
Labastie, P., and Whetten, R.L. (1990). *Phys. Rev. Lett.* **65,** 1567.
Lifshitz, C., Louage, F., Aviyente, V., and Song, K. (1991). *J. Phys. Chem.* **95,** 9298.
Lorquet, J.C., and Takeuchi, T. (1990). *J. Phys. Chem.* **94,** 2279.

Luo, X., and Rizzo, T.R. (1991). *J. Chem. Phys.* **94,** 889.
Marcus, R.A. (1966). *J. Chem. Phys.* **45,** 4493.
Marcus, R.A., and Coltrin, M.E. (1977). *J. Chem. Phys.* **67,** 2609.
McCoy, A.B., Burleigh, D.C., and Sibert, E.L. (1991). *J. Chem. Phys.* **95,** 7449.
McCulloh, K.E., and Dibeler, V.H. (1976). *J. Chem. Phys.* **64,** 4445.
Meisels, G.G., Verboom, G.M.L., Weiss, M.J., and Hsieh, T.C. (1979). *J. Am. Chem. Soc.* **101,** 7189.
Mies, F.H. (1969). *J. Chem. Phys.* **51,** 798.
Miller, W.H. (1976). *J. Chem. Phys.* **65,** 2216.
Miller, W.H. (1979). *J. Am. Chem. Soc.* **101,** 6810.
Miller, W.H., Handy, N.C., and Adams, J.E. (1980). *J. Chem. Phys.* **72,** 99.
Miller, W.H. (1983). *J. Phys. Chem.* **87,** 21.
Miller, W.H. (1987). *Chem. Rev.* **87,** 19.
Miller, W.H., Hernandez, R., Handy, N.C., Jayatilaka, D., and Willetts, A. (1990). *Chem. Phys. Lett.* **172,** 62.
Minkin, V.I., Simkin, B.Ya., and Minyaev, R.M. (1990). *Quantum Chemistry of Organic Compounds*. Springer-Verlag, Berlin.
Moiseyev, N., Lipkin, N., Farrelly, D., Atabek, O., and Lefebvre, R. (1989). *J. Chem. Phys.* **91,** 6246.
Mondro, S.L., Vande Linde, S.R., and Hase, W.L. (1986). *J. Chem. Phys.* **84,** 3783.
Morrow, J.C., and Baer, T. (1988). *J. Phys. Chem.* **92,** 6567.
Muller-Dethlefs, K., and Schlag, E.W. (1991). *Ann. Rev. Phys. Chem.* **42,** 109.
Nadler, I., Pfab, J., Reisler, H., and Wittig, C. (1984). *J. Chem. Phys.* **81,** 653.
Neumark, D.M. (1992). *Ann. Rev. Phys. Chem.* **43,** 153.
Neusser, H.J. (1987). *Int. J. Mass Spectrom. Ion. Proc.* **79,** 141.
Neusser, H.J. (1989). *J. Phys. Chem.* **93,** 3897.
Nikitin, E.E. (1965). *Theor. Exp. Chem.* **1,** 90.
Nikitin, E.E. (1974). *Theory of Elementary Atomic and Molecular Processes in Gases*. Clarendon Press, Oxford.
Nose, S. (1984a). *Mol. Phys.* **52,** 255.
Nose, S. (1984b). *J. Chem. Phys.* **81,** 511.
Nyman, G., Nordholm, S., and Schranz, H.W. (1990). *J. Chem. Phys.* **93,** 6767.
Olzmann, M., and Troe, J. (1992). *Ber. Bunsenges. Phys. Chem.* **96,** 1327.
Orchard, S.W., and Thrush, B.A. (1974). *Proc. R. Soc. Lond.* **A337,** 257.
Pechukas, P., Light, J.C., and Rankin, C. (1966). *J. Chem. Phys.* **44,** 794.
Pechukas, P. (1981). *Ann. Rev. Phys. Chem.* **32,** 159.
Pechukas, P., and Light, J.C. (1965). *J. Chem. Phys.* **42,** 3281.
Peslherbe, G.H., and Hase, W.L. (1994). *J. Chem. Phys.* **101,** 8535.
Polik, W.F., Moore, C.B., and Miller, W.H. (1988). *J. Chem. Phys.* **89,** 3584.
Polik, W.F., Guyer, D.R., Miller, W.H., and Moore, C.B. (1990a). *J. Chem. Phys.* **92,** 3471.
Polik, W.F., Guyer, D.R., and Moore, C.B. (1990b). *J. Chem. Phys.* **92,** 3453.
Potter, E.D., Gruebele, M., Khundkar, L.R., and Zewail; A.H. (1989). *Chem. Phys. Lett.* **164,** 463.
Pratt, S.T., and Chupka, W.A. (1981). *Chem. Phys.* **62,** 153.
Quack, M., and Troe, J. (1974). *Ber. Bunsenges. Phys. Chem.* **78,** 240.
Quack, M., and Troe, J. (1981). *Int. Rev. Phys. Chem.* **1,** 97.
Rai, S.N., and Truhlar, D.G. (1983). *J. Chem. Phys.* **79,** 6046.
Reddy, K.V., and Berry, M.J. (1979). *Chem. Phys. Lett.* **66,** 223.
Reinhardt, W.P. (1990). *J. Mol. Struct.* **223,** 157.
Reisner, D.E., Field, R.W., Kinsey, J.L., and Dai, H.L. (1984). *J. Chem. Phys.* **80,** 5968.
Rosenstock, H.M., Stockbauer, R., and Parr, A.C. (1980). *J. Chem. Phys.* **73,** 773.

Rosenstock, H.M., Dannacher, J., and Liebman, J.F. (1982). *Radiat. Phys. Chem.* **20,** 7.
Scherer, N.F., and Zewail, A.H. (1987). *J. Chem. Phys.* **87,** 97.
Schneider, F.W., and Rabinovitch, B.S. (1962). *J. Am. Chem. Soc.* **84,** 4215.
Schranz, H.W., Nordholm, S., and Nyman, G. (1991). *J. Chem. Phys.* **94,** 1487.
Segall, J., and Zare, R.N. (1988). *J. Chem. Phys.* **89,** 5704.
Skodje, R.T., Schwenke, D.W., Truhlar, D.G., and Garrett, B.C. (1984). *J. Phys. Chem.* **88,** 628.
Stein, S.E., and Rabinovitch, B.S. (1973). *J. Chem. Phys.* **58,** 2438.
Steinfeld, J.I., Francisco, J.S., and Hase, W.L. (1989). *Chemical Kinetics and Dynamics.* Prentice-Hall, Englewood Cliffs, NJ.
Swamy, K.N., and Hase, W.L. (1986). *J. Chem. Phys.* **84,** 361.
Townes, C.H., and Schalow, A.L. (1955). *Microwave Spectroscopy*. McGraw-Hill, New York.
Troe, J. (1977). *J. Chem. Phys.* **65,** 4758.
Troe, J. (1979). *J. Phys. Chem.* **83,** 114.
Troe, J. (1981). *J. Chem. Phys.* **75,** 226.
Troe, J. (1983). *J. Chem. Phys.* **79,** 6017.
Troe, J. (1984). *J. Phys. Chem.* **88,** 4375.
Troe, J. (1986). In *Tunneling,* Jortner, J., and Pullmann, B., Eds. Reidel, Dordrecht, pp. 149–164.
Troe, J. (1988). *Ber. Bunsenges. Phys. Chem.* **92,** 242.
Troe, J. (1991). In *Mode Selective Chemistry,* Jortner, J., Levine, R.D., and Pullman, B., Eds. Kluver Academic, Dordrecht, pp. 241–259.
Truhlar, D.G., Isaacson, A.D., Skodje, R.T., and Garrett, B.C. (1982). *J. Phys. Chem.* **86,** 2252.
Truhlar, D.G., and Garrett, B.C. (1980). *Acc. Chem. Res.* **13,** 440.
Truhlar, D.G., and Kuppermann, A. (1971). *J. Am. Chem. Soc.* **93,** 1840.
Vande Linde, S.R., Mondro, S.L., and Hase, W.L. (1987). *J. Chem. Phys.* **86,** 1348.
Varshni, Y.P. (1957). *Rev. Mod. Phys.* **29,** 664.
Wagner, A.F., Kiefer, J.H., and Kumaran, S.S. (1992). In *Twenty-Fourth Symposium on Combustion.* The Combustion Institute, pp. 613–619.
Waite, B.A., Gray, S.K., and Miller, W.H. (1983). *J. Chem. Phys.* **78,** 259.
Waite, B.A., and Miller, W.H. (1981). *J. Chem. Phys.* **74,** 3910.
Waller, I.M., Kitsopoulus, T.N., and Neumark, D.M. (1990). *J. Phys. Chem.* **94,** 2240.
Wang, H., Zhu, L., and Hase, W.L. (1994). *J. Phys. Chem.* **98,** 1608.
Wardlaw, D.M., and Marcus, R.A. (1984). *Chem. Phys. Lett.* **110,** 230.
Wardlaw, D.M., and Marcus, R.A. (1985). *J. Chem. Phys.* **83,** 3462.
Wardlaw, D.M., and Marcus, R.A. (1986). *J. Phys. Chem.* **90,** 5383.
Wardlaw, D.M., and Marcus, R.A. (1988). *Adv. Chem. Phys.* **70,** 231.
Weerasinghe, S., and Amar, F.G. (1993). *J. Chem. Phys.* **98,** 4967.
Weitzel, K.M. (1991). *Chem. Phys. Lett.* **186,** 490.
Weitzel, K.M., Booze, J.A., and Baer, T. (1991). *Chem. Phys.* **150,** 263.
Whitten, G.Z., and Rabinovitch, B.S. (1963). *J. Chem. Phys.* **38,** 2466.
Whitten, G.Z., and Rabinovitch, B.S. (1964). *J. Chem. Phys.* **41,** 1883.
Yu, J., and Klippenstein, S.J. (1991). *J. Phys. Chem.* **95,** 9882.
Zhu, L., Chen, W., Hase, W.L., and Kaiser, E.W. (1993). *J. Phys. Chem.* **97,** 311.
Zhu, L., and Hase, W.L. (1990). *Chem. Phys. Lett.* **175,** 117.

8

Dynamical Approaches to Unimolecular Rates

8.1 INTRODUCTION

In the previous chapters theories were discussed for calculating the unimolecular rate constant as a function of energy and angular momentum. The assumption inherent in these theories is that a microcanonical ensemble is maintained during the unimolecular reaction and that every state in the energy interval $E \rightarrow E + dE$ has an equal probability of decomposing. Such theories are viewed as statistical since the unimolecular rate constant is found from a statistical counting of states in the microcanonical ensemble. A dynamical description of unimolecular decomposition is concerned with properties of individual states of the energized molecule. Of interest are the decomposition probabilities for the states as well as the rate of transitions between the states.

Dynamical theories of unimolecular decomposition deal with the properties of vibrational/rotational energy levels, state preparation and intramolecular vibrational energy redistribution (IVR). Thus, the presentation in this chapter draws extensively on the previous chapters 2 and 4. Unimolecular decomposition dynamics can be treated using quantum and classical mechanics, and both perspectives are considered here. The role of nonadiabatic electronic transitions in unimolecular dynamics is also discussed.

8.2 DESCRIPTION OF QUANTUM RESONANCE STATES

A molecule which can dissociate does not, strictly speaking, have a discrete energy spectrum. The relative motion of the product fragments is unbounded and, in this sense the motion of the unimolecular system is infinite, and hence the energy spectrum is continuous. However, it may happen that the dissociation probability of the molecule is sufficiently small that one can introduce the concept of quasi-stationary states. Such states are commonly referred to as resonances since the energy of the unimolecular fragments in the continuum is in resonance with (i.e., matches) the energy of a vibrational/rotational level of the unimolecular reactant. For unimolecular reactions there are two types of resonance states. The simplest type, a shape resonance, occurs when a molecule is temporarily trapped by a fairly high and wide potential energy barrier. The second type of resonance, called a Feshbach or compound-state resonance, occurs when energy is initially distributed between vibrational/rotational degrees of freedom of the molecule which are not strongly coupled to the fragment relative motion, so that there is a time lag for unimolecular dissociation. In a time-dependent picture, resonances can be viewed as localized wave packets composed of a superposition of continuum wave functions, which for a time qualitatively resemble bound states.

Compound-state resonances are important in quantal theories of unimolecular decomposition. They are prepared in low-energy atom (molecule)-molecule collisions when part of the relative kinetic energy of the motion becomes temporarily converted into excitation of the internal (rotational and/or vibrational) degrees of freedom of either partner. When this excitation occurs, the molecular system has insufficient energy in its relative motion to separate. One can also prepare compound-state resonances by using electromagnetic radiation (e.g., a laser) to excite the molecule. Thus, it is proper to view these resonances as the natural extension of the bound vibrational/rotational eigenstates into the dissociative continuum.

For a quasi-stationary resonance state the unimolecular reactant moves within the potential energy well for a considerable period of time, leaving it only when a fairly long time interval τ has elapsed; τ may be called the lifetime of the almost stationary resonance state. The energy spectrum of these states will be quasi-discrete; it consists of a series of broadened levels with Lorentzian line-shapes [recall Eq. (4.35)], whose full-width at half-maximum Γ is related to the lifetime by $\Gamma = \hbar/\tau$.

In chapter 2 solutions to Schrödinger's equation for bound vibrational/rotational states were obtained with a boundary condition requiring the finiteness of the wave function at infinity. This type of solution is inappropriate for resonance states (Landau and Lifshitz, 1965). Instead, one looks for solutions which represent an outgoing spherical wave at infinity. This corresponds to the fragments separating as the molecule dissociates. Since such a boundary condition is complex, we cannot assert that the eigenvalues of the energy must be real. On the contrary, by solving Schrödinger's equation, we obtain a set of complex values, which we write in the form

$$E_n^{\circ} = E_n - i\Gamma_n/2, \tag{8.1}$$

where E_n and Γ_n are constants, which are positive. The constant E_n, the real component to the eigenvalue, gives the position of the resonance in the spectrum.

It is easy to see the physical significance of complex energy values. The time factor in the wave function of a quasi-stationary state is of the form

$$\exp\,[-(i/\hbar)E_n^{\circ}t] = \exp\,[-(i/\hbar)E_n t]\,\exp\,[-(\Gamma_n/2\hbar)t]. \tag{8.2}$$

Hence, all probabilities given by the squared modules of the wave function decrease with time as $\exp[-(\Gamma_n/\hbar)t]$, that is,

$$|\psi_n(t)|^2 = |\psi_n(0)|^2 \exp\,[-(\Gamma_n/\hbar)t]. \tag{8.3}$$

In particular, the probability of finding the unimolecular reactant within its potential energy well decreases according to this law. Thus, Γ_n determines the lifetime of the state and the state specific unimolecular rate constant is

$$k_n = \Gamma_n/\hbar = 1/\tau_n, \tag{8.4}$$

where τ_n is the state's lifetime.

The resonance states are said to be isolated if the widths of their levels are small compared with the distances between them; that is,

$$\Gamma_n << 1/\rho(E), \tag{8.5}$$

where $\rho(E)$ is the density of states for the energized molecule. As the line widths broaden and/or the number of resonance states in an energy interval increases, the spectrum may no longer be quasi-discrete since the resonance lines may overlap; that is,

$$\Gamma_n >> 1/\rho(E). \tag{8.6}$$

The theory of isolated resonances is well understood and is discussed below. Some initial work has been done on the theory of overlapping resonances (Remacle et al., 1989; Desouter-Lecomte and Culot, 1993; Someda et al., 1994a,b) and its relation to experiment (Reid et al., 1994). Much of the research of overlapping resonances has its origins in nuclear physics, where the dissociation of a compound nucleus is treated (Ericson, 1960, 1963; Satchler, 1990; Rotter, 1991). For example, fluctuations in product state populations, called Ericson fluctuations (Satchler, 1990; Rotter, 1991), may arise from coherent excitation of overlapping resonances. However, more work needs to be done to develop a complete theory of overlapping resonances and this topic is not discussed here. Mies and Krauss (1966, 1969) and Rice (1971) were pioneers in treating unimolecular rate theory in terms of the decomposition of isolated Feshbach resonances.

8.3 ISOLATED RESONANCES

A possible absorption spectrum for a molecule near its unimolecular dissociation threshold E_o is shown in figure 8.1. Below E_o the absorption lines for the molecular eigenstates are very narrow and are only broadened by interaction of the excited molecule with the radiation field. However, above E_o the excited states leak toward product space, which gives rise to characteristic widths for the resonances in the spectrum. Since the line widths do not overlap, the resonances are isolated. Each

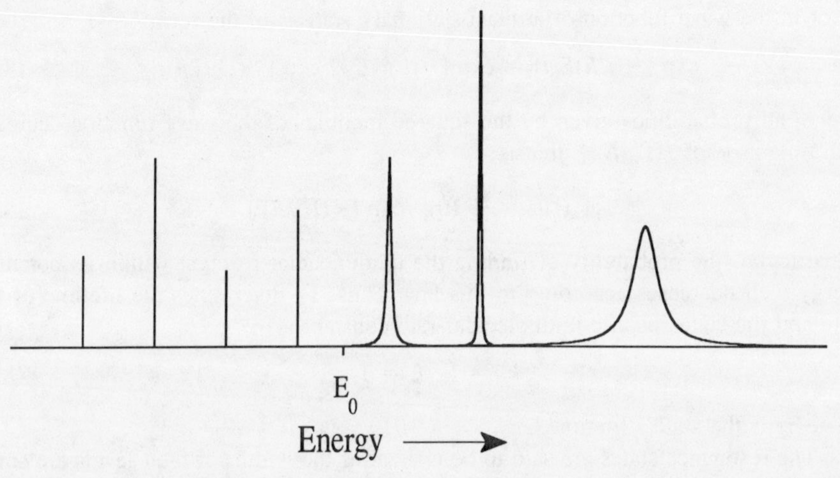

Figure 8.1 Possible absorption spectrum for a molecule which dissociates via isolated compound-state resonances. E_o is the unimolecular threshold.

resonance has its own characteristic width (i.e., lifetime). The areas of the lines in the spectrum are determined by the transition moment μ_{if} between the initial state ψ_i and final state ψ_f, that is, $\mu_{if} = \langle\psi_i|\hat{\mu}|\psi_f\rangle$. In the study of a resonance one is interested in its position given by its energy E_n and its width Γ_n.

8.3.1 Calculation of Resonance Positions and Widths

8.3.1.1 One-Dimensional Semiclassical Tunneling Model

To illustrate the calculation of resonance positions and widths it is instructive to consider tunneling through a one-dimensional potential energy barrier (fig. 8.2). The states above the unimolecular dissociation threshold (E_o) are shape resonances as a result of tunneling through the potential energy barrier. The energies for both the bound states and resonance states are given quite accurately by applying the semiclassical quantization condition, Eq. (2.72), to the motion inside the potential energy well. For the resonance states this gives E_n in Eq. (8.1). The semiclassical unimolecular rate constant for the resonance is (Miller, 1975; Waite and Miller, 1980):

$$k_n = \nu_n \kappa_n, \tag{8.7}$$

where ν_n is the classical vibrational frequency for the resonance state inside the potential energy well and κ_n is the one-dimensional semiclassical tunneling probability through the potential barrier. κ_n is given by

$$\kappa_n = \exp(-2\theta_n)/[1 - \exp(-2\theta n)], \tag{8.8a}$$

$$\theta_n = \int_{r^+}^{r^-} dr\sqrt{2m[V(r) - E_n]}\,. \tag{8.8b}$$

Here r^- and r^+ are the inner and outer turning points for the classically forbidden region inside the potential energy barrier, m is the mass for the one-dimensional

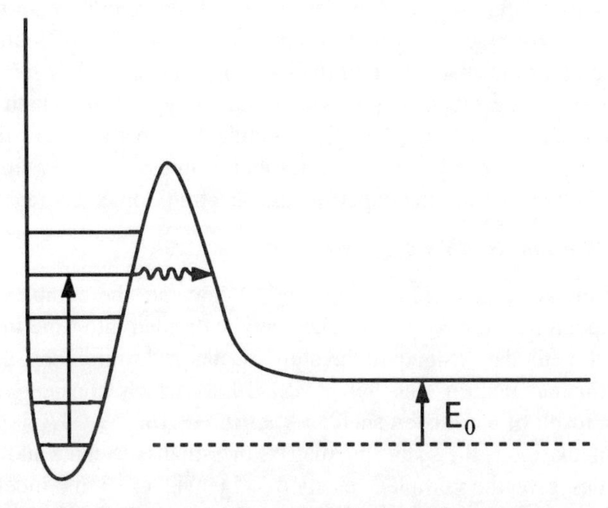

Figure 8.2 Energy levels for a one-dimensional potential energy curve. The levels are shape resonances and are broadened by tunneling through the potential energy barrier.

motion, $V(r)$ is the potential energy curve, and E_n is the resonance state's real energy. The tunneling rates increase with increase in n, since the barrier is easier to penetrate. Thus, the resonance width Γ_n increases with n.

8.3.1.2 Stabilization Graph

In general it is much easier to calculate resonance positions than widths (Truhlar, 1984). A particularly useful approach for calculating resonance positions, and one that is instructive, is the stabilization method pioneered by Hazi and Taylor (1970). This method is based upon the principle that wave functions for resonance states are large inside the bound region of the potential energy surface and much smaller outside this region. In contrast the wave functions for nonresonant states are larger outside the bound region. In figure 8.2, the resonant states are those trapped in the potential energy well and the nonresonant states are those which scatter off the barrier from the right and those with energies greater than the barrier maximum (one should note that the concept of resonant and nonresonant states is more general than this, and applies to a system with Feshbach resonances).

Consider the effect of first placing an infinite potential energy barrier at a position of the reaction coordinate which is outside the bound region of the potential energy surface and then moving this barrier along the reaction coordinate, keeping it out of the bound region. The effect of this barrier is to create bound states out of both the resonant and nonresonant states. As the barrier is moved the energies of the resonant states are stable, nearly independent of the barrier position. However, since the wave functions for the nonresonant states are not localized in the bound region, the particle in the box problem tells us that their energies should drop as the potential barrier is moved out along the reaction coordinate.

To implement the stabilization method one chooses a square integrable (L^2) basis set for the Hamiltonian centered in the bound region of the potential energy surface and performs a variational calculation as described in section 2.3.2. To mimic the effect of moving an infinite potential energy barrier along the reaction coordinate, the spatial size of the basis functions is scaled by a parameter α. By performing a variational calculation versus α, eigenvalues (i.e., energies) for the Hamiltonian are determined versus α. Energies for resonance states are nearly insensitive to α, while the nonresonant energies drop as the spatial size of the basis becomes larger by increasing α. This gives rise to numerous avoided crossings between energy levels, which yields a stabilization graph as shown in figure 8.3. The stabilization method has been used extensively for finding resonance energies (Eastes and Marcus, 1973; Swamy et al., 1986; Bai et al., 1983) and interpreting experimental spectra (Gomez Llorente et al., 1989).

8.3.1.3 Calculational Methods

Different theoretical methods have been used to calculate the complex energies, Eq. (8.1), for compound-state resonances. They can be divided into time-independent and time-dependent methods. A standard quantum mechanical time-independent method is a close-coupling calculation (Stechel et al., 1978) which considers resonant state formation as a result of a collision such as $A + BC \rightarrow ABC^* \rightarrow AB + C$. Determined from such a calculation is the scattering matrix, or **S**-matrix (Schatz and Ratner, 1993), whose properties give the complex resonance eigenvalues. Time-independent variational quantum-mechanical methods (Truhlar and Mead, 1990) have also been used in which the scattering wave functions are expanded in terms of a square-integrable (L^2)

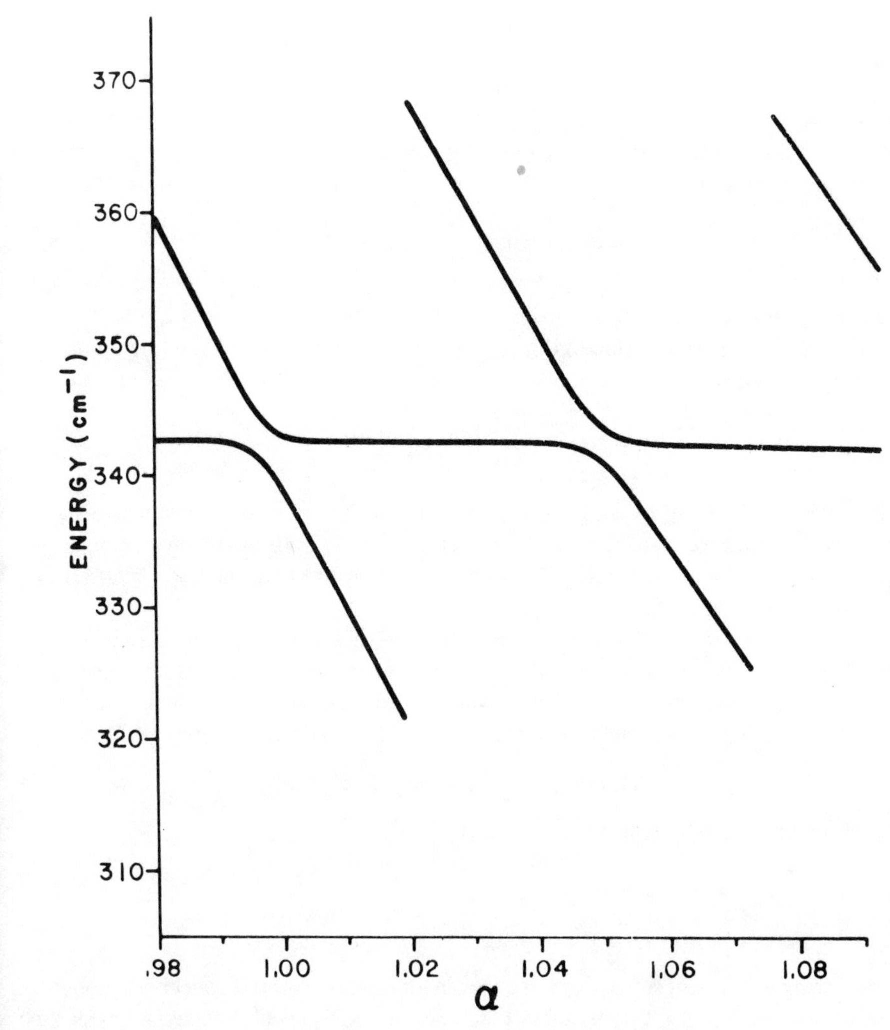

Figure 8.3 Stabilization graph of a meta-stable state for a van der Waals complex (Bačić and Simons, 1982).

finite basis set. In this approach, which include the secular equation (Grabenstetter and LeRoy, 1979), complex coordinate rotation (Reinhardt, 1982), stabilization graph (Hazi, 1978; McCurdy and McNutt, 1983), and golden rule (Tucker and Truhlar, 1988) methods, the complex resonance energies are determined by calculating wave functions for the resonance states in the bound region of the potential energy surface. Multidimensional semiclassical methods have also been used to calculate complex resonance energies when the classical motion is quasiperiodic and decomposition occurs by tunneling through a potential barrier (Waite and Miller, 1980; Gray and Child, 1984). If the decomposition is not by tunneling and/or the classical motion is chaotic, the semiclassical calculation becomes considerably more difficult (Marcus, 1973).

Time-dependent approaches (Feit and Fleck, 1982, 1983; Kosloff, 1988; Schinke,

1993; Kosloff, 1994) have also been used to find the complex energies for compound-state resonances. A localized wave packet [i.e., a coherent superposition state, Eq. (4.7)], $\Psi(0)$ is initially placed in the bound region of the potential energy surface and propagated in time to give $\langle\Psi(0)|\Psi(t)\rangle$, which is $C(t)$ in Eq. (4.16). If $\Psi(0)$ is a superposition of resonant states, it can be considered a zero-order state (see chapter 4) and can be written as

$$\Psi(0) = \sum_n c_n \psi_n , \tag{8.9}$$

where ψ_n are the wave functions for the compound state resonances in the bound region of the potential energy surface. Using Eq. (8.9) and Eq. (4.7) and including the unimolecular decay of the resonance state one obtains

$$C(t) = \langle\Psi(0)|\Psi(t)\rangle = \sum_n c_n^2 \exp(-iE_n t/\hbar)\exp(-\Gamma_n t/2\hbar) , \tag{8.10}$$

which is the same as Eq. (8.2), except there now is a superposition of resonance states. If $C(t)$ is followed for a sufficiently long period of time, both the resonance positions E_n and widths Γ_n can be evaluated (Gray, 1992b) by Prony's method (Marple, Jr., 1987) of spectral analysis.

Assume that the initial wave packet $\Psi(0)$ is not a superposition state but an individual resonance state. The energy E_n and wave function ψ_n for the resonance could be determined by performing a stabilization graph calculation as described above. If the initial wave-packet is ψ_n, Eq. (8.10) becomes the same as (8.2), that is,

$$C(t) = \exp(-iE_n t/\hbar) \exp(-\Gamma_n t/2\hbar) \tag{8.11}$$

and following $C(t)$ versus time gives Γ_n.

8.3.2 Recurrences and Quantum Beats

By preparing a coherent superposition of resonant states one could conceivably observe recurrences, section 4.4.1.1, which are damped by unimolecular decay. Consider the situation where there are only two resonances in the superposition state so that Eq. (8.10) becomes

$$\begin{aligned} C(t) = {} & c_1^2 \exp(-iE_1 t/\hbar) \exp(-\Gamma_1 t/2\hbar) \\ & + c_2^2 \exp(-iE_2 t/\hbar) \exp(-\Gamma_2 t/2\hbar). \end{aligned} \tag{8.12}$$

This equation should be compared to Eq. (4.20) which describes the time dependence of a superposition state for the double-well tunneling problem. The two equations are identical except in Eq. (8.12), $\Psi(t)$ is multiplied by $\Psi(0) = c_1\psi_1 + c_2\psi_2$ and includes damping from unimolecular decay, that is, the $\exp(-\Gamma_n t/2\hbar)$ terms. The probability of populating the superposition state versus time, that is, $P(t) = |C(t)|^2$, is given by

$$\begin{aligned} P(t) = {} & c_1^4 \exp(-\Gamma_1 t/\hbar) + c_2^4 \exp(-\Gamma_2 t/\hbar) \\ & + 2c_1^2 c_2^2 \cos\left(\frac{(E_1 - E_2)t}{\hbar}\right) \exp[-(\Gamma_1 + \Gamma_2)t/2\hbar] , \end{aligned} \tag{8.13}$$

where Euler's relation, $e^{i\alpha} = \cos\alpha + i \sin\alpha$, has been used to represent the complex exponential terms in Eq. (8.12). $P(t)$ in Eq. (8.13) displays oscillatory dynamics, with frequency $\omega_{12} = (E_1 - E_2)/\hbar$, which results from motion from the initial zero-order state to another zero-order state also formed by a linear combination between ψ_1 and ψ_2. This is an example of quantum beats, as discussed in section 4.4.1.5. There the damping was by fluorescence, here it is by unimolecular decomposition. Both are first-order processes.

8.3.3 Mode Specificity

The observation of decomposition from isolated compound-state resonances does not necessarily imply mode-specific unimolecular decomposition. Nor is mode specificity established by the presence of fluctuations in state-specific rate constants for resonances within a narrow energy interval. What is required for mode-specific unimolecular decomposition is a distinguishable and, thus, assignable pattern (or patterns) in the positions of resonance states in the spectrum. Identifying such patterns in a spectrum allows one to determine which modes in the molecule are excited when forming the resonance state. It is, thus, possible to interpret particularly large or small state-specific rate constants in terms of mode-specific excitations. Therefore, mode specificity means there are exceptionally large or small state-specific rate constants depending on which modes are excited.

The ability to assign a group of resonance states, as required for mode-specific decomposition, implies that the complete Hamiltonian for these states is well approximated by a zero-order Hamiltonian with eigenfunctions $\phi_i(\mathbf{m})$; see section 2.3.2 and section 4.4.2. The ϕ_i are product functions of a zero-order orthogonal basis for the reactant molecule and the quantity $\mathbf{m}$ represents the quantum numbers defining ϕ_i. The wave functions ψ_n for the compound state resonances are given by

$$\psi_n = \sum_i c_{in}\phi_i(\mathbf{m}) . \tag{8.14}$$

Resonance states in the spectra, which are assignable in terms of zero-order basis $\phi_i(\mathbf{m})$, will have a predominant expansion coefficient c_{in}. Hose and Taylor (1982) have argued that for an assignable level $c_{in}^2 > 0.5$ for one of the expansion coefficients.

In contrast to resonance states which can be assigned quantum numbers and which may exhibit mode-specific decomposition are states which are intrinsically unassignable. Because of extensive couplings, a zero-order Hamiltonian and its basis set cannot be found to represent the wave functions ψ_n for these states. The spectrum for these states is irregular without patterns, and fluctuations in the k_n are related to the manner in which the ψ_n are randomly distributed in coordinate space. Thus, the states are intrinsically unassignable and have no good quantum numbers apart from the total energy and angular momentum. Energies for these resonance states do not fit into a pattern, and states with particularly large or small rate constants are simply random occurrences in the spectrum. For the most statistical (i.e., nonseparable) situation, the expansion coefficients in Eq. (8.14) are random variables, subject only to the normalization and orthogonality conditions

$$\sum_n c_{in}^2 = 1 \quad \text{and} \quad \sum_i c_{in}c_{im} = 0. \tag{8.15}$$

The distribution for the expansion coefficients is a normalized Gaussian (Brody et al., 1981; Carmeli, 1983; Polik et al., 1988):

$$P(\mathbf{c}) = \prod_{n=1}^{m} \sqrt{f/2\pi}\,\mathrm{esp}(-mc_n^2/2)\,, \tag{8.16}$$

where m is the number of expansion coefficients.

To conclude this section, for many reactant molecules it is expected that a microcanonical ensemble of resonance states will contain states which exhibit mode-specific decay and can be identified by patterns (i.e., progressions) in the spectrum, as well as unassignable states with random ψ_n and, thus, state-specific rate constants with random fluctuations. In general, it is not expected that the ψ_n, which form a microcanonical ensemble, will have identical k_n which equal the RRKM $k(E)$.

8.3.4 Distributions of State-Specific Rate Constants

If all the resonance states which form a microcanonical ensemble have random ψ_n, as defined by Eqs. (8.15) and (8.16), and are thus intrinsically unassignable, a situation arises which we will refer to as *statistical state-specific* behavior. Since the wave function coefficients of the ψ_n are Gaussian random variables when projected onto the ϕ_i basis functions for any zero-order representation (Polik et al., 1990b), the distribution of the state-specific rate constants k_n will be as statistical as possible. If these k_n within the energy interval $E \rightarrow E + dE$ form a continuous distributions, Levine (1987) has argued that the probability of a particular k is given by the Porter-Thomas (1956) distribution

$$P(k) = \frac{\nu}{2\bar{k}} \left(\frac{\nu k}{2\bar{k}} \right)^{\frac{\nu-2}{2}} \frac{\exp(-\nu k/2\bar{k})}{\Gamma(\frac{1}{2}\nu)} \tag{8.17}$$

where $\bar{k}$ is the average state-specific unimolecular rate constant within the energy interval $E \rightarrow E + dE$,

$$\bar{k} = \int_0^\infty kP(k)\,dk \tag{8.18}$$

and ν is the "effective number of decay channels." Equation (8.17) is derived in statistics as the probability distribution

$$\chi_\nu^2 = x_1^2 + x_2^2 + \ldots + x_\nu^2, \tag{8.19}$$

where the ν x_i's are each independent Gaussian distributions (Polik et al., 1990b). Equation (8.17) has also been derived with maximal entropy arguments by Levine (1988) and with direct integration of probability distributions (Polik et al., 1988).

Increasing the effective number of decay channels ν reduces the variance of the distribution $P(k)$, as can be seen in figure 8.4. If ν equals 1 or 2 the most probable k is a value of zero. For $\nu > 2$ there is a maximum in $P(k)$ located at (Lu and Hase, 1989a):

$$k_{\max} = \frac{\nu - 2}{\nu}\bar{k}. \tag{8.20}$$

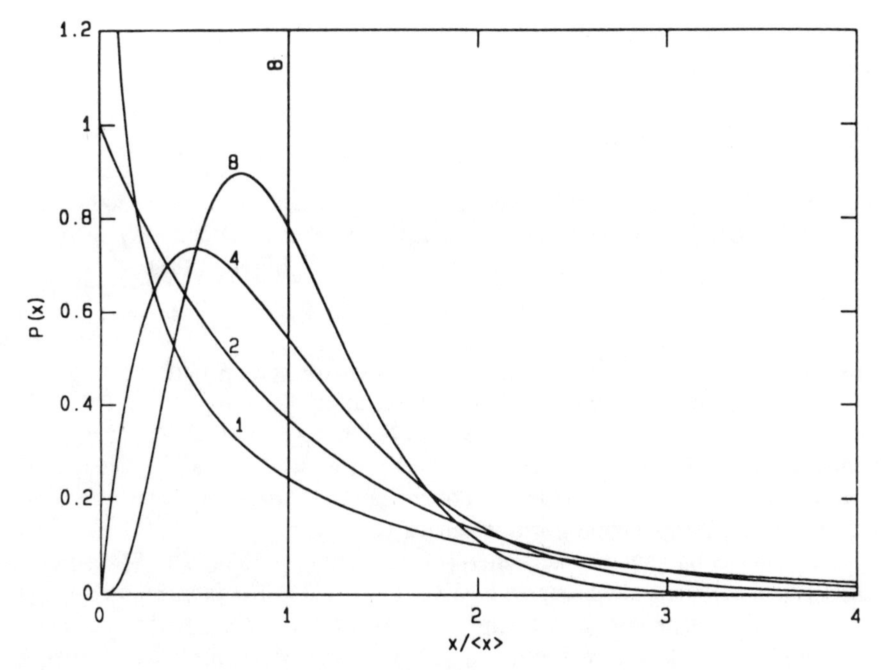

Figure 8.4 Porter-Thomas distribution of state specific rate constants, Eq. (8.15), for ν = 1, 2, 4, 8, and ∞. In these plots $x = k$ and $\langle x \rangle = \bar{k}$ (Polik et al., 1990b).

The extent of fluctuations in k for $\nu > 2$ may be studied by considering the second moment (Hase et al., 1989):

$$\langle [(k - k_{\max})/k_{\max}]^2 \rangle = \int_0^\infty [(k - k_{\max})/k_{\max}]^2 P(k)\, dk$$
$$= \frac{2(\nu + 1)}{(\nu - 2)^2} . \tag{8.21}$$

As ν becomes large this moment approaches zero.

Possible forms of $P(k)$ have not been established for groups of resonance states which are mode-specific or groups which contain both mode specific and non-mode-specific resonances. More work is required to determine $P(k)$ for systems that are not statistical-state-specific. However, as discussed below, the $P(k)$ in Eq. (8.17) may also fit systems which are not statistical-state-specific.

8.3.5 Statistical State Specificity and RRKM Theory

A microcanonical ensemble of isolated resonances excited incoherently decays according to

$$\mathbb{N}(t, E) = \sum_n \exp(-k_n t) , \tag{8.22}$$

where the k_n are state-specific rate constants. If the state-specific rate constants are assumed continuous, Eq. (8.22) can be written as (Lu and Hase, 1989a,b)

$$\mathbb{N}(t, E) = \mathbb{N}_o \int_0^\infty \exp(-kt) P(k)\, dk\,, \tag{8.23}$$

where $P(k)$ is the probability of a state-specific rate constant and $\mathbb{N}_o$ is the total number of molecules in the microcanonical ensemble. For the Porter-Thomas $P(k)$ distribution in Eq. (8.17), $\mathbb{N}(t, E)$ becomes (Lu and Hase, 1989a,b; Miller, 1988)

$$\mathbb{N}(t, E)/\mathbb{N}_o = (1 + 2\bar{k}t/\nu)^{-\nu/2}. \tag{8.24}$$

Thus, for state-specific decay and the most statistical (or nonseparable) case, a microcanonical ensemble does *not* decay exponentially as predicted by RRKM theory. It is worthwhile noting that when $\nu/2$ becomes very large, the right-hand side of Eq. (8.24) approaches $\exp(-\bar{k}t)$ (Miller, 1988), since $\lim (1 + x/n)^{-n} = \exp(-x)$, when $n \rightarrow \infty$. Other distributions for $P(k)$, besides the Porter-Thomas distribution, have been considered and all give $\mathbb{N}(t, E)$ expressions which are nonexponential (Lu and Hase, 1989b).

The connection between the Porter-Thomas nonexponential $\mathbb{N}(t, E)$ distribution and RRKM theory is made through the parameters $\bar{k}$ and ν. The average of the statistical state-specific rate constants $\bar{k}$ is expected to be similar to the RRKM rate constant $k(E)$. This can be illustrated (Waite and Miller, 1980) by considering a separable (uncoupled) two-dimensional Hamilton $H = H_x + H_y$, whose decomposition path is tunneling through a potential energy barrier along the x-coordinate. The complex energy for a resonance level is

$$E^o_{n_x,n_y} = E_{n_x} - i\Gamma_{n_x}/2 + E_{n_y}, \tag{8.25}$$

where only the energy of the x-coordinate is complex, since the Hamiltonian is uncoupled and tunneling occurs along the x-coordinate. The state-specific rate constant for the resonance is

$$k_{n_x,n_y} = \Gamma_{n_x}/\hbar. \tag{8.26}$$

Figure 8.2 provides an illustration of the E_{n_x} energy levels and the potential energy barrier.

Consider the limit of many closely spaced energy levels and the *average* rate constant at energy E:

$$\bar{k} \equiv \frac{\sum\limits_{n_x,n_y} k_{n_x,n_y}\, \delta(E - E_{n_x,n_y})}{\sum\limits_{n_x,n_y} \delta(E - E_{n_x,n_y})}. \tag{8.27}$$

If the semiclassical approximation for Γ_{n_x} in Eq. (8.26) is used, that is,

$$\Gamma_{n_x} = \frac{\kappa(E_{n_x})}{2\pi} \left(\frac{dE_{n_x}}{dn_x} \right), \tag{8.28}$$

where $\kappa(E_{n_x})$ is the one-dimensional tunneling probability for the x-coordinate [see Eqs. (8.7) and (8.8)], and one uses the continuum limit

$$\sum_{n_x} \rightarrow \int dn_x \,, \tag{8.29}$$

Eq. (8.27) can be written as

$$\bar{k} = \frac{\sum_{n_y} \kappa(E - E_{n_y}) \left[\int dE_{n_x}\, \delta(E - E_{n_y} - E_{n_x}) \right]}{h \sum_{n_x, n_y} \delta(E - E_{n_y} - E_{n_x})} \,. \tag{8.30}$$

If the δ function is broadened to include the energy range $E \rightarrow E + dE$, it is straightforward to show that Eq. (8.30) becomes

$$k(E) = \sum_{n_y} \kappa(E - E_{n_y})/h\rho(E) \tag{8.31}$$

which is the RRKM rate expression including tunneling (Miller, 1979). [When the δ function is broadened the denominator of Eq. (8.27) becomes $\rho(E)dE$.] The numerator is the sum of the tunneling probabilities. A comparison of Eqs. (8.27) and (8.31) shows that for decomposition by tunneling the average of the state-specific rate constants for a microcanonical ensemble $\bar{k}$ is the same as the RRKM rate constant $k(E)$.

Equation (8.31) suggests that the average quantum mechanical rate for a microcanonical ensemble of resonance states is the RRKM rate. However, this is not a general result and the above analysis only pertains to state-specific decomposition by tunneling. For this case the tunneling barrier defines the dividing surface, with no recrossings, which is needed to derive RRKM from classical (not quantum) mechanical principles (see chapter 6). For state-specific decomposition which does not occur by tunneling, a dividing surface cannot be constructed for a quantum calculation as is done for the above case. Therefore, it cannot be shown that $\bar{k}$, the microcanonical average of the state-specific rate constants, equals the RRKM $k(E)$. However, the above analysis is highly suggestive that $\bar{k}$ may be a good approximation to the RRKM $k(E)$.

The parameter ν in Eq. (8.17) has also been related to RRKM theory. Polik et al. (1990b) have shown that for decomposition by quantum mechanical tunneling

$$\nu = \left[\sum_{n} \kappa(E - E^{\ddagger}_{\mathbf{n}}) \right]^2 \Big/ \sum_{n} \kappa(E - E^{\ddagger}_{\mathbf{n}})^2 \,, \tag{8.32}$$

where $\kappa(E - E^{\ddagger}_{\mathbf{n}})$ is a one-dimensional tunneling probability through a potential barrier and $E^{\ddagger}_{\mathbf{n}}$ is the vibrational energy in the $3N - 7$ modes orthogonal to the tunneling coordinate. $\kappa(E - E^{\ddagger}_{\mathbf{n}})$ is the extension of $\kappa(E - E_{n_y})$ in Eq. (8.31) to many vibrational degrees of freedom. If the energy is sufficiently low that all the tunneling probabilities are <<1 and one makes a parabolic approximation to the tunneling probabilities (Polik et al., 1990b; Miller et al., 1990), Eq. (8.32) becomes

$$\nu = \prod_{k=1}^{3N-7} \coth(\pi\omega_k^{\ddagger}/\omega_b) \,, \tag{8.33}$$

where the $\omega_k^{\ddagger}$ are the $3N - 7$ frequencies for the modes orthogonal to the tunneling coordinate and ω_b is the barrier frequency. The interesting aspect of Eq. (8.33) is that it shows ν to be energy-independent in the tunneling region. On the other hand, for energies significantly above the barrier so that $\kappa(E - E_{\mathbf{n}}^{\ddagger}) = 1$, it is easy to show (Polik et al., 1990b, Miller et al., 1990) that

$$\nu = N^{\ddagger}(E), \tag{8.34}$$

where $N^{\ddagger}(E)$ is the sum of states for the transition state. Thus, in this energy region, ν rapidly increases with increase in energy and the $P(k)$ in Eq. (8.17) becomes more narrowly peaked, and $\mathbb{N}(t, E)$ in Eq. (8.24) more closely approximates the exponential decay predicted by RRKM theory. The second moment of the distribution of rate constants (i.e., fluctuations) is determined from Eqs. (8.20) and (8.21) by equating ν to $N^{\ddagger}(E)$.

8.3.6 Calculational Results

Theoretical studies (Beswick and Shapiro, 1982; Chu and Datta, 1982; Hutson et al., 1983; Ashton et al., 1983; Schatz et al., 1988) have shown that many van der Waals molecules, such as He---I_2 and Ar---HCl, dissociate via isolated resonances with state-specific rate constants. The wave functions for the resonances are found to be assignable, so that the unimolecular decomposition is mode specific. However, for the van der Waals molecules $ArCl_2$ (Halberstadt et al., 1992), ArI_2 (Gray, 1992a) and those formed by rare-gas atoms attached to aromatic molecules (Semmes et al., 1990) there is substantial coupling between zero-order states in forming the resonance states.

In contrast to the studies for van der Waals molecules, there have been substantially fewer theoretical studies of state-specific decomposition for molecules with covalent intramolecular potentials. A survey of the theoretical work is given in table 8.1. For the most part the studies are two-dimensional (i.e., one coordinate coupled to a reaction coordinate) and involve model potentials. An important counterexample are the studies for the formyl radical; that is, H—C=O → H + C≡O (Gazdy et al., 1991; Cho et al., 1992a; Gray, 1992b; Dixon, 1992; Wang and Bowman, 1994; Werner et al., 1995).

Mode-specific unimolecular decomposition has been observed in much of the theoretical work listed in table 8.1. In some cases more than one zero-order Hamiltonian is necessary to assign resonance states for an excited molecule. For the Hénon-Heiles Hamiltonian two zero-order Hamiltonians are used to identify assignable resonances as either restricted precessors (Q^{I}) or quasiperiodic liberators (Q^{II}) (Bai et al., 1983; Hose and Taylor, 1982). Similarly, Manz and co-workers (Bisseling et al., 1987; Joseph et al., 1988) assigned many stretching resonances of *ABA* molecules as either hyperspherical mode or local mode states. At the same energy, the Q^{II} states for the Hénon-Heiles Hamiltonian decay about an order of magnitude faster than the Q^{I} states.

Table 8.1. Quantum Mechanical Studies of Positions and Widths of Isolated Unimolecular Resonances.[a]

System	Reference
Two-dimensional studies	
ABC → AB + C	Eastes and Marcus, 1973; Christoffel and Bowman, 1983; Basilevsky and Ryaboy, 1983
Barbanis Hamiltonian	Waite and Miller, 1980
Hénon-Heiles Hamiltonian	Waite and Miller, 1981; Christoffel and Bowman, 1982; Bai et al., 1983
model isomerization	Bowman, et al., 1979
ABA → AB + A (H_2O, CH_2)	Hedges and Reinhardt, 1983; Bisseling et al., 1987; Joseph et al., 1988
DCN → D + CN	Chuljian et al., 1984
$H_2CO \rightarrow H_2 + CO$	Waite et al., 1983; Gray and Child, 1984; Aquilanti and Cavalli, 1987
H—C—C → H + C═C (model alkyl)	Moiseyev and Bar-Adon, 1984; Swamy et al., 1986
$H_2C{=}C \rightarrow HC{\equiv}CH$	Carrington et al., 1984
Three or more dimensions	
HCO → H + CO	Cho et al., 1992a; Gray, 1992b; Dixon, 1992; Wang and Bowman, 1994; Werner et al., 1995
$H_3^+ \rightarrow H^+ + H_2$	Pollak and Schlier, 1989
H_2O and CH_2O decomposition	Hartke et al., 1992
$HO_2 \rightarrow H + O_2$	Dobbyn et al., 1995

[a]The calculations are for systems with covalent intramolecular potentials.

At low excitation energies resonance stretching states of *ABA* molecules identified as hyperspherical modes decay at slower rates than do the local mode resonances (Bisseling et al., 1987). However, for higher energies such a rule cannot be established for the *ABA* resonances. Mode-specific decomposition was observed in both two- and three-dimensional models of H_2O and CH_2O dissociation (Hartke et al., 1992).

Levine (1988) has found that the *ABA* resonances studied by Manz and co-workers (Bisseling et al., 1985, 1987) can be fit by Eq. (8.17) with $\nu = 1.8$. These *ABA* resonances include a large number of mode-specific states and the *ABA* system is certainly not statistical state specific. The inferences to be made, in light of this result, is that the ability to fit a collection of resonance widths to Eq. (8.17) does not prove the system is statistical-state-specific. As discussed above, the evidence for statistical state specificity is the absence of *any* patterns in the positions of the resonances in the spectrum so that all the resonance states are intrinsically unassignable. This will be the case when the expansion coefficients, for the resonance wave functions ψ_n, are Gaussian random variables for any zero-order basis set (Polik et al., 1990b).

Mode-specific decomposition has been illustrated for the H—C—C → H + C═C model Hamiltonian (Swamy et al., 1986). Lifetimes for different resonance states are given in table 8.2. The states are listed according to the quantum numbers for the HC and CC stretch modes. The CC stretch progression of resonance states with zero quantum in the HC stretch has the longest lifetimes, while the progression with two quanta in the HC stretch has the shortest lifetimes. Such a finding is characteristic of mode-specific decomposition. A similar mode-specific decomposition is observed for formyl radical decomposition, HCO → H + CO (Wang and Bowman, 1994; Werner et

Table 8.2. Resonance Positions and Widths for the H—C—C→ H + C═C Model Alkyl System

State[a]	Energy (cm^{-1})[b]	Lifetime(s)
0,12	16195 (16200)	1.28(−10)
0,13	17221 (17220)	1.50(−8)
0,14	18223 (qp)	1.19(−8)
0,15	19205 (qp)	4.33(−10)
0,16	20167 (qp)	4.03(−10)
1,9	15803 (15800)	4.90(−11)
1,10	16894 (16890)	1.36(−10)
1,11	17963 (17960)	5.38(−11)
1,12	19009 (19010)	1.33(−10)
1,13	20032 (20030)	2.48(−11)
2,7	16209 (ch)	1.34(−12)
2,8	17357 (ch)	2.13(−13)
2,9	18419 (ch)	1.90(−13)
2,10	19536 (ch)	1.55(−13)

[a]In the identification of the state(n_1,n_2), n_1 and n_2 are the HC and CC stretch quantum numbers, respectively.

[b]The position (real energy) of the resonance. Semiclassical values for the resonance positions are given in parenthesis; qp means that the trajectory for the resonance is quasi-periodic, but was not quantized; ch means that the resonance trajectory is chaotic. The semiclassical energies are accurate to four significant figures.

al., 1995). Resonance states, with zero quantum in the HC stretch and HCO bend and excess quanta in the CO stretch, are particularly long-lived. Theoretical calculations of the widths of these resonance states are compared in table 8.3.

Three-dimensional quantum mechanical calculations have been performed to determine the unimolecular rate constants for the resonances in $HO_2 \rightarrow H + O_2$ dissociation (Dobbyn et al., 1995). The resonances are not assignable and the fluctuations in the resonance rate constants can be represented by the Porter-Thomas distribution, Equation 8.17. Thus, the unimolecular dissociation of HO_2 is apparently statistical

Table 8.3. Quantum Mechanical Calculations of Linewidths $\Gamma(cm^{-1})$ for HCO |0, v, 0> Resonances[a]

v	(Werner et al., 1995)	(Dixon, 1992)	(Wang and Bowman, 1994)
3	0.28(−3)		0.74(−2)
4	0.23(−1)	0.1	0.1
5	0.186	0.1	0.15
6	0.363	0.6	0.53
7	0.607	1.5	1.86
8	0.68(−1)	3.7	
9	1.24		
10	0.297		

[a]v is the CO stretch quantum number. Numbers in parentheses indicate powers of ten.

state specific. The HO_2 vibrational energy level spacings are well approximated by the Wigner surmise, Equation 2.81, which indicates the resonance state eigenfunctions are strongly mixed and, thus, unassignable.

8.3.7 Experimental Results

Compared to the large number of experimental studies of state-specific decomposition for van der Waals molecules, there is a paucity of such experimental data for the unimolecular decomposition of covalently bound molecules. This is because, for the latter class of molecules, it is often the case that the molecule's density of states is sufficiently large and its unimolecular lifetime sufficiently short that there is extensive overlapping of the resonance line widths. Experimental studies of state specific unimolecular decomposition are listed in table 8.4. In the following, experimental studies of D_2CO, HFCO, and NO_2 state-specific decomposition are reviewed.

As discussed in section 6.2.3.2 and illustrated in Figure 6.12, extensive fluctuations are observed in the unimolecular rate constant for D_2CO decomposition on the ground electronic state, S_0 (Polik et al., 1990b). For D_2CO with 28,300 cm^{-1} excitation, it is found that neighboring resonances have lifetimes that fluctuate by over two orders of magnitude. However, careful analysis of the resonance properties indicate that these fluctuations are not a result of mode specificity, but arise from random fluctuations in the state specific rate constants. The Porter-Thomas function, Eq. 8.17, with $\nu = 3.8$ provides an excellent fit to the distribution of S_0 unimolecular rate constants in the limit of a zero electric field, and the average rate constant for the distribution, as defined by Eq. 8.18 equals the RRKM rate constant (Polik et al., 1990b). Also, the energy level spacings between resonant states are well approximated by the Wigner surmise, Eq. 2.81, which indicates the resonant states are intrinsically unassignable. Thus, analyses of the experiments indicate formaldehyde unimolecular decomposition is *statistical state specific*. However there is still some uncertainty concerning complete mixing of the $2J + 1$ K-levels (Hernandez et al., 1993).

Stimulated emission pumping (SEP) has been used to prepare resonance states for HFCO near the dissociation threshold for HF + CO formation (Choi and Moore 1991, 1992). A difference between the H_2CO and HFCO potential energy surfaces, is that the HFCO → HF + CO dissociation threshold has been estimated as 49 ±4 kcal/mol and is significantly smaller than the $H_2CO \rightarrow H_2 + CO$ threshold of ~85 kcal/mol.

Table 8.4. Experimental Studies of State Specific Unimolecular Decomposition

Reaction	Reference
$D_2CO \rightarrow D_2 + CO$	Polik et al., 1988, 1990a
$HFCO \rightarrow HF + CO$	Choi and Moore, 1991, 1992
$H_2O_2 \rightarrow 2OH$	Luo and Rizzo, 1990, 1991, 1992
$NO_2 \rightarrow NO + O$	Miyawaki et al., 1993; Hunter et al., 1993; Reid et al., 1993, 1994
$CH_3O \rightarrow H + H_2CO$	Geers et al., 1993
$DCO \rightarrow D + CO$	Tobiasen et al. (1995a)
$HCO \rightarrow H + CO$	Tobiasen et al. (1995b); Adamson et al. (1993); Neyer et al. (1995)

Franck-Condon factors for the $S_1 \rightarrow S_0$ stimulated transitions result in S_0 resonance states in the 13,000 to 23,000 cm^{-1} energy range which are extremely high overtones of the C—H out-of-plane bending mode (ν_6) in combination with the C=O stretching mode (ν_2). The extent of state mixing in the wave functions Eq. (8.14), for the resonance states is strongly mode dependent. For vibrational states with almost the same total vibrational energy, resonance states with the most quanta in ν_6 show the least mixing between zero-order states, that is, fewer c_{in} of appreciable size in Eq. (8.14). The extent of state mixing as a function of quanta in ν_2 and ν_6 is depicted in figure 8.5. Consistent with this state mixing is the finding that for states with approximately the same total energy those with higher excitation in the out-of-plane bending mode v_6 dissociate more slowly than others. This is the type of mode specificity observed in the quantum calculations discussed above.

In the experimental studies of state specific NO_2 unimolecular dissociation (Miyawaki et al., 1993; Hunter et al., 1993; Reid et al., 1994, 1993), NO_2 is first vibrationally/rotationally cooled to ~1 K by supersonic jet expansion. Ultraviolet excitation is then used to excite a NO_2 resonance state which is an admixture of the optically active 1^2B_2 and the ground $\tilde{X}^2A_1$ electronic states. [It should be noted that in the subpicosecond experiments by Ionov et al. (1993a) discussed in section 6.2.3.1, a superposition of resonance states is prepared instead of a single resonance state.] The NO product states are detected by laser-induced fluorescence. Both lifetime and product energy distributions for individual resonances are measured in these experiments. A stepwise increase in the unimolecular rate constant is observed when a new product channel opens. Fluctuations in the product state distributions, depending on the resonance state excited, are observed. The origin of the dynamical results is not clearly understood, but it apparently does not arise from mode specificity, since analyses of

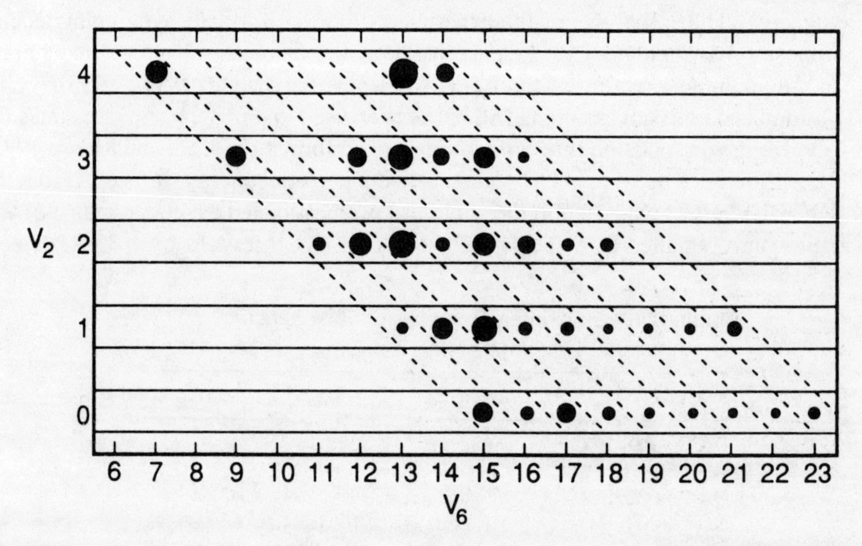

Figure 8.5 The dependence of vibrational state mixing on ν_2, ν_6, and total energy determined by high-resolution SEP spectroscopy. The vibrational states between two dashed lines are nearly isoenergetic. The diameter of the filled circles is proportional to the extent of mixing (Choi and Moore, 1991).

nearest-neighbor level spacings and spectral intensity distributions suggests chaotic dynamics and intrinsically unassignable resonance states. The fluctuations in the product energy distributions may arise from fluctuations in the properties of the resonance states and adiabaticity as the system moves from the transition state to products.

8.3.8 Relationship between State Specificity and Macroscopic Unimolecular Rate Constants

As discussed by Slater (1959) and Bunker (1964), the lifetime distribution $P(t, E)$ provides a link between microscopic and macroscopic unimolecular kinetics. The lifetime distribution is related to $\mathbb{N}(t, E)$, Eq. (8.22), according to

$$P(t,E) = -\frac{1}{\mathbb{N}_o}\frac{d\mathbb{N}(t,E)}{dt} \tag{8.35a}$$

$$= \sum_n k_n \exp(-k_n t)\left(\sum_n 1\right)^{-1}, \tag{8.35b}$$

where $\mathbb{N}_o = \Sigma_n 1$ is the number of resonances in the energy interval $E \rightarrow E + dE$.

Studies of unimolecular reactions involving chemical activation (Rabinovitch and Setser, 1964), radiationless transitions (Hippler et al., 1983), and overtone excitation (Crim, 1984) are often interpreted by the mechanism (Rabinovitch and Setser, 1964)

$$A^* \xrightarrow{k(\omega,E)} \text{decomposition products } (D)$$

$$\xrightarrow{\omega} \text{stabilized reactant } (S),$$

where A^* is the monoenergetically excited reactant and ω is the collisision frequency. The phenomenological collision-averaged monoenergetic unimolecular rate constant $k(\omega, E)$ is given by

$$k(\omega, E) = \omega D/S \tag{8.36}$$

and is often determined experimentally from a Stern-Volmer plot. If one makes the strong-collision assumption of unimolecular rate theory, D is given by (Forst, 1973; Hase, 1976)

$$D = \int_0^\infty W(t)P(t,E)\,dt, \tag{8.37}$$

where $W(t) = \exp(-\omega t)$ is the probability the reactant avoids a collision for time t (Slater, 1959). S is simply equal to $1 - D$. Inserting these expressions for D and S into Eq. (8.36) yields

$$k(\omega,E) = \mathbb{N}_o \Big/ \left(\int_0^\infty \mathbb{N}(t, E)\exp(-\omega t)\,dt\right) - \omega. \tag{8.38}$$

The $\mathbb{N}(t, E)$ for the isolated compound-state resonances, Eq. (8.22) gives

$$k(\omega,E) = \mathbb{N}_o \bigg/ \left\{ \sum_n 1/(k_n + \omega) \right\} - \omega\,. \tag{8.39}$$

The $\omega \rightarrow \infty$ and $\omega \rightarrow 0$ limiting values of $k(\omega, E)$ are

$$k(\infty,E) = \sum_n k_n/\mathbb{N}_o = \langle k_n \rangle \tag{8.40}$$

and

$$k(0,E) = \mathbb{N}_o \bigg/ \sum_n (1/k_n) = \langle 1/k_n \rangle^{-1} \tag{8.41}$$

Thus, in the high-pressure limit $k(\omega\ ,E)$ equals the average of the state-specific rate constants within the energy interval $E \rightarrow E + dE$, while $k(\omega, E)$ for the low-pressure limit is one divided by the average of the inverse of the state-specific rate constants. If all the k_i are equal, $k(\infty, E) = k(0, E)$ and normal RRKM behavior is observed. However, for *statistical state specificity*, where there are random fluctuations in the k_n, $k(\omega, E)$ will be pressure dependent.

The $\omega \rightarrow \infty$ and $\omega \rightarrow 0$ limiting values of $k(\omega, E)$ have been determined for the Porter-Thomas $P(k)$ distribution (Lu and Hase, 1989a,b). For the $\omega \rightarrow \infty$ limit, $k(\omega, E)$ is independent of ν and equals $\bar{k}$, as given by Eq. (8.40). At the $\omega \rightarrow 0$ limit, $k(\omega, E)$ depends upon the value of ν and the following expressions are obtained:

$$\nu = 1 \text{ and } 2: \qquad k(0, E) = 0\,, \tag{8.42}$$

$$\nu > 2 \text{ and finite:} \qquad k(0, E) = \left(\frac{\nu - 2}{\nu} \right) \bar{k}\,. \tag{8.43}$$

It should be noted that $k(0, E)$ in Eq. (8.43) is the same as k_{max} in Eq (8.20). That is, for $\nu > 2$ and finite, the value of $k(\omega, E)$ in the $\omega \rightarrow 0$ limit is the same as the value of k where the $P(k)$ distribution has its maximum. For $\nu = 1$ and 2, where there is not a maximum in $P(k)$, $k(0, E)$ equals the value of k where $P(k)$ has its largest value; that is, $k = 0$. In the experimental studies of Polik et al. (1990b) the state-specific rate constants for $D_2CO \rightarrow D_2 + CO$ dissociation are well-described by the Porter-Thomas distribution with $\nu = 3.8$. For this case $k(\omega, E)$ varies from $\bar{k}$ to $\bar{k}/2$ between the $\omega \rightarrow \infty$ and $\omega \rightarrow 0$ limits, respectively. This is not a substantial change in $k(\omega, E)$.

The Lindemann-Hinshelwood unimolecular rate constant $k_{uni}(\omega, E)$ is related to the lifetime distribution $P(t, E)$ according to (Slater, 1959; Bunker, 1964)

$$k_{uni}(\omega, E) = \omega D = \omega \int_0^\infty W(t)P(t, E)dt\,. \tag{8.44}$$

By comparing Eqs. (8.36) and (8.44), it is seen that the relationship between $k_{uni}(\omega, E)$ and $k(\omega, E)$ is given by

$$k_{uni}(\omega, E) = \frac{\omega k(\omega, E)}{k(\omega, E) + \omega}\,. \tag{8.45}$$

If Eq. (8.35) is used for $P(t, E)$, $k_{uni}(\omega, E)$ in Eq. (8.44) becomes

$$k_{\text{uni}}(\omega, E) = \frac{\omega}{\mathbb{N}_o} \sum_n \frac{k_n}{k_n + \omega}, \tag{8.46}$$

where the summation is over the resonance states within $E \rightarrow E + dE$. If the energy E can be assumed to be continuous, one obtains the thermal Lindemann-Hinshelwood unimolecular rate constant by Boltzmann weighting the $k_{\text{uni}}(\omega, E)$ given by Eq. (8.45), that is,

$$k_{\text{uni}}(\omega, T) = \int_0^{\infty} \frac{k_{\text{uni}}(\omega, E)\,\rho(E)\exp(-E/k_B T)\,dE}{Q}$$

$$= \frac{\omega}{Q}\int_0^{\infty} \frac{k(\omega, E)\,\rho(E)\exp(-E/k_B T)\,dE}{k(\omega, E) + \omega}, \tag{8.47}$$

where $\rho(E)$ and Q are the density of states and partition function for the reactant molecule's internal degrees of freedom. Equation (8.47) is a further extension of the standard thermal Lindemann-Hinshelwood unimolecular rate constant (Robinson and Holbrook, 1972; Forst, 1973). If energy cannot be assumed to be continuous, one obtains $k_{\text{uni}}(\omega, T)$ by Boltzman weighting the $k_{\text{uni}}(\omega, E)$ in Eq. (8.46) to give

$$k_{\text{uni}}(\omega, T) = \frac{\omega \sum_n k_n \exp(-E_n/k_B T)/(k_n + \omega)}{Q}. \tag{8.48}$$

Here, the summation is over all the resonance states, not only those within $E \rightarrow E + dE$. An equation similar to Eq. (8.48) has been used to interpret thermal unimolecular rate constants for the formyl radical (Wagner and Bowman, 1987).

The Porter-Thomas distribution for $\mathbb{N}(t, E)$, Eq. (8.24), can be inserted into Eq. (8.35) to obtain the Porter-Thomas lifetime distribution:

$$P(t, E) = \bar{k}(1 + 2\bar{k}t/\nu)^{-[(\nu/2)+1]}, \tag{8.49}$$

which becomes the RRKM exponential form $P(t, E) = \bar{k}\exp(-\bar{k}t)$ for $\nu \rightarrow \infty$. When Eq. (8.49) for $P(t, E)$ is combined with Eq. (8.44), the Lindemann-Hinshelwood $k_{\text{uni}}(\omega, E)$ is obtained, which results from the Porter-Thomas $P(k)$ distribution in Eq. (8.17). As shown in figure 8.6, the resultant $k_{\text{uni}}(\omega, E)$ depends on the value of ν in the Porter-Thomas distribution. This result suggests that $k_{\text{uni}}(\omega, T)$ in Eqs. (8.47) and (8.48) may be sensitive to the distribution of state-specific rate constants $P(k)$ (Miller, 1988).

8.4 NONRANDOM ENERGY-SELECTED EXPERIMENTS

In the previous section excitation of a single, isolated resonance and its ensuing unimolecular decomposition was considered. However, unimolecular dynamics has also been investigated by exciting a superposition of resonance states, which is initially localized in one part of the molecule, for example, a C—H bond. If this superposition contains all the resonance states in the energy width ΔE of the excitation process, statistical unimolecular decomposition might be expected after complete IVR for the

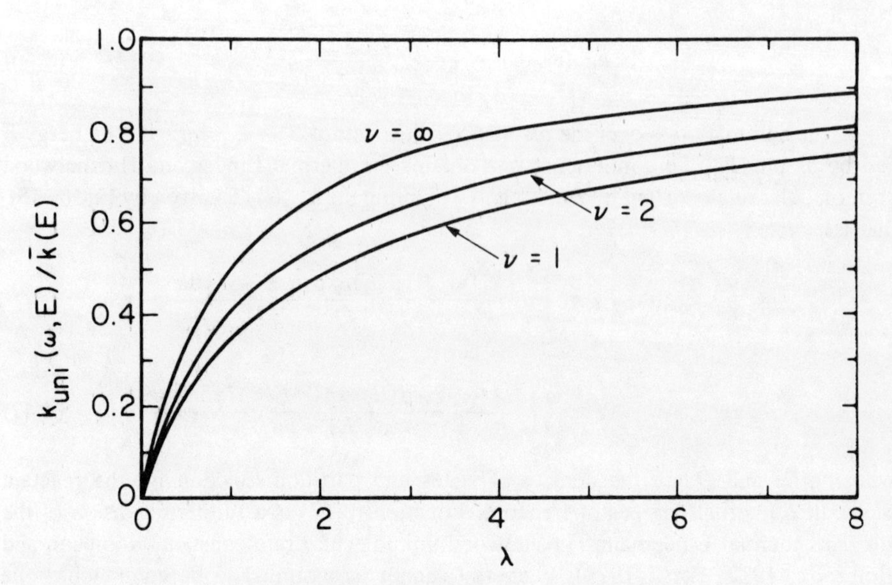

Figure 8.6 Microcanonical pressure-dependent rate constant (normalized to the average microcanonical rate $\bar{k}(E)$, as a function of the reduced collision frequency $\lambda \equiv \omega/\bar{k}(E)$. ν is the "number of channels" which characterizes the Porter-Thomas distribution; $\nu = \infty$ is the function $\lambda/(\lambda + 1)$, the standard textbook result (Miller, 1988).

initial superposition state. Of interest is to detect the nature of the decomposition process before complete IVR. As discussed in the next section, initial nonstatistical unimolecular decomposition, arising from the nature of the excitation process, is called *apparent* non-RRKM behavior.

To detect the initial apparent non-RRKM decay in a photoactivation or chemical activation experiment, one has to monitor the reaction at short times. This can be performed by studying the unimolecular decomposition at high pressures, where collisional stabilization competes with the rate of IVR. The first successful detection of apparent non-RRKM behavior was accomplished by Rynbrandt and Rabinovitch (1971) who used chemical activation to prepare vibrationally excited hexafluorobicyclopropyl-d_2 as described in chapter 1 (Eq. 1.21). Similar studies were also performed (Meagher et al., 1974) on the series of chemically activated fluoroalkyl cyclopropanes:

F F
F —R, $R = CF_3, C_3F_7, C_5F_{11}$. (8.50)
D D

The chemically activated molecules are formed by reaction of ${}^1CD_2$ with the appropriate fluorinated alkene. In all these cases apparent non-RRKM behavior was observed. As displayed in figure 8.7, the measured unimolecular rate constants are strongly dependent on pressure. However, at low pressures each rate constant approaches the RRKM value.

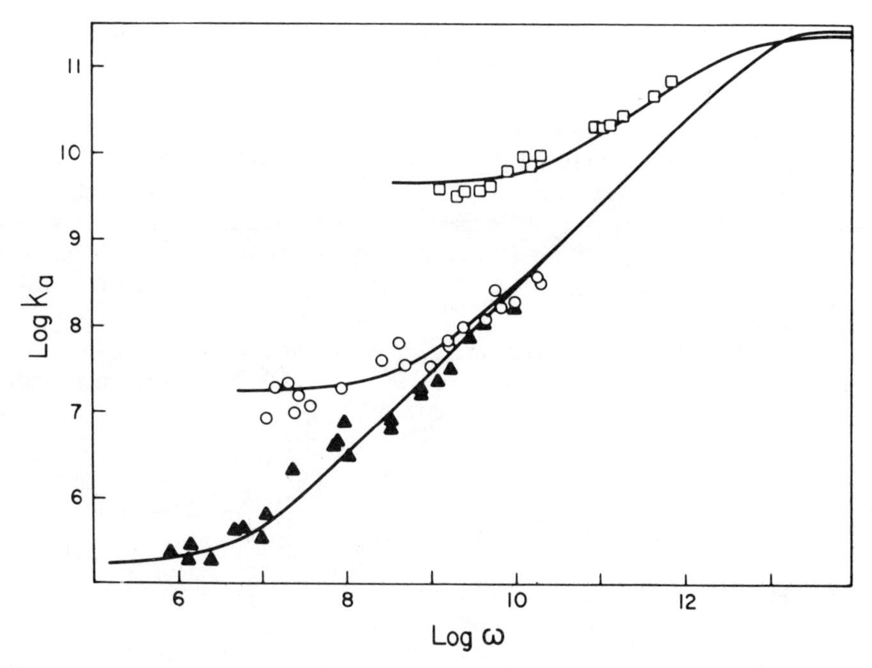

Figure 8.7 Chemical activation unimolecular rate constants vs. ω for fluoroalkyl cyclopropanes; see Eq. (8.50) (Meagher et al., 1974). The □, ○, and ▲ points are for $R = CF_3$, C_3F_7, and C_5F_{11}, respectively.

Vibrational overtone excitation of C—H and O—H bonds has been used to excite molecules, which have an initial thermal distribution of energy (Reddy and Berry, 1977; Crim, 1984; Jasinski et al., 1983). Unimolecular rate constants are determined either directly by measuring the time evolution of the products or indirectly from Stern-Volmer plots. Because of the thermal energy distribution of the molecules to be excited, the experiments are not particularly sensitive to state (or mode)-specific unimolecular decomposition (Hase, 1985). Experiments of this type for the isomerization of allyl isocyanide were initially thought to exhibit mode-specific rate constants (Reddy and Berry, 1979), but later experiments (Segall and Zare, 1988) showed the isomerization could be accurately modeled by RRKM theory. However, two experiments involving Stern-Volmer plot analyses seem to show non-RRKM behavior. In the dissociation of *t*-butyl hydroperoxide by O—H overtone excitation, it has been suggested (Chandler et al., 1982; Gutow et al., 1988) that at high pressures the dissociation of a molecule with nonrandomized energy competes with IVR. This is an apparent non-RRKM effect, like that discussed above for reactions (8.50). Also, the isomerization of CHD_2NC exhibits a K quantum number dependence, which cannot be explained by RRKM theory (Gutow and Zare, 1992). Finally in pump-probe experiments of $H_2O_2 \rightarrow 2OH$ dissociations, performed in the time-domain so that collisions are unimportant, the rate of product formation is nonexponential and different rotational states of the OH product are formed at different rates (Scherer and Zewail, 1987). The same type of experiment performed for $NCNO \rightarrow CN + NO$ dissociation (Khundkar et al., 1987) shows similar behavior. However, it is not clear whether the results are due to non-RRKM dynamics or reflect the thermal energy distribution of the preexcited reactant molecules and the distribution of resonance states excited by the broad laser bandwith.

Real-time experiments (Khundkar and Zewail, 1990; Zewail, 1991) with a sub-picosecond resolution have probed the unimolecular dynamics of $NO_2 \rightarrow NO + O$ (Ionov et al., 1993a) and $H + CO_2 \rightarrow HOCO \rightarrow HO + CO$ (Scherer et al., 1987, 1990; Ionov et al., 1993b). The NO_2 experiment is described and discussed in section 6.2.3.1 (p. 196). The $H + CO_2$ reaction and ensuing formation of HOCO is initiated by photodissociation of HI in the HI---CO_2 van der Waals complex (Fig. 8.8). A sub-picosecond laser pulse is used to initiate the reaction while a second laser pulse probes the product formation. The reactants are vibrationally and rotationally cold prior to excitation, and the experiments demonstrate that the $H + CO_2$ reaction proceeds

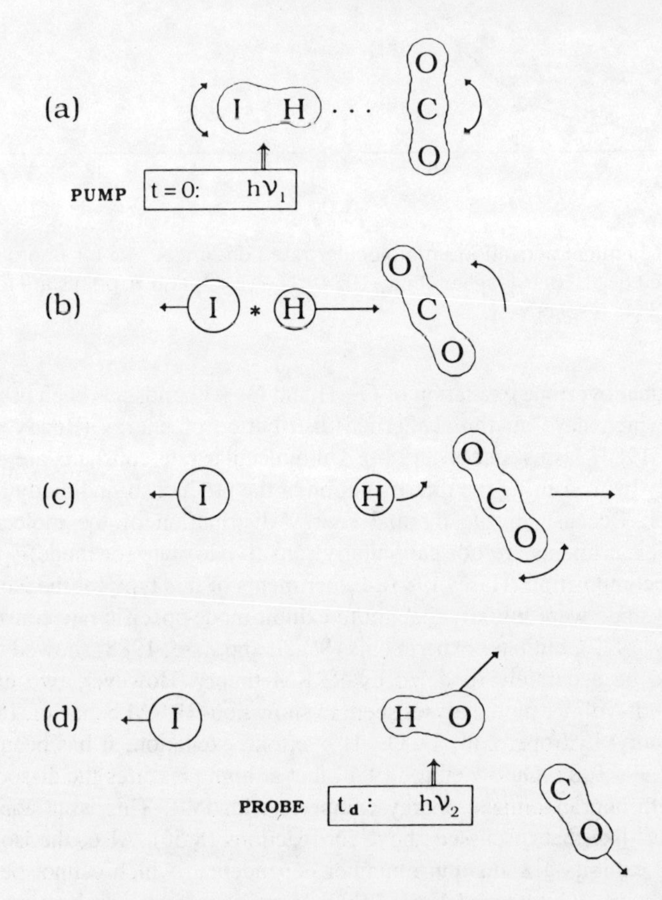

Figure 8.8 Experimental description: (a) Zero-of-time ($t = 0$) established by pump laser pulse (λ_1), photon energy $h\nu_1$, which photodissociates the HI component of the HI·CO_2 *vdW* molecule. (b) The translationally hot H atom is ejected toward CO_2. (c) The H atom forms the HOCO complex, which then undergoes intramolecular vibrations and unimolecular dissociation to yield (d) OH (and CO), detected at a time t_d by the probe laser pulse (λ_2) photon energy $h\nu_2$ (Scherer et al., 1990).

through an HOCO collision complex. While the HOCO rate constants are in accord with RRKM theory, the HOCO molecules which decay to yield OH in the $J = 6$ rotational state do so significantly faster than those which yield $J = 1$ (Scherer et al., 1990). This appears to be a non-RRKM result. Finally, it is of interest that an initial apparent non-RRKM decay is not observed in either the experiments for NO_2 or HOCO dissociation. Apparently IVR is very rapid for these reactive systems.

8.5 CLASSICAL DYNAMICS

8.5.1 Intrinsic and Apparent Non-RRKM Behavior

Classical trajectory studies of unimolecular decomposition have helped define what is meant by RRKM and non-RRKM behavior (Bunker, 1962, 1964; Bunker and Hase, 1973; Hase, 1976, 1981). RRKM theory assumes that the phase space density of a decomposing molecule is uniform. A microcanonical ensemble exists at $t = 0$ and rapid intramolecular processes maintain its existence during the decomposition [fig. 8.9(a), (b)]. The lifetime distribution, Eq. (8.35a), is then

$$P(t) = k(E)\exp[-k(E)t)], \tag{8.51}$$

where $k(E)$ is the RRKM rate constant. This property has been called *intrinsic* RRKM behavior (Bunker and Hase, 1973). As discussed in section 4.4.3, trajectory calculations have identified two distinct types of intramolecular motion: chaotic and quasiperiodic. Intrinsic RRKM behavior is observed if *all* the classical motion is sufficiently chaotic so that the intramolecular dynamics is ergodic on the time scale of the unimolecular decomposition. This ensures that a microcanonical ensemble is maintained as phase space points cross the transition state (dividing surface) from excited reactant to products. Molecules, for which intrinsic RRKM behavior has been observed in trajectory studies, are listed in table 8.5.

Intrinsic non-RRKM behavior occurs when an initial microcanonical ensemble decays nonexponentially or exponentially with a rate constant different from that of RRKM theory. The former occurs when there is a bottleneck (or bottlenecks) in the classical phase space so that transitions between different regions of phase space are less probable than that for crossing the transition state [fig. 8.9(e)]. Thus, a microcanonical ensemble is not maintained during the unimolecular decomposition. A limiting case for intrinsic non-RRKM behavior occurs when the reactant molecule's phase space is metrically decomposable into two parts, for example, one part consisting of chaotic trajectories which can decompose and the other of quasiperiodic trajectories which are trapped in the reactant phase space (Hase et al., 1983). If the chaotic motion gives rise to a uniform distribution in the chaotic part of phase space, the unimolecular decay will be exponential with a rate constant k_{ch} given by

$$k_{\text{ch}} = \frac{N^{\ddagger}(E)}{h\rho_{\text{ch}}(E)}. \tag{8.52}$$

where $\rho_{\text{ch}}(E)$ is the density of states for the chaotic region of phase space. Thus k_{ch} is related to the RRKM rate constant by

$$k_{\text{RRKM}} = f_{\text{ch}}\, k_{\text{ch}}, \tag{8.53}$$

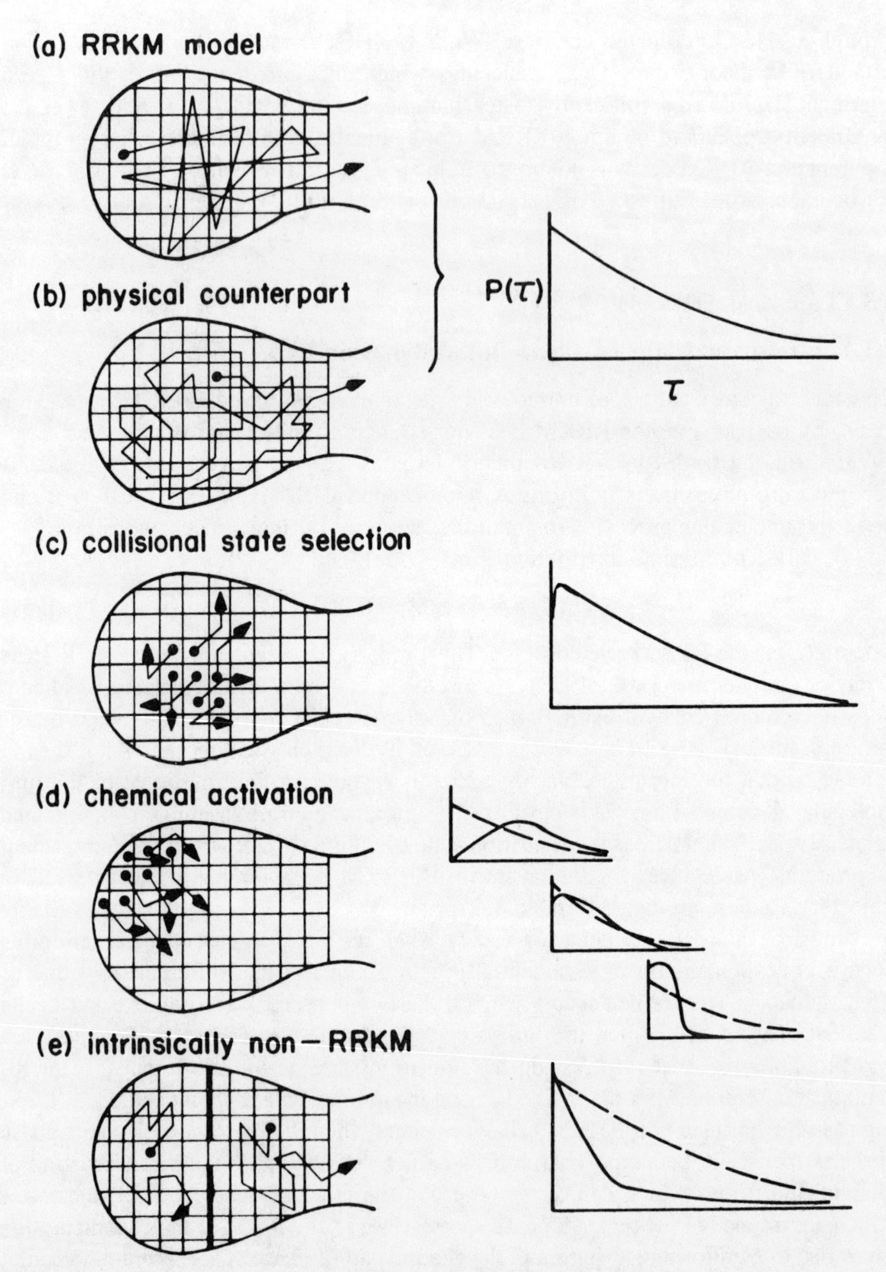

Figure 8.9 Relation of state occupation (schematically shown at constant energy) to lifetime distribution for the RRKM theory and for various actual situations. Dashed lines in lifetime distributions for (d) and (e) indicate RRKM behavior. (a) RRKM model. (b) Physical counterpart of RRKM model. (c) Collisional state selection. (d) Chemical activation. (e) Intrinsically non-RRKM (Bunker and Hase, 1973).

where $f_{ch} = \rho_{ch}(E)/\rho(E)$ is the fraction of phase space which consists of chaotic motion. Since there is a microcanonical ensemble at $t = 0$, the $t = 0$ intercept of $P(t)$ in Eq. (8.35a) is k_{RRKM} so that

$$P(t) = \kappa_{RRKM} \exp(-k_{ch}t). \tag{8.54}$$

A list of trajectory studies identifying intrinsic non-RRKM decomposition is given in table 8.5.

In many experiments such as photoactivation and chemical activation, the molecular vibrational/rotational states are excited nonrandomly. Regardless of the pattern of the initial energizing, the RRKM model requires the phase space distribution to become random in a negligibly short time. Three different possible lifetime distributions are represented by figure 8.9(d). As shown in the middle, the lifetime distribution may be similar to that of RRKM theory. In other cases, the probability of a short lifetime with respect to reaction may be enhanced or reduced, depending on the location of the initial excitation within the molecule. These are examples of *apparent* non-RRKM behavior arising from the initial nonrandom excitation. If there are very strong internal couplings, the lifetime distribution will become that of RRKM theory after rapid intramolecular vibrational energy redistribution. The systems studied by Rabinovitch and co-workers (Rynbrandt and Rabinovitch, 1971; Meagher, et al., 1974) are classic examples of apparent non-RRKM behavior.

8.4.2 Slater Theory

The classical mechanical model of unimolecular decomposition developed by Slater (1959) is based on the normal mode harmonic oscillator Hamiltonian, Eq. (2.47).

Table 8.5. Classical Trajectory Studies Identifying Intrinsic RRKM and non-RRKM Unimolecular Decomposition.

Reaction	Reference
Intrinsic RRKM decomposition	
$NO_2 \rightarrow NO + O$, $O_3 \rightarrow O_2 + O$	Bunker, 1962, 1964
$C_2HCl \rightarrow C_2H + Cl$	Hase and Feng, 1974, 1976; Sloane and Hase, 1977
$C_2H_5 \rightarrow H + C_2H_4$	Hase et al., 1979; Hase and Buckowski, 1982; Hase et al., 1983
$C_2H_4Cl \rightarrow C_2H_4 + Cl$	Sewell and Thompson, 1990; Sewell et al., 1991
$CH_4 \rightarrow H + CH_3$	Hu and Hase, 1991
$SiH_2 \rightarrow$ products	Schranz et al., 1991c
$Ar_n \rightarrow Ar_{n-1} + Ar$ $(12 \leq n \leq 14)$	Weerasinghe and Amar, 1993
$Al_3 \rightarrow Al_2 + Al$	Peslherbe and Hase, 1994
Intrinsic non-RRKM decomposition	
Model $HCC \rightarrow H + CC$	Bunker, 1964; Wolf and Hase, 1980a, 1980b; Hase and Wolf, 1981a, 1981b
$CH_3NC \rightarrow CH_3CN$	Bunker and Hase, 1973; Sumpter and Thompson, 1987
Model one-dimensional chains	Schranz et al., 1986
$HNC \rightarrow HCN$	Smith et al., 1987
$Si_2H_6 \rightarrow$ products	Schranz et al., 1991a; Sewell et al., 1991
1, 2-$C_2H_2F_2 \rightarrow$ products	Schranz et al., 1991b; Sewell et al., 1991
Cl^-—$CH_3Cl \rightarrow Cl^- + CH_3Cl$	Vande Linde and Hase, 1990a, 1990b; Cho et al., 1992b

Solving the classical equations of motion, Eq. (2.9), for this Hamiltonian gives rise to quasiperiodic motion in which each normal mode coordinate Q_i varies with time according to

$$Q_i = Q_i^\circ \cos(2\pi \nu_i t + \delta_i), \tag{8.55}$$

where Q_i° is the amplitude and δ_i the phase of the motion. Reaction is assumed to have occurred if a particular internal coordinate q, such as a bond length, attains a critical extension $q^\ddagger$. In the normal mode approximation, the displacement $\mathbf{d}$ of internal coordinates and normal mode coordinates $\mathbf{Q}$ are related through the linear transformation

$$\mathbf{d} = \mathbf{L}\,\mathbf{Q}. \tag{8.56}$$

The transformation matrix $\mathbf{L}$ is obtained from a normal mode analysis performed in internal coordinates, namely Eq. (2.49). Thus, as the evolution of the normal mode coordinates versus time is evaluated from Eq. (8.55), displacements in the internal coordinates and a value for q are found from Eq. (8.56). The variation in q with time results from a superposition of the normal modes. At a particular time the normal mode coordinates may phase together, figure 8.10, so that q exceeds the critical extension $q^\ddagger$, which defines the occurrence of decomposition.

If a microcanonical ensemble is chosen at $t = 0$, Slater theory gives an initial

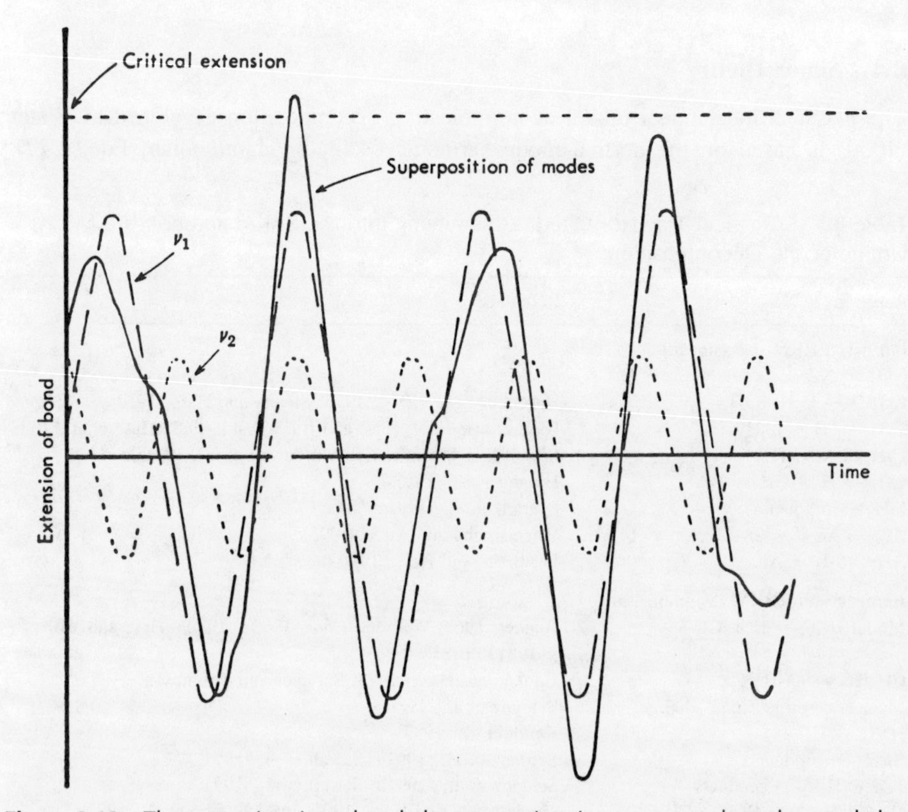

Figure 8.10 The extension in a bond due to motion in two normal modes, and the resultant extension (Laidler, 1965).

decay rate which agrees with the RRKM value. However, Slater theory gives rise to intrinsic non-RRKM behavior (Bunker, 1962, 1964, 1966). The trajectories are quasiperiodic and each trajectory is restricted to a particular type of motion and region of phase space. Thus, as specific trajectories react, other trajectories cannot fill up unoccupied regions of phase space. As a result, a microcanonical ensemble is not maintained during the unimolecular decomposition. In addition, some of the trajectories may be unreactive and trapped in the reactant region of phase space.

A severe shortcoming of Slater theory is the assumption that the intramolecular motion of a highly excited molecule is represented by the normal mode Hamiltonian. As discussed in section 2.3, at high degrees of excitation anharmonic couplings between the normal modes become important. It is found that the molecular phase space is not metrically decomposable (Bunker, 1964) and the chaotic motion of the trajectories may maintain a microcanonical ensemble during the unimolecular decomposition. More complete unimolecular rate theories, like the one described in the next section, are needed to describe the transition from normal mode quasiperiodic Slater dynamics at low energies to chaotic RRKM dynamics at high energies.

8.4.3 Phase Space Bottlenecks and Rate Constants

In research similar to that described in section 4.3.1, phase space structures and phase space bottlenecks have been used to analyze unimolecular reaction dynamics (Davis and Gray, 1986; Gray et al., 1986b; Gray and Rice, 1987; Zhao and Rice, 1992; Jain et al., 1993; DeLeon, 1992a,b; Davis and Skodje, 1992). Important phase space structural properties are illustrated in figure 8.11, for the one-dimensional pendulum Hamiltonian (Lichtenberg and Lieberman, 1991):

$$H = \frac{1}{2}Gp^2 - F\cos\phi = E. \tag{8.57}$$

For $E > E_{sx}$ the motion is rotational, while the motion is librational for $E < E_{sx}$. For $E = E_{sx}$ the oscillation period becomes infinite. The p, ϕ phase space curve for $E = E_{sx}$ in figure 8.11(b) is called a *separatrix*, because it separates different types of motion, in this case librational from rotational. There are singular points for $p = 0$. One at the origin $\phi = 0$ with $U = 0$ and the other where two branches of the separatrix cross at $\phi = \pm\pi$. The first point is a stable or *elliptic* singular point, while the second is an unstable or *hyperbolic* singular point. Near an elliptic singular point a trajectory remains in its neighborhood, while a trajectory near a hyperbolic point diverges from the point.

Phase space curves, associated with a one-dimensional attractive potential such as that for He-I, are illustrated in figure 8.12. The separatrix (i.e., dashed curve) is called a "reaction separatrix," because it is a boundary between bound unreactive motion and unbound reactive motion. Regardless of the length of time the trajectory is evaluated it will remain on this curve and, therefore, the separatrix is called an invariant curve. It is plotted in figure 8.13(a).

If the He-I one-dimensional potential is coupled to other degrees of freedom as in HeI_2, the separatrix is no longer an invariant curve (Davis and Gray, 1986; Davis and Skodje, 1992). Trajectories that initially do not have sufficient energy in the He-I coordinate to cross the separatrix and dissociate can acquire the needed energy by

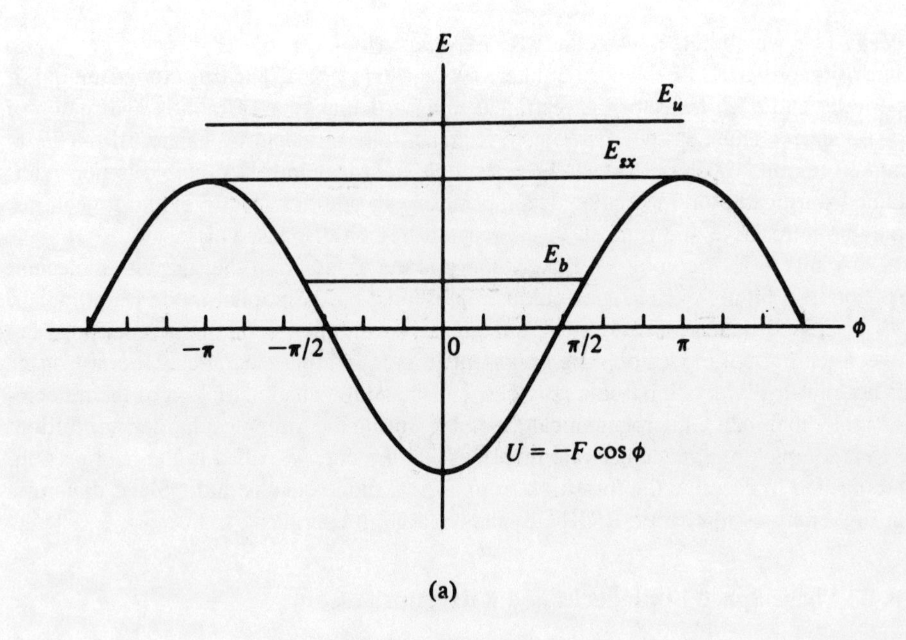

(a)

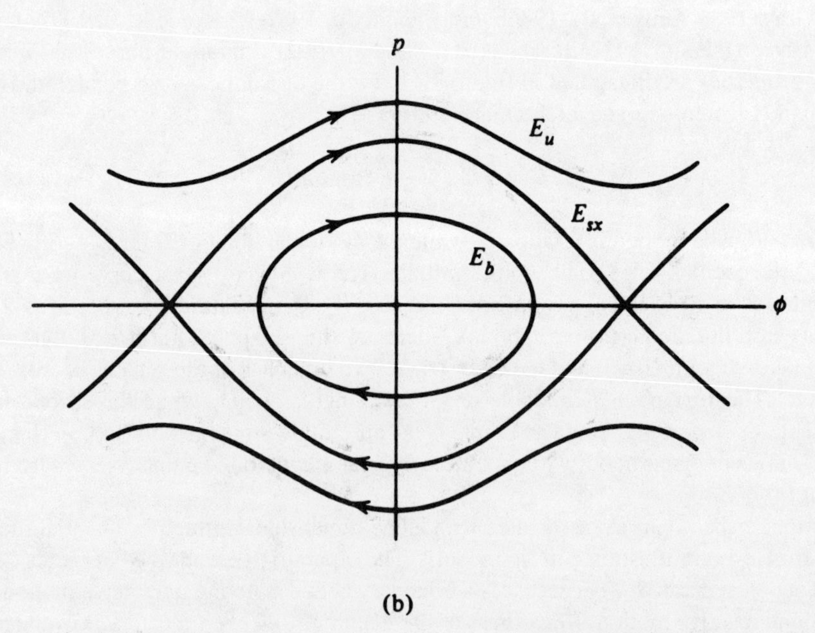

(b)

Figure 8.11 Correspondence between (a) an energy diagram and (b) a phase space diagram for the pendulum Hamiltonian (Lichtenberg and Lieberman, 1992).

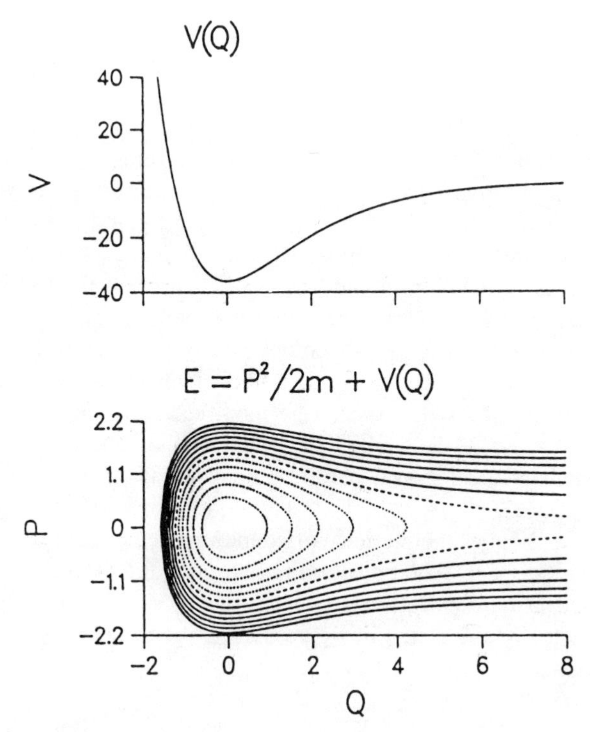

Figure 8.12 One-dimensional potential for the He—I coordinate (top) and phase space curves for this potential (bottom) (Davis and Skodje, 1992).

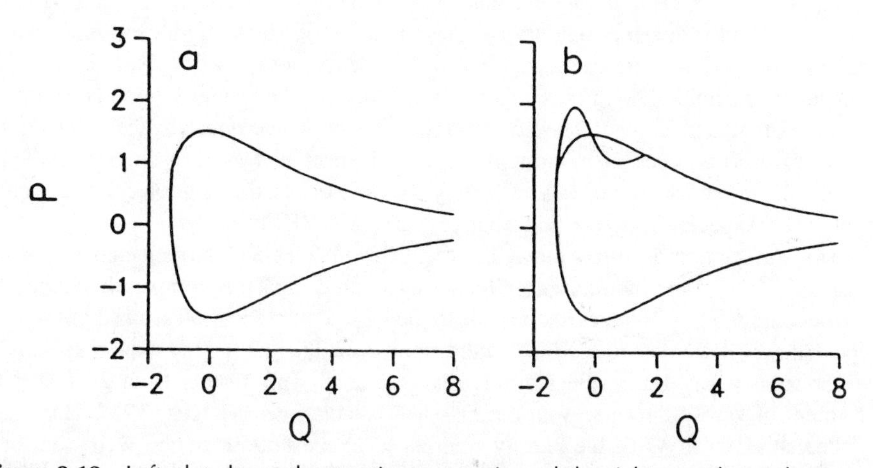

Figure 8.13 Left plot shows the reactive separatrix and the right one shows the propagated version of the separatrix plotted on top of the original separatrix, for $HeI_2 \rightarrow He + I_2$ (Davis and Skodje, 1992).

transfer of energy from other degrees of freedom. Since the reactive separatrix is no longer an invariant curve, if it is propagated for an additional period of time a new separatrix curve results. It has a figure-eight "turnstile," with regions outside and inside the original separatrix [fig. 8.13(b)]. These two regions are called "flux in" and "flux out," respectively. A collision between He and I_2 forms a HeI_2 collision complex by passing through the flux in half of the turnstile and then dissociates to He + I_2 by passing through the flux out half. This is illustrated in figure 8.14. Due to Liouville's theorem, the area bounded by the original and propagated versions of the separatrix are equal and the areas of the flux in and flux out are also equal. If there are no intramolecular bottlenecks for energy transfer within the collision complex, the unimolecular rate constant for dissociation of the complex is proportional to the area of the flux out region of the turnstile divided by the area of the interaction region. It is worthwhile noting that classical RRKM theory can be modified such that the separatrix is the transition state. This was done numerically with exact separatrices by Davis and Gray (1986). An actual statistical theory, requiring no exact dynamics, was then developed by Gray et al. (1986a).

Figure 8.15 illustrates the presence of an intramolecular bottleneck in the interaction region of phase space. The transition rate through the turnstile in this bottleneck can be calculated using concepts described in section 4.3.1. An intramolecular bottleneck, such as the one depicted in figure 8.15, is expected to give rise to intrinsic non-RRKM behavior.

8.5.4 Comparisons with Quantum Dynamics

In comparing unimolecular rate constants calculated by classical and quantum dynamics, it is useful to consider three different types of intramolecular motion for the dissociating molecule (Hase, 1986; Hase et al., 1989). For the first, the motion throughout the phase space is chaotic and classical RRKM theory correctly describes the classical unimolecular rate constant. The dynamics is intrinsically RRKM. The second case is for quasiperiodic trajectories. These trajectories are locked in a regular type motion and will not dissociate even though the total energy exceeds the unimolecular threshold. The third case arises when the intramolecular motion is so highly correlated that a trajectory undergoes infrequent (and hence nonstatistical) transitions between different types of motion which may themselves be nearly quasiperiodic or chaotic. Here the molecule will ultimately dissociate, but the unimolecular rate constant is not expected to agree with that of classical RRKM theory.

An association is often made between chaotic classical motion and statistical behavior. Thus, if the motion within the classical phase space is completely chaotic so that decomposing molecules can be described by a microcanonical ensemble, it is expected that RRKM theory will be valid. However, the classical and quantal RRKM rate constants may be in considerable disagreement. This results from an incorrect treatment of zero-point energy in the classical calculations (Marcus, 1977; Hase and Buckowski, 1982). With the energy referenced at the bottom of the well, the total internal energy of the dissociating molecule is $E = E^* + E_z^*$, where E^* is the internal energy of the molecule and E_z^* is its zero-point energy. The classical dissociation energy is D_e, and the energy available to the dissociating molecule at the classical

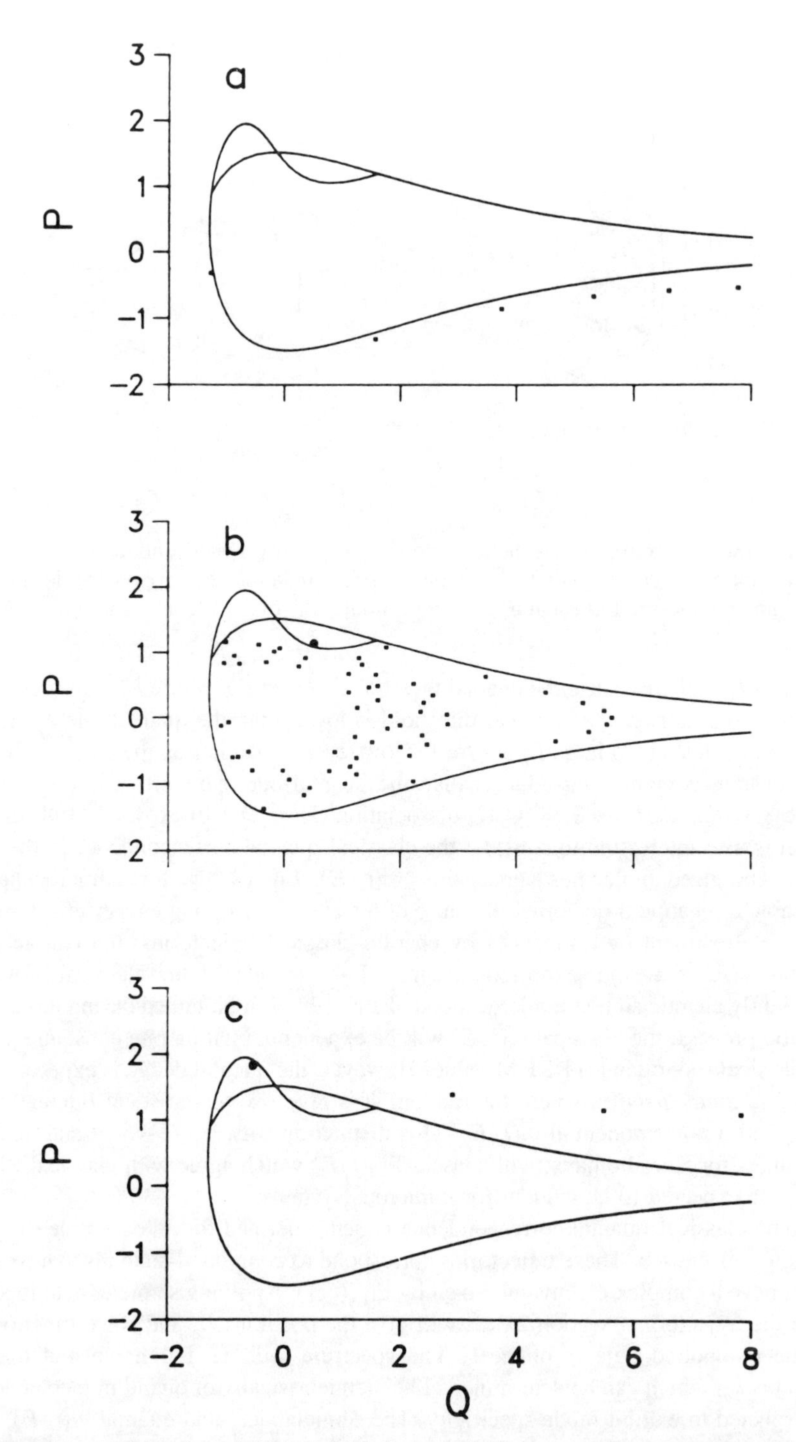

Figure 8.14 A He + I_2 trajectory which forms a collision complex. The intermolecular bottleneck is plotted along with the turnstile which leads to dissociation (left side) or complex formation (right side). (a) The approach of the trajectory, (b) the trajectory while it is a collision complex, and (c) the dissociative part of the trajectory. In (b) and (c) the first point on each plot is shown as a larger dot. Note the trajectory in (b) enters through the flux in region and leaves in (c) through the flux out half (Davis and Gray, 1986).

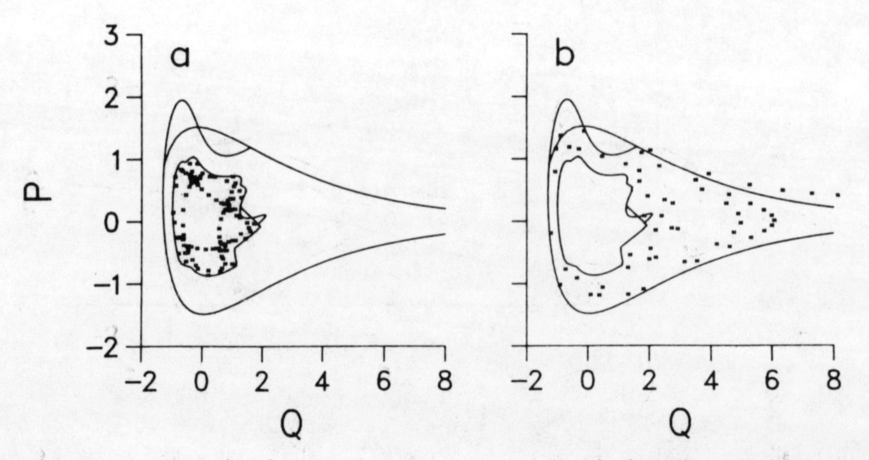

Figure 8.15 An example of a HeI_2 + He + I_2 trajectory which undergoes dissociation with an intramolecular bottleneck. In part (a) the trajectory is trapped inside an intramolecular bottleneck and escapes in (b) and finally dissociates (Davis and Gray, 1986).

barrier is $E - D_e$. Because the quantal threshold is $D_e + E_z^‡$, where $E_z^‡$ is the zero-point energy at the barrier, the classical threshold is lower than the quantal one by $E_z^‡$. For large molecules with a large $E_z^‡$ and/or for low levels of excitation the classical RRKM rate constant is significantly larger than the quantal one. This effect is illustrated in figure 8.16 for $C_2H_5 \rightarrow H + C_2H_4$ dissociation (Hase and Buckowski, 1982).

It is also interesting to consider the classical/quantal correspondence in the number of energized molecules versus time $\mathbb{N}(t, E)$, Eq. (8.22), for a microcanonical ensemble of chaotic trajectories. Because of the above zero-point energy effect and the improper treatment of resonances by chaotic classical trajectories, the classical and quantal $\mathbb{N}(E, t)$ are not expected to agree. For example, if the classical motion is sufficiently chaotic so that a microcanonical ensemble is maintained during the decomposition process, the classical $\mathbb{N}(t, E)$ will be exponential with a rate constant equal to the classical (not quantal) RRKM value. However, the quantal decay is expected to be *statistical state specific*, where the random ψ_n's give rise to statistical fluctuations in the k_n and a nonexponential $\mathbb{N}(t, E)$. This distinction between classical and quantum mechanics for Hamiltonians, with classical $\mathbb{N}(t, E)$ which agree with classical RRKM theory, is expected to be evident for numerous systems.

The classical/quantal correspondence is better defined for an ensemble of quasiperiodic trajectories. These trajectories correspond to compound-state resonance states which have a complex eigenvalue given by Eq. (8.1). Applying semiclassical mechanics to the trajectories, section 2.3.3, will give the positions E_n and wave functions ψ_n for the compound-state resonances. The spectrum will be assignable and the rate constants k_n, which can be determined either semiclassically or quantum mechanically, are expected to exhibit mode specificity. The semiclassical and quantal $\mathbb{N}(t, E)$, for a microcanonical ensemble of these resonances, are expected to be nonexponential and in good agreement. However, since the trajectories are quasiperiodic and trapped in the reactant region of phase space, the classical microcanonical ensemble will not decompose, that is, the classical rate is zero.

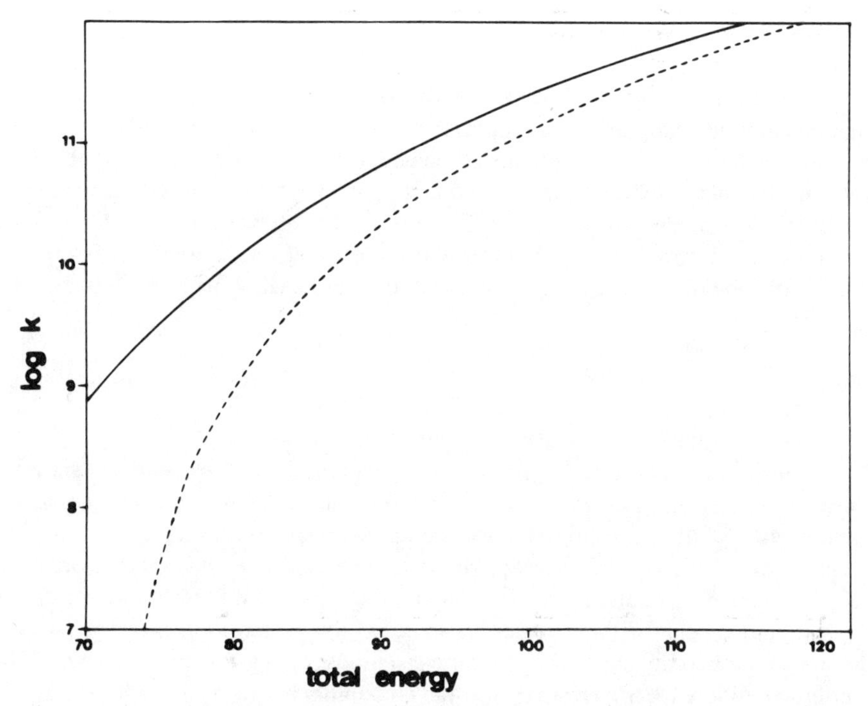

Figure 8.16 Harmonic RRKM $C_2H_5 \rightarrow H + C_2H_4$ unimolecular rate constants: classical state counting (solid curve), quantal state counting (dashed curve). Rate constant is in units of sec^{-1} and energy in kcal/mol (Hase and Buckowski, 1982).

The above are two extreme cases. Many examples are known where the complete microcanonical ensemble of trajectories is chaotic, but the situation where all the microcanonical motion above the classical threshold is quasiperiodic is more of an idealization, which is more likely if the unimolecular process is barrier tunneling. A common classical microcanonical ensemble is one which consists of both quasiperiodic and chaotic trajectories (Hase, 1981; Hase and Wolf, 1981b). There also may be "vague tori" trajectories (Jaffe and Reinhardt, 1982; Shirts and Reinhardt, 1982) which, in some cases, may yield correct quantum k_n for resonance states. For example, the classical lifetimes determined by Woodruff and Thompson (1979) for HeI_2 are in excellent agreement with the approximate quantal lifetimes of Beswick and Jortner (1978). For the mixed van der Waals clusters HeI_2He, NeI_2Ne, HeI_2Ne, and $HeNeI_2$, Schatz et al. (1983) find mode-specific behavior in classical trajectories and suggest that these classical results qualitatively represent the exact quantum dynamics. The unimolecular lifetimes of several vibrational states of $n\pi^*$ excited DCN have been determined via both classical and quantum dynamical means (Chuljian et al., 1984), and the two sets of rate constants are in good agreement. The above studies indicate that the nonstatistical, correlated classical motion found for some systems may accurately reflect the quantum dynamics. However, the point to be emphasized is that, if the quantum mechanical decomposition occurs via isolated compound-state resonances, only in special cases are the classical and quantal $\mathbb{N}(t, E)$ the same.

8.6 ELECTRONICALLY NONADIABATIC DYNAMICS

In the previous sections it has been implicitly assumed that the unimolecular reaction is electronically adiabatic and, thus, occurs on a single potential energy surface. Electronically excited states (i.e., multiple potential energy surfaces) for unimolecular reactions was discussed in chapter 3 and it is assumed that the reader has read and is familiar with this material (Nikitin, 1974; Hirst, 1985; Steinfeld et al., 1989). Transitions between electronic states are particularly important for the unimolecular decomposition of ions. For example, the following two dissociation paths:

$$AB^+ \longrightarrow A^+ + B$$
$$AB^+ \longrightarrow B^+ + A$$

must occur on different potential energy surfaces.

To begin, consider a curve crossing for a diatomic molecule, which is shown in figure 3.10 in both diabatic and adiabatic representations. A transition between the two potential energy curves, in either representation, involves two traversals of the avoided crossing region, one as the atoms approach each other and a second as they separate. In order for the diatom to end up on a potential energy surface different from the one on which it started, one of these traversals must involve a transition between potential curves with probability P and the other traversal must not have such a transition. This latter probability is $1 - P$. The net transition probability between curves 1 and 2 is then (Nikitin, 1974)

$$p_{12} = 2\,P(1 - P). \tag{8.58}$$

If the transition probability between adiabatic curves is denoted as P_a and that between diabatic curves P_d, it can be shown (Nikitin, 1974) that $P_a = 1 - P_d$. For a diatomic molecule initially in the diabatic state (see fig. 3.10), the predissociation rate constant is

$$k(E, J) = \nu\, p_{12}, \tag{8.59}$$

where ν is the frequency at which the diabatic crossing point is reached, E is the energy of the diatom relative to its minimum, and J is the angular momentum.

In the Landau-Zener approximation for the transition probability (Landau and Lifshitz, 1965; Zener, 1933; Nikitin, 1974; Hirst, 1985; Nesbitt and Hynes, 1986), $P_a = 1 - P_d$ is given by

$$P_{LZ} = \exp\left(\frac{-2\pi V_{12}^2}{\hbar v|s_1 - s_2|}\right), \tag{8.60}$$

where v is the relative velocity at the diabatic crossing, s_1 and s_2 are the slopes of the diabatic curves at the crossing point, and V_{12} is one-half the splitting between the adiabatic potential curves at the crossing. The Landau-Zener model has been used in trajectory studies of nonadiabatic transitions (Tully and Preston, 1971; Tully, 1976; Zahr et al., 1975).

For the Landau-Zener model the velocity and coupling term V_{12} are assumed to be constant in the interaction region. The latter assumption breaks down at high energies, while the former is poor at low energies. If the coupling between the potential energy curves is weak, additional formulas can be used to calculate the transition probability (Nikitin, 1974; Delos, 1973; Child, 1970; Bandrauk and Laplante, 1976; Desouter-Lecomte and Lorquet, 1979).

For a polyatomic molecule with n degrees of freedom, the crossing point is replaced by a crossing "seam" in a $(n - 1)$-dimensional space or, for certain symmetric systems, is replaced by canonical or Jahn-Teller intersection (see chapter 3). Such nonadiabatic interactions for polyatomics have been studied extensively by Lorquet and co-workers (Desouter-Lecomte et al., 1979; Dehareng et al., 1983; Desouter-Lecomte et al., 1985; Barbier et al., 1984).

The Fermi Golden Rule (Merzbacher, 1970) is often used to interpret rate constants for electronically nonadiabatic transitions in polyatomic molecules. Figure 8.17 depicts vibrational/rotational levels for two electronic states 1 and 2. Unimolecular decomposition occurs on the ground electronic state 1. When the system is initially prepared in the electronically excited state, the complete unimolecular rate constant depends on both the rate constant k_{12} for the electronic transition $1 \leftarrow 2$ and the unimolecular rate constant for the ground electronic state. If a single vibrational/rotational state of electronic state 2 is initially excited, the Fermi Golden Rule expression for k_{12} is

$$k_{12} = \frac{2\pi}{\hbar} V_{12}^2 \rho_1(E) , \tag{8.61}$$

where V_{12} is the rms coupling between the initial vibrational/rotational state of electronic state 2 and the vibrational/rotational states of electronic state 1, and $\rho_1(E)$ is density of vibrational/rotational states for electronic state 1.

The electronic transition rate constant in Eq. (8.60) for a diatomic molecule can

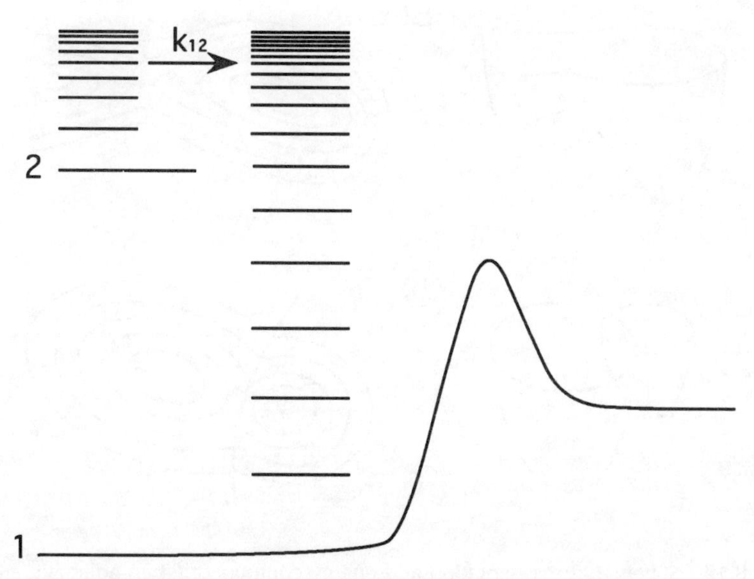

Figure 8.17 Depiction of a nonadiabatic electronic transition from electronic state 2 to electronic state 1.

be generalized to give a RRKM-like equation for a nonadiabatic unimolecular transition (Tully, 1974; Zahr et al., 1975; Praet et al., 1982; Heller and Brown, 1983; Desouter-Lecomte et al., 1983; Lorquet and Leyh-Nihant, 1988; Lorquet and Takeuchi, 1990). The assumption is made that, in a particular set of coordinates, the seam between the two crossing surfaces (fig. 8.18) is a plane perpendicular to the reaction coordinate r, which has the value r_c at the seam. It is also assumed that the position of the crossing point r_c along the reaction coordinate remains constant when degrees of freedom perpendicular to the reaction coordinate are excited. This later assumption holds when both potential energy surfaces are locally identical at $r = r_c$. In particular, this requires that degrees of freedom other than the reaction coordinate have the same equilibrium positions and force constants in both electronic states. With these assumptions the Landau-Zener one-dimensional surface crossing model can be used to calculate the predissociation rate constant (Lorquet and Leyh-Nihant, 1988), that is,

$$k(E, J) = \frac{2\int_0^{E-E_c} dE_u \rho_u^{\ddagger}(E_u, J) P_{LZ}(E - E_c - E_u)}{h\rho(E, J)} . \tag{8.62}$$

In Eq. (8.62), E and J are the total energy and angular momentum of the predissociating molecule, E_c is the potential energy of the crossing point relative to the potential minimum of the predissociating molecule, $\rho^{\ddagger}(E_u, J)$ is the density of vibrational/rotational states at the crossing point for the degrees of freedom perpendicular to the reaction coordinate, P_{LZ} is the Landau-Zener transition probability from Eq. (8.60), and $\rho(E, J)$ is the density of vibrational/rotational states for the predissociating molecule. The additional factor of 2 in Eq. (8.62), in comparison to RRKM theory,

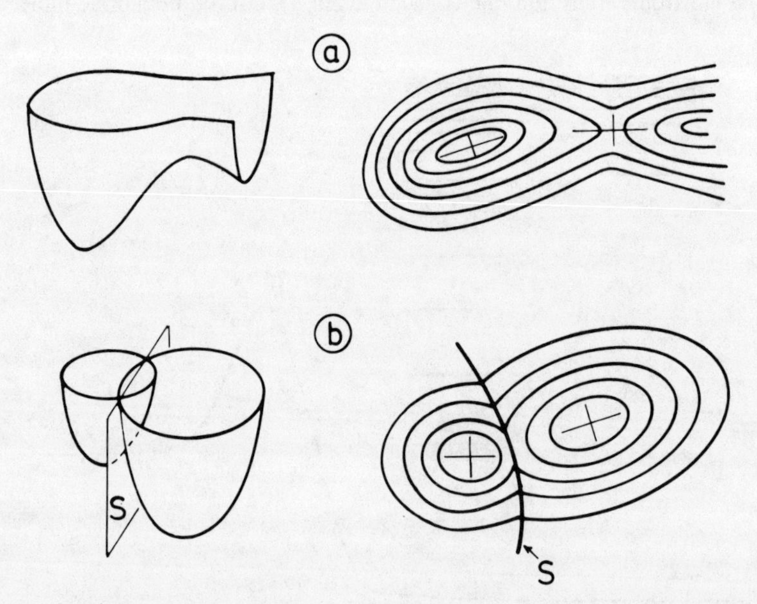

Figure 8.18 Schematic representation and energy contours of (a) an adiabatic and (b) a nonadiabatic reaction. S is the seam between the two crossing surfaces (Lorquet and Leyh-Nihant, 1988).

arises from the fact that a crossing of the seam in either a positive or negative direction can bring about a nonadiabatic transition, while in RRKM theory only the positive direction is effective.

Experiments of the photodissociation dynamics of CH_3SH (Jensen et al., 1993) at 248, 222, and 193 nm illustrate the interesting nonadiabatic interactions that can occur between adiabatic potential energy surfaces (also see chapter 3, section 3). Excitation of CH_3SH at the lower energies of 248 and 222 nm excites the molecule to a repulsive region of the S_1 potential energy surface which preferentially dissociates to $CH_3S + H$, even though the S—H bond is stronger than the C—S bond. At 193 nm the transition is to the second excited singlet state S_2 which is bound with respect to both $CH_3S + H$ and $CH_3 + SH$ dissociation. However, S_2 has good Franck-Condon overlap with S_1 when the C—S bond is extended and, thus, exciting S_2 preferentially forms $CH_3 + SH$.

REFERENCES

Adamson, G.W., Zhao, X., and Field, R.W. (1993). *J. Mol. Spectrosc.* **160,** 11.

Aquilanti, V., and Cavalli, S. (1987). *Chem. Phys. Lett.* **133,** 538.

Ashton, C.J., Child, M.S., and Hutson, J.M. (1983). *J. Chem. Phys.* **78,** 4025.

Bačić, Z., and Simons, J. (1982). *J. Phys. Chem.* **86,** 1192.

Bai, Y.Y., Hose, G., McCurdy, C.W., and Taylor, H.S. (1983). *Chem. Phys. Lett.* **99,** 342.

Bandrauk, A.D., and Laplante, J.P. (1976). *J. Chem. Phys.* **65,** 2592.

Barbier, C., Galloy, C., and Lorquet, J.C. (1984). *J. Chem. Phys.* **81,** 2975.

Basilevsky, M.V., and Ryaboy, V.M. (1983). *Chem. Phys. Lett.* **98,** 221.

Beswick, J.A., and Jortner, J. (1978). *J. Chem. Phys.* **69,** 512.

Beswick, J.A., and Shapiro, M. (1982). *J. Chem. Phys.* **64,** 333.

Bisseling, R.H., Kosloff, R., Manz, J., and Schor, H.H.R. (1985). *Ber. Bunsenges. Phys. Chem.* **89,** 270.

Bisseling, R.H., Kosloff, R., Manz, J., Mrugala, F., Römelt, J., and Weichselbaumer, G. (1987). *J. Chem. Phys.* **86,** 2626.

Bowman, J.M., Christoffel, K., and Tobin, F. (1979). *J. Phys. Chem.* **83,** 905.

Brody, T.A., Flores, J., French, J.B., Mello, P.A., Pandey, A., and Wong, S.S. (1981). *Rev. Mod. Phys.* **53,** 385.

Bunker, D.L. (1962). *J. Chem. Phys.* **37,** 393.

Bunker, D.L. (1964). *J. Chem. Phys.* **40,** 1946.

Bunker, D.L. (1966). *Theory of Elementary Gas Reaction Rates.* Pergamon, New York.

Bunker, D.L., and Hase, W.L. (1973). *J. Chem. Phys.* **59,** 4621.

Carmeli, M. (1983). *Statistical Theory and Random Matrices.* Marcel Dekker, New York.

Carrington, Jr., T., Hubbard, L.M., Schaefer, III, H.F., and Miller, W.H. (1984). *J. Chem. Phys.* **80,** 4347.

Chandler, D.W., Farneth, W.E., and Zare, R.N. (1982). *J. Chem. Phys.* **77,** 4447.

Child, M.S. (1970). *J. Mol. Spectrosc.* **33,** 487.

Cho, S.W., Wagner, A.F., Gazdy, B., and Bowman, J.M. (1992a). *J. Chem. Phys.* **96,** 2799.

Cho, Y.J., Vande Linde, S.R., Zhu, L., and Hase, W.L. (1992b). *J. Chem. Phys.* **96,** 8275.

Choi, Y.S., and Moore, C.B. (1991). *J. Chem. Phys.* **94,** 5414.

Choi, Y.S., and Moore, C.B. (1992). *J. Chem. Phys.* **97,** 1010.

Christoffel, K.M., and Bowman, J.M. (1982). *J. Chem. Phys.* **76,** 5370.

Christoffel, K.M., and Bowman, J.M. (1983). *J. Chem. Phys.* **78,** 3952.

Chu, S.O., and Datta, K.K. (1982). *J. Chem. Phys.* **64,** 5307.

Chuljian, D.T., Ozment, J., and Simons, J. (1984). *J. Chem. Phys.* **80,** 176.
Crim, F. (1984). *Ann. Rev. Phys. Chem.* **35,** 657.
Davis, M.J., and Gray, S.K. (1986). *J. Chem. Phys.* **84,** 5389.
Davis, M.J., and Skodje, R.T. (1992). In *Advances in Classical Trajectory Methods, Vol. 1, Intramolecular and Nonlinear Dynamics,* Hase, W.L., Ed. Plenum, New York, p. 77.
Dehareng, D., Chapuisat, X., Lorquet, J.C., Galloy, C., and Raseev, G. (1983). *J. Chem. Phys.* **78,** 1246.
DeLeon, N. (1992a). *J. Chem. Phys.* **96,** 285.
DeLeon, N. (1992b). *Chem. Phys. Lett.* **189,** 371.
Delos, J.B. (1973). *J. Chem. Phys.* **59,** 2365.
Desouter-Lecomte, M., and Culot, F. (1993). *J. Chem. Phys.* **98,** 7819.
Desouter-Lecomte, M., and Lorquet, J.C. (1979). *J. Chem. Phys.* **71,** 4391.
Desouter-Lecomte, M., Dehareng, D., Leyh-Nihant, B., Praet, M. Th., Lorquet, A.J., and Lorquet, J.C. (1985). *J. Phys. Chem.* **89,** 214.
Desouter-Lecomte, M., Galloy, C., Lorquet, J.C., and Vaz Pires, M. (1979). *J. Chem. Phys.* **71,** 3661.
Desouter-Lecomte, M., Sannen, C., and Lorquet, J.C. (1983). *J. Chem. Phys.* **79,** 894.
Dixon, R.N. (1992). *J. Chem. Soc. Faraday Trans.* **88,** 2575.
Dobbyn, A.J., Stumpf, M., Keller, H.-M., Hase, W.L., and Schinke, R. (1995). *J. Chem. Phys.* **102,** 5867.
Eastes, W., and Marcus, R.A. (1973). *J. Chem. Phys.* **59,** 4757.
Ericson, T. (1960). *Phys. Rev. Lett.* **5,** 430.
Ericson, T. (1963). *Ann. Phys.* **23,** 390.
Feit, M.D., and Fleck, Jr., J.A. (1983). *J. Chem. Phys.* **78,** 301.
Feit, M.D., Fleck, Jr., J.A., and Steiger, A. (1982). *J. Comp. Phys.* **47,** 412.
Forst, W. (1973). *Theory of Unimolecular Reactions.* Academic Press, New York.
Gazdy, B., Bowman, J.M., Cho, S.-W., and Wagner, A.F. (1991). *J. Chem. Phys.* **94,** 4192.
Geers, A., Kappert, J., Temps, F., and Wiebrecht, J.W. (1993). *J. Chem. Phys.* **99,** 2271.
Gomez Llorente, J.M., Zakrzewski, J., Taylor, H.S., and Kulander, K.C. (1989). *J. Chem. Phys.* **90,** 1505.
Grabenstetter, J.E., and LeRoy, R.J. (1979). *Chem. Phys.* **42,** 41.
Gray, S.K. (1992a). *Chem. Phys. Lett.* **197,** 86.
Gray, S.K. (1992b). *J. Chem. Phys.* **96,** 6543.
Gray, S.K., and Child, M.S. (1984). *Mol. Phys.* **53,** 961.
Gray, S.K., and Rice, S.A. (1987). *J. Chem. Phys.* **86,** 2020.
Gray, S.K., Rice, S.A., and Davis, M.J. (1986a). *J. Phys. Chem.* **90,** 3470.
Gray, S.K., Rice, S.A., and Noid, D.W. (1986b). *J. Chem. Phys.* **84,** 3745.
Gutow, J.H., and Zare, R.N. (1992). *J. Phys. Chem.* **96,** 2535.
Gutow, J.H., Klenerman, D., and Zare, R.N. (1988). *J. Phys. Chem.* **92,** 172.
Halberstadt, N., Serna, S., Roncero, O., and Janda, K.C. (1992). *J. Chem. Phys.* **97,** 341.
Hartke, B., Janza, A.E., Karrlein, W., Manz, J., Mohan, V., and Schreier, H.-J. (1992). *J. Chem. Phys.* **96,** 3569.
Hase, W.L. (1976). In *Dynamics of Molecular Collisions, Part B,* W.H. Miller, Ed. Plenum, New York, p. 121.
Hase, W.L. (1981). In *Potential Energy Surfaces and Dynamics Calculations,* Truhlar, D.G., Ed. Plenum, New York, p. 1.
Hase, W.L. (1985). *Chem. Phys. Lett.* **116,** 312.
Hase, W.L. (1986). *J. Phys. Chem.* **90,** 365.
Hase, W.L., and Buckowski, D.G. (1982). *J. Comp. Chem.* **3,** 335.
Hase, W.L., and Feng, D.-F. (1974). *J. Chem. Phys.* **61,** 4690.
Hase, W.L., and Feng. D.-F. (1976). *J. Chem. Phys.* **64,** 651.

Hase, W.L., and Wolf, R.J. (1981a). *J. Chem. Phys.* **75,** 3809.
Hase, W.L., and Wolf, R.J. (1981b). *Potential Energy Surfaces and Dynamics Calculations,* Truhlar, D.G., Ed. Plenum Press, New York, p. 37.
Hase, W.L., Buckowski, D.G., and Swamy, K.N. (1983). *J. Phys. Chem.* **87,** 2754.
Hase, W.L., Cho, S.-W., Lu, D.-h., and Swamy, K.N. (1989). *Chem. Phys.* **139,** 1.
Hase, W.L., Wolf, R.J., and Sloane, C.S. (1979). *J. Chem. Phys.* **70,** 2911.
Hazi, A.U. (1978). *J. Phys. B.* **11,** 2259.
Hazi, A.U., and Taylor, H.S. (1970). *Phys. Rev. A* **1,** 1109.
Hedges, Jr., R.M., and Reinhardt, W.P. (1983). *J. Chem. Phys.* **78,** 3964.
Heller, E.J., and Brown, R.C. (1983). *J. Chem. Phys.* **79,** 3336.
Hernandez, R., Miller, W.H., Moore, C.B., and Polik, W.F. (1993). *J. Chem. Phys.* **99,** 950.
Hippler, H., Luther, K., Troe, J., and Wendelken, H.J. (1983). *J. Chem. Phys.* **79,** 129.
Hirst, D.M. (1985). *Potential Energy Surfaces. Molecular Structure and Reaction Dynamics.* Taylor and Francis, London.
Hose, G., and Taylor, H.S. (1982). *J. Chem. Phys.* **76,** 5356.
Hu, X., and Hase, W.L. (1991). *J. Chem. Phys.* **95,** 8073.
Hunter, M., Reid, S.A., Robie, D.C., and Reisler, H. (1993). *J. Chem. Phys.* **99,** 1093.
Hutson, J.M., Ashton, C.J., and LeRoy, R.J. (1983). *J. Phys. Chem.* **87,** 2713.
Ionov, S.I., Brucker, G.A., Jaques, C., Chen, Y., and Wittig, C. (1993a). *J. Chem. Phys.* **99,** 3420.
Ionov, S.I., Brucker, G.A., Jaques, C., Valachovic, L., and Wittig, C. (1993b). *J. Chem. Phys.* **99,** 6553.
Jaffe, C., and Reinhardt, W.P. (1982). *J. Chem. Phys.* **77,** 5191.
Jain, S., Bleher, S., and Dumont, R.S. (1993). *J. Chem. Phys.* **99,** 7793.
Jasinski, J.M., Frisoli, J.K., and Moore, C.B. (1983). *J. Chem. Phys.* **79,** 1312.
Jensen, E., Keller, J.S., Waschewsky, G.C.G., Stevens, J.E., Graham, R.L., Freed, K.F., and Butler, L.J. (1993). *J. Chem. Phys.* **98,** 2882.
Joseph, T., Manz, J., Mohan, V., and Schreier, H.-J. (1988). *Ber. Bunsenges. Phys. Chem.* **92,** 397.
Khundkar, L.R., Knee, J.L., and Zewail, A.H. (1987). *J. Chem. Phys.* **87,** 77.
Khundkar, L.R., and Zewail, A.H. (1990). *Ann. Rev. Phys. Chem.* **41,** 15.
Kosloff, R. (1988). *J. Phys. Chem.* **92,** 2087.
Kosloff, R. (1994). *Ann. Rev. Phys. Chem.* **45,** 145.
Laidler, K.J. (1965). *Chemical Kinetics.* McGraw-Hill, New York.
Landau, L.D., and Lifshitz, E.M. (1965). *Quantum Mechanics.* Pergamon, New York.
Levine, R.D. (1987). *Adv. Chem. Phys.* **70,** 53.
Levine, R.D. (1988). *Ber. Bunsenges. Phys. Chem.* **92,** 222.
Lichtenberg, A.J., and Lieberman, M.A. (1992). *Regular and Chaotic Dynamics, 2nd ed.* Springer-Verlag, New York.
Lorquet, J.C., and Leyh-Nihant, B. (1988). *J. Phys. Chem.* **92,** 4778.
Lorquet, J.C., and Takeuchi, T. (1990). *J. Phys. Chem.* **94,** 2279.
Lu, D.H., and Hase, W.L. (1989a). *J. Phys. Chem.* **93,** 1681.
Lu, D.H., and Hase, W.L. (1989b). *J. Chem. Phys.* **90,** 1557.
Luo, X., and Rizzo, T.R. (1990). *J. Chem. Phys.* **93,** 8620.
Luo, X., and Rizzo, T.R. (1991). *J. Chem. Phys.* **94,** 889.
Luo, X., and Rizzo, T.R. (1992). *J. Chem. Phys.* **96,** 5129.
Marcus, R.A. (1973). *Faraday Discuss. Chem. Soc.* **55,** 34.
Marcus, R.A. (1977). *Ber. Bunsenges. Phys. Chem.* **81,** 190.
Marple, Jr., S. (1987). *Digital Spectral Analysis with Applications.* Prentice-Hall, Englewood Cliffs, NJ.
McCurdy, C.W., and McNutt, J.F. (1983). *Chem. Phys. Lett.* **94,** 306.

Meagher, J.F., Chao, K.J., Barker, J.R., and Rabinovitch, B.S. (1974). *J. Phys. Chem.* **78,** 2535.
Merzbacher, E. (1970). *Quantum Mechanics*. Wiley, New York.
Mies, F.H. (1969). *J. Chem. Phys.* **51,** 798.
Mies, F.H., and Krauss, M. (1966). *J. Chem. Phys.* **45,** 4455.
Miller, W.H. (1975). *Adv. Chem. Phys.* **30,** 77.
Miller, W.H. (1979). *J. Am. Chem. Soc.* **101,** 6810.
Miller, W.H. (1988). *J. Phys. Chem.* **92,** 4261.
Miller, W.H., Hernandez, R., and Moore, C.B. (1990). *J. Chem. Phys.* **93,** 5657.
Miyawaki, J., Yamanouchi, K., and Tsuchiya, S. (1993). *J. Chem. Phys.* **99,** 254.
Moiseyev, N., and Bar-Adon, R. (1984). *J. Chem. Phys.* **80,** 1917.
Nesbitt, D.J., and Hynes, J.T. (1986). *J. Chem. Phys.* **84,** 1554.
Neyer, D.W., Luo, X., Burak, I., and Houston, P.L. (1995). *J. Chem. Phys.* **102,** 1645.
Nikitin, E.E. (1974). *Theory of Elementary Atomic and Molecular Processes in Gases*. Clarendon, Oxford.
Peslherbe, G.H., and Hase, W.L. (1994). *J. Chem. Phys.* **101,** 1179.
Polik, W.F., Guyer, D.R., and Moore, C.B. (1990a). *J. Chem. Phys.* **92,** 3453.
Polik, W.F., Guyer, D.R., Miller, W.H., and Moore, C.B. (1990b). *J. Chem. Phys.* **92,** 3471.
Polik, W.F., Moore, C.B., and Miller, W.H. (1988). *J. Chem. Phys.* **89,** 3584.
Pollak, E., and Schlier, Ch. (1989). *Acc. Chem. Res.* **22,** 223.
Porter, C.E., and Thomas, R.G. (1956). *Phys. Rev.* **104,** 483.
Praet, M.Th., Lorquet, J.C., and Raseev, G. (1982). *J. Chem. Phys.* **77,** 4611.
Rabinovitch, B.S., and Setser, D.W. (1964). *Adv. Photochem.* **3,** 1.
Reddy, K.V., and Berry, M.J. (1977). *Chem. Phys. Lett.* **52,** 111.
Reddy, K.V., and Berry, M.J. (1979). *Chem. Phys. Lett.* **66,** 223.
Reid, S.A., Brandon, J.T., Hunter, M., and Reisler, H. (1993). *J. Chem. Phys.* **99,** 4860.
Reid, S.A., Robie, D.C., and Reisler, H. (1994). *J. Chem. Phys.* **100,** 4256.
Reinhardt, W.P. (1982). *Ann. Rev. Phys. Chem.* **33,** 223.
Remacle, F., Desouter-Lecomte, M., and Lorquet, J.C. (1989). *Chem. Phys.* **91,** 4155.
Rice, O.K. (1971). *J. Chem. Phys.* **55,** 439.
Robinson, P.J., and Holbrook, K.A. (1972). *Unimolecular Reactions*. Wiley-Interscience, London.
Rotter, I. (1991). *Rep. Prog. Phys.* **54,** 635.
Rynbrandt, J.D., and Rabinovitch, B.S. (1971). *J. Phys. Chem.* **75,** 2164.
Satchler, G.R. (1990). *Introduction to Nuclear Reactions*. Oxford University, New York, pp. 252–256.
Schatz, G.C., Buch, V., Ratner, M.A., and Gerber, R.B. (1983). *J. Chem. Phys.* **79,** 1808.
Schatz, G.C., Gubes, R.B., and Ratner, M.A. (1988). *J. Chem. Phys.* **88,** 3709.
Schatz, G.C., and Ratner, M.A. (1993). *Quantum Mechanics in Chemistry.* Prentice-Hall, Englewood Cliffs, NJ.
Scherer, N.F., and Zewail, A.H. (1987). *J. Chem. Phys.* **87,** 97.
Scherer, N.F., Khundkar, L.R., Bernstein, R.B., and Zewail, A.H. (1987). *J. Chem. Phys.* **87,** 1451.
Scherer, N.F., Sipes, C., Bernstein, R.B., and Zewail, A.H. (1990). *J. Chem. Phys.* **92,** 5239.
Schinke, R. (1993). *Photodissociation Dynamics*. Cambridge, New York.
Schranz, H.W., Nordholm, S., and Freasier, B.C. (1986). *Chem. Phys.* **108,** 93.
Schranz, H.W., Raff, L.M., and Thompson, D.L. (1991a). *J. Chem. Phys.* **95,** 106.
Schranz, H.W., Raff, L.M., and Thompson, D.L. (1991b). *Chem. Phys. Lett.* **182,** 455.
Schranz, H.W., Raff, L.M., and Thompson, D.L. (1991c). *J. Chem. Phys.* **94,** 4219.
Segall, J., and Zare, R.N. (1988). *J. Chem. Phys.* **89,** 5704.
Semmes, D.H., Baskin, J.S., and Zewail, A.H. (1990). *J. Chem. Phys.* **92,** 3559.

Sewell, T.D., Schranz, H.W., Thompson, D.L., and Raff, L.M. (1991). *J. Chem. Phys.* **95,** 8089.

Sewell, T.D., and Thompson, D.L. (1990). *J. Chem. Phys.* **93,** 4077.

Shirts, R.B., and Reinhardt, W.P. (1982). *J. Chem. Phys.* **77,** 5204.

Slater, N.B. (1959). *Theory of Unimolecular Reactions.* Cornell University Press, Ithaca, NY.

Sloane, C.S., and Hase, W.L. (1977). *J. Chem. Phys.* **66,** 1523.

Smith, R.S., Shirts, R.B., and Patterson, C.W. (1987). *J. Chem. Phys.* **86,** 4452.

Someda, K., Nakamura, H., and Mies, F.H. (1994a). *Prog. Theor. Phys. Suppl.* **116,** 443.

Someda, K., Nakamura, H., and Mies, F.H. (1994b). *Chem. Phys.* **187,** 195.

Stechel, E.B., Walker, R.B., and Light, J.C. (1978). *J. Chem. Phys.* **69,** 3518.

Steinfeld, J.I., Francisco, J.S., and Hase, W.L. (1989). *Chemical Kinetics and Dynamics.* Prentice-Hall, Englewood Cliffs, NJ.

Sumpter, B.G., and Thompson, D.L. (1987). *J. Chem. Phys.* **87,** 5809.

Swamy, K.N., Hase, W.L., Garrett, B.C., McCurdy, C.W., and McNutt, J.F. (1986). *J. Phys. Chem.* **90,** 3517.

Tobiason, J.D., Dunlop, J.R., and Rohlfing, E.A. (1995a). *Chem. Phys. Lett.* **235,** 268.

Tobiason, J.D., Dunlop, J.R., and Rohlfing, E.A. (1995b). *J. Chem. Phys.* **103,** 1448.

Truhlar, D.G., and Mead, C.A. (1990). *Phys. Rev. A.* **42,** 2593.

Truhlar, D.G., Ed. (1984). *Resonances.* American Chemical Society, Washington, D.C.

Tucker, S.C., and Truhlar, D.G. (1988). *J. Chem. Phys.* **88,** 3667.

Tully, J.C. (1974). *J. Chem. Phys.* **61,** 61.

Tully, J.C. (1976). In *Dynamics of Molecular Collisions, Part B,* Miller, W.H., Ed. Plenum Press, New York.

Tully, J.C., and Preston, R.K. (1971). *J. Chem. Phys.* **55,** 562.

Vande Linde, S.R., and Hase, W.L. (1990a). *J. Chem. Phys.* **93,** 7962.

Vande Linde, S.R., and Hase, W.L. (1990b). *J. Phys. Chem.* **94,** 6148.

Wagner, A.F., and Bowman, J.M. (1987). *J. Phys. Chem.* **91,** 5314.

Waite, B.A., and Miller, W.H. (1980). *J. Chem. Phys.* **73,** 3713.

Waite, B.A., and Miller, W.H. (1981). *J. Chem. Phys.* **74,** 3910.

Waite, B.A., Gray, S.K., and Miller, W.H. (1983). *J. Chem. Phys.* **78,** 259.

Wang, D., and Bowman, J.M. (1994). *J. Chem. Phys.* **100,** 1021.

Weerasinghe, S., and Amar, F.G. (1993). *J. Chem. Phys.* **98,** 4967.

Werner, H.-J., Bauer, C., Rosmus, P., Keller, H -M., Stumpf, M., and Schinke, R. (1995). *J. Chem. Phys.* **102,** 3593.

Wolf, R.J., and Hase, W.L. (1980a). *J. Chem. Phys.* **72,** 316.

Wolf, R.J., and Hase, W.L. (1980b). *J. Chem. Phys.* **73,** 3779.

Woodruff, S.B., and Thompson, D.L. (1979). *J. Chem. Phys.* **71,** 376.

Zahr, G.E., Preston, R.K., and Miller, W.H. (1975). *J. Chem. Phys.* **62,** 1127.

Zener, C. (1933). *Proc. Roy. Soc. London A* **140,** 660.

Zewail, A.H. (1991). *Faraday Discuss. Chem. Soc.* **91,** 1.

Zhao, M., and Rice, S.A. (1992). *J. Chem. Phys.* **97,** 943.

9

Product Energy Distributions

The measurement of product translational and rotational energies, and in some cases vibrational energy, is often more readily accomplished than the measurement of the dissociation rate. As a result there exists a considerable body of experimental information about product energy distributions (PED) for many classes of reactions. The only simple model for treating these PED is the statistical one; however, there is a considerable diversity in its application. In the dissociation of large molecules at moderate to large excess energies, the translational, rotational, and vibrational energy distributions can be treated as continuous functions. On the other hand, in the dissociation of triatomic molecules, it is often possible to measure the quantized rotational energy distribution for specific vibrational energy levels of the diatomic product.

Just as in the determination of the dissociation rates, product energy partitioning is highly sensitive to the potential energy surface. If there is no reverse activation barrier, the product energies are often distributed statistically. That is, the distributions depend only upon the product phase space and are independent of the detailed shape of the potential energy surface. On the other hand, for reactions with a "tight" transition state located at the top of a reverse activation barrier, statistical redistribution of the product energies is often not possible. After passing through the transition-state region, the products move down the repulsive wall and rapidly dissociate with little chance to exchange and equilibrate the available energy. Often, such products are ejected with considerable translational energy. This happens in large as well as small molecules or ions. The resulting product energy partitioning is then highly nonstatistical, even though the dissociation rate is perfectly predicted by RRKM theory. That is, the dissociation rate and product energy partitioning are separate and uncoupled events. The rate is governed early in the reaction history by the structure of the transition state, while product energy partitioning is determined late in the reaction and is governed by the shape of the potential energy surface at large internuclear distances.

The most effective model for treating product energy distributions (PEDs) of reactions with no reverse activation barriers is the statistical theory. Several models are available for predicting PEDs of reactions with potential barriers. The simplest ones do not require a knowledge of the potential energy surface. They avoid this by ignoring exit channel effects beyond the transition state. Schinke (1992) treats several of these models in detail. The simplest, and probably the least accurate, one is the impulsive model, which does not permit energy flow among the normal modes beyond the transition state. A more sophisticated model involves mapping of the transition state wave function on to the product energy distributions. In this chapter, we discuss the statistical theories in detail and treat some of the nonstatistical theories such as the impulsive model as well as the use of classical trajectories only briefly.

The statistical theory for product energy distributions is based on the phase space theory, originally developed by Light and co-workers (Pechukas and Light, 1965; Light, 1964; Light and Lin, 1965) and applied to unimolecular reactions by Nikitin (1965, 1974) and Klots (1971, 1972, 1976). It is a loose transition-state model which assumes that all product states are populated with equal probability. It has been applied to product energy distributions at several levels of complexity.

(*a*) *The prior distribution:* This model is simple to apply, but does not explicitly conserve linear or angular momentum. However, for the dissociation of medium and large molecules, it gives results that are very close to the more complicated phase space theory.

(*b*) *The phase space theory (PST) distribution:* This approach conserves angular momentum, but ignores centrifugal barriers associated with the orbital angular momentum of the departing fragments. It is independent of the product interaction potential and thus is applicable to infinitely attractive interaction potentials. Densities of states for vibrational and rotational degrees of freedom refer to the free products.

(*c*) *The orbiting transition state phase space theory (OTS/PST) distribution:* This approach conserves angular momentum and takes into account the centrifugal barrier. As such, it depends upon the assumed interaction potential between the separating fragments. Usually, isotropic interaction potentials of the form $-ar^{-n}$ are assumed. Densities of states for vibrational and rotational degrees of freedom refer to the species at the centrifugal barrier.

(*d*) *The statistical adiabatic channel model (SACM):* This approach is based on model vibrational/rotational diabatic potential energy curves connecting the unimolecular reactant and products. The energy levels of the reactant are assumed to be populated statistically.

(*e*) *The separate statistical ensemble (SSE) distribution:* This approach adds additional constraints on energy transfer by treating the conserved and disappearing modes differently. Complete equilibration of all degrees of freedom is assumed only up to the centrifugal barrier. The resulting PEDs exhibit a lower rotational temperature than obtained with PST or OTS/PST.

(*f*) *Variational RRKM distribution with exit channel couplings:* A distinction is made between conserved and disappearing modes. It is assumed that between the variational TS and the loose orbiting TS the conserved modes are decoupled from the disappearing modes. However free energy flow among the latter is still maintained. Beyond the orbiting TS, the disappearing modes are also adiabatic.

9.1 THE MICROCANONICAL PRIOR DISTRIBUTION OF PRODUCT ENERGIES

For a single dissociation event with a given excess energy, E, we write with no assumptions that

$$E = E_t + E_{r1} + E_{r2} + E_{v1} + E_{v2} \,, \tag{9.1}$$

where the terms on the right are the translational, rotational, and vibrational energies, respectively, and the numbers in the subscript refer to products 1 or 2. The energy, E, is measured relative to the final products in their 0 K ground states. This energy available for redistribution among the product modes is often denoted as E_{avl}. When the results of many events are averaged, the same relation can be written in terms of the average quantities

$$E = \langle E \rangle_t + \langle E \rangle_{r1} + \langle E \rangle_{r2} + \langle E \rangle_{v1} + \langle E \rangle_{v2} \,. \tag{9.2}$$

In a later section, we will show how these average energies can be very simply calculated. But first we consider the distribution of each of these energies. In order to simplify the equations, thc individual rotational energies are combined into the total rotational energy, E_r.

The prior distribution is determined by the available phase space of the products. The fraction of the products that are formed with translational energy between E_t and $E_t + dE_t$, with total rotational energy between E_r and $E_r + dE_r$, and with vibrational energy of each product between E_{vi} and $E_{vi} + dE_{vi}$, $i = 1$ or 2, is given by the expression:

$$P(E_t, E_r, E_1, E_2 \mid E)dP = \frac{1}{N(E)} \rho_t(E_t)\rho_r(E_r)\rho_1(E_1)\rho_2(E_2)\, dE_1 dE_2 dE_r dE_t \,. \tag{9.3}$$

In order to simplify the notation, E_{v1} and E_{v2} have been replaced by E_1 and E_2, respectively. The horizontal bar separating E from the other energies indicates that the total energy, E, is a constant. As a result, we can express one of the energies in Equation (9.3) in terms of the total excess energy, E. When the vibrational energy, E_2, has been arbitrarily chosen to be replaced by: $E_2 = E - E_t - E_r - E_1$, the normalization integral in the denominator, $N(E)$, is given by

$$N(E) = \int_0^E \int_0^{E-E_t} \int_0^{E-E_t-E_r} \rho_t(E_t)\rho_r(E_r)\rho_1(E_1)\rho_2(E - E_t - E_r - E_1)\, dE_1 dE_r dE_t \,. \tag{9.4}$$

Equation (9.4) is just a convolution integral which combines the several density of states into the sum of states at a total available energy E.

Experimental data seldom provide as much detail as is contained in Equation (9.3). More typically, we are interested in the distribution function for one particular energy. Thus, to obtain for instance the translational energy distribution, Equation (9.3) must be integrated over E_1 and E_r. The translational energy distribution is then given by

$$P(E_t|E)dE_t = \frac{\rho_t(E_t)dE_t}{N(E)} \int_0^{E-E_t} \int_0^{E-E_t-E_r} \rho_r(E_r)\rho_1(E_1)\rho_2(E - E_t - E_r - E_1)\, dE_1 dE_r \,. \tag{9.5}$$

In which all variables except for E_t have been integrated out. This equation is equivalent to

$$P(E_t|E)dE_t = \frac{\rho_t(E_t)\,\rho_{rv}(E - E_t)dE_t}{N(E)} \,, \tag{9.6}$$

where ρ_{rv} is the ro-vibrational density for both products which is given by the double integral in Eq. (9.5). [The translational energy distribution is often referred to as the kinetic energy release distribution (KERD). This terminology is unfortunate because kinetic energy includes rotational energy as well. We thus avoid its use here.]

In an analogous manner, the distribution of vibrational energy in fragment 1 can be determined by the following integral:

$$P(E_1|E)dE_1 = \frac{\rho_1(E_1)dE_1}{N(E)} \int_0^{E-E_1} \int_0^{E-E_t-E_1} \rho_t(E_t)\rho_r(E_r)\rho_2(E - E_t - E_r - E_1)dE_r dE_t \,. \quad (9.7)$$

The average energy of any of the energies can also be obtained from Eq. (9.3). For instance, the average rotational energy is expressed as

$$\langle E_r \rangle = \int\int\int E_r \, P(E_t, E_r, E_1, E_2|E)dE_1 dE_2 dE_r dE_t \,. \quad (9.8)$$

In all but Eq. (9.6), the desired quantities are expressed in terms of several integrals. Although this "formal" approach illustrates the basic principles involved in obtaining product energy distributions and averages energies, it is often advantageous to eliminate as many of these integrals as possible by using combined density of states as was done in Eq. (9.6). These combined density of states are readily calculated with the steepest descent program listed in the appendix.

9.1.1 Degrees of Freedom in the Prior and Phase Space Theories

For large molecules at significant excess energies it is convenient and adequate to treat both the translational and rotational density of states classically. The classical density for translations and rotations is proportional to $E^{(x-2)/2}$, where x is the number of translational or rotational degrees of freedom.

So far, no account has been taken of the conservation of angular and linear momentum. In the absence of these constraints there are three degrees of translation and the full number of rotational degrees of freedom (i.e., two for a diatom and three for a nonlinear polyatomic fragment). This is precisely the numbers assumed in the usual formulation of the prior distribution (Ben-Shaul, 1977; Levine and Kinsey, 1979; Steinfeld et al., 1989). However, it is not correct because it does not conserve angular or linear momentum. When these conservation laws in a global sense are taken into account as is done in transition state theory, the translational degrees of freedom are reduced to two. The number of rotational degrees of freedom depends upon the total angular momentum of the molecule. In the $J = 0$ limit, the rotational degrees of freedom are also reduced by one. However, in the high J limit, this is not the case.

The number of translational degrees of freedom in TST can be rationalized as follows. According to Eyring (1935) and Evans and Polanyi (1935), motion of the reacting system in the vicinity of the transition state is given by a one-dimensional translational motion of the separating fragments and its state density is given by the one-dimensional density of translational states. However, in TST and phase space theory, the final products are populated according to the *flux* along the reaction coordinate. This yield function is proportional to the translational-state density at the TS multiplied by the velocity in the reaction coordinate. The radial velocity in the reaction coordinate, $u^\ddagger$, is

$$u^\ddagger = \sqrt{\frac{2E_t}{\mu}} \quad (9.9)$$

so that the one-dimensional density of translational states [$\rho_t \propto E_t^{-1/2}$] is multiplied by $E_t^{1/2}$. This effectively converts the translational motion in PST into a two-dimensional density of states, which is just a constant. This reduction in the translational degrees of freedom from the prior (3) to the PST (2) value can be attributed to the conservation of linear momentum. It is interesting that an alternative derivation of the TST rate expression, treats the reaction coordinate as a very loose vibration which becomes a two-dimensional distribution. These two derivations are compared by Laidler (1987).

The other difference between the usual prior distribution and PST is that the former includes all of the product rotational degrees of freedom. Because of detailed angular momentum conservation in PST the rotational product energy distributions are no longer so simple. The details are explained in the section on PST. However, it is useful to summarize the results here. In the limit of the total $J \rightarrow \infty$, the full product rotational degrees of freedom become available. This is the prior distribution limit. On the other hand, if the reactants are prepared in low J states, the product rotational degrees of freedom are effectively reduced by 1. While this is somewhat simplified, it can be used to convert the prior distribution into a more correct distribution.

In summary, the prior distribution can be made more realistic by reducing the translational degrees of freedom from 3 to 2, and by reducing the product rotational degrees of freedom by one if the system is rotationally cold. If the dissociation results in a molecule plus an atom, there is no ambiguity about which molecule's rotational degrees should be reduced by one. If the products are two di- or polyatomics, but we are interested only in the total rotational energy distribution, the rotational density of states is again given by $\rho(E_r) = E^{(r-3)/2}$, where r is the sum of the product rotational degrees of freedom. In this case we are not forced to specify from which product the one degree of freedom has been taken. However, it is less evident which rotational state densities are to be used when we are interested in the rotational product energy distribution of the products individually. Klots (1994b) states that the joint product energy distribution over j_1 and j_2 (the angular momentum quantum number of the two products) will be given by

$$P(j_1, j_2) = \frac{1}{2J_\ell + 1} P_{\text{thermo}}(j_1)\, P_{\text{thermo}}(j_2)\,, \tag{9.10}$$

where P_{thermo} is the thermodynamic or canonical rotational energy distribution at the temperature of the dissociated products. This temperature will be discussed shortly. J_ℓ is the larger of j_1 and j_2. Division by $2J_\ell + 1$ makes one of the degrees of freedom appear missing.

9.1.2 The Concept of the Product Temperature

Experimental product rotational and translational energy distributions derived from energy-selected dissociation reactions can frequently be characterized by a temperature which implies that the distribution is a canonical one. This is found even when rotationally cold reactants are prepared in a state-selective manner. How can this be? We illustrate the origin of these canonical distributions by calculating the rotational and vibrational distributions for a system of classical harmonic oscillators.

The product translational energy distribution is given by

$$P(E_t|E)dE_t = \frac{\rho_t(E_t)dE_t}{N(E)} \int_0^{E-E_t} (E - E_t - E_v)^{(r-3)/2} \rho_v(E_v)dE_v \,, \qquad (9.11)$$

where the expression raised to the $(r - 3)/2$ is the rotational density of states. (Recall that one rotational degree of freedom has been removed.) A numerical constant associated with this density of states is omitted because the same constant will appear in the sum of states expression in the denominator and thus will cancel. By conservation of energy, the rotational energy has been replaced by $E - E_t - E_v$, and the vibrational density of states is the combined vibrational density for both products.

Consider the loss of an atom from a polyatomic molecule (e.g. $C_4H_9I \rightarrow C_4H_9 + I$) for which $r = 3$ so that the rotational density of states is just a constant. The translational density of states is also a constant because of the two translational degrees of freedom. In order to solve the problem analytically, we use the classical harmonic oscillator density of states which is

$$\rho_v(E_v) = \frac{E_v^{s-1}}{(s - 1)! \prod h\nu_i} \,. \qquad (9.12)$$

The sum of states [denominator in Eq. (9.11)] is given by

$$N(E) = \frac{1}{(s - 1)! \prod h\nu_i} \int_0^E \int_0^{E-E_t} E_v^{s-1} \, dE_v dE_t = \frac{E^{s+1}}{(s + 1)! \prod h\nu_i} \,, \qquad (9.13)$$

where we have also ignored the constants associated with the rotational and translational densities of states. The normalized translational PED can now be expressed as

$$P(E_t|E) = \frac{(s + 1)! \prod h\nu_i}{E^{s+1}} \int_0^{E-E_t} \frac{E_v^{s-1}}{(s - 1)! \prod h\nu_i} dE_v = \frac{s + 1}{E^{s+1}} (E - E_t)^s . \qquad (9.14)$$

This is the microcanonical distribution of translational energies for the unimolecular dissociation of a system of s *classical* harmonic oscillators. It cannot be emphasized enough that this is not an accurate expression because the approximation of classical harmonic oscillator density of states is highly inaccurate. It is used here only to provide us with a convenient analytical expression and to illustrate certain points about the PED.

The microcanonical prior distribution in Eq. (9.14) can be compared to a two-dimensional canonical translational energy distribution which is given by

$$P(E_t|T)dE_t = \frac{1}{k_B T} e^{-E_t/k_B T} dE_t \,. \qquad (9.15)$$

Georgiou and Wight (1988) have pointed out the connection between these two distributions by expanding them in a power series. Equations (9.14) and (9.15) then become

$$P(E_t|E) = \frac{s + 1}{E} \left(1 - \frac{s}{E} E_t + \frac{s(s - 1)}{2!E^2} E_t^2 - \ldots \right) \qquad (9.16)$$

and

$$P(E_t|T) = \frac{1}{k_B T} \left[1 - \frac{E_t}{k_B T} + \left(\frac{E_t}{k_B T} \right)^2 - \ldots \right) . \qquad (9.17)$$

If the number of vibrational degrees of freedom is large then the translational energy will be small and the series in Eq. 9.16 can be terminated after the first few terms. By comparing the two series term by term, the k_BT quantity can be related to E and the number of oscillators. Thus from the zeroth-, first-, and second-order terms, k_BT is found to be equal to $E/(s+1)$, E/s, and $E/[s(s+1)]^{1/2}$, respectively. When s is large, these values differ little from each other. For simplicity, we thus choose the value of $k_BT = E/s$. Because k_BT is just the average energy of a two-dimensional translational or rotational distribution, we can assign immediately an average canonical or thermodynamic energy to the microcanonical energy distributions. Had we carried out this analysis for the rotational energy distribution for a reaction that produces a nonlinear plus a diatomic fragment, the total rotational degrees of freedom would have been 5 which would result in a rotational density of states that varies linearly with the energy, that is, a four-dimensional distribution. The average rotational energy would then be given by $2k_BT$.

The microcanonical [Eq. (9.14)] and the canonical [Eq. (9.15)] translational energy distributions are compared graphically in figure 9.1 for the case of three molecules with 3, 5, and 25 classical harmonic oscillators at a constant product temperature, $k_BT^* = 400\ \text{cm}^{-1}$ (ca. 580 K). Because of the different product heat capacities, the excess energy varies for the three different reactions. It is evident that the largest molecule with 25 product oscillators results in microcanonical and canonical distributions that are nearly indistinguishable. As the size of the molecule decreases, the discrepancy between the two distributions increases. The same trends would be evident had we used the more correct quantum vibrational density of states for the microcanonical translational PEDs.

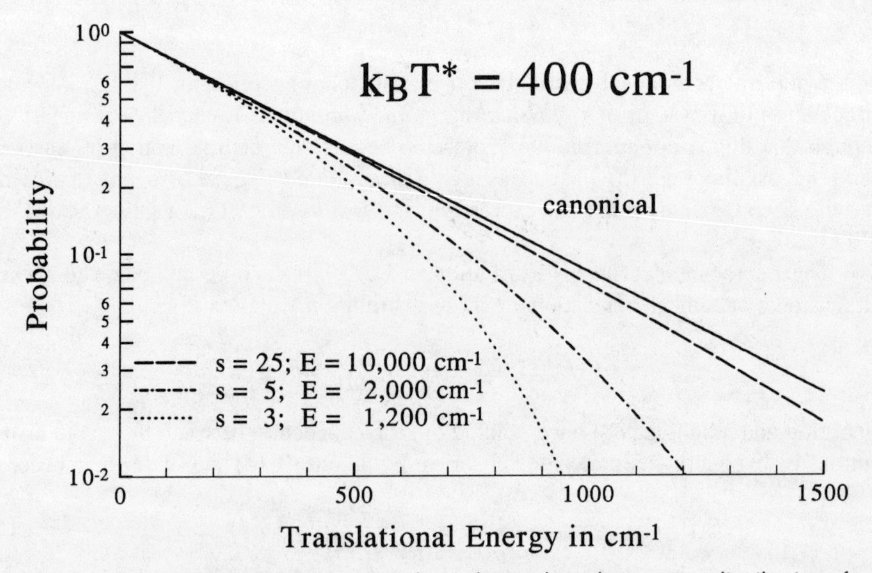

Figure 9.1 The calculated microcanonical translational product energy distributions for a system of s classical oscillators at the indicated excess energies. The equivalent canonical PED at a system energy $k_BT^* = 400\ \text{cm}^{-1}$ is a simple exponential function, $\exp(-E_t/k_BT^*)$.

So far we have shown that the translational and rotational product distributions are described in first order by a canonical distribution at some temperature T^*, which we will identify shortly. The question still remains: Why should a microcanonical system dissociate and give a canonical product energy distribution? The answer lies in the size of the vibrational heat bath. Clearly the dissociation of a diatomic molecule into two atoms will not give rise to a canonical distribution of translational energy. However, if the vibrational heat bath is sufficiently large (see fig. 9.2), the small translational and rotational energy sink exchanges energy with the much larger vibrational heat bath as though it were a canonical system. This difference will be made clearer in the following section when we illustrate and compare the translational, rotational, and vibrational distributions obtained for the dissociation of the C_8H_8 molecule.

Based on the results of figure 9.1 it is evident that the product translational and rotational distributions are well described by their canonical counterpart. Since the dimensionality is readily determined, the only as yet unknown parameter is the temperature, T^*. This can be obtained from Eq. (9.2) by replacing the average energies by their canonical expressions. This will certainly be a good approximation for the translational and rotational energy, and it can be hoped that it will work equally well for the average vibrational energy. With this approximation, Eq. (9.2) can be re-written as

$$E = k_B T^* + \frac{r^\dagger}{2} k_B T^* + \sum_{i=1}^{s_1} \frac{h\nu_i}{\exp[h\nu_i/k_B T^*] - 1} + \sum_{i=1}^{s_2} \frac{h\nu_i}{\exp[h\nu_i/k_B T^*] - 1}, \tag{9.18}$$

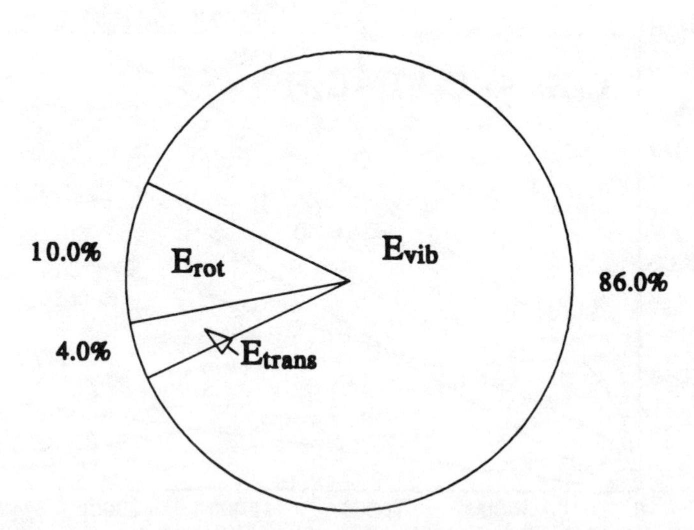

Figure 9.2 Average product energy partitioning for the system $C_8H_8 \rightarrow C_6H_6 + C_2H_2$. Note the large vibrational heat bath which insures that the translational and rotational product energy distributions are well described by a canonical distribution at a temperature T^* which is the product system temperature.

where s_1 and s_2 are the number of vibrational modes in products 1 and 2 and $r^\dagger$ is the number of product rotational degrees of freedom. Recall that this should be the full product rotational degrees of freedom if the total J is large, but should be reduced by one if the sample is rotationally cold. Because the average vibrational energy in Eq. (9.18) is expressed in its quantized form, this equation can be applied to the dissociation of real molecules.

Equation (9.18), first proposed by Klots (1973), is one of the most useful equations dealing with product energies. The only parameters required for its evaluation are the product vibrational frequencies and the total excess energy. The four terms give, respectively, the average translational, rotational, and vibrational (for each fragment) energies. Although it is not possible to solve this equation explicitly for $k_B T^*$, E can easily be generated as a function of $k_B T^*$.

9.1.3 The Dissociation of $C_8H_8 \rightarrow C_6H_6 + C_2H_2$

The average energies predicted by Eq. (9.18) for the dissociation of styrene to benzene and acetylene are shown in figure 9.3 as a function of the excess energy, E. At low energies, the classical rotational and translational energies increase rapidly because the vibrations are not fully active. However, as the vibrations reach their equipartition limit, the average translational and rotational energies level off. At 1600 K, the vibrational energy sink has already reached 95% of its equipartition limit. In this limit, Eq. (9.18) can be written as

$$E = k_B T^* + \frac{r^\dagger}{2} k_B T^* + s_1 k_B T^* + s_2 k_B T^* = \left(\frac{r^\dagger + 2}{2} + s_1 + s_2 \right) k_B T^*. \qquad (9.19)$$

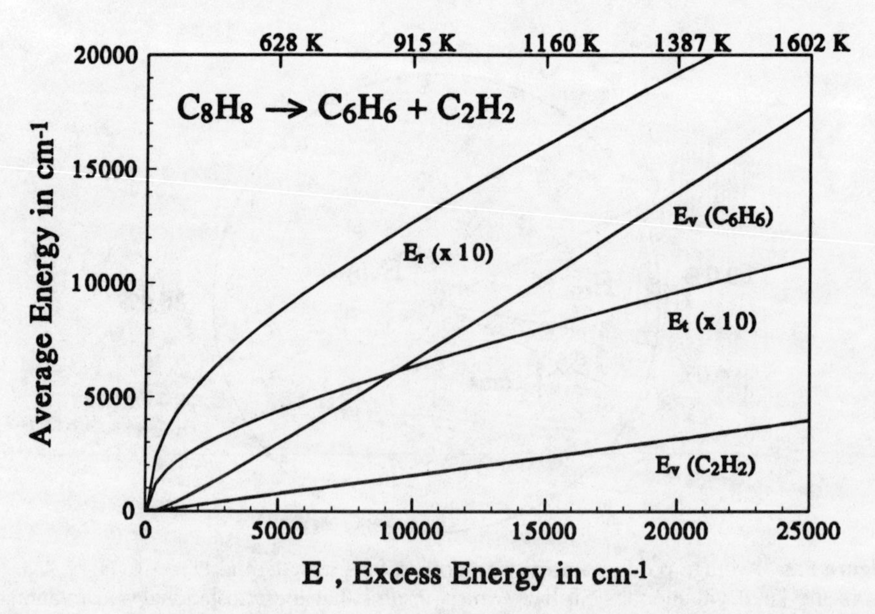

Figure 9.3 The average energy release as a function of the excess energy for the $C_8H_8 \rightarrow C_6H_6 + C_2H_2$ reaction. These were calculated with the Klots Equation (9.18).

So far we have shown that the translational PED can be approximated by a canonical distribution at a certain product temperature, T^*, defined by Eq. (9.18). What about the distributions for the other energies? The calculated microcanonical energy distributions, using the correct quantum vibrational density of states, for the dissociation of C_8H_8 to benzene and acetylene, are shown in figure 9.4. The rotational energy distribution is very close to a four-dimensional canonical distribution, $E_r \exp(-E_r/k_BT)$. Because of the high vibrational frequencies of the acetylene molecule, its distribution is highly structured, but looks nevertheless, somewhat like a canonical energy distribution. Finally, the bulk of the product energy is taken by the C_6H_6 vibrational energy. It is, in fact, the major heat bath for the dissociating system. Its internal energy distribution is compared to that of a canonical distribution in figure 9.5. As suggested by Klots, the two distributions have the same *average* energies. However, the two distributions are different. The canonical energy distribution is given by

$$P(E_v|T) = \rho_v(E_v)e^{-E_v/k_BT} . \tag{9.20}$$

The two distributions were calculated for a rather high excess energy of 42,000 cm^{-1} (5.2 eV) at which energy the temperature is about 2,300 K. The microcanonical distribution is more sharply peaked and drops sensibly to zero as the energy reaches the total excess energy of 5.2 eV. On the other hand, the canonical distribution has no such constraint and extends to much higher energies. This large discrepancy between the microcanonical and the canonical vibrational product energy distributions arises because there is no large energy sink with which the vibrations can exchange their

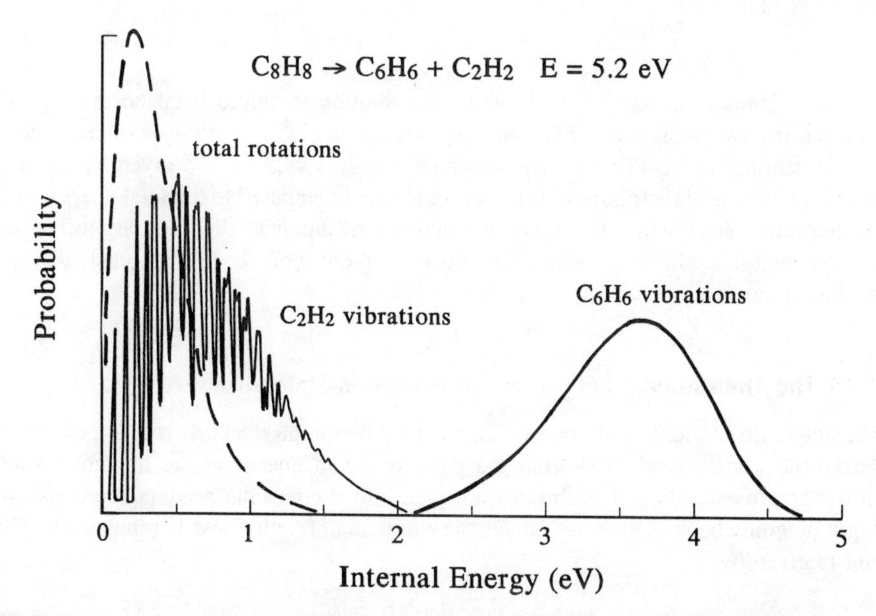

Figure 9.4 Microcanonical rotational and vibrational prior distributions for the $C_8H_8 \rightarrow C_6H_6 + C_2H_2$ reaction. The structure in the C_2H_2 vibrational distribution is due to the sparse vibrational density of states and the use of exact density of states calculations. The rotational and C_2H_2 vibrational distributions are well described by canonical distributions.

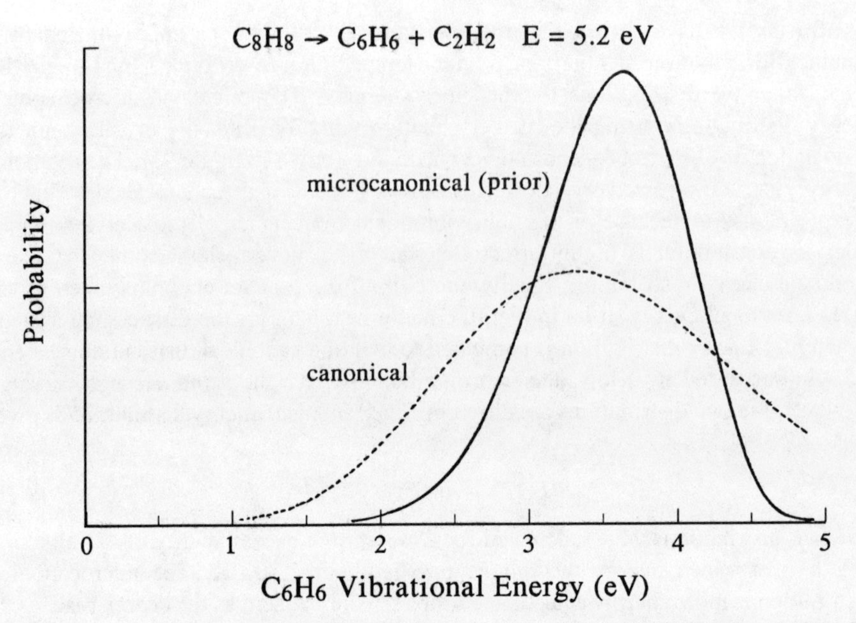

Figure 9.5 A comparison of the microcanonical and canonical C_6H_6 vibrational PEDs for the $C_8H_8 \rightarrow C_6H_6 + C_2H_2$ reaction. The two distributions have the identical average energy. Note that the correct microcanonical distribution goes sensibly to zero at the excess energy of 5.2 eV, while the canonical distribution extends to infinity. These two distribu- differ because there is no large heat sink with which the C_6H_6 system can equilibrate its energy (see fig. 9.2).

energies. Unlike the translational energy distribution in figure 9.1, the discrepancy between the two vibrational distributions increases with s, the number of oscillators.

In summary, the PED of any subset of energy levels will be given by a corresponding canonical distribution if the subset is small compared to the total energy sink. Furthermore, even when the canonical and microcanonical distributions differ, the average energies will be the same. This justifies the use of Eq. (9.18) to calculate the product temperature.

9.1.4 The Translational PED from the Transition-State Structure

The above discussions have focused exclusively on the products. It is also possible to derive the translational PED from the partitioning of energy at the transition state. However, this simple approach does not take into account the product rotations. We begin by going back to the rate constant as a function of both E and E_t (chapter 6). This rate is given by

$$k(E,E_t) = \frac{\sigma\, \rho^{\ddagger}(E - E_t)}{h\rho(E + E_o)} . \tag{9.21}$$

Recall that the energy in this chapter is defined relative to the final product energies. If it is assumed that the translational energy in the transition state remains as translational energy of the products, the translational PED will be proportional to $\rho^{\ddagger}(E - E_t)$.

Expressing this distribution in terms of the classical harmonic oscillator density of states permits a comparison of this translational PED with the prior PED of Eq. (9.14). This classical translational PED, based on the transition state energy partitioning, is given by

$$P^{\ddagger}(E_t|E) \propto (E - E_t)^{s^{\ddagger}-1} . \tag{9.22}$$

This distribution looks very similar to that of Eq. (9.14). However, the s and $s^{\ddagger}$ are different. The latter is the number of oscillators in the transition state, which is just one less than the number of oscillators in the molecule, while s is the number of oscillators in the product vibrations. For the case of the $CH_3I \rightarrow CH_3 + I$ reaction, s in Eq. (9.14) is 6, while the $s^{\ddagger}$ in Eq. (9.22) is 8. Hence, these two approaches predict slightly different (6 vs. 7) power dependencies of the energy. This difference comes about because vibrational modes are converted into rotational modes during the course of reaction. While each vibrational mode has a classical energy of kT^*, the corresponding rotational energy is only $kT^*/2$. However, these differences are small and would be difficult to distinguish experimentally, especially for large molecules where the fractional difference between s and $s + 1$ is very small.

9.1.5 Quantum Version of the Prior Distribution

When the experiment can resolve the individual quantum states of the dissociated products, it is necessary to express the product energy distribution in a quantized manner. This is necessary primarily when dealing with small molecules. Consider the dissociation of NO_2:

$$NO_2(E, J, K) \rightarrow NO(^2\Pi_\Omega; v; J_{NO}) + O\ (^3P, J_O) . \tag{9.23}$$

The O atom can be formed in several spin orbit states with $J = 0, 1, 2$, with energies, of 0, 68.7, and 227 cm^{-1} (Radzig and Smirnov, 1985); while Ω for NO is either 1/2 or 3/2, the two states being split by 119.82 cm^{-1} (Huber and Herzberg, 1979).

In expressing the product energy distribution, we now use the degeneracies rather than the density of states and sum over the quantum numbers, rather than integrate over the energies. The prior PED for the NO + O products is thus given by:

$$P(v, J_{NO}, J_O, E_t|E) \propto (2J_{NO} + 1)(2J_O + 1)E_t^x , \tag{9.24}$$

in which the rotational angular momentum quantum number of NO is J_{NO}, and the electronic angular momentum of the oxygen atom is J_O. As before, the exponent for E_t can be either 1/2 (a three-dimensional translational energy distribution) or 0 (a two-dimensional distribution consistent with PST). The full product rotational degrees of freedom have been used in Eq. (9.24). By energy conservation

$$E = E_v(NO) + E_r(NO) + E_\Omega(NO) + E_J(O) + E_t . \tag{9.25}$$

Thus, to obtain the rotational energy distribution, for a given NO vibrational level, but for a distribution over the oxygen J_O levels as well as the two NO spin orbit levels, we perform the following sum:

$$P(v, J_{NO}|E) = (2J_{NO} + 1) \sum_{\Omega=1/2;3/2} \sum_{J_O=0}^{2} (2J_O + 1)E_t^x , \tag{9.26}$$

where

$$E_t = E - E_v(\mathrm{NO}) - J_{\mathrm{NO}}(J_{\mathrm{NO}} + 1)B_{\mathrm{NO}} - E_\Omega(\mathrm{NO}) - E_J(\mathrm{O}) \qquad (9.27)$$

and B_{NO} is the NO rotational constant. Recall that the two spin orbit states of NO have the same degeneracy of 2. This function for an excess energy of 1949 cm^{-1} is plotted in figure 9.6 as solid lines for both the three-dimensional and two-dimensional translational energy distributions. (The excess energy is just below that necessary to produce NO in $v = 1$.) Interestingly, the rotational energy distribution for $x = 0$ is simply proportional to $(2J_{\mathrm{NO}} + 1)$ and thus gives a straight line.

The translational energy distribution can be obtained in a similar manner. However, it is evident that if the data are highly correlated, the translational PED may consist simply of several delta functions rather than a smooth distribution. If, for instance, the translational energy distribution is determined by Doppler spectroscopy on rotationally and vibrationally selected NO products, the translational PED, for a given v, J_{NO}, and Ω, will consist simply of three delta functions which correspond to the three O atom J_{O} levels. That is,

$$P(E_t|E, v, J_{\mathrm{NO}}, \Omega) = (2J_{\mathrm{NO}} + 1)E_t^x \sum_{J_{\mathrm{O}}=0}^{2} (2J_{\mathrm{O}} + 1)\,, \qquad (9.28)$$

where $E_t = E - E_v(\mathrm{NO}) - E_r(\mathrm{NO}) - E_\Omega - E_J(\mathrm{O})$. On the other hand, if the data are not correlated, which would be the case if the NO velocity had been measured for all product quantum numbers, then the PED would take the following form:

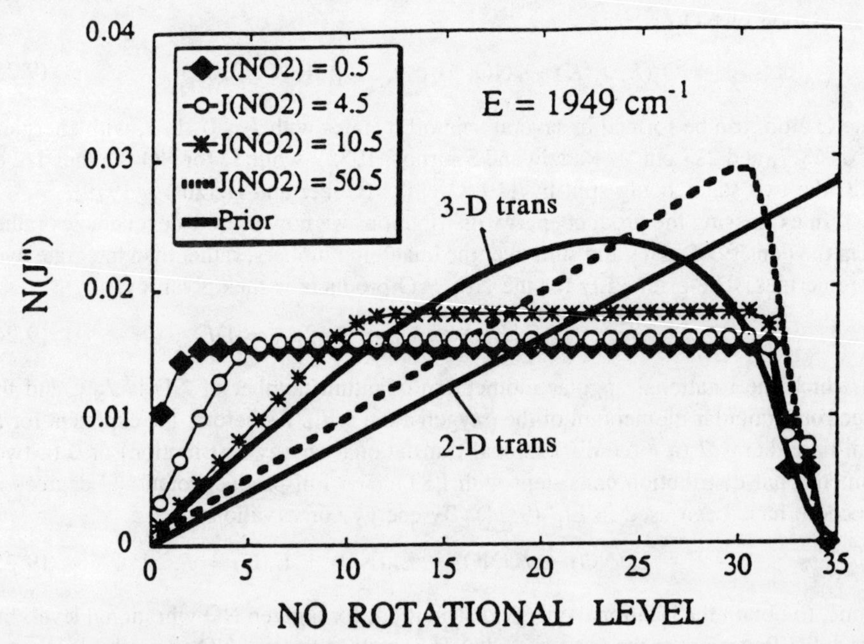

Figure 9.6 The effect of NO_2 parent total angular momentum (J_{NO_2}) on PST calculations of NO product rotational distributions at $E = 1949\ \mathrm{cm}^{-1}$. Two prior distributions (solid lines) are included which differ by their translational degrees of freedom (x in Eq. 9.26 is ½ or 0 for 3-D or 2-D distributions, respectively). This figure was adapted from Reisler and coworkers (Hunter et al., 1993).

$$P(E_t|E) = E_t^x \sum_{J_{NO}=0}^{J_{max}} \sum_{\Omega=1/2}^{3/2} \sum_{J_O=0}^{2} (2J_{NO} + 1)(2J_O + 1) \qquad (9.29)$$

where

$$J_{max}(J_{max} + 1)B_{NO} \leq E - E_{\Omega} - E_J(O) - E_v(NO) - E_t \, .$$

This translational energy distribution is not well characterized by a temperature because there are an insufficient number of degrees of freedom. Because the electronic energies of the NO and O products are very small, the total energy is very nearly the sum of E_t and E_r. Thus, when the excess energy is below the $v = 1$ level as is the case for the calculation shown in figure 9.6, $P(E_t|E) \approx 1 - P(E_r|E)$.

9.2 PHASE SPACE THEORY (PST) PRODUCT ENERGY DISTRIBUTIONS

PST as developed by Light and co-workers (Pechukas et al., 1966; Light, 1967; Pechukas and Light, 1965) as well as Klots (1971, 1972, 1976), Nikitin (1965), and others (Troe, 1983; Olzmann and Troe, 1992) conserves angular momentum. The following pages treat the theory for products that are combinations of atoms, linear species, and spherical tops. We begin by relating the unimolecular dissociation with the reverse recombination reaction. By microscopic reversibility, the cross section for a bimolecular (σ_b) recombination step with reactants in a given angular momentum state, j, and a given orbital angular momentum, ℓ, which produces a molecule with total angular momentum, J, is related to the rate constant for the reverse dissociation reaction, k_u, which forms the products in the same quantum states by

$$k_u(E, J \rightarrow j,\ell) = u\sigma_b(E, \ell, j \rightarrow E, J) \qquad (9.30)$$

where u is the velocity of the colliding pair. This then offers a prescription for determining the rotational degeneracies for product formation in specific angular momentum states.

9.2.1 Rotational Degeneracy for Molecule Plus Atom Products

We consider first a reaction that produces a molecule with angular momentum j and an atom with no angular momentum. The probability that a particular j and ℓ combines to form a total angular momentum, J is given by (Light, 1967)

$$P(j + \ell \rightarrow J) = \frac{(2J + 1)}{(2j + 1)(2\ell + 1)} \, . \qquad (9.31)$$

However, the cross section for association of the fragments with orbital angular momentum, ℓ, is proportional to $(2\ell + 1)$. In addition, the product rotor has a rotational degeneracy of $2j + 1$ if it is a linear product or $(2j + 1)^2$ in case of a spherical top. Thus the cross section for $j + \ell \rightarrow J$ is proportional to

$$\sigma_b(j + \ell\, J) \propto \frac{(2J + 1)}{(2j + 1)(2\ell + 1)} (2\ell + 1)(2j + 1) = (2J + 1) \quad \text{(linear)}, \qquad (9.32)$$

$$\sigma_b(j + \ell + J) \propto \frac{(2J + 1)}{(2j + 1)(2\ell + 1)} (2\ell + 1)(2j + 1)^2 = (2J + 1)(2j + 1) \quad \text{(sphere)}. \qquad (9.33)$$

Even though these cross sections do not themselves depend upon the value of ℓ, they must be summed over all the allowed values of ℓ so that

$$\sigma_b(j \to J) = \sum_{\ell} \sigma_b(\ell + j \to J) = \sigma_b(\ell + j \to J) \sum_{\ell} 1 . \tag{9.34}$$

The orbital angular momentum is restricted by conservation of angular momentum by the vector equation

$$\mathbf{J} = \mathbf{j} + \boldsymbol{\ell} \tag{9.35}$$

The restricted domain of j and ℓ is shown in the ℓ vs. j graph in figure 9.7. It is evident that the allowed values of ℓ depend upon the relative values of j and J. Hence when the products consist of a linear molecule plus an atom the cross section, or the rotational degeneracy, $\Phi(j|J)$, is given by

$$\sigma_b(j \to J) \propto \Phi(j|J) = (2J + 1)(2j + 1), \qquad j \leq J, \tag{9.36a}$$

$$\sigma_b(j \to J) \propto \Phi(j|J) = (2J + 1)^2, \qquad j \geq J, \tag{9.36b}$$

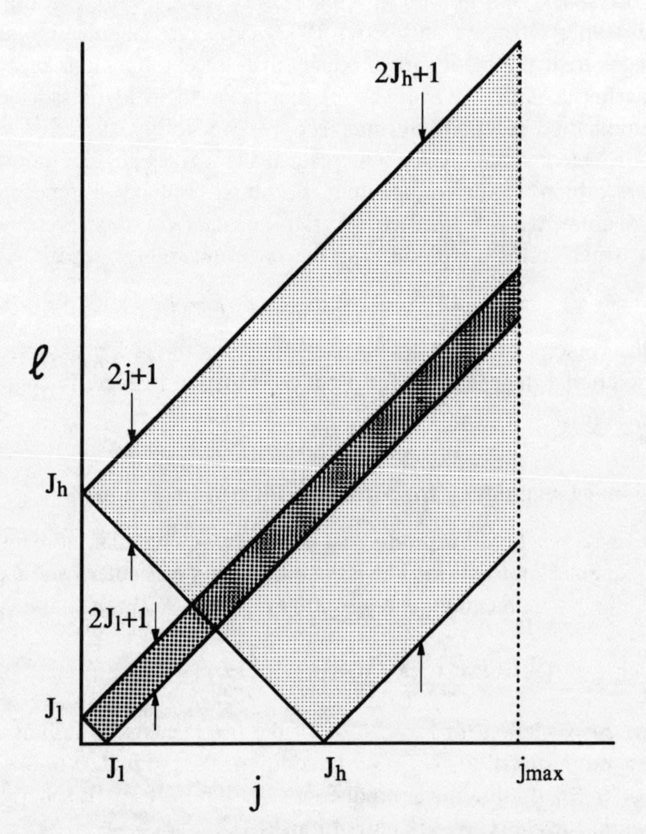

Figure 9.7 The constraint of angular momentum conservation and the allowed values of the orbital angular momentum (ℓ) and the product rotational quantum number, j, imposed by PST. The total, and conserved, angular momentum is J. The case of low J (J_l) and high J (J_h) are illustrated.

while for products that consist of a spherical top plus an atom the rotational degeneracy is given by

$$\sigma_b(j \rightarrow J) \propto \Phi(j|J) = (2J + 1)(2j + 1)^2, \qquad j \leq J, \tag{9.37a}$$

$$\sigma_b(j \rightarrow J) \propto \Phi(j|J) = (2J + 1)^2(2j + 1), \qquad j \geq J. \tag{9.37b}$$

We note now the difference between the prior and the PST rotational degeneracies. For the case of the NO_2 dissociation, the prior probability in Eq. (9.24) for production of J_{NO} was $(2J_{NO} + 1)$. This is identical to the PST result when $j \leq J$ which is the high J limit case [Eq. (9.36a)]. On the other hand, for cold samples in which $J_{NO_2} \rightarrow 0$, the PST result predicts a probability that is independent of J_{NO} [Eq. (9.36b)]. A similar reduced dimensionality is apparent in the case of the spherical top plus an atom.

9.2.2 Rotational Degeneracies for a Sphere Plus a Diatom or Sphere

When both products can be produced in different angular momentum states, the situation is more complicated. Various schemes can be used to combine j_1, j_2, and ℓ to produce a total J. If the products are a spherical top with angular momentum, j_1, and a diatom with angular momentum, j_2, the two angular momenta are combined into a resultant j. This j is then combined with ℓ to yield J. The rotational degeneracies are obtained by joining the two steps. The cross section for combining the spherical rotor angular momentum quantum number j_1 with the diatomic quantum number j_2, yields $\sigma(j_1, j_2 \rightarrow j) = (2j + 1)(2j_1 + 1)$. Combining j with the orbital angular momentum ℓ yields $\sigma(j, \ell \rightarrow J) = (2J + 1)/(2j + 1)$. The rotational degeneracies are then obtained by multiplying these two cross sections and summing over the orbital angular momenta as well as over the intermediate angular momentum j as shown in Eq. (9.38) (Troe, 1983):

$$\Phi(j_1, j_2|J) = (2J + 1)(2j_1 + 1) \sum_j \sum_\ell 1 \, , \tag{9.38}$$

where the limits for ℓ and j are

$$|J - j| \leq \ell \leq (J + j), \tag{9.39}$$

$$|j_1 - j_2| \leq j \leq (j_1 + j_2). \tag{9.40}$$

Of course, j_1 and j_2 are limited by energy conservation by the following inequality:

$$j_1(j_1 + 1)B_1 + j_2(j_2 + 1)B_2 = E_r = E - E_v - E_t \leq E - E_v \, . \tag{9.41}$$

Finally, when both products are spherical tops, the rotational degeneracy is

$$\Phi(j_1, j_2|J) = (2J + 1)(2j_1 + 1)(2j_2 + 1) \sum_j \sum_\ell 1 \, . \tag{9.42}$$

Although the $(2J + 1)$ factor is unnecessary when dealing with J-selected systems, it is important when the reactant is in a thermal distribution of rotational energies, and the PEDs must be averaged over this thermal distribution of J.

9.2.3 The Product Energy Distributions

Now that we have expressions for the rotational degeneracies, the PST PEDs are obtained in the same manner as those for the prior distributions. That is the PED for products having angular momenta j_1 and j_2 is given simply by

$$P(E_v, j_1, j_2, E_t | E, J) \propto \rho(E_v)\Phi(j_1, j_2 | J)\rho_t(E_t)$$
$$= \frac{1}{N(E, J)} \rho(E_v)\Phi(j_1, j_2 | J)\rho_t(E_t) , \qquad (9.43)$$

where $N(E, J)$ in the denominator is the sum of states in which quantum numbers or energies have been summed over all of their allowed values, consistent with angular momentum and energy conservation. This equation is analogous to Eq. (9.3) for the prior distribution.

Suppose that the dissociation products are a polyatomic linear molecule and an atom (e.g., BrC≡C· + Br), and that the available energy is well above the vibrational energies of the BrC≡C· fragment. Then the vibrational density of states must be included in the PED. If the upper Br atom spin-orbit state is ignored, the PED for a given E_v, j, and E_t, is

$$P(E_v, j, E_t | E, J) \propto (2J + 1)\rho(E_v)(2j^{<} + 1) , \qquad (9.44)$$

where $j^{<}$ is the lesser of j and J, or

$$j^{<} = \begin{matrix} j & \text{if } j \leq J \\ J & \text{if } j \geq J, \end{matrix} \qquad (9.45)$$

and where j is the rotational quantum number for the BrC≡C· radical. As before, the two-dimensional translational density of states is just a constant. The normalization factor which converts the proportionality in Eq. (9.44) into an equality is the total phase space volume which is obtained by summing over all of the quantum states as was done in Eq. (9.4). However, now the sums of the rotational quantum numbers are restricted as shown in figure (9.7).

In order to obtain the rotational PED for BrC≡C· in a distribution of vibrational levels, it is necessary to integrate over the vibrational energy. The rotational PED for BrC≡C· in any vibrational state will be given by

$$P(j | E, J) \propto (2J + 1)(2j^{<} + 1)N_v(E - E_r) , \qquad (9.46)$$

where $N_v(E_r)$ is the sum of the vibrational states up to the energy $E_r = j(j + 1)B$. The integration is necessary because the available energy $(E - E_r)$ can be partitioned between the vibrational and translational energies.

The vibrational PED is obtained by summing over all the product j states as

$$P(E_v | E, J) \propto (2J + 1)\rho(E_v) \sum_{j=0}^{j_{max}} (2j^{<} + 1) , \qquad (9.47)$$

where $j_{max} = \{(E - E_v)/B\}^{1/2}$. The sum over the rotational levels is again called for because the energy, $E - E_v$, can be partitioned between the rotational and translational energies.

All of the above equations can be written in terms of continuous J functions and thus converted to E_r. Convenient expressions for the translational energy release distributions have been summarized (Klots, 1971).

9.2.4 PST for $NO_2(J) \rightarrow NO(J_{NO}) + O(J_O)$

The rotational PEDs for the dissociation of state-selected NO_2 have been measured and analyzed using both the prior model and PST (Robie et al., 1992; Hunter et al., 1993). A convenient approximation is to assume that the product energy distributions are independent of the NO_2 M and K quantum numbers. Three product angular momenta, J_{NO}, J_O, and ℓ must be combined to add up to the total angular momentum, J_{NO_2}. For convenience, the NO Ω value is included in the J_{NO} term thereby adding either 0.5 or 1.5 to the rotational quantum number. The angular momenta are combined in two steps. The intermediate $\mathbf{J}'$ is introduced which is defined by the vector addition $\mathbf{J}' = \mathbf{J}_{NO} + \boldsymbol{\ell}$ such that

$$|J_{NO} - J'| \le \ell \le (J_{NO} + J') . \quad (9.48)$$

This is followed by combining J' with J_O to yield the vector sum, $\mathbf{J}_{NO_2} = \mathbf{J}' + \mathbf{J}_O$, so that

$$|J_{NO_2} - J_O| \le J' \le (J_{NO_2} + J_O) . \quad (9.49)$$

The PST rotational degeneracy thus becomes

$$\Phi(J_{NO}, J_O | J_{NO_2}) = (2J_{NO_2} + 1) \sum_{J'} \sum_{\ell} 1 , \quad (9.50)$$

where the limits are given by the two sets of inequalities in Eqs. (9.48) and (9.49). Summing over ℓ for a given J' yields

$$P(J_{NO}, J_O, E_t | E, J_{NO_2}) \propto (2J_{NO_2} + 1) \sum_{J'} (2J^{<} + 1) , \quad (9.51)$$

where the sum over J' is between the two limits in Eq. (9.48) and

$$J^{<} = \begin{matrix} J_{NO} & \text{if } J_{NO} \le J' \\ J' & \text{if } J_{NO} \ge J' . \end{matrix} \quad (9.52)$$

A major difference between the prior and PST product energy distributions is that in PST the rotational degeneracies of the O atom and NO are intimately intertwined. On the other hand, in the prior distribution, the two terms are independent of each other so that the rotational degeneracies of the O atom and NO are simply multiplied together as in Eq. (9.24).

A second difference between the prior and PST distributions is in the $2J^{<} + 1$ term in Eq. (9.51). Figure 9.7 shows the allowed angular momenta for the case of high and low J. (It is helpful to think of J' simply as J_{NO_2} since in any case, adding or subtracting J_O will change it only slightly.) If $J' > J_{NO}$, then $2J^{<} + 1$ is simply $2J_{NO} + 1$, that is, the same as in the prior distribution. However, once $J_{NO} > J'$, the degeneracy becomes a constant $2J' + 1$.

For the case of a cold sample, J' is very small so that even for low J_{NO}, the $J^{<}$ degeneracy is simply the constant, $2J' + 1$. The rotational PED for various initial J_{NO_2}

values and a total energy of 1949 cm^{-1} have been calculated by Hunter et al. (1993) and are shown in figure 9.6. In each case, the onset of the constant $P(J_{NO})$ distribution corresponds to J_{NO} becoming larger than J_{NO_2}. The conservation of angular momentum has a profound effect on the rotational distribution for the case of a cold sample. However, as J_{NO_2} increases, the PST and prior distributions converge.

The experimentally obtained NO rotational distribution for the dissociation of cold NO_2 molecules is shown in figure 9.8. This figure is in the form of a Boltzmann plot ln $\{N(j)/(2j+1)\}$ vs. E_r/k_BT. Figure 9.8 shows the NO($v = 0$) rotational distribution at 1949 cm^{-1} excess energy. The results were analyzed with the usual prior model using two rotational degrees of freedom for the NO product, as well as by PST. Note that the prior distribution converges to a straight line at low rotational energies. This is expected because a thermal two-dimensional rotational distribution would yield a straight line with a slope of $-1/k_BT$. The departure from the straight line at high rotational energies is a result of energy conservation which demands that the Boltzmann plot go to zero at 1949 cm^{-1}. It is evident that PST agrees much better with the experimental results, especially at low rotational energies. The effectively reduced dimensionality of the PST is clearly evident at low rotational energies, where it is much closer to a one-dimensional distribution of rotational energies. Because of the strong

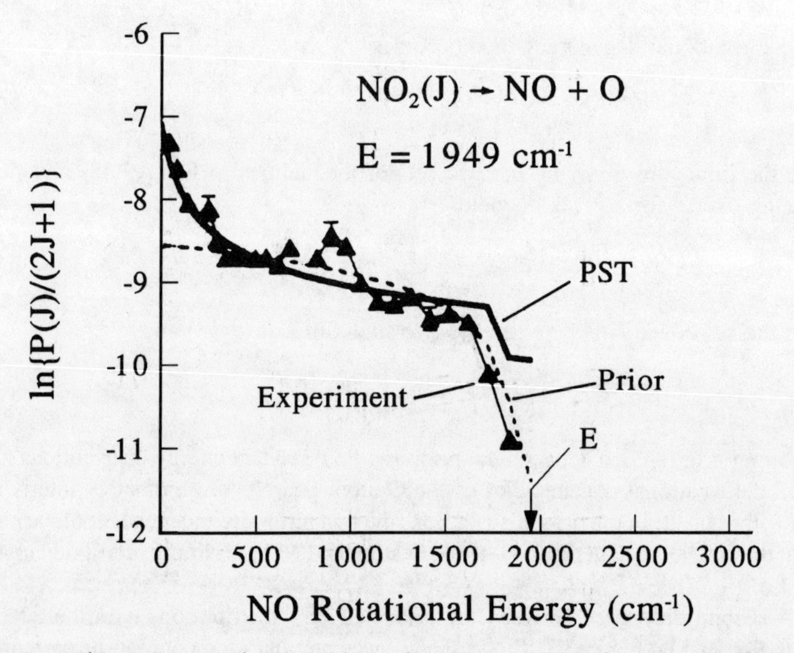

Figure 9.8 The experimental rotational PED for NO($v = 0$) from the $NO_2 \rightarrow$ NO($v = 0$) + O reaction at an excess energy of 1949 cm^{-1}. The lines through the experimental points are the PST and the prior distributions. The prior distribution assumes two rotational degrees of freedom. Thus a plot of $[N(J)/(2J+1)]$ yields a straight line at low energies. The reduced degrees of freedom due to angular momentum conservation in the PST causes the deviation and the much better agreement with experiment. Taken with permission from Reisler and co-workers (Hunter et al. 1993).

long range attractive forces, the PST and the orbiting transition state theory (to be discussed) yield the same rotational PEDs.

9.3 ORBITING TS THEORY (OTS/PST)

Both PST and orbiting transition state theory (OTS/PST) conserve angular momentum. However, in simple PST the product interaction potential and thus the centrifugal barrier is ignored, and densities of states for vibrational and rotational degrees of freedom refer to the free products. In orbiting transition state theory the additional constraints imposed by the interaction potential and the resulting centrifugal barrier are considered. These are:

1. Orbital angular momentum must be less than the critical value of ℓ_{max} which is determined by the total energy of the products and the long-range intermolecular potential. For values of ℓ larger than ℓ_{max}, the fragments cannot surmount the centrifugal barrier.
2. Densities of states for vibrational and rotational degrees of freedom are calculated at the centrifugal barrier.
3. Orbital angular momentum is assumed to be conserved from the centrifugal barrier to products so that the centrifugal potential is converted to product relative translational energy.

In common usage, the terms phase space theory and orbiting transition state are often and erroneously used interchangeably. However, the distinction is an important one. While the PST and OTS/PST PEDs are nearly indistinguishable for many reactions, they differ markedly for reactions involving H-atom loss. Because of the considerable complexity introduced by the centrifugal barrier, Klots has developed approximate analytic formulas for certain limiting cases (1971, 1972, 1976). More recently, the distributions have been related to those of a thermal sample (1994b). This has resulted in a particularly convenient equation for simple product systems such as a spherical top plus an atom. Chesnavich and Bowers (1976, 1977a,b, 1979, 1976) developed rigorous expressions valid over a large range of angular momenta and applicable to various product geometries. We first present the simple formulas of Klots and then summarize the relationships developed by Chesnavich and Bowers.

9.3.1 The Klots Equations for the Translational PED

Klots has developed a range of relationships between microcanonical and canonical systems which are extremely useful in extracting the microcanonical properties (rates and PEDs) from their corresponding canonical forms. The connection between the microcanonical and canonical treatment is based on the assumption that the molecule provides a sufficiently large heat bath to equilibrate the whole system at some temperature, which we call T^*. The derived product energy distributions are treated as continuous functions, just as are the density of states derived from the inversion of the partition function. Some of the basic aspects of the Klots approach are treated in the last part of chapter 10. Further information can be obtained from a variety of papers by Klots (1989, 1990a,b, 1992, 1993a,b, 1994a–c).

We have already established that for moderately large molecules, the translational and rotational degrees of freedom can be treated as a canonical distribution because they are in equilibrium with the much larger vibrational heat bath. That is, there exists a temperature, T^* such that

$$P(E_t|E) = P(E_t|T^*) \tag{9.53}$$

Thus, the problem is reduced to determining the canonical translational energy distribution at the temperature T^*, $P(E_t|T^*)$. In simple PST, this is a two-dimensional canonical distribution whose temperature is determined by an equation similar to Eq. (9.18). A more accurate approximation of the final temperature in Eq. (9.53) is $E = \frac{1}{2}(r^\dagger)k_BT^* + \langle E_v \rangle$ (Klots, 1994c). That is, a unit of k_BT^* should be subtracted from Eq. (9.18). In fact, for intermediate to large-sized molecules, the difference of a k_BT^* factor is negligible.

The translational energy distribution [Eq. (9.53)] must now be corrected for the effect of the centrifugal barrier. In *PST*, the $P(E_t|T^*)$ function is not that of a free particle in a heat bath. It is best to consider the distribution in terms of the flux of the reverse association reaction. By microscopic reversibility, this will be the distribution of product energies in the reaction. The flux for the reverse association reaction is given by

$$\text{flux}(E_t, T^*) \propto [E_t^{1/2} \exp(-E_t/k_BT^*)][\sigma(E_t)\, u] \tag{9.54}$$

The term in the first bracket is just the three-dimensional thermal Maxwell-Boltzmann distribution of translational energies. The second bracket contains the reaction cross section and the relative product velocities. The velocity (u) is proportional to $E_t^{1/2}$.

The bimolecular reaction cross section can be written in terms of the impact parameter as $\sigma = \pi b_{\max}^2$. In OTS/PST the maximum orbital angular momentum is related to the impact parameter by $\hbar\ell_{\max} = \mu u b_{\max}$ (see fig. 7.7). Combining these two yields the following expression for the reaction cross section (Light, 1967):

$$\sigma(\ell_{\max}) = \pi\left(\frac{\hbar}{\mu u}\right)^2 \sum_{\ell=0}^{\ell_{\max}} (2\ell + 1) = \pi\left(\frac{\hbar}{\mu u}\right)^2 (\ell_{\max} + 1)^2 \,. \tag{9.55}$$

This accounts for the centrifugal barrier in OTS/PST. To this must now be added angular momentum conservation. For a given initial J only certain combinations of product j's and ℓ's can be formed. We thus want the cross section for the association reaction which produces a product in a given J state. As in PST, the product angular momentum quantum numbers are limited by

$$|\ell - j| \leq J \leq |\ell + j| \,. \tag{9.56}$$

For simplicity we consider the case of products consisting of a spherical top with total angular momentum, j, plus an atom. The probability of finding this product sphere in the angular momentum level, j is

$$p(j) = (2j + 1)^2 \exp(-Bj^2/k_BT^*) \tag{9.57}$$

Furthermore, the probability that a given j combines with an orbital angular momentum, ℓ to give J is given by

$$(2J + 1)/(2\ell + 1)(2j + 1) \tag{9.58}$$

The reaction cross section, summed over all allowed values of ℓ, thus becomes

$$\sigma(E_t, j+\ell \to J) = \pi\left(\frac{\hbar}{\mu u}\right)^2 \sum_{\ell=0}^{\ell_{max}} \frac{(2J+1)}{(2\ell+1)(2j+1)} (2j+1)^2 \exp\left(-\frac{Bj^2}{k_B T^*}\right). \quad (9.59)$$

This cross section can be readily evaluated for the case of $J \approx 0$, that is, a rotationally cold sample. When the sum in Eq. (9.59) is replaced by an integration and the approximation that $J \approx 0$ so that $j \approx \ell$ is made, the flux function [Eq. (9.54)] reduces to

$$P(E_t, T^*) \propto \exp\left(-\frac{E_t}{k_B T^*}\right)\left[1 - \exp\left(-\frac{B\ell_{max}^2}{k_B T^*}\right)\right]. \quad (9.60)$$

Note that the velocities, u, have canceled the $E_t^{1/2}$ preexponential term in the flux equation thereby converting the translational PED into a two-dimensional distribution, albeit modified by the second term in Eq. (9.60). This is valid for low J samples and all spherically symmetric interaction potentials. The effect of the assumed potential function is contained in the ℓ_{max} term. The expression for ℓ_{max} can be obtained from V_{max} in Eq. (7.13b). This energy is the maximum of the centrifugal barrier. All of this energy is assumed to be converted to product translational energy. Thus upon replacing V_{max} by E_t and substituting $(\hbar\ell_{max})$ for the angular momentum P_Θ, we obtain

$$(\hbar\ell_{max})^2 = 4\mu(E_t a)^{1/2} \qquad \text{for } n = 4\ , \quad (9.61)$$

$$(\hbar\ell_{max})^2 = 3\mu(2a)^{1/3}E_t^{2/3} \qquad \text{for } n = 6\ . \quad (9.62)$$

where n in the interaction potential $V = -ar^{-n}$ is equal to 4 and 6 for ions and neutrals, respectively. The parameter $a = \alpha q^2/2$ and $2\epsilon r_e^6$, for ionic and neutral reactions, respectively. Klots has pointed out that more complex forms of the spherically symmetric interaction potentials can also be used (1994a). For instance, a hard sphere potential in which the colliding species have an average diameter of d yields $(\hbar\ell_{max})^2 = 2\mu E_t d^2$, while a combination of a hard sphere with an ion-induced dipole $-ar^{-4}$ interaction yields (at high kinetic energies)

$$(\hbar\ell_{max})^2 = 2\mu d^2(E_t + a/d^4)\ . \quad (9.63)$$

Finally, the total available energy E is related to the temperature of the products by Eq. (9.18). This equation is not exactly correct because the average translational energy can vary between $k_B T^*$ and $2k_B T^*$ (Klots, 1994c). However, this difference has a negligible effect on the calculated T^* for all but very small molecules. In the latter case, the whole concept of expressing the translational or rotational product energy distributions as a canonical one breaks down in any case.

Figure 9.9 shows examples of calculated translational PEDs for the following two reactions at an excess energy of 4000 cm^{-1} ($\approx$0.5 eV):

$$C_2H_6 \to CH_3 + CH_3\ , \quad (9.64)$$

$$C_2H_6 \to C_2H_5 + H\ . \quad (9.65)$$

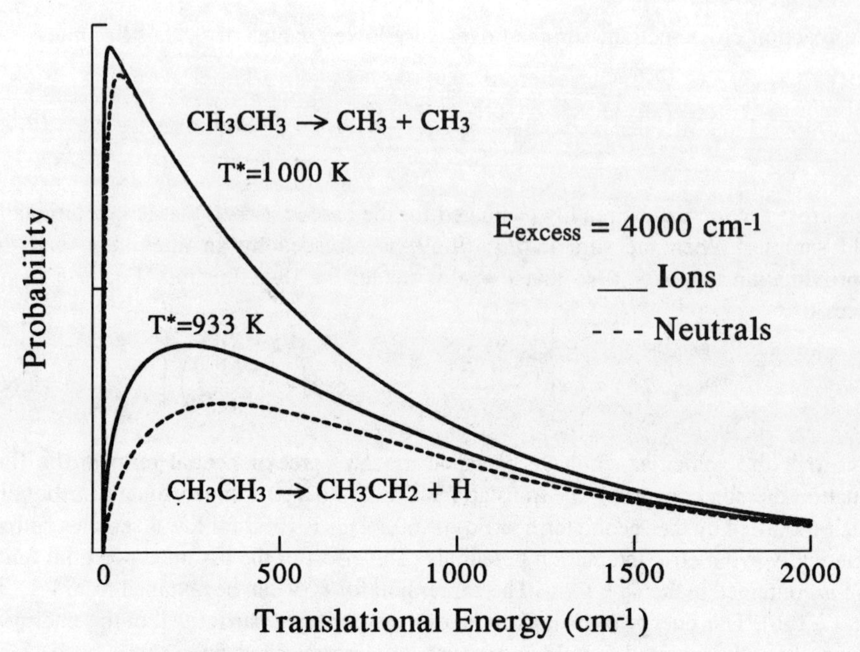

Figure 9.9 Calculated translational PEDs for the two indicated reactions with the OTS/PST Equation (9.60). The reactants are assumed to be prepared rotationally cold so that the total $J = 0$. The much larger centrifugal potential for the H atom loss reactions, results in the depletion of low translational energies.

Both the neutral and ionic reactions have already been discussed in chapter 7 (Eq. 7.15) in which all of the parameters necessary for the translational PED calculation [Eq. (9.60)] are also listed. As suggested by Klots (1994b), the rotational constant B for the production of two methyl radicals is given by $B = \hbar/2(I_1 + I_2)$. The two reactions yield strikingly different PEDs. Most reactions which loose a heavy atom or molecular unit have translational PEDs similar to that of CH_3 loss from ethane. The H-atom loss PED is different because of the small reduced mass of the H—C_2H_5 unit which results in a large centrifugal barrier. As a result, the translational energies are depleted at low energies. The slightly more attractive surface for the ionic reaction tends to lower the centrifugal barrier (see fig. 7.8) which accounts for the difference between the neutral and ionic H atom loss results.

It is evident that the translational PED for CH_3 loss has the general shape of a two-dimensional kinetic energy distribution [$\exp(-E_t/k_BT^*)$], as suggested by PST. In fact, Eq. (9.60) is simply the product of the PST PED and a correction factor (second term in brackets) which incorporates the effect of the centrifugal barrier for cases of low initial J. If the centrifugal barrier is neglected, $B\ell^2_{max}$ = total available energy. Because the average translational energy, $k_BT^* << E$, the second term in Eq. (9.60) disappears.

The translational PED for the I-atom loss from CH_3I^+ is a classic example of a reaction with statistical PEDs. Figure 9.10 shows the translational PEDs at three excess energies as measured by photoelectron photoion coincidence (PEPICO) using a room-

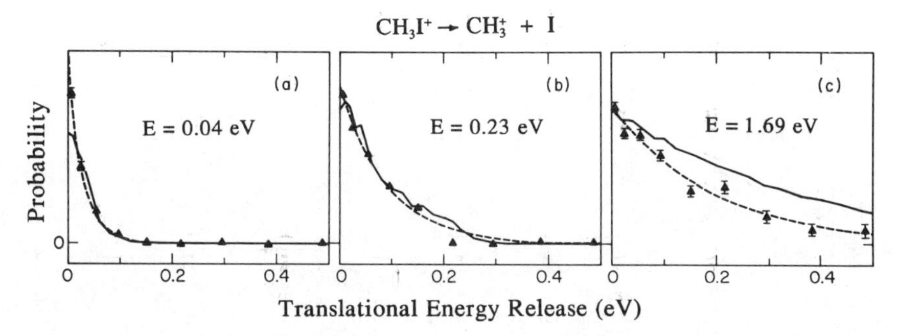

Figure 9.10 The translational PEDs for the $CH_3I^+ \rightarrow CH_3^+ + I$ reaction obtained by photoelectron photoion coincidence at the indicated excess energies. The solid lines are OTS/PST calculations while the dashed lines are the best fit exponential curves (PST). Taken with permission from Mintz and Baer (1976).

temperature methyliodide sample (Mintz and Baer, 1976). Because of the strongly attractive interaction between CH_3^+ and I, the PST and the OTS/PST distributions are nearly identical. Only at the very lowest energies does the OTS/PST deviate from a simple exponential function.

Figure 9.10(c) shows a marked deviation from the expected PST result. The departure from the PST appears to coincide with the accessibility of the $CH_3 + I^+$ channel. Thus a possible explanation for this large deviation is that the reaction initially proceeds along the I^+ channel. At some time in the course of dissociation, the electron hops to the I^+ yielding the final $CH_3^+ + I$ product. But by this time, energy exchange between CH_3^+ and I is no longer possible so that the final translational energy is less than expected.

Because these PEDs are so well described by a simple exponential function, Powis and co-workers (Johnson et al., 1981; Powis, 1987) suggested that the PED can be scaled by the excess or average energy. When the data in figure 9.10 plus data of Powis (1983) are plotted versus the reduced translational energy, $E_t/\langle E_t \rangle$, a single line shown in figure 9.11 is obtained. With these combined data, the depletion at low energies due to the centrifugal barrier becomes more apparent.

A reaction in which the scaled translational PED is particularly instructive is Br loss from $C_2H_5Br^+$ shown in figure 9.12 (Miller and Baer, 1984). At low energies the PED is statistical. However, when the excited electronic state is excited (see the inserted photoelectron spectrum), the translational PED suddenly becomes nonstatistical. This indicates that the excited state is repulsive which results in the large, and nonstatistical release of translational energy.

Equation (9.60) and figure 9.9 are for reactions with initially little angular momentum. Equation (9.60) can be applied even to room-temperature data especially when the reaction involves the loss of a heavy unit such as a Br atom. However, for H-atom loss reactions or for the dissociation of bimolecular reaction complexes, which contain large amounts of angular momentum, the assumptions used to derive Eq. (9.60) are no longer valid. However, Klots (1994b) has generalized his results to the case of large J. The integrals are more complicated, but still easily calculated.

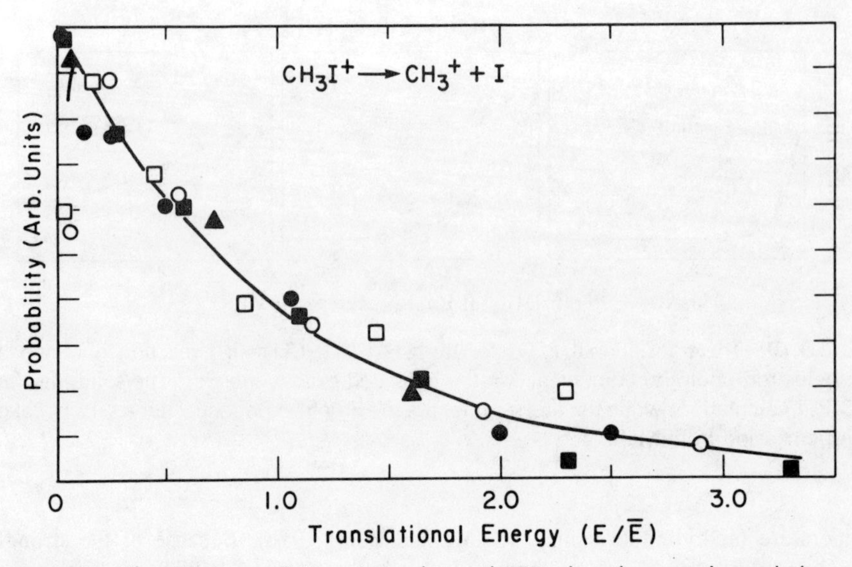

Figure 9.11 The $CH_3I^+ \rightarrow CH_3^+ + I$ translational PED plotted versus the scaled energy $E_t/\langle E_t \rangle$. The data are from Mintz and Baer (1976) and from Powis (1983). Each symbol represents data taken at a different excess energy which ranges from 0.04 eV to 3 eV. Taken with permission from Baer (1986).

9.3.2 The Chesnavich-Bowers Treatment of PED with OTS/PST

Chesnavich and Bowers (1976, 1977a,b, 1979) developed rigorous expressions for the product energy distributions, which are valid over a large range of angular momenta and are applicable to various product geometries.

Consider a molecule dissociating to products which have angular momenta, j_1 and j_2. As before, j_1 and j_2 are first coupled to a resultant j, which is then coupled to the orbital angular momentum, ℓ, to yield the total J. In PST, Eqs. (9.39) and (9.40) limit the values to the indicated regions in figure 9.7. This will now be modified by the centrifugal barrier.

9.3.2.1 The Centrifugal Barrier

In OTS/PST an additional condition is imposed by the centrifugal barrier and the relative product translational energy. Recall [Eq. (7.13)] that for an interaction potential given by $-ar^{-n}$, this barrier has a maximum, $V_{\max}$ given by

$$V_{\max} = \left(\frac{P_\theta^2}{\mu}\right)^{n/(n-2)} \frac{1}{(na)^{2/(n-2)}} \left(\frac{1}{2} - \frac{1}{n}\right), \tag{9.66}$$

where P_Θ is the magnitude of the orbital angular momentum. The constant a is defined as in Eqs. (9.61) and (9.62). Because all of the centrifugal energy is converted to product translational energy, $E_t \geq V_{\max}$. Or stated in another way, for a given orbital angular momentum, there is a minimum translational energy, $E_t^\ddagger$, which is given by

$$E_t^\ddagger = \frac{\ell^{2n/(n-2)}}{\left(\frac{n\mu}{\hbar^2}\right)^{n/(n-2)} a^{2/(n-2)} \left(\frac{2}{n-2}\right)} = \frac{\ell^{2n/(n-2)}}{\Lambda}, \tag{9.67}$$

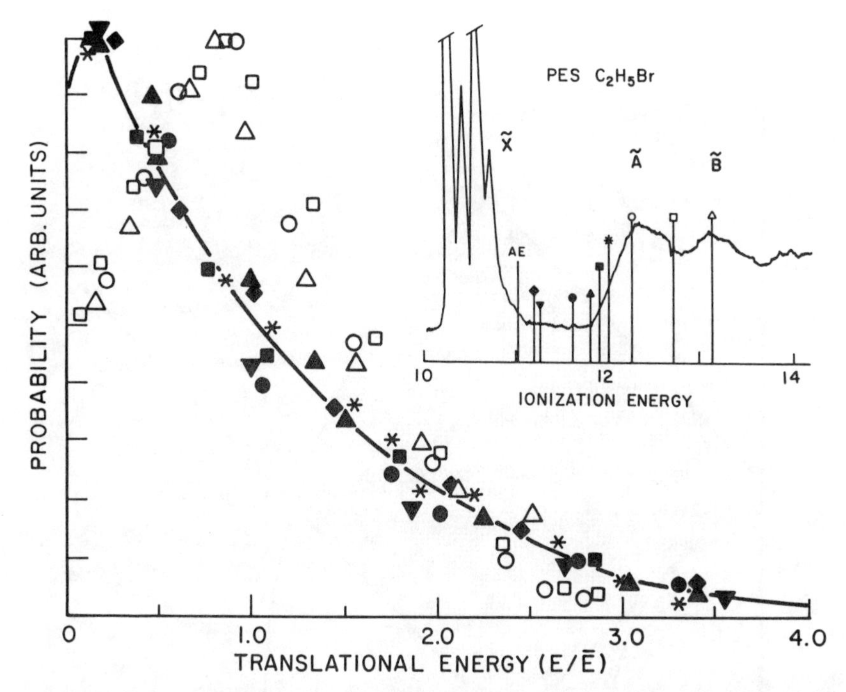

Figure 9.12 The scaled translational PED for the $C_2H_5Br^+ \rightarrow C_2H_5^+ + Br$ reaction. The insert is the photoelectron spectrum of ethylbromide which shows the ground states (two spin-orbit states) as well as the excited $\tilde{A}$ and $\tilde{B}$ states. When the ion is prepared in the ground states, the PED is statistical (solid line). The open points correspond to data from the excited states which yield a nonstatistical PED. This indicates that the $\tilde{A}$ state does not undergo an internal conversion to the ground state, but dissociates directly along a repulsive surface. Taken with permission from Baer (1986).

where Λ is the expression in the denominator. The superscript $\ddagger$ on any variable indicates that this is a minimum value for that variable. In converting Eq. (9.66) into Eq. (9.67), P_Θ was replaced by $\hbar[\ell(\ell + 1)]^{1/2} \approx \hbar\ell$.

A similar minimum value for the combined translational and rotational energy can be developed by noting that $E_{tr} = E_t + E_r = E_t + B_r j^2 \geq E_t + B_r(J - \ell)^2$. This yields

$$E_{tr}^{\ddagger} = \frac{\ell^{2s/(s-2)}}{\Lambda} + B_r(J - \ell)^2 \,. \tag{9.68}$$

When this equation is inverted, the maximum orbital angular momentum quantum number, ℓ_{max}, for a given translational/rotational energy, is obtained:

$$\ell_{max} = [\Lambda E_t]^{(n-2)/2n} = \Lambda^{(n-2)/2n}\,(E_{tr} - B_r j^2)^{(n-2)/2n} \,. \tag{9.69}$$

This equation is plotted for a given E_{tr} on the ℓ vs. j graph of figure 9.13. It provides an upper limit to the orbital angular momentum for a given value of E_{tr} and j. The corresponding PST line (shown in figure 9.7) which ignores the centrifugal barrier is vertical because it is limited only by energy conservation. Thus the deviation of the ℓ_{max} line from the vertical is a measure of the centrifugal barrier. In the extreme case of H loss reaction the ℓ_{max} line cuts across the ℓ vs. j plot almost horizontally as was indeed found by Powis (1979) for the case of H loss from CH_4^+. The resulting transla-

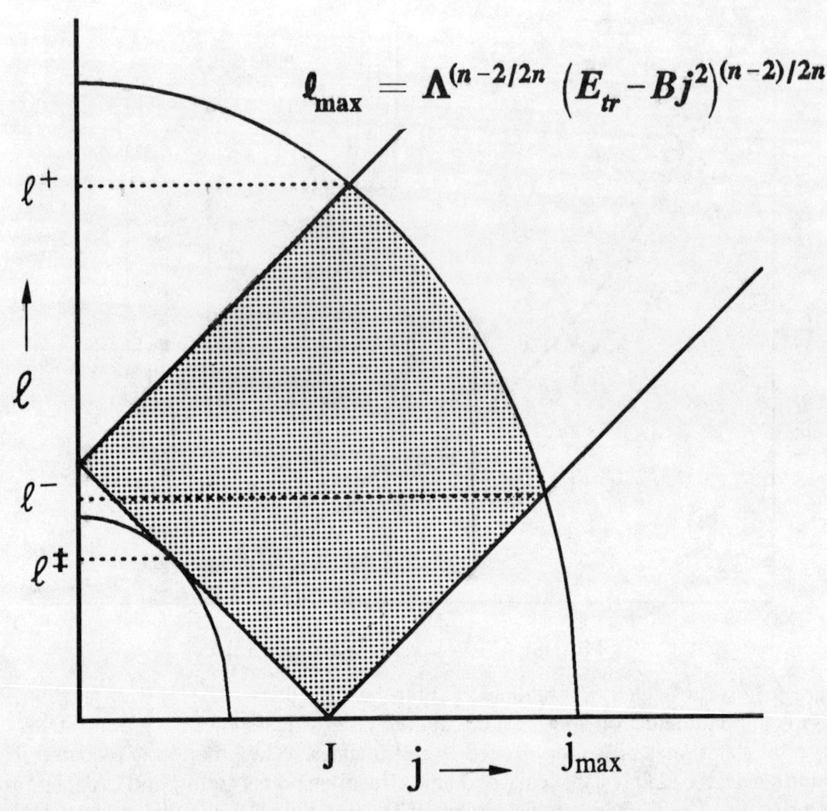

Figure 9.13 The ℓ vs. j plot for a reaction with total angular momentum quantum number J and a fixed $E_{tr} = j^2_{max}B$, where B is the effective product rotational constant. j_{max} is the maximum product rotational angular momentum. The ℓ_{max} line deviates from the vertical because of the centrifugal barrier. The shaded area is the range of integration for the $\Gamma(E_r, j)$ function to yield $\Gamma_{ro}(E_{tr}, J)$ [e.g., Eqs. (9.81)–(9.83)].

tional distribution is then close to $E_t \exp[-E_t/k_B T^*]$ which is a four-dimensional distribution (Klots, 1988). Examples are the H loss reactions in figure 9.9.

A minimum $\ell^\ddagger$ can also be established from figure 9.13 (Chesnavich and Bowers, 1976). For a reaction producing neutral products with $n = 6$,

$$\ell^\ddagger_{\text{neutral}} = \frac{B_r \Lambda}{3} \left[\left(1 + \frac{6J}{B_r \Lambda} \right)^{1/2} - 1 \right], \tag{9.70}$$

while for a reaction that produces an ion and a neutral product, $n = 4$, in which case

$$\ell^\ddagger_{\text{ion}} = \left(\frac{B_r \Lambda}{2} \right)^{1/3} [(S + J/2)^{1/3} - (S - J/2)^{1/3}], \tag{9.71}$$

where

$$S = \left(\frac{J^2}{4} + \frac{B_r \Lambda}{54} \right)^{1/2}. \tag{9.72}$$

B_r in these equations is the reduced rotational constant of the two products, which is given by

$$B_r = \frac{B_1 B_2}{B_1 + B_2}. \quad (9.73)$$

9.3.2.2 The Total Sum of States in OTS/PST

The OTS/PST procedure for determining PEDs is similar to that of PST except that the translational and rotational density and sums of states are treated in a manner consistent with constraints (1) and (3) above. In order to do so, it is necessary to identify and carefully distinguish a number of terms shown in table 9.1 (Chesnavich and Bowers, 1979).

The OTS/PST total sum of states [the analog of Eq. (9.4) in the prior treatment] is given by

$$N(E, J) = \int_{E_{tr}^{\ddagger}}^{E} \rho_v(E - E_o - E_{tr})\, \Gamma(E_{tr}, J) dE_{tr}\,. \quad (9.74)$$

where the lower limit, $E_{tr}^{\ddagger}$, in Eq. (9.74) is the minimum value of the translational and rotational energy needed to surmount the centrifugal energy barrier [Eq. (9.68)]. Note that if the total angular momentum (J) is zero, the lower limit of integration, $E_{tr}^{\ddagger} = \ell^{\ddagger} = 0$, which is identical to the lower limit used in the corresponding prior integral.

The rotational-orbital sum of states function $\Gamma(E_{tr}, J)$ depends upon the nature of the products, for example, two spheres, a sphere plus a prolate top, etc. Expressions for it and other rotational sums and densities will be given later. First we present the expressions for the product energy distributions.

9.3.2.3 Product Energy Distributions (PED)

The differential rate constant for forming products with translational/rotational kinetic energy, E_{tr}, is given by Eq. (7.57). This rate constant is proportional to the probability $P_{tr}(E_{tr}|J, E)$ of forming products with energy E_{tr}. If the products of the unimolecular reaction are an atom and a polyatomic, the probability for vibrational energy, E_v, in the polyatomic is simply

Table 9.1. Glossary of Important Terms.

J	Total system angular momentum.
$\rho(E_r,j)$	Rotational density of states at a rotational energy E_r, and a total rotational angular momentum j.
$\Gamma(E_r,j)$	Rotational sum of states at a rotational energy $\leq E_r$, and a total rotational angular momentum of j.
$\rho(E_t,E_r,J)$	Rotational-orbital density of states at a translational energy of E_t, rotational energy of E_r, and a total angular momentum of J.
$\Gamma(E_{tr},J)$	Rotational-orbital sum of states at a translational and rotational energy sum of E_{tr}, and total angular momentum of J.
$\rho(E,E_t,J)$	Total phase space density of states at energy E, of which E_t is translational energy and at a total angular momentum of J.
$N(E,J)$	Total phase space sum of states at energy E and total angular momentum of J.

$$P_v(E_v|J, E) = 1 - P_{tr}(E_{tr}|J, E) \tag{9.75}$$

since the total energy available to the products is $E = E_{tr} + E_v$. Note that the energy reference in this chapter is the dissociated products, while in chapter 7 it was the ground state of the dissociating molecule.

9.3.2.3.a The Translational PED. In order to obtain the distribution of relative translational energy, E_t, the differential rate constant for forming products with energy E_{tr}, Eq. (7.57), must be integrated over all values of E_r with E_t held constant. The resulting translational PED is

$$P_t(E_t|J, E)dE_t = \frac{dE_t}{N(E, J)} \int_{E_r^\ddagger}^{E-E_t} \rho_v(E - E_t - E_r)\rho(E_t\ , E_r, J)dE_r\ , \tag{9.76}$$

where we have again expressed the excess or available energy as E. Note that this is a distribution for a fixed E and J. The lower limit, $E_r^\ddagger$, is the minimum rotational energy which can generate a $\mathbf{j} = \mathbf{j}_1 + \mathbf{j}_2$ which is large enough to couple with ℓ to produce J. Its value depends upon the relative magnitudes of J and the maximum available orbital angular momentum, ℓ^*. They are given by

$$E_r^\ddagger = 0 \qquad \text{when } \ell_{max} > J\ , \tag{9.77}$$

$$E_r^\ddagger = B_r(J - \ell_{max})^2 \qquad \text{when } J \geq \ell_{max}\ . \tag{9.78}$$

An important difference between this and the corresponding prior and PST expressions is that the rotational and translational density of states are no longer separable.

9.3.2.3.b Product Rotational Energy Distributions. The product rotational energy distribution is determined much like the translational PED. The integral is

$$P_r(E_r|J, E)dE_r = \frac{dE_r}{N(E, J)} \int_{E_t^\ddagger}^{E-E_t} \rho_v(E - E_t - E_r)\ \rho(E_t, E_r, J)dE_t\ , \tag{9.79}$$

where instead of integrating over E_r, we integrate over E_t.

9.3.2.4 Rotational and Rotational-Orbital Sums and Densities of States

In order to evaluate the above product energy distributions, it is necessary to know the appropriate rotational-orbital sums of states $\Gamma(E_{tr}, J)$ and density of states $\rho(E_t, E_r, J)$. The term $\Gamma(E_{tr}, J)$ is required for obtaining the total sum of states, $N(E, J)$ in Eq. (9.74). The density of states is needed in evaluating the product energy distributions in Eqs. (9.76) and (9.79). What complicates the issue is that these terms depend on the particular products formed and their evaluation presents one of the major hurdles in applying OTS/PST to reaction dynamics. Chesnavich and Bowers have evaluated many of these quantities for several types of products (1976, 1977a, 1979). The types of possible products are: atoms, linear molecules, spheres, prolate and oblate symmetric tops, and asymmetric tops. Rather than treating all possible combinations, we focus on a subset of important ones. Studies have shown that negligible errors are introduced by approximating asymmetric and symmetric tops with spheres

(Chesnavich and Bowers, 1977a). Because this approximation simplifies considerably the calculations, we do not discuss the symmetric or asymmetric tops. The rotational constant of an asymmetric top treated as a spherical top is just:

$$B_r = (B_a B_b B_c)^{1/3} \tag{9.80}$$

The sums of rotational states reported by Chesnavich and Bowers (1979) for various combinations of products are listed in table 9.2. These rotational sums of states can now be used to generate all of the rotational-orbital sums and densities as needed.

9.3.2.5 $\Gamma_{ro}(E_{tr}, J)$ for Linear-Atom and Sphere-Atom Systems

To determine $\Gamma_{ro}(E_{tr}, J)$, the rotational-orbital sum of states, the rotational sum of states, $\Gamma(E_r, j)$, in table 9.2 should be integrated in the ℓ–j plane as shown in figure 9.13:

$$\Gamma_{ro}(E_{tr}, J) = \int\int \Gamma(E_r, j)\,dj\,d\ell\;. \tag{9.81}$$

Case J = 0. For both linear and spherical products, the rotational energy is $E_r = Bj^2$, where B is the rotational constant. Since for $J = 0, j = \ell$, the double integral over j and ℓ in Eq. (9.81) can be replaced by a single integral over j or ℓ, so that $\Gamma_{ro}(E_{tr},J)$ reduces to

$$\Gamma_{ro}(E_{tr}, J) = \int_0^{j_{max}(E_{tr})} \Gamma(E_r, j)\,dj\,. \tag{9.82}$$

For these systems, the rotational sum of states at fixed j, $\Gamma(E_r, j)$, is just the rotational degeneracy, which is unity and $2j$ for the linear and spherical products, respectively (table 9.2). Inserting these expressions into Eq. (9.82) gives $\Gamma_{ro}(E_{tr}, J)$ of $j_{max}(E_{tr})$ and $[j_{max}(E_{tr})]^2$ for the two respective products. These terms could be written as $[j_{max}(E_{tr})]^{2s}$, where s is 1/2 or 1 for the linear and sphere, respectively (Klots, 1972). The term $j_{max}(E_{tr})$ is found from Eq. (9.68) by replacing both ℓ and $J - \ell$ with j, and solving for j_{max} versus E_{tr}.

Case J ≠ 0. To find $\Gamma_{ro}(E_{tr}, J)$ for the linear atom and sphere-atom systems, with $J \neq 0$, one first applies Green's theorem to reduce the double integral in Eq. (9.81) to a single integral over ℓ. $\Gamma_{ro}(E_{tr}, J)$ takes the general form

$$\Gamma_{ro}(E_{tr}, J) = \int_0^{\ell^-} (J + \ell)^{2s} d\ell + \int_{\ell^+}^{0} |J - \ell|^{2s} d\ell + \int_{\ell^-}^{\ell^+} \left(\frac{E_{tr}}{B} - \frac{\ell^\alpha}{\Lambda B}\right)^s d\ell\;, \tag{9.83}$$

where $\alpha = 2n/(n - 2)$, and $s = 1/2$ and 1 for the linear and sphere systems, respectively. ℓ^+ and ℓ^- are determined from figure 9.13 and depend on the position of the curved boundary. The solution of Eq. (9.83) depends upon whether $E_{tr} \geq BJ^2$ or $E_{tr} \leq BJ^2$, and whether $\ell^+ \geq J$ or $E_{tr} \leq J^\alpha/\Lambda$.

Equation (9.83) is not strictly valid in the limit as $J \to 0$ because we approximated the correct rotational degeneracies of $2j + 1$ (for the sphere) with $2j$. When these degeneracies are treated correctly, the sums of states in the limit of $J = 0$ are the same as given above.

Table 9.2. Rotational Sums of States.[a]

System	$\Gamma(E_r)$	$\Gamma_z(E_r)$	$\Gamma(E_r,j)$	Boundaries
Linear-atom	$\left(\frac{E_r}{B_1}\right)^{1/2}$	$\left(\frac{E_r}{B_1}\right)^{1/2}$	1	$E_r \geq B_1 j^2$
Sphere-atom	$\frac{4}{3}\left(\frac{E_r}{B_s}\right)^{3/2}$	$\frac{E_r}{B_s}$	$2j$	$E_r \geq B_s j^2$
Linear-linear	$\frac{E_r^2}{B_1 B_2}$	$\frac{2}{3}\,\frac{\sqrt{B_1}+\sqrt{B_2}-\sqrt{w}}{B_1 B_2}E_r^{3/2}$	$q\sin^{-1}\left(\frac{jz}{wq}\right) - \frac{jz}{w}$	$B_1 j^2 \geq E_r \geq B_r j^2$
			$q\sin^{-1}\left(\frac{2\sqrt{B_1 B_2}}{w}\right) - \frac{B_1 j^2}{w}$	$B_2 j^2 \geq E_r \geq B_1 j^2$
			$q\sin^{-1}\left(\frac{jz}{wq}\right) + \frac{jz}{w} - j^2$	$E_r \geq B_2 j^2$
Sphere-linear	$\frac{8E_r^{5/2}}{15B_1 B_s^{3/2}}$	$\frac{E_r^2}{2B_1^{1/2}B_s^{3/2}}\sin^{-1}\sqrt{\frac{B_s}{w}}$	$\frac{4}{3B_s w^2}z^3$	$B_s j^2 \geq E_r \geq B_r j^2$
			$\left(\frac{2j}{w^2}\right)\left[wE_r - B_s\left(B_1 + \frac{B_s}{3}\right)j_2\right]$	$E_r \geq B_s j^2$
Sphere-sphere	$\frac{\pi E_r^3}{6(B_1 B_2)^{3/2}}$	$\frac{8E_r^{5/2}}{15B_1 B_2 w^{1/2}}$	$\frac{8jz^3}{3w^3}$	$E_r \geq B_r j^2$

[a] Taken from Chesnavich and Bowers (1979).

B_s and B_l are the spherical top and linear rotational constants. B_r is the reduced rotational constant of the rigid rotor pair. For near spherical tops use $B_s = (ABC)^{1/3}$. For linear-linear system, $q = E_r/(2\sqrt{B_1 B_2}$; $z = (\omega E_r - B_1 B_2 j^2)^{1/2}$; $\omega = B_1 + B_2$, and $B_1 < B_2$. For sphere-linear system, $z = (\omega E_r - B_s B_1 j^2)^{1/2}$, and $\omega = B_s + B_1$. For sphere-sphere system, $z = (\omega E_r - B_1 B_2 j^2)^{1/2}$, and $\omega = B_1 + B_2$

9.3.2.6 $\rho(E_t, E_r, J)$ for Linear-Atom and Sphere-Atom Systems

In order to get the rotational-orbital density of states $\rho(E_t, E_r, J)$ it is necessary to integrate the $\rho(E_r, J)$ rotational density of states in the ℓ–j plane, that is,

$$\rho(E_t, E_r, J) = \int\int \rho(E_r, j)\,dj\,d\ell\ . \tag{9.84}$$

The rotational density of states is given by

$$\rho(E_r, j) = \frac{\partial\Gamma(E_r, j)}{\partial E_r}\ . \tag{9.85}$$

The limits for the integral in Eq. (9.84) are different from those for the integral in Eq. (9.81). Now the boundaries of j and ℓ are rectangles and only depend upon E_r and E_t. For the general case, this integration must be carried out as shown in figure 9.14 where the boundaries are

$$j_{max} = (E_r/B)^{1/2} \quad \text{and} \quad \ell_{max} = (\Lambda E_t)^{(n-2)/2n}\ . \tag{9.86}$$

Case $J = 0$. Since there is only one fragment with j fixed by $E_r = Bj^2$ for the linear-atom and sphere-atom systems, their rotational density of states is $(2j)^{2s-1}$ times the delta function $\delta(E_r - Bj^2)$, where $s = 1/2$ and 1 for the linear and sphere systems,

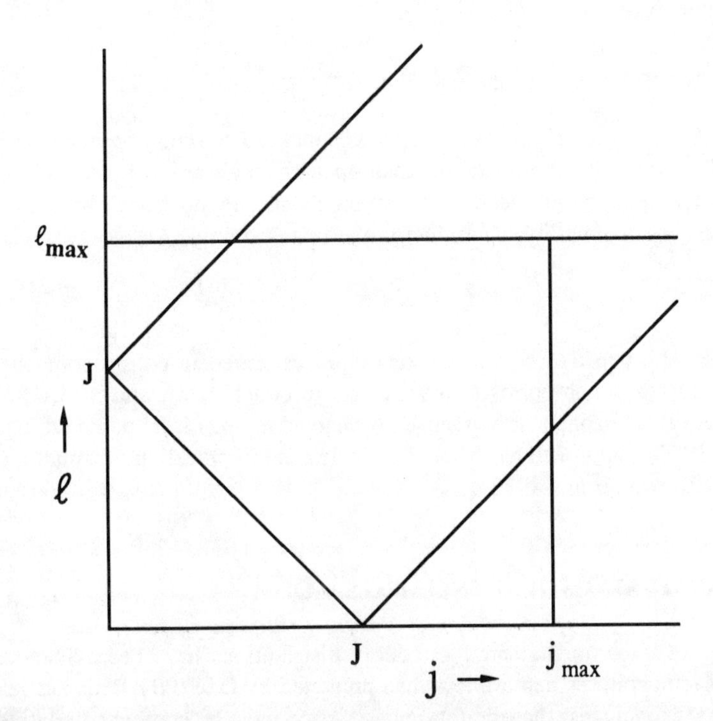

Figure 9.14 The ℓ vs. j plot for a reaction with total angular momentum quantum number J and fixed $E_r = j^2_{max}B$, and fixed E_t. The horizontal line is the fixed E_t boundary, while the vertical line is the fixed E_r boundary. The area bounded by those lines and the five lines is the range of integration for the $\rho(E_r, j)$ function to yield $\rho(E_t, E_r, J)$.

respectively. Thus, because $J = 0$, integration in Eq. (9.84) is carried out at the point $j = \ell = (E_r/B)^{1/2}$. The resulting expression for the rotational-orbital density of states $\rho(E_t, E_r, J)$ is $(2j)^{2s-1}$. Finally, since $E_t \geq \ell^\alpha/\Lambda$ [see Eq. (9.67)], $j = \ell \leq (\Lambda E_t)^{1/\alpha}$.

Case $J \neq 0$. For this situation j is fixed by $E_r = Bj^2$ and the rotational densities of states $\rho(E_r, j)$ are the same as described above. Since the total angular momentum is a vector quantity, and j and ℓ are magnitudes, the range of integration for ℓ in Eq. (9.85) is $|J - j|$ to $J + j$, with the constraint [Eq. (9.69)] that $\ell_{max} = (\Lambda E_t)^{1/\alpha}$. There are four solutions for $\rho(E_t, E_r, J)$, which depend on the relative magnitudes of j and ℓ_{max} with respect to J. For $E_r = Bj^2 \geq BJ^2$ and $E_t = (\ell_{max})^\alpha/\Lambda \geq J^\alpha/\Lambda$, one obtains

$$\rho(E_t, E_r, J) = (2j)^{2s-1}\,[\min\{\ell_{max}, J + j\} - (j - J)]. \tag{9.87}$$

9.4 STATISTICAL ADIABATIC CHANNEL MODEL

In chapter 7 the statistical adiabatic channel model (SACM) (Quack and Troe, 1974, 1975) was described for calculating unimolecular reaction rates. This theory assumes the reaction system remains on the same diabatic potential energy curve while moving from reactant to products. Two parameters, α and β are used to construct model diabatic potential curves. The unimolecular rate constant, at fixed E and J, for forming products with specific energy E_i (e.g., a specific vibrational energy in one of the fragments) is

$$k(E, J; E_i) = \frac{N(E, J; E_i)}{h\,\rho(E, J)}, \tag{9.88}$$

where $N(E, J; E_i)$ is the number of open channels for forming products with internal energy, E_i, for a given E and J. Since all open channels are assumed to be populated with equal weight and all reactant and product states are connected by diabatic potential curves, the probability of product properties is simply

$$P(E, J; E_i) = \frac{N(E, J; E_i)}{N(E, J)}. \tag{9.89}$$

The major effort in SACM calculations is concentrated in constructing the diabatic potential curves which correlate reactant and product states, and in determining the number of open channels at a given E. A channel is open if its potential maximum is less than E. This approach has been effectively used in modeling the rotational PED of HOOH (Brouwer et al., 1987).

9.5 SEPARATE STATISTICAL ENSEMBLES MODEL

A number of experimental product energy distributions have been characterized by rotational temperatures that are less than predicted by OTS/PST. This has lead Wittig et al. (1985) to suggest that the total phase space must be contracted relative to PST. This model incorporates some of the ideas contained in the SACM and VTST but is much simpler to implement. Consider the fate of the normal modes of the ethane molecule which dissociates to 2 CH_3 (table 7.4). Of the 18 normal modes, 12 are conserved which means that during the course of reaction, they are converted from ethane vibrations into the two methyl radical vibrations. The remaining six are disappearing modes which end up as either product translation or rotation. At the PST transition state, these disappearing modes have already evolved into rotations with their correspondingly very large densities of states. However, in VTST the TS is closer in and is thus much "tighter" than the PST TS. That is, PST overestimates the rotational density of states. The hypothesis of the SSE model is that energy flow is complete up to the inner TS, but that beyond that point energy flow between the conserved vibrational modes and the disappearing modes ceases. Thus, the partitioning of the energy between the conserved vibrational modes (which end up as product vibrations) and the disappearing modes (which end up as rotations and translations) should be calculated with a reduced contributions from the rotors. After determining the fraction of the energy in the product vibrations, the energy allocated to the disappearing modes is then distributed statistically among the rotational and translational degrees of freedom using the loose PST transition state. The net result is that less energy is apportioned to the rotational and translational degrees of freedom in SSE model than in PST.

In practice, the SSE model distributes the vibrational energy by using one degree of freedom less than phase space theory. Consider the reaction, $MAB \rightarrow M + AB$, where M is a polyatomic and AB is a diatomic fragment. The vibrational energy distribution in AB is given by

$$P(E_v(AB)|E) = \frac{\displaystyle\int_0^{E-E_v(AB)} \rho_{tr}[E - E_v(AB) - E_v(M)]\, \rho_v[E_v(M)] dE_v(M)}{\displaystyle\int_0^{E}\int_0^{E-E_v(AB)} \rho_{tr}[E - E_v(AB) - E_v(M)]\, \rho_v[E_v(M)]\, dE_v(M)\, dE_v(AB)}. \tag{9.90}$$

Now the key feature in SSE is the number of degrees of freedom associated with the ρ_{tr}. In PST for a rotationally cold sample, this would be $r + 1$, where r is the total number of rotational degrees of freedom of the products (recall that we have subtracted one degree of freedom from both the translations and the rotations). In SSE, one additional degree of freedom is subtracted in this step of the calculation. Thus for the dissociation of NCNO, ρ_{tr} is a four-dimensional density of states, which is proportional to [E-E_v(NO)-E_v(NC)] (Wittig et al., 1985).

In the case of the previously mentioned $NO_2 \rightarrow O + NO$ reaction, the distribution of NO vibrational levels (ignoring the three low lying O electronic states) is given by

$$P_v(\text{NO}) = \frac{(E - E_v)^x}{\sum_v (E - E_v)^x}, \tag{9.91}$$

where the sum is over the allowed NO vibrational levels and $E_{tr} = E - E_v$. However, for this reaction, $x = 0$ so that P_v(NO) is $1/(v_{max} + 1)$, where v_{max} is the highest NO vibrational level consistent with energy conservation. Thus the vibrational distribution of the NO products is uniform. For instance above the $v = 1$ level, the two vibrational states should be equally populated. Figure 9.15 shows that the data for the NO_2 dissociation fall in between the SSE model and the VTST model (to be discussed). The big advantage of the SSE model is that it is much easier to implement than the VTST model.

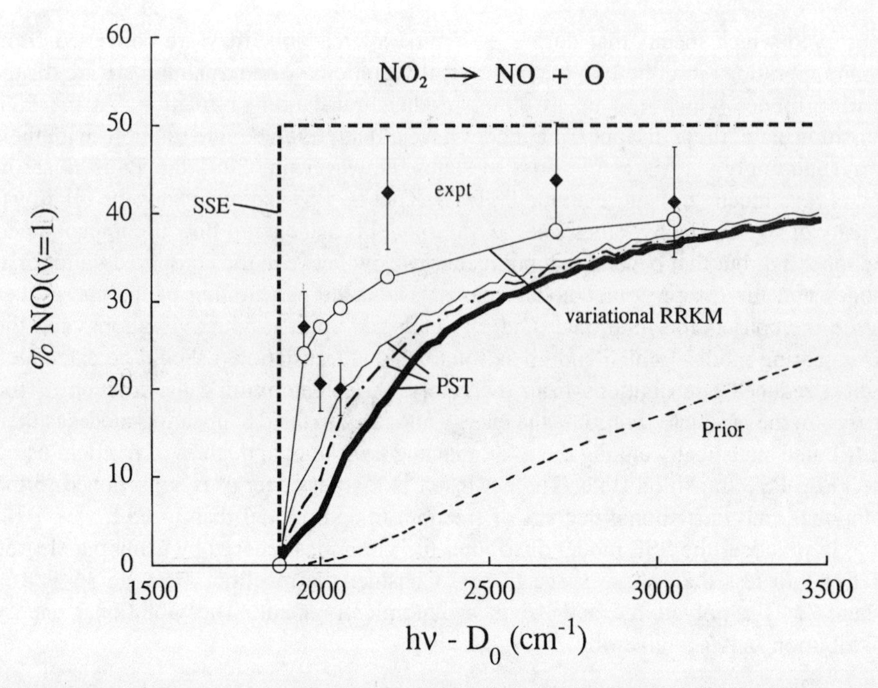

Figure 9.15 The measured and calculated NO vibrational population for the $NO_2 \rightarrow NO + O$ reaction at a range of excess energies in the vicinity of the NO $v = 1$ threshold. The measured distributions are sums of separate $^2\Pi_{1/2}$ and $^2\Pi_{3/2}$ NO distributions. The three PST curves were obtained with various assumptions about the ratios of the O (^3P_J) and NO spin–orbit populations. The variational RRKM calculation was taken from Klippenstein and Radivoyevitch (1993). Taken with permission from Hunter et al. (1993).

A final example of the difference in the PST and SSE models is in the rotational distribution for the NCNO → CN + NO reaction. Qian et al. (1985) studied the rotational distribution of CN and NO for various product vibrational states. Of particular interest is that at energies below which product vibrations can be excited, the experimental rotational distributions agrees with OTS/PST. At these energies the SSE and OTS/PST models are identical. However, at energies in excess of 1876 cm^{-1}, which is the vibrational energy of the CN fragment, the two models differ with the SSE predictions agreeing better with the experiment than the OTS/PST results (Wittig et al., 1985). Figure 9.16 shows the CN($v = 0$) rotational distribution at an excess energy of 2348 cm^{-1}, which is well above the CN $v = 1$ threshold. It is clear that the rotational temperature at this energy is lower than predicted by OTS/PST, and that the SSE model accounts for this reduced rotational energy.

9.6 VARIATIONAL RRKM THEORY WITH EXIT CHANNEL COUPLINGS

Aubanel et al. (1991) have discussed the application of variational RRKM theory to product energy distributions. What follows is a summary of this review.

A wide range of transition state locations is found in the application of variational

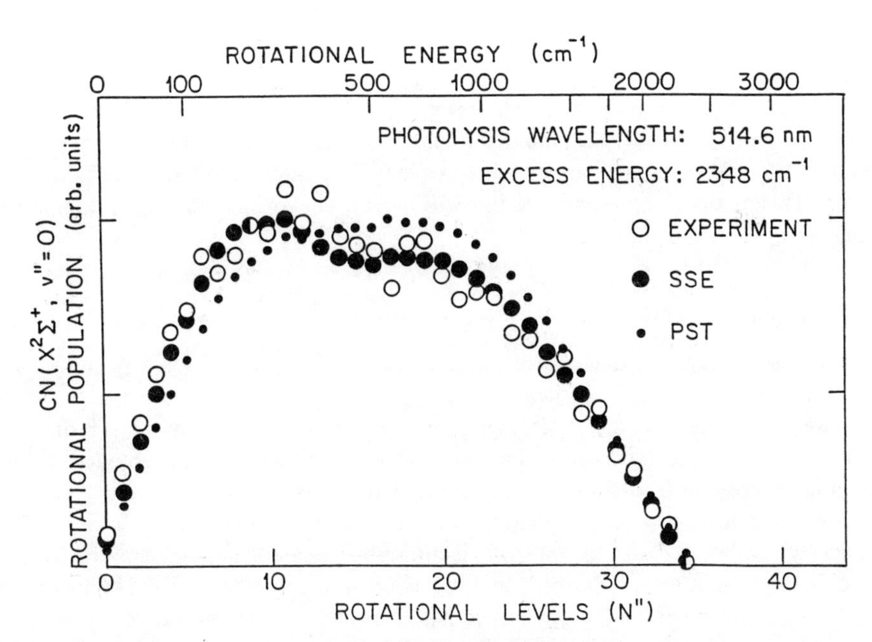

Figure 9.16 The CN ($v = 0$) product rotational energy distribution from the dissociation of rotationally cold and state selected NCNO at an energy above the CN, $v = 1$ threshold [Data from Qian et al. (1985)]. The PST predicts a distribution that is shifted toward higher energies, relative to the experimental and SSE model distributions. Taken with permission from Wittig et al. (1985).

RRKM theory to dissociation reactions having no barrier to the reverse process of association. Accordingly one must, in the general case, allow for the influence of exit channel couplings in order to predict the properties of separated products based on the transition state distributions embodied in $N(E, J)$. Two ways to accomplish this in approximate fashion are outlined below. One models the effect of exit channel dynamics within the framework of the flexible transition states discussed in chapter 7 and the other handles the exit channel dynamics explicitly using classical dynamics. The former model is referred to as variational RRKM theory with exact channel couplings (VRRKM/ECC).

9.6.1 Flexible Transition State Model

A VRRKM/ECC model for product vibrational and rotational distributions was introduced by Wardlaw and Marcus (1988). Subsequently, Marcus (1988) constructed a refined version which successfully describes rotational quantum number distributions of products arising from the decomposition of NCNO (Klippenstein et al., 1988) and CH_2CO (Klippenstein and Marcus, 1989). In the latter model, the conserved modes are assumed vibrationally adiabatic (as in SACM) after passage through the transition state and, consequently, the distribution of vibrational quantum numbers for the products is the same as it is at the transition state. The transitional modes are assumed nonadiabatic between the variationally determined TS and a loose TS located at the centrifugal barrier. [These are the same two transition states associated with the TS switching model of Chesnavich and Bowers (Chesnavich et al., 1981).] Beyond this loose TS, the transitional modes are treated as adiabatic. For energies below those needed to yield vibrationally excited products, the phase space theory distribution of product rotational states is obtained. At energies above the vibrational excitation thresholds, the product rotational state distributions are generally expected to differ from those of phase space theory. This model is thus similar to the SSE model, but considerably more complex to implement.

9.6.2 Trajectory Model

RRKM theory assumes that the relaxation dynamics of the molecule is sufficiently rapid to maintain a microcanonical distribution of states at the transition state. Initial trajectory conditions could, in principle, be selected uniformly from this TS distribution and an ensemble of trajectories propagated to the product region, provided that the potential energy surface from the transition state region to the asymptotic region is available. Product property distributions would then be readily extracted from the final trajectory conditions. Such an approach is difficult to implement in general, because of the complexities of sampling from a microcanonical distribution in a region intermediate between reactants and products. It has, however, been applied to a two-atom system in a study of post barrier effect on energy exchange (Rynefors, 1982). An alternative approach has been proposed by Hamilton and Brumer (1985) who chose to initiate a microcanonical set of trajectories in the product region and propagating this ensemble toward the parent molecule. The initial conditions for these 'time reversal' trajectories are readily determined via standard quasi-classical techniques and represent a phase space theory distribution. One retains those trajectories which reach a (predetermined) TS and discards the remainder, leaving an "exit channel corrected" ensemble which is implicitly assumed to correspond to a microcanonical distribution at the TS. Product internal energy distributions are then extracted from this corrected phase space theory set of initial trajectory conditions. The well-known problem of violation of zero-point energy conservation of classical trajectories (Bowman et al., 1989; Miller et al., 1989; Peslherbe and Hase, 1994) (see also chapter 8) is ignored in this model, although it is not expected to be a serious one for the direct process (i.e., short-time dynamics) involved here (Untch et al., 1993).

9.7 NONSTATISTICAL MODELS

The statistical models discussed in the previous section are for unimolecular reactions without a potential energy barrier for the reverse recombination reaction. For reactions with such potential energy barriers, these statistical models are probably inappropriate. This is because the release of the barrier's potential energy in forming the products is expected to be nonstatistical. Although the energy distribution at the potential energy barrier (i.e., at the TS) may be statistical as assumed by RRKM theory, the final product energy distribution may be highly nonstatistical. Exceptions arise when a secondary minimum associated with a bound isomer is located after the tight TS. If the lifetime in that state is sufficiently long to equilibrate the molecule's modes, statistical PEDs can still be attained. This happens quite frequently in ionic reactions which produce dipolar products with their attendant long range ion–dipole interactions which are typically 10–15 kcal/mole deep (Weitzel, 1991). [A notable exception is the Sn2 reaction $Cl^- \cdots CH_3Cl$ in which the reaction coordinate has the form of a double well potential (Graul and Bowers, 1994). Because of the weak interaction between the closed shell anion and the molecule, and the attendant vibrational frequency mismatch, coupling among the vibrational degrees of freedom is apparently hindered (Cho et al., 1992).] In the absence of wells, molecules which dissociate along a repulsive potential will have nonstatistical PEDs. Most often they are characterized by significantly higher than statistical translational energies. Schinke, 1992 discusses these classes of reactions in detail.

Two models for determining PEDs for dissociation along repulsive potentials are discussed here. The impulsive model is a very crude classical model whose major virtue is its simplicity. A more sophisticated model is the transition-state mapping model. Finally, we mention the use of classical trajectories in determining PEDs.

9.7.1 The Impulsive Model

The impulsive model (Holdy et al., 1970; Busch and Wilson, 1972) serves as the epitome of the nonstatistical model. It is most readily applied to the dissociation of triatomic molecules or ions which dissociate directly on a repulsive surface with lifetimes on the order of femtoseconds. Consider the molecule *ABC* in figure 9.17 which dissociates to *A* + *BC* with a total available energy of *E*. According to the impulsive model, the entire available energy is initially released as relative kinetic energy between atoms *A* and *B*. Atom *C* acts simply as a spectator. Once *A* and *B* are separated (energy flow between them ceases) the kinetic energy of the *B* atom is shared among the translational, rotational, and vibrational degrees of freedom of the *BC* fragment. Conservation of momentum and energy permits us to write the following equations (Holdy et al., 1970):

$$|P_a| = |P_b| = |P_{bc}| \, , \tag{9.92}$$

$$E = \frac{P_a^2}{2M_a} + \frac{P_b^2}{2M_b} \, , \tag{9.93}$$

$$E_t = \frac{P_a^2}{2M_a} + \frac{P_{bc}^2}{2M_{bc}} \, , \tag{9.94}$$

$$\frac{E_t}{E} = \frac{M_a M_b/(M_a + M_b)}{M_a M_{bc}/(M_a + M_b + M_c)} = \frac{\mu_{a,b}}{\mu_{a,bc}} \, , \tag{9.95}$$

where E_t is the total translational energy carried off by the atom and the diatom and E_t/E is the fraction of the total energy released as translational energy.

Once the translational energy is accounted for, the remaining energy is partitioned

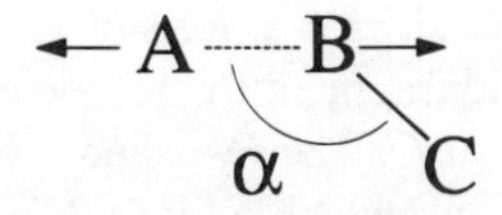

Figure 9.17 Impulsive model for the dissociation of *ABC* → *A* + *BC*. The dissociation energy is initially directed only at *A* and *B*, while *C* is a spectator atom. The fraction of rotational and vibrational energy imparted to *BC* is a function of the angle and the reduced masses.

between vibrational and rotational energy of the BC fragment. This is determined by the bond angle and is given by

$$E_v = E\left(1 - \frac{\mu_{a,b}}{\mu_{a,bc}}\right)\cos^2\alpha\ , \tag{9.96}$$

$$E_r = E\left(1 - \frac{\mu_{a,b}}{\mu_{a,bc}}\right)\sin^2\alpha\ . \tag{9.97}$$

A linear molecule is thus not expected to produce any rotationally excited products. This model predicts for the case of the NO_2 dissociation, that 71.5% of the available energy is released as translational energy compared to the observed 60% when NO_2 is photodissociated at 28,800 cm^{-1} (Busch and Wilson, 1972). Such discrepancies are not unexpected in view of its simplicity. As pointed out by Schinke (1992) the impulsive model would be strictly applicable only if the "dissociative potential energy surface depends so weakly on the bond angle that the torque $-\partial V/\partial\alpha$ can be safely neglected." This is almost never true.

One modification to the impulse model can be made by varying the position along the diatom bond at which the force is exerted (Daniels and Wiesenfeld, 1993). An example in which this fitting parameter was used to force the average measured rotational energy to agree with the predicted rotational energy is in the photodissociation of O_3 (Daniels and Wiesenfeld, 1993). However, this modification only fits the average energy. In fact, the impulsive model with a stationary reactant molecule yields a single and unique product rotational energy. Experimentally one observed a distribution of rotational energies. In the context of the impulsive model, these arise from initial thermal energy of the molecule as well as a distribution of bond angles due to zero-point vibrational (bending) energy. When this is taken into account (Levene and Valentini, 1987) a distribution is indeed obtained. The ozone photodissociation represents a particularly good example of the failure of the statistical PST and the value of the impulsive model (figure 9.18). Less good agreement have been noted for other reactions, including four-atom reactions such as H_2CO (Carleton et al., 1990).

9.7.2 Transition State Mapping of Rotational Distributions

More sophisticated models use calculated transition-state wave functions to predict product energy distributions, an approach that can be justified as long as the potential energy surface is a simple repulsive one. Direct dissociation of neutral species are thus better candidates for these techniques than ionic dissociations because of the long range interactions associated with the ion-induced dipole forces. Good agreement between theory and experiment has been noted for the rotational product energy distribution in the photodissociation of formaldehyde (Schinke, 1985, 1986).

We illustrate this approach with a model that has been developed (Qian et al., 1990; Ogai et al., 1992; Solter et al., 1992; Reisler et al., 1994) to interpret rotational state distributions of the diatomic fragment BC following the dissociation of resonance states in the triatomic molecule ABC. The resonances (see chapter 8) are identified by the quantum numbers v_1, v_2, and v_3, which refer to BC stretch, AB stretch, and bending vibrations, respectively. In the model, very weak coupling is assumed between the A—BC bending mode and the BC vibration. the decay is viewed as a direct dissociation process which starts at the transition state, with the stationary wave function for the

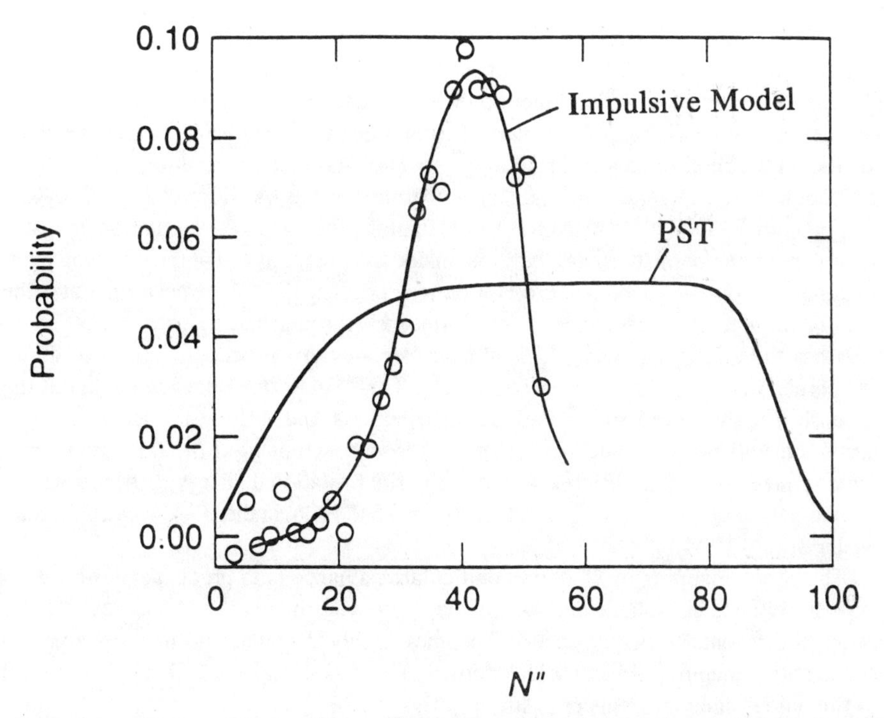

Figure 9.18 The $O_2(^3\Sigma_g^-, v = 12)$ product rotational distribution obtained by LIF from the photodissociation of O_3 at 248 nm. The solid lines are the calculated statistical PST distribution and the impulsive model prediction. Taken and adapted with permission from Daniels and Wiesenfeld (1993).

resonance at the transition states, Ψ^{TS}, determining the "initial conditions" for the fragmentation step. Then, the final rotational state distribution "maps" or "reflects," in a dynamical sense, the shape of $|\Psi^{TS}|^2$.

The dissociation of ClNO(T_1) and FNO(S_1) to NO exemplify the general features of the above model (Qian et al., 1990; Ogai et al., 1992; Solter et al., 1992). The NO vibration is treated as an adiabatic (i.e., uncoupled) degree of freedom and the X + NO separation defines the reaction coordinate. Thus, the only relevant coordinate for the transition state wavefunction is the X—NO bending angle, γ. Trajectories are initialized at the transition state for different γ and followed to products. This calculation defines the relation between the final rotational angular momentum quantum number j (a continuous variable in classical mechanics) and the initial bending angle, γ_o. The final distribution of j, for a particular bending quantum number, v_3, at the transition state, is given by

$$P_{v_3}(j) \propto \sum \sin\gamma_o \, (\Psi_{v_3}^{TS}(\gamma_o))^2 \left(\left| \frac{dJ}{d\gamma_o} \right| \right)^{-1}, \tag{9.98}$$

where the relation between the initial angle γ_o and final angular momentum j, the rotational excitation function, is defined by

$$J(\gamma_o) = j\,. \tag{9.99}$$

An application of the above model to FNO(S_1) dissociation (Ogai et al., 1992) is illustrated in figure 9.19. The left-hand side of this figure depicts the rotational excitation function and the weighting function $|\Psi_{v_3}^{TS}(\gamma_o)|^2$, for four bending resonance states. The rotational state distributions obtained from a quantum mechanical calculation are depicted on the right-hand side of figure 9.19. The rotational distribution is essentially a reflection of the quantum mechanical probability at the transition state $|\Psi_{v_3}^{TS}(\gamma_o)|^2$.

For both FNO(S_1) and ClNO(T_1) dissociation, the shape of the stationary wave function and its nodal pattern are the same inside the potential and at the TS. However, this is not always the case. If there is a narrow bottleneck at the transition state, the wave function for the resonance has to "squeeze" through the tight transition state, which has the effect of reducing the number of nodes in the wave function for directions perpendicular to the reaction coordinate. This "squeezing" does not rule out the relationship between the final rotational distributions and $|\Psi_{v_3}^{TS}(\gamma_o)|^2$. However, this distribution will not be related to v_3 for the resonance, but v_3 at the transition state. Such a "squeezing" effect has been seen in the final rotational energy distributions for the dissociation of HCO in the ground electronic state HCN in the C $^1A'$ electronic state (Reisler et al., 1994).

The above mapping of Ψ^{TS} for a particular resonance onto product states requires that the resonance be isolated and assignable. However, in the case of classical chaotic motion, the resonance wave function becomes highly irregular and unassignable, so that the above mapping scheme breaks down. The dissociation of NO_2 appears to fall into this latter category (Reisler et al., 1994).

9.7.3 Classical Trajectories

Trajectories are widely used to study the unimolecular and intramolecular dynamics of highly excited molecules (Hase, 1981). However, as discussed in chapter 8, because of the improper treatment of zero-point energy motion by classical trajectories, these calculations, in general, do not give accurate product energy distributions. Trajectories allow the potential energy barrier to be crossed without zero-point energy in the modes orthogonal to the reaction coordinate and products to be formed without zero-point energy. A way to circumvent these difficulties and study the effect of the potential energy barrier on product energy partitioning is to initiate the trajectories at the barrier. The quasi-classical model (Truhlar and Muckerman, 1979) can be used to select the trajectories so that the vibrational modes are populated in accord with quantum RRKM theory. This will insure that each vibrational mode has zero-point energy. Because of the direct short time dynamics in moving from the barrier to products, the unphysical transfer of zero-point energy between modes may not be a serious problems (Untch et al., 1993).

Trajectory calculations of this type have been performed for $CH_2OH^+ \rightarrow CHO^+ + H_2$ (Uggerud and Helgaker, 1992), and for $H_2CO \rightarrow H_2 + CO$ (Chang et al., 1992; Chen et al., 1994) dissociations. These calculations used direct dynamics (Wang and Karplus, 1973) in which *ab initio* or semi-empirical molecular orbital calculations are used directly in evaluating the classical trajectories. The vibrational distributions calculated by *ab initio* direct dynamics for the CO and H_2 products of H_2CO dissociation are compared with the experimental results (Debarre et al., 1985) in table 9.3. There is very good agreement between theory and experiment.

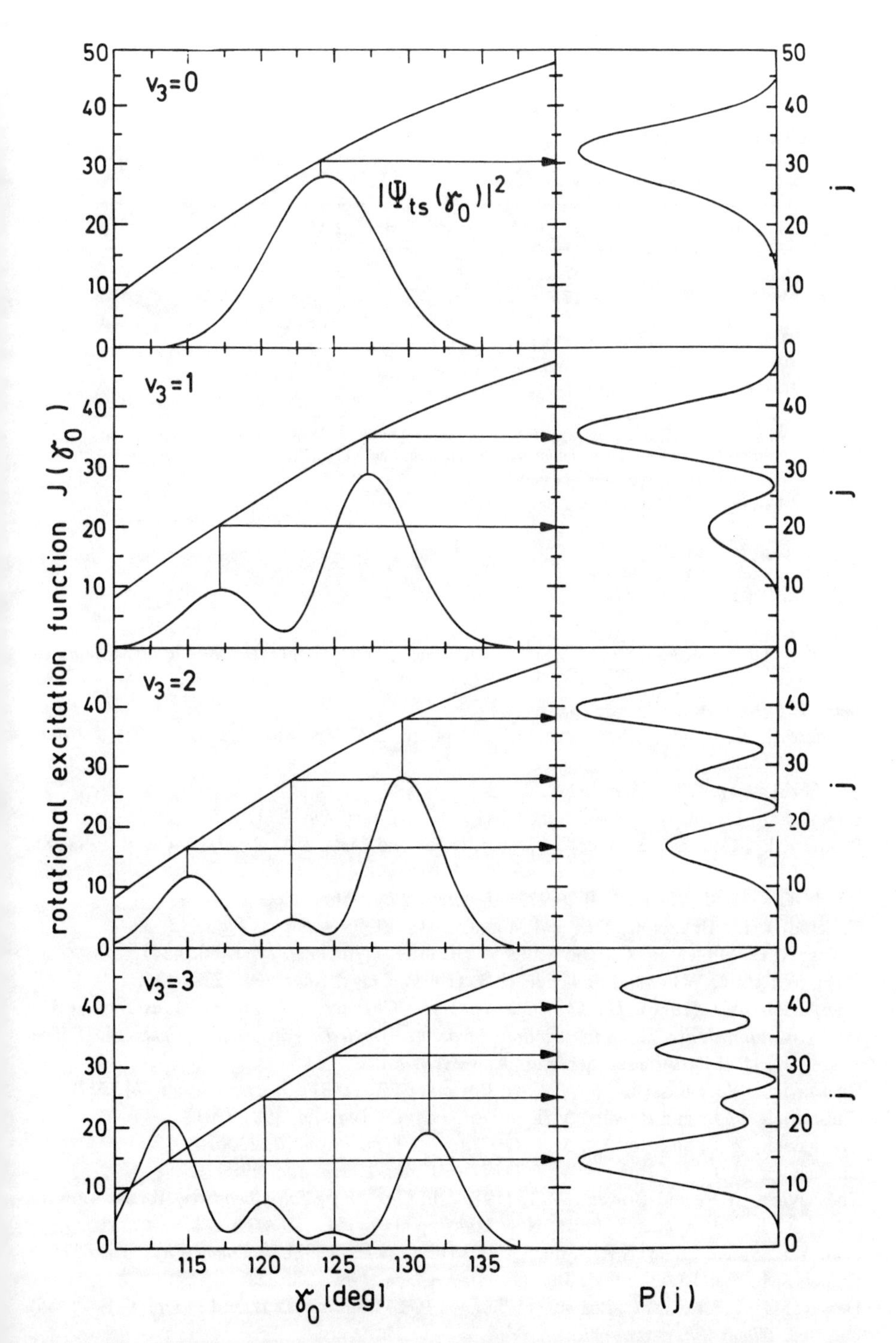

Figure 9.19 The product rotational quantum number distributions, $P(j)$, of FNO obtained by mapping the transition state wave function $|\Psi^{TS}_{v_3}(\gamma_o)|^2$ onto the rotational excitation function, $J(\gamma_o)$. Examples are for four different bending quantum numbers, v_3, at the transition state of a triatomic reactant. γ_o is the bending angle. The arrows illustrate the reflection principle. Taken with permission from Ogai et al. (1992).

Table 9.3. Vibrational Distributions for CO and H_2 Formed by H_2CO Dissociation[a]

	CO			H_2		
v	HF/3–21G	HF/6–31G**	Exp[b]	HF/3–21G	HF/6–31G**	Exp[c]
0	83.2	82.2	88	11.0	22.8	24.2
1	16.8	17.8	12	30.7	36.5	41.3
2				29.5	27.0	24.6
3				21.7	11.2	8.6
4				6.8	2.5	0.3
5				0.3	0.0	0.0

[a]The distributions are expressed in %. The experimental results are from Debarre et al. (1985), while the trajectory studies on the *ab initio* surfaces are from Chen et al. (1994).

[b]Experimental result are from Bamford et al. (1985).

[c]Experimental results are from Debarre et al. (1985).

REFERENCES

Aubanel, E.E., Wardlaw, D.M., Zhu, L., and Hase, W.L. (1991). *Int. Rev. Phys. Chem.* **10,** 249.

Baer, T. (1986). *Adv. Chem. Phys.* **64,** 111.

Bamford, D.J., Filseth, S.V., Foltz, M.F., Hepburn, J.W., and Moore, C.B. (1985). *J. Chem. Phys.* **82**, 3032.

Ben-Shaul, A. (1977). *Chem. Phys.* **22,** 341.

Bowman, J.M., Gazdy, B., and Sun, Q. (1989). *J. Chem. Phys.* **91,** 2859.

Brouwer, L., Cobos, C.J., Troe, J., Dubal, H.R., and Crim, F.F. (1987). *J. Chem. Phys.* **86,** 6171.

Busch, G.E., and Wilson, K.R. (1972). *J. Chem. Phys.* **56,** 3626.

Carleton, K.L., Butenhoff, T.J., and Moore, C.B. (1990). *J. Chem. Phys.* **93,** 3907.

Chang, Y.T., Minichino, C., and Miller, W.H. (1992). *J. Chem. Phys.* **96,** 4341.

Chen, W., Hase, W.L., and Schlegel, H.B. (1994). *Chem. Phys. Lett.* **228,** 436.

Chesnavich, W.J. (1976). *The Statistical Theory of Chemical Kinetics: Rigorous angular Momentum Conservation in Reactions Involving Symmetric Polyatomic Species*. Ph.D. Thesis, UMI Dissertation Services. Ann Arbor, MI.

Chesnavich, W.J., Bass, L., Su, T., and Bowers, M.T. (1981). *J. Chem. Phys.* **74,** 2228.

Chesnavich, W.J., and Bowers, M.T. (1976). *J. Am. Chem. Soc.* **98,** 8301.

Chesnavich, W.J., and Bowers, M.T. (1977a). *J. Chem. Phys.* **66,** 2306.

Chesnavich, W.J., and Bowers, M.T. (1977b). *J. Am. Chem. Soc.* **99,** 1705.

Chesnavich, W.J., and Bowers, M.T. (1979). In *Gas Phase Ion Chemistry, Vol. 1*. Bowers, M.T., Ed. Academic Press, New York, pp. 119–151.

Cho, Y.J., Vande Linde, S.R., Zhu, L., and Hase, W.L. (1992). *J. Chem. Phys.* **96,** 8275.

Daniels, M.J., and Wiesenfeld, J.R. (1993). *J. Chem. Phys.* **98,** 321.

Debarre, D., Lefebvre, M., Pealat, M., Taran, J.P.E., Bamford, D.J., and Moore, C.B. (1985). *J. Chem. Phys.* **83,** 4476.

Evans, M.G., and Polanyi, M. (1935). *Trans. Faraday Soc.* **31,** 875.

Eyring, H. (1935). *J. Chem. Phys.* **3,** 107.

Georgiou, S., and Wight, C.A. (1988). *J. Chem. Phys.* **88,** 7418.

Graul, S.T., and Bowers, M.T. (1994). *J. Am. Chem. Soc.* **116,** 3875.

Hamilton, I., and Brumer, P. (1985). *J. Chem. Phys.* **82,** 595.

Hase, W.L. (1981). In *Potential Energy Surfaces and Dynamics Calculations,* Truhlar, D.G., Ed. Plenum Press, New York, pp. 1–35.
Holdy, K.E., Klotz, L.C., and Wilson, K.R. (1970). *J. Chem. Phys.* **52,** 4588.
Huber, K.P., and Herzberg, G. (1979) *Molecular Spectra and Molecular Structure IV. Constants of Diatomic Molecules.* Van Nostrand Reihhold, New York.
Hunter, M., Reid, S.A., Robie, D.C., and Reisler, H. (1993). *J. Chem. Phys.* **99,** 1093.
Johnson, K., Powis, I., and Danby, C.J. (1981). *Chem. Phys.* **63,** 1.
Klippenstein, S.J., Khundkar, L.R., Zewail, A.H., and Marcus, R.A. (1988). *J. Chem. Phys.* **89,** 4761.
Klippenstein, S.J., and Marcus, R.A. (1989). *J. Chem. Phys.* **91,** 2280.
Klippenstein, S.J., and Radivoyevitch, T. (1993). *J. Chem. Phys.* **99,** 3644.
Klots, C.E. (1971). *J. Phys. Chem.* **75,** 1526.
Klots, C.E. (1972). *Z. Naturforsch.* **27a,** 553.
Klots, C.E. (1973). *J. Chem. Phys.* **58,** 5364.
Klots, C.E. (1976). *J. Chem. Phys.* **64,** 4269.
Klots, C.E. (1988). *Acc. Chem. Res.* **21,** 16.
Klots, C.E. (1989). *J. Chem. Phys.* **90,** 4470.
Klots, C.E. (1990a). *J. Chem. Phys.* **93,** 2513.
Klots, C.E. (1990b). *J. Chem. Phys.* **93,** 6585.
Klots, C.E. (1992). *J. Phys. Chem.* **96,** 1733.
Klots, C.E. (1993a). *J. Chem. Phys.* **98,** 206.
Klots, C.E. (1993b). *J. Chem. Phys.* **98,** 1110.
Klots, C.E. (1994a). *J. Chem. Phys.* **100,** 1035.
Klots, C.E. (1995). *J. Phys. Chem.* **99,** 1748.
Klots, C.E. (1994c). In *Unimolecular and Bimolecular Ion Molecule Reaction Dynamics,* Baer, T., Ng, C.Y., and Powis, I., Eds. Wiley, New York, pp. 311–335.
Laidler, K.J. (1987). *Chemical Kinetics, 3rd ed.* Harper & Row, New York.
Levene, H.B., and Valentini, J.J. (1987). *J. Chem. Phys.* **87,** 2594.
Levine, R.D., and Kinsey, J.L. (1979). In *Atom-Molecule Collision Theory—A Guide for the Experimentalist.* Bernstein, R.B., Ed. Plenum Press, New York, pp. 693–746.
Light, J.C. (1964). *J. Chem. Phys.* **40,** 3221.
Light, J.C. (1967). *Faraday Discuss. Chem. Soc.* **44,** 14.
Light, J.C., and Lin, J. (1965). *J. Chem. Phys.* **43,** 3209.
Marcus, R.A. (1988). *Chem. Phys. Lett.* **144,** 208.
Miller, B.E., and Baer, T. (1984). *Chem. Phys.* **85,** 39.
Miller, W.H., Hase, W.L., and Darling, C.L. (1989). *J. Chem. Phys.* **91,** 2863.
Mintz, D.M., and Baer, T. (1976). *J. Chem. Phys.* **65,** 2407.
Nikitin, E.E. (1965). *Theor. Exp. Chem.* **1,** 90.
Nikitin, E.E. (1974). *Theory of Elementary Atomic and Molecular Processes in Gases.* Clarendon Press, Oxford.
Ogai, A., Brandon, J., Reisler, H., Suter, H.U., Huber, J.R., von Dirke, M., and Schinke, R. (1992). *J. Chem. Phys.* **96,** 6643.
Olzmann, M., and Troe, J. (1992). *Ber. Bunsenges. Phys. Chem.* **96,** 1327.
Pechukas, P., Light, J.C., and Rankin, C. (1966). *J. Chem. Phys.* **44,** 794.
Pechukas, P., and Light, J.C. (1965). *J. Chem. Phys.* **42,** 3281.
Peslherbe, G.H., and Hase, W.L. (1994). *J. Chem. Phys.* **100,** 1179.
Powis, I. (1979). *J. Chem. Soc. Faraday Trans.* **75,** 1294.
Powis, I. (1983). *Chem. Phys.* **74,** 421.
Powis, I. (1987). *Acc. Chem. Res.* **20,** 179.
Qian, C.X.W., Noble, M., Nadler, I., Reisler, H., and Wittig, C. (1985). *J. Chem. Phys.* **83,** 5573.

Qian, C.X.W., Ogai, A., Iwata, L., and Reisler, H. (1990). *J. Chem. Phys.* **92,** 4296.
Quack, M., and Troe, J. (1974). *Ber. Bunsenges. Phys. Chem.* **78,** 240.
Quack, M., and Troe, J. (1975). *Ber. Bunsenges. Phys. Chem.* **79,** 469.
Radzig, A.A., and Smirnov, B.M. (1985). *Reference Data on Atoms, Molecules, and Ions.* Springer-Verlag, Berlin.
Reisler, H., Keller, H.M., and Schinke, R. (1994). *Comments At. Mol. Phys.* **30,** 191.
Robie, D.C., Hunter, M., Bates, J.L., and Reisler, H. (1992). *Chem. Phys. Lett.* **193,** 413.
Rynefors, K. (1982). *J. Chem. Phys.* **77,** 6051.
Schinke, R. (1985). *Chem. Phys. Lett.* **120,** 129.
Schinke, R. (1986). *J. Chem. Phys.* **84,** 1487.
Schinke, R. (1992). *Photodissociation Dynamics.* Cambridge University Press, Cambridge.
Solter, D., Werner, H.J., von Dirke, M., Untch, A., Vegiri, A., and Schinke, R. (1992). *J. Chem. Phys.* **97,** 3357.
Steinfeld, J.I., Francisco, J.S., and Hase, W.L. (1989). *Chemical Kinetics and Dynamics.* Prentice-Hall, Englewood Cliffs, NJ.
Troe, J. (1983). *J. Chem. Phys.* **79,** 6017.
Truhlar, D.G., and Muckerman, J.T. (1979). In *Atom-Molecule Collision Theory*, Bernstein, R.B., Ed. Plenum, Press, New York, pp. 505–566.
Uggerud, E., and Helgaker, T. (1992). *J. Am. Chem. Soc.* **114,** 4265.
Untch, A., Schinke, R., Cotting, R., and Huber, J.R. (1993). *J. Chem. Phys.* **99,** 9553.
Wang, I.S.Y., and Karplus, M. (1973). *J. Am. Chem. Soc.* **95,** 8160.
Wardlaw, D.M., and Marcus, R.A. (1988). *Adv. Chem. Phys.* **70,** 231.
Weitzel, K.M. (1991). *Chem. Phys. Lett.* **186,** 490.
Wittig, C., Nadler, I., Reisler, H., Noble, M., Catanzarite, J., and Radhakrishnan, G. (1985). *J. Chem. Phys.* **83,** 5581.

10

The Dissociation of Small and Large Clusters

10.1 THE DEFINITION OF CLUSTERS

Clusters are aggregates of loosely bonded molecules, in which each of the units retains the structure that it has as a free molecule. Because of the weak interactions among the molecules, clusters are stable only in cold environments such as are found in molecular beams. The weak intermolecular bonds provide an interesting testing ground for theories of intramolecular vibrational energy redistribution (IVR) and thus for theories of unimolecular dissociation. In addition, clusters constitute the bridge between the gas and liquid phases. Such phenomena as solvation, heat capacity, and phase transitions, which are ill defined for small clusters, become progressively more precise as the cluster size increases.

Typical binding energies for neutral clusters are below 1000 cm^{-1}. Ionic clusters, because of their ion-induced dipole forces, tend to be more strongly bonded with binding energies in excess of 5000 cm^{-1}. Not infrequently, a neutral van der Waals dimer such as Ar_2 with its binding energy of about 100 cm^{-1} (Tang and Toennies, 1986) changes its character upon ionization. The equilibrium bond distance is reduced from about 4 Å to 2.43 Å (Huber and Herzberg, 1979; Ma et al., 1993) and the binding energy increases to 10,000 cm^{-1} (Norwood et al., 1989; Furuya and Kimura, 1992). Clearly, the Ar_2^+ ion no longer meets our definition of a dimer. Rather, the neutral dimer is converted into a stable ion with a bond order of 1/2. A molecule that is frequently referred to as a cluster is C_{60}. However, it is held together neither by weak bonds, nor is it composed of a collection of monomers. It is thus better classified as a large covalently bonded molecule. Table 10.1 summarizes some binding energies for various classes of dimers.

When clusters comprise several loosely bound molecules, the atoms within each molecule are held together by strong bonds while the molecules themselves are attracted to neighboring molecules by weak bonds. This discrepancy in forces translates into disparities in the respective vibrational frequencies. In the case of the HF dimer the two H—F vibrational frequencies are ca. 4000 cm^{-1} while the four van der Waals frequencies are 480, 400, 210, and 150 cm^{-1} (Quack and Suhm, 1991). The consequence of these different frequencies and forces is that energy cannot flow as freely among constituent units of the cluster as it can within each molecule. That is, there are bottlenecks in the phase space which means that IVR is inhibited. Thus, cluster dissociations, especially when energized by infrared radiation, provide some of the rare example of nonstatistical decay.

Table 10.1. Dimer Binding Energies

Dimer	Binding Energy (cm^{-1})	Reference
$(C_2H_4)_2$	374	(Tsuzuki & Tanabe, 1992)
$(C_2H_4)_2^+$	6,400	(Jungwirth & Bally, 1993)
$(HF)_2$	1,062	(Bohac et al., 1992a)
$Ar—CO_2$	166	(Bohac et al., 1992b)
He—HF	7.1	(Lovejoy & Nesbitt, 1990)
$(C_6H_6)_2$	560	(Krause et al., 1992)
$(C_6H_6)_2^+$	5300	(Grover et al., 1987)
$(NO)_2$	710	(Hetzler et al., 1991)
$C_6H_6—Ar$	425	(van der Avoird, 1993)

10.1.2 How Are Clusters Observed?

Clusters are generally observed in molecular beam expansions. The low-temperature characteristic of this environment permits the condensation of molecules into dimers, trimers and, if the cooling is sufficient, veritable snowballs with thousands of monomer units. These broad distributions of clusters in a beam cause major problems in detection. While it is possible to control the expansion so that dimers are the dominant cluster (in the presence of a large excess of monomers), the trimers can only be studied in the presence of an excess of dimers, etc. Thus a major experimental problem in the study of neutral clusters is the detection of a particular cluster in the presence of many other sized clusters.

The most common method for detecting clusters is by mass spectrometry. The molecular beam is ionized (often with a laser) one or more centimeters from the nozzle and the resulting ions are mass analyzed. At very low temperatures achieved by seeding the sample in a large excess of a rare gas, or when the clusters are strongly bound because of dipolar interactions, very large clusters can be produced. An example of a mass spectrum of antimony clusters is shown in figure 10.1. Both singly and doubly charged cluster ions are observed.

Because the stabilities and structures of ions and neutrals are often very different, vertical ionization produces vibrationally excited ion cores whose energies often exceed the monomer binding energy. Hence, dissociative ionization processes result in ion cluster distributions that can be very different from the distribution in the neutral beam. The Sb^+ and Sb^{+2} distributions in figure 10.1 are a result of several competing processes including evaporation of single neutral atoms and fission of doubly charged ions into two singly charged ions (Brechignac et al., 1995). Fission of the doubly charged ions into singly charged ions is energetically favored for the lower cluster sizes, while evaporation of neutral Sb atoms is favored for the higher cluster sizes. Hence, the doubly charged ions are observed only for the higher cluster sizes.

Other examples of dissociative ionization, but at lower excitation energies, are shown for the clusters of methanol and methylchloride in figure 10.2 (Booze and Baer, 1992). Both of these "dimer" ions were produced by photoionization at the lowest possible energies consistent with a measurable signal level. Yet, dissociative ionization is apparently the dominant process as is evident from the broad TOF peaks (see chapter 5). Only the small sharp TOF peaks in figure 10.2 are the result of nondissociative

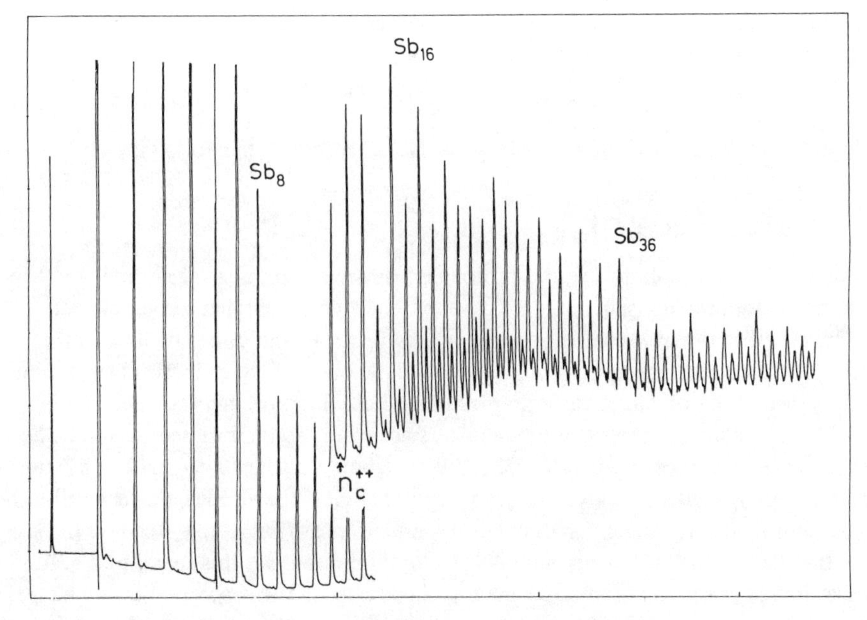

Figure 10.1 The mass spectrum of antimony (Sb) clusters formed by electron impact ionization of the cluster beam several centimeters from the nozzle. Both singly and doubly charged cluster ions are observed. Taken with permission from Brechignac et al. (1995).

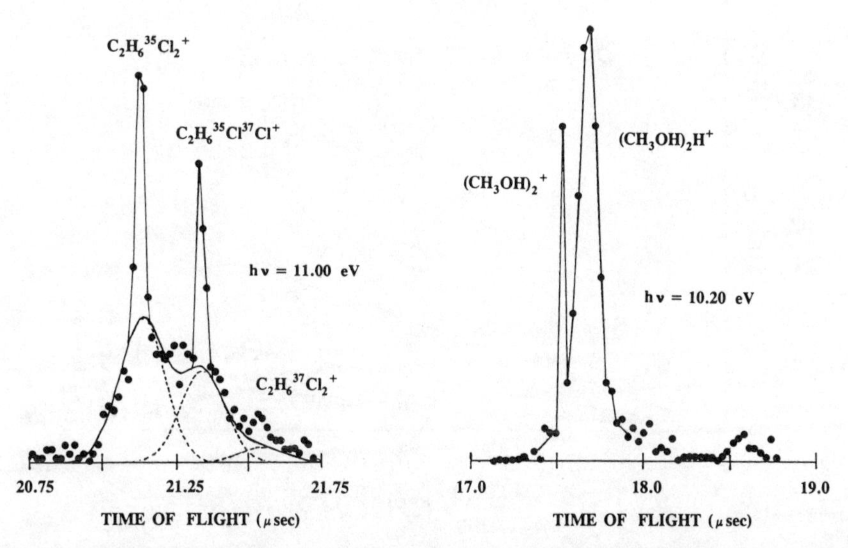

Figure 10.2 Photoelectron photoion coincidence (PEPICO) time of flight spectra of methanol and methylchloride clusters. Sharp peaks are observed if a cluster ionizes without dissociation. The broad peaks are a result of kinetic energy release in dissociative ionization. Taken with permission from Booze and Baer (1992).

ionization. The broad $(CH_3OH)_2H^+$ peak is a result of an intermolecular proton transfer reaction in the methanol trimer:

$$(CH_3OH)_3 + h\nu \rightarrow (CH_3OH)_2H^+ + CH_2OH + e^- \tag{10.1}$$

while the broad methylchloride ion peak is the result of the following dissociative ionization process:

$$(CH_3Cl)_3 + h\nu \rightarrow (CH_3Cl)_2{}^+ + CH_3Cl + e^-. \tag{10.2}$$

Although it is possible to use the protonated dimer signal as a measure of the trimer neutral (assuming no sequential decays of energy rich higher order clusters), the methylchloride dimer signal is a combination of trimer and dimer neutral so that care must be exercised by distinguishing the two on the basis of their kinetic energy. Similar results have been obtained for argon clusters (Furuya and Kimura, 1992).

An approach for separating the clusters according to their masses has been developed by Buck and co-workers (1988, 1990). When a cluster beam with a mixture of cluster sizes is scattered with a He beam at right angles, the various clusters will have a maximum scattering angle that depends upon their mass (fig. 10.3). Because the signal as a function of the scattering angle has a slight maximum at this maximum scattering angle, it is possible to differentiate neutral cluster sizes. The separation is much cleaner than appears in figure 10.4 because time of flight analysis separates the lower from the higher mass clusters. That is, at an angle of for instance 9°, the methanol dimer will appear as a single TOF peak while the monomers appear as forward and backward scattered TOF peaks which are situated to either side of the monomer peak. By this method it was first established that the methanol clusters, upon electron impact ioniza-

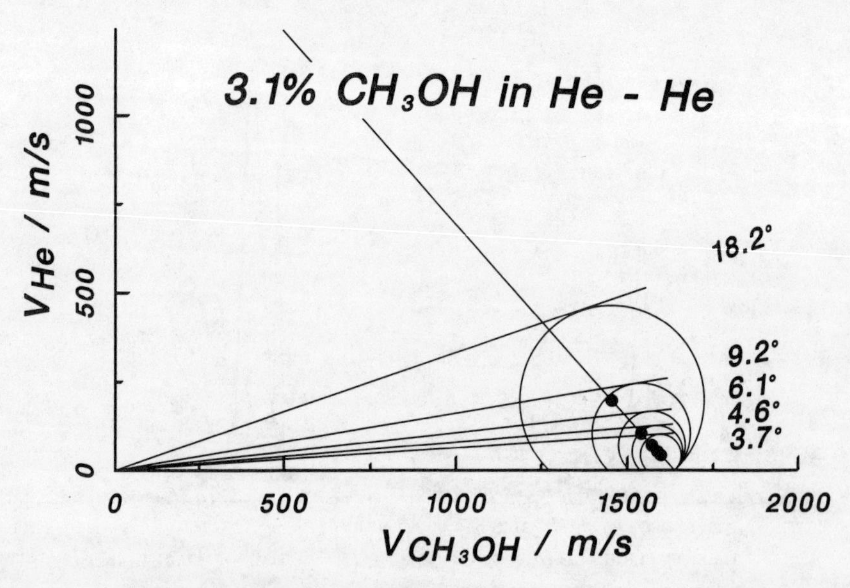

Figure 10.3 The Newton diagram for scattering of a $(CH_3OH)_n$ cluster beam with He. The positions of the elastically scattered clusters $n = 1$–5 are indicated by the limiting angles in the laboratory system. The limiting angles are 18.2° for CH_3OH monomers, 9.2° for dimers, etc. Taken with permission from Buck et al. (1990).

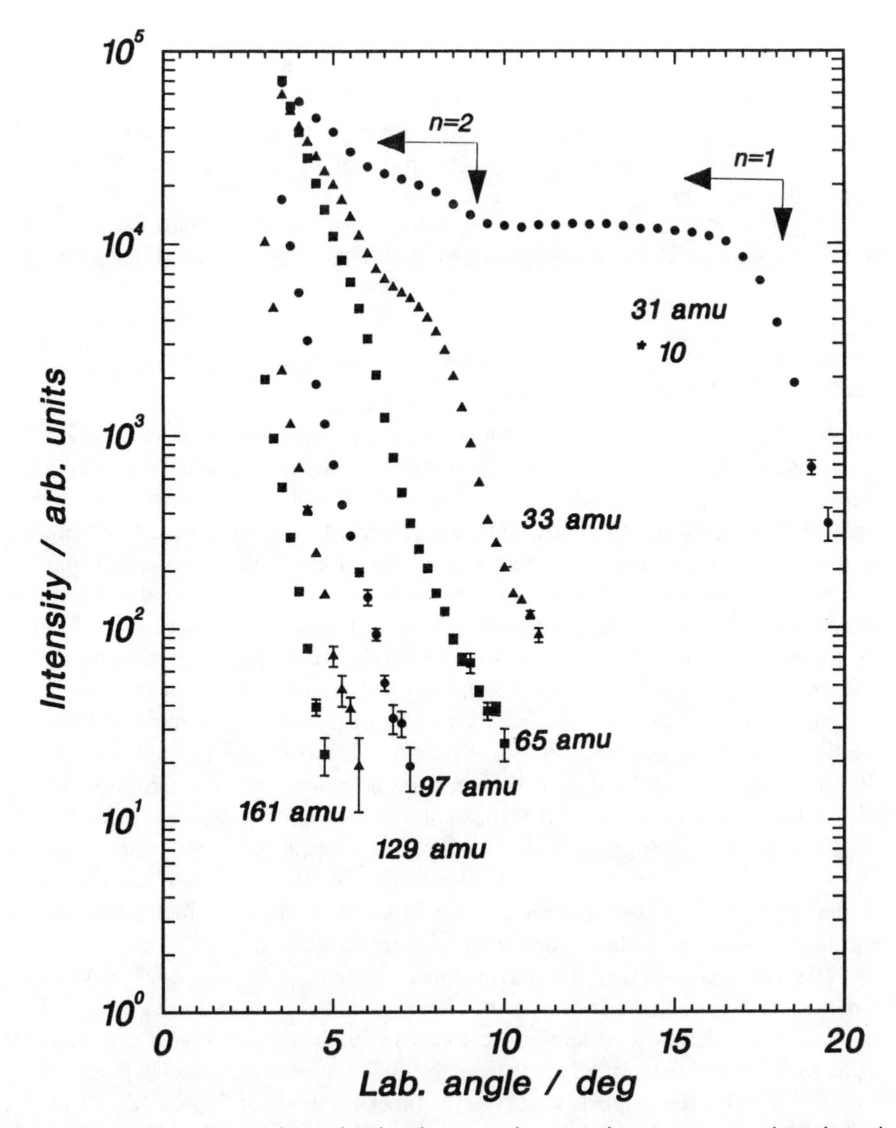

Figure 10.4 The measured angular distributions of scattered ion masses resulting from the electron impact ionization of the scattered neutral methanol clusters. The onset of the H loss fragment ion from the methanol monomer is shown at the limiting angle for the monomer (18.2°). Similarly, the dimer onset, producing the protonated monomer is shown at 9.2°, while the protonated dimer peak produced by reaction (1) is evident with a weak onset at 6.1°. Taken with permission from Buck et al. (1990).

tion, mostly dissociate (fig. 10.4). In particular, trimer ionization resulted predominantly in the production of the protonated dimer ion, in agreement with the PEPICO results of figure 10.2 and Eq. (10.1).

The best method to differentiate various sized clusters is spectroscopically. Not only can different sized clusters be selected, but even different conformations of the same mass cluster can be investigated. This approach presupposes that the resolution is sufficiently high and that the cluster spectra have been interpreted so that the spectrum provides a positive signature of the structure. In this fashion structures have been determined for various conformations by microwave spectroscopy (Nelson et al., 1985; Fraser et al., 1986; Klots et al., 1989; Ruoff et al., 1990). The structure and lifetimes of selected dimers and trimers have also been measured by infrared (Miller, 1990; Dayton et al., 1990; Riehn et al., 1992b; Nesbitt and Lovejoy, 1990; Farrell et al., 1992) and electronic spectroscopy (Mons et al., 1990; Brutschy, 1990; Riehn et al., 1992a).

10.1.3 How Do Clusters Dissociate?

Many small cluster have been investigated by IR photodissociation in which the chromophore is a high-frequency mode of one of the monomer units. Such spectra of small dimers and trimers are characterized by sharp absorption peaks, which is a consequence of the long cluster lifetimes. Thus it is possible to prepare the clusters in highly localized and well characterized states which lie well above their dissociation limits. The lifetimes of numerous IR excited dimers have been measured from the absorption peak widths. The interpretation of peak widths in terms of dissociation rates rather than, for instance, IVR is not without its problems because both processes give rise to broadened absorption peaks (Miller, 1990; Uzer and Miller, 1991). However, there are some characteristic signs of IVR, which can be observed at low energies (infrared excitation) in systems with few degrees of freedom. If broadened lines are the result of IVR, "dark" states should appear with intensities borrowed from the zero-order bright states. Although such dark states have been observed in a few cases with IR excitation (Jucks and Miller, 1988; Lovejoy et al., 1986), they appear to be rare in small dimers such as HCN—CO_2 (Miller, 1990). Furthermore, the line widths are not correlated with the appearance or nonappearance of the dark states. Thus, the dark states are not the primary cause for the line broadening in these small system.

In the electronic excitation of larger dimers, such as *p*-difluorobenzene-Ar (Hye-Keun et al., 1988; Kelley and Bernstein, 1986), or stilbene-He (Semmes et al., 1990) there is evidence that IVR is rapid and competes with the dissociation rate so that real time measurements are required to measure the dissociation rate. In fact in the stilbene-He dimer, the absorption peak widths gave dimer lifetimes of 2 psec (Zwier et al., 1983; DeHaan et al., 1989) whereas the real-time lifetime measurements are about 40 psec (Semmes et al., 1990).

Excited dimers, with the energy placed initially in monomer vibrational modes, decay via one of three mechanisms: (1) direct vibrational predissociation (VP) at a rate k_{vp} in which the energy from the bright state is transferred directly to the van der Waals bond followed by rapid dissociation; (2) sequential IVR (rate $= k_{ivr}$) in which the energy becomes equilibrated among the dimer normal modes including the van der Waals modes, followed by statistical dissociation of the equilibrated complex via the rate k_d; and (3) a combination of IVR and VP in which the two proceed in a parallel fashion:

$$(1)\ AB + h\nu \rightarrow AB^* \xrightarrow{k_{vp}} A + B \qquad \text{predissociation (VP)}$$

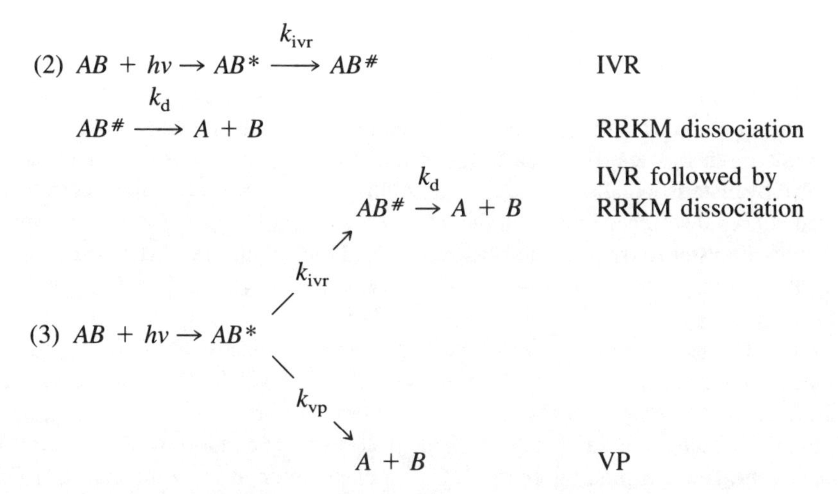

The use of the term vibrational predissociation in the above sense is not universal. Herzberg (1967) and others (Kelley and Bernstein, 1986) use VP and statistical RRKM dissociation interchangeably. The suggested use here makes the connection between vibrational and electronic predissociation stronger. In both cases, the reaction can be modeled via coupling of only two or three potential energy surfaces. The full density of vibrational states is not considered in such predissociation models.

10.2 EXAMPLES OF VAN DER WAALS DIMERS DISSOCIATION

It is useful to categorize the various classes of dimers and clusters before discussing the theoretical approaches and models for understanding the cluster dissociation rates and energy partitioning. Clusters fall into three broad classes:

Class A: clusters of small neutral atoms and molecules
[HCl-Ar, $(N_2O)_2$, HF-HCCH, $(H_2O)_2$, $(HF)_2$, I_2-He, etc.];
Class B: clusters containing large polyatomic molecules
[*p*-difluorobenzene-*X*, aniline-X, stilbene-*X*, where *X* = rare gas. CH_4];
Class C: ionic clusters
[$Cs^+(CH_3OH)_n$, $[N_2O\text{-}NO]^-$, $(CO_2)_n^+$, Ar-$CH_3COCH_3^+$, etc].

It is helpful in certain cases to further subdivide these clusters into closed-shell and open-shell species. This is generally not an issue in neutral clusters so far studied because most are closed shell, except for some notable exceptions such as: ICl—Ne (Drobits and Lester, 1988), OH—Ar (Berry et al., 1990), and benzyl-*X* (Disselkamp and Bernstein, 1993). On the other hand, it is important in ionic clusters because an ionic cluster generated from a neutral cluster is an open-shell species. These tend to rearrange into stable clusters with strong bonds that often do not resemble their neutral counterparts. By contrast, ionic closed shell clusters such as $Cs^+(X)_n$ are expected to be similar in structure and energy to the isoelectronic neutral cluster $Xe(X)_n$.

Class A clusters are characterized by weak bonds (D_o = 200–1000 cm^{-1}) and low density of states. When the high-frequency monomer modes are excited with infrared radiation these dimers tend to decay via vibrational predissociation. IVR is inhibited because of the low density of states in the high-frequency molecular modes and the weak coupling between the molecular and the van der Waals normal modes.

Class B clusters are also weakly bound, but their constituent molecular units have large density of states with low vibrational frequencies which approach those of the van der Waals modes. Thus, these clusters are much more likely to decay via rapid IVR followed by a statistical dissociation. At least this is their behavior on the electronically excited potential surface. Studies in the infra-red have not been carried out. Among the well studied systems are *p*-difluorobenzene—*X* (Tiller et al., 1989; Hye-Keun et al., 1988; Butz et al., 1986) and stilbene—*X* dimers, where *X* is a rare gas (Semmes et al., 1990; Khundkar et al., 1983; DeHaan et al., 1989).

Class C clusters (ions) tend to have strong bonds in excess of 2000 cm^{-1} if they are open shell. An example is the benzene neutral dimer which is bound by 565 cm^{-1}, while the benzene ionic dimer is bound by 5300 cm^{-1} (Grover et al., 1989; Krause et al., 1991). On the other hand, closed-shell species are much more weakly bound. For instance, the average binding energy of a methanol monomer to the $Cs^+(CH_3OH)_n$ cluster is below 1000 cm^{-1} (Liu and Lisy, 1988; Draves et al., 1990).

The following examples illustrate a number of the interesting properties of dimer dissociations.

10.2.1 $HCN\text{-}C_2H_2$

HCN—acetylene dimer is formed when the monomers and a seed gas (typically in the ratio of 1:1:98 in order to avoid higher order clusters) are expanded through a small nozzle. The analysis of high-resolution IR spectra showed the presence of two conformations of the HCN--HCCH dimer (Block et al., 1989):

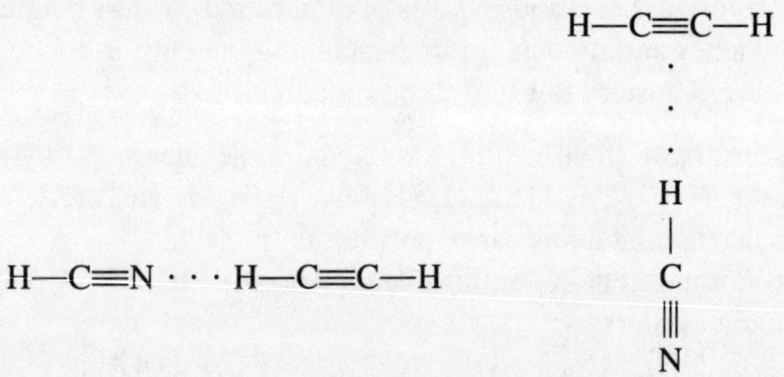

H—C≡N · · · H—C≡C—H

H—C≡C—H
⋮
H
|
C
|||
N

The vibrational frequencies, determined by *ab initio* molecular orbital calculations, of these two dimers along with the frequencies of the monomer unit are shown in table 10.2 (Block et al., 1989).

Because of anharmonicity, these calculated frequencies (6-31G**/MP2 level) are about 5% higher on the average than the experimentally determined frequencies. Because this error is fairly constant, the calculated shifts between the monomer and dimer frequencies (shown in parenthesis) are accurate to within a few wavenumbers. For instance, the measured shifts for the HCN and acetylene stretch frequencies in the monomer and dimer are -1 and -24 cm^{-1} ($+1$ and -27 cm^{-1} calculated) for the linear isomer; and -47 and -11 (-44 and -10 cm^{-1} calculated) for the T-shaped isomer.

The table of frequencies demonstrates a number of the characteristics of van der

Table 10.2. Calculated[a] Vibrational Frequencies of Linear and T Shaped HCCH-HCN Dimers and Monomers [from Bløck et al. (1989)]

Linear Dimer	T-Shaped Dimer	HCN	HCCH
41	67		
41	104		
105	105		
125	135		
125			
500 (+42)[b]	473 (+15)[b]		458
500 (+42)	483 (+25)		458
733 (−6)	755 (+16)	739	
733 (−6)	769 (+30)	739	
842 (+87)	842 (+87)		755
842 (+87)	851 (+96)		755
1997 (−48)	1998 (−47)	2045	
2054 (+51)	2047 (+44)		2003
3476 (−27)	3493 (−10)		3503[c]
3534 (+1)	3489 (−44)	3533[c]	
3577 (−16)	3583 (−10)		3593

[a]Calculated by *ab initio* calculations with 6–31G**/MP2 basis set. Uncorrected for anharmonicity effects.

[b]Frequency shifts between dimer and monomer.

[c]These are the active IR frequencies excited.

Waals complexes. One is that the vibrational frequencies of each monomer change little upon dimer formation. A similar statement holds for the bond distances and angles. Second, there is often a considerable gap between the frequencies of the monomer units and the van der Waals frequencies. Finally, it is interesting to compare the C—H frequencies and their shifts in the linear and T-shaped dimers. In the linear isomer, the C—H frequency of the HCN unit is "free," while the acetylene C—H unit is "bound" as it forms part of the van der Waals bond. This difference in the C—H environment is reflected in the frequency shifts. That is, the HCN frequency is shifted by only 1 cm^{-1}, while the "bound" acetylene frequency is shifted by 27 cm^{-1}. In the T-shaped isomer, it is the other way around in that the acetylene C—H stretch is "free" while the HCN stretch is "bound." (Apparently, the acetylene C—H stretch is not totally free in the T-shaped dimer as its shift is still 10 cm^{-1}.)

The binding energies of the HCN--HCCH complexes have been calculated to be 775 and 813 cm^{-1} for the linear and T-shaped isomers, respectively (Block et al., 1989). Thus the excitation of the C—H stretch at about 3300 cm^{-1} is more than sufficient to dissociate the dimer. An interesting property of this dimer is that the C—H stretch frequencies for the acetylene and HCN unit are very similar so that the complex can be excited in a mode selective manner with essentially the same total energy content. The results are shown in table 10.3.

The fact that the dimer lifetimes, or dissociation rates, are mode dependent is the first indication that this dissociation violates one of the basic assumptions of the statistical theory, namely that the decay rate depends only upon the total energy and

Table 10.3. Lifetimes for Mode-Selected HCN-HCCH Dimers.

C-H Mode Excited	Spectral Shift[a]	Dimer Lifetime[b]	RRKM[c]
Free HCN (linear)	−1.2	200 μsec< τ >160 nsec	2 psec
Bound HCCH (linear)	−24.5	1.14 nsec	2 psec
Bound HCN (T-shaped)	−46.9	1.06 nsec	2 psec
Free HCCH (T-shaped)	−10.8	200 μsec< τ >160 nsec	2 psec

[a]Measured shifts of CH stretch in the 3300 cm^{-1} region (Block et al., 1989).

[b]Lifetimes measured from absorption peak widths (Block et al., 1989).

[c]RRKM lifetime based on $k(E)$ at 3300 cm^{-1} with D_o = 775 cm^{-1}.

angular momentum, and not upon the manner of state preparation. A second discrepancy is between the statistically expected (RRKM) and the observed lifetimes shown in table 10.3. The RRKM rate was calculated by using the set of vibrational frequencies in table 10.2 (multiplied by a common factor of 0.95 to take into account the consistently high calculated values). The transition-state frequencies were assumed to be the same as the dimer frequencies except that one frequency (119 cm^{-1}) was dropped. The transition state was thus somewhat "tight." No reasonable adjustment of dissociation limits or vibrational frequencies can bridge this gap between theory and experiment which differs by orders of magnitude. Thus, we conclude that the dimer dissociation rates cannot be explained in terms of RRKM theory.

The reason for the inapplicability of the statistical theory goes to the very heart of the statistical assumption, namely that the energy flow in the dimer is not random. That is, there is evidently a dynamical barrier that prevents free energy flow from one mode to another. It is not a matter of an insufficient density of states. Even ignoring the rotations, at 3300 cm^{-1} the density of states is on the order of 10^5 /cm^{-1} and the sum of states of the transition state is 10^6 states. These numbers exceed those of H_2CO (near its dissociation limit) which is characterized by a statistical dissociation (Polik et al., 1988) (see chapter 6). Nor is the calculated rate too fast to be statistical. There are many examples of statistical dissociations in which the rates are in the ps region. This is truly a case of a long-lived molecule with nonstatistical energy flow. Whether this reaction dissociates by vibrational predissociation (VP) or by IVR followed by a fast statistical dissociation is not known with certainty because product state analysis has not been done. However, the absence of any dark states in the spectrum from the van der Waals modes suggests that IVR is not taking place. This reaction has all of the characteristics associated with direct vibrational predissociation, that is, mechanism 1.

Similar nonstatistical decays have been observed for numerous neutral dimers, including the $(NO)_2$ dissociation which was investigated in real time by a pump-probe scheme (Casassa et al., 1986). The dimer lifetime increased from 39 to 880 psec as the photodissociation energy increased from 1789 to 1870 cm^{-1}. Because of the limitations imposed by the picosecond pulses, J levels could not be resolved. Most findings of nonstatistical decay have been in neutral systems, in part because dimer ions have not been investigated with as much precision. One example of an ionic cluster dissociation which may be nonstatistical is $(N_2)_n^+$ (Leisner et al., 1988). It has been observed that meta-stable N_2 cluster ions dissociate by evaporating three or four neutral nitrogen molecules. The explanation is that vibrational energy residing in the N_2 monomers is transferred directly to the van der Waals modes which then leads to the evaporation of three or four monomer units. The implication is that energy in the N_2 vibrational modes is slow to transfer to the van der Waals modes. The lifetime of these clusters is then determined by the rate of energy transfer from the N_2 vibrations to the rest of the cluster. Whether this is a case of VP (mechanism 1) or IVR followed by a rapid RRKM like dissociation (mechanism 2) is not known.

10.2.2 The *Trans*-Stilbene—He Complex

Uncomplexed *trans*-stilbene is an important molecule in the field of unimolecular reactions, because it is one of the few reactions which takes place on an excited potential energy surface and which is statistical. This is the famous *trans* → *cis* isomerization which is a very rapid reaction on the excited potential energy surface with a greatly reduced barrier of 1300 cm^{-1}. The rates have been measured by several workers (Troe, 1985; Syage et al., 1982; Majors et al., 1984) in the energy range from 1300 to 3000 cm^{-1}, and found to be approximately consistent with the statistical RRKM theory (see figure 5.18). In addition, IVR in the excited S_1 state has been extensively investigated (Felker and Zewail, 1988; Felker et al., 1985; Perry et al., 1983). Dissipative IVR becomes significant at about 1230 cm^{-1} which is very near the barrier for isomerization. Below 600 cm^{-1} IVR is essentially absent in pure *trans*-stilbene (Felker et al., 1985).

The binding energy of the He atom to *trans*-stilbene in the excited state is only 47 cm^{-1}, so the dimer dissociation occurs at an energy well below the *trans–cis* isomerization reaction and well below the onset of IVR in the stilbene molecule (DeHaan et al., 1989). The weakness of the binding of He to stilbene is reflected in the $S_o \rightarrow S_1$ transition origin which is shifted by only -6.5 cm^{-1} relative to the monomer stilbene (DeHaan et al., 1989). Despite this weak interaction IVR, although incomplete, is much more pronounced in the dimer than in the stilbene monomer, and takes place even at very low energies of excitation. This is particularly the case when out of plane bending frequencies in the stilbene molecule are excited. This is understood in terms of the dimer geometry in which the He atom is located over the phenyl rings and can move freely between the two rings. The three low van der Waals mode frequencies of 9, 19, and 48 cm^{-1} are strongly coupled to the out of plane bends which causes inhomogeneous broadening of the transitions, some as wide as 3 cm^{-1} (Semmes et al., 1990).

The dissociation rates of stilbene—He have been measured in real time as a function of the energy above the dissociation limit (Semmes et al., 1987, 1990). The states accessed are various combinations of the following three vibrational modes: ν_{25}(198 cm^{-1}), ν_{36}(35 cm^{-1}), and ν_{37}(48 cm^{-1}). They are nearly the same value in the monomer stilbene as in the stilbene—He complex. However, some of them overlap with the monomer bands so that dispersed fluorescence spectra are not pure. Figure 10.5 shows several time resolved fluorescence spectra of both reagent and product states. The reagent in this case is the dimer, while the product is the pure stilbene. State to state dissociation rates are obtained by choosing the excitation and emission frequency. Because stilbene and the stilbene—He dimer have slightly shifted spectra, the reaction can be followed either by monitoring the decay of the reagent (top two traces) or the appearance of the product (lower traces). It was found that the rates, which do not vary smoothly with excitation energy, are highly mode specific. The decay rate

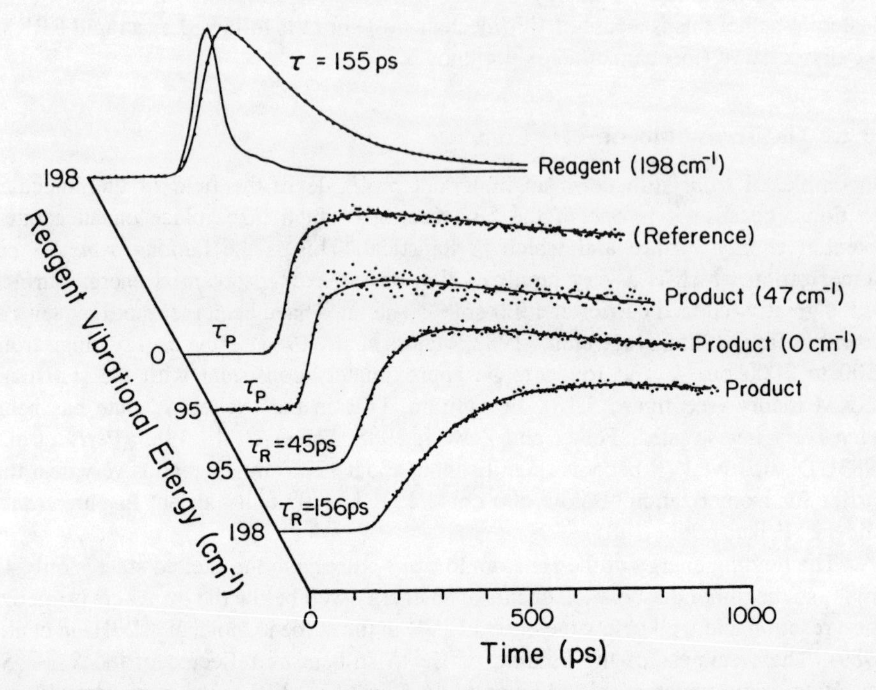

Figure 10.5 Picosecond time resolved emission after photodissociation of *trans*-stilbene-He dimers. Excitation at 198 cm^{-1} is followed by both decay of the reagent (top trace) and the appearance of product stilbene (bottom trace). The reference trace at zero energy is of the dimer, which is characterized by an instrumentally limited rise time of 20 ps and the fluorescence decay (2.67 ns) of the dimer. At an excitation energy of 95 cm^{-1}, the rate is monitored by the appearance of ground state stilbene and vibrationally excited stilbene. The excited product is formed more rapidly than can be measured, while the ground state product is formed slowly (45 ps). Furthermore, the rates are faster at 95 cm^{-1} than they are at 198 cm^{-1} excitation energies. These findings show that this reaction is not statistical. Taken with some modification from Semmes et al. (1987).

upon excitation of $2\nu_{37}$ at an energy of 95 cm^{-1} is three times greater than when the ν_{25} mode is excited at 198 cm^{-1}.

What determines this mode specificity? Is this rate limited by the IVR rate, or by the dissociation rate? Detailed LIF studies of this molecule have shown that IVR and dissociation are strongly coupled (DeHaan et al., 1989). Because of the low dissociation energy of 47 cm^{-1}, IVR is not complete and is strongly mode dependent. The observation of different rates for the production of ground and vibrationally excited stilbene at an excitation energy of 95 cm^{-1} is particularly telling. It implies that the reagent decay would be bi-exponential. However, as a result of spectral congestion, the reagent decay could not be measured (Semmes et al., 1990). Nevertheless, the 95 cm^{-1} data demonstrate that the reaction proceeds by branching toward two paths which do not communicate during the course of the reaction (see section 7.7.3). The mechanism that can account for these findings is probably mechanism 3, in which the VP and IVR are in competition. If IVR is irreversible, then a two component decay of the reagent can result.

An RRKM calculation of the stilbene-He decay rate gives lifetimes at 95, 198, and 396 cm^{-1} of 10, 3, and 2 psec (Semmes et al., 1990). These rates, although somewhat faster, are nevertheless within an order of magnitude of the measured rates. Thus, we see that the stilbene—He decay is complex in which IVR, vibrational predissociation, and possibly statistical RRKM dissociation following IVR may all be important in determining the dissociation rates.

Somewhat less information is available for the complex between stilbene and Ne. However, it is interesting that appearance of the stilbene product after excitation of the ν_{25} state at 198 cm^{-1}, can only be fitted by assuming a 233-psec rising component and a 63-psec decaying component. This is rather convincing evidence that the sequential mechanism 2 is operative. The slow IVR [$k_{ivr} = 1/(255 \text{ psec})$] is followed by a faster dissociation rate [$k_d = 1/(65 \text{ psec})$], which agrees well with an RRKM calculation (Semmes et al., 1990).

10.3 THEORY OF VIBRATIONAL PREDISSOCIATION IN SMALL DIMERS

A number of simple models for dimer dissociation rates have been developed, among them the momentum gap model (Ewing, 1987) and the Miller-LeRoy frequency shift model (Miller, 1988; LeRoy et al., 1991). Since they are based on approximations of the more exact theory, we begin with the latter and show how the two simple models are related.

The dimer is approximated as a strongly bonded diatomic molecule which is loosely bonded to an atom. The diatom vibration is generally treated as a harmonic oscillator, while the van der Waals bond is described by various functional forms including a simple Morse potential (Ewing, 1987). However, in order to account for the importance of product rotations, three-dimensional, anisotropic interaction potentials must be used. These are usually model potentials in which the parameters have been fitted to measured experimental quantities such as the spectrum of the binary complex, pressure broadened absorption lines, second virial coefficients in mixtures of the two gases, and/or molecular beam differential scattering cross sections. Figure 10.6 shows both a three-dimensional and an equipotential contour plot for the "T"-shaped Ar···HCCH van der Waals potential with the acetylene unit in its ground vibrational state (Bemish et al., 1993). This dimer has only two van der Waals modes, which are the stretch and the bend. The potential surface in the two-dimensional cartesian coordinates is very shallow and shows that the Ar atoms executes wide amplitude motion which extends to within 30° of the linear conformation. The double minimum appears to be characteristic of all "T"-shaped systems. In the case of Ar···HCCH, the zero-point level lies above the barrier between the two wells.

Three-dimensional theoretical treatments which take into account the angular dependence on the dimer dissociation rate have been carried out (Truhlar, 1990; Halberstadt et al., 1987, 1992; Roncero et al., 1990; Tiller et al., 1989; Clary, 1992). However, considerable physical insight can be derived by ignoring the angular dependence, and treating the dissociation in terms of just two coordinates, R (the van der

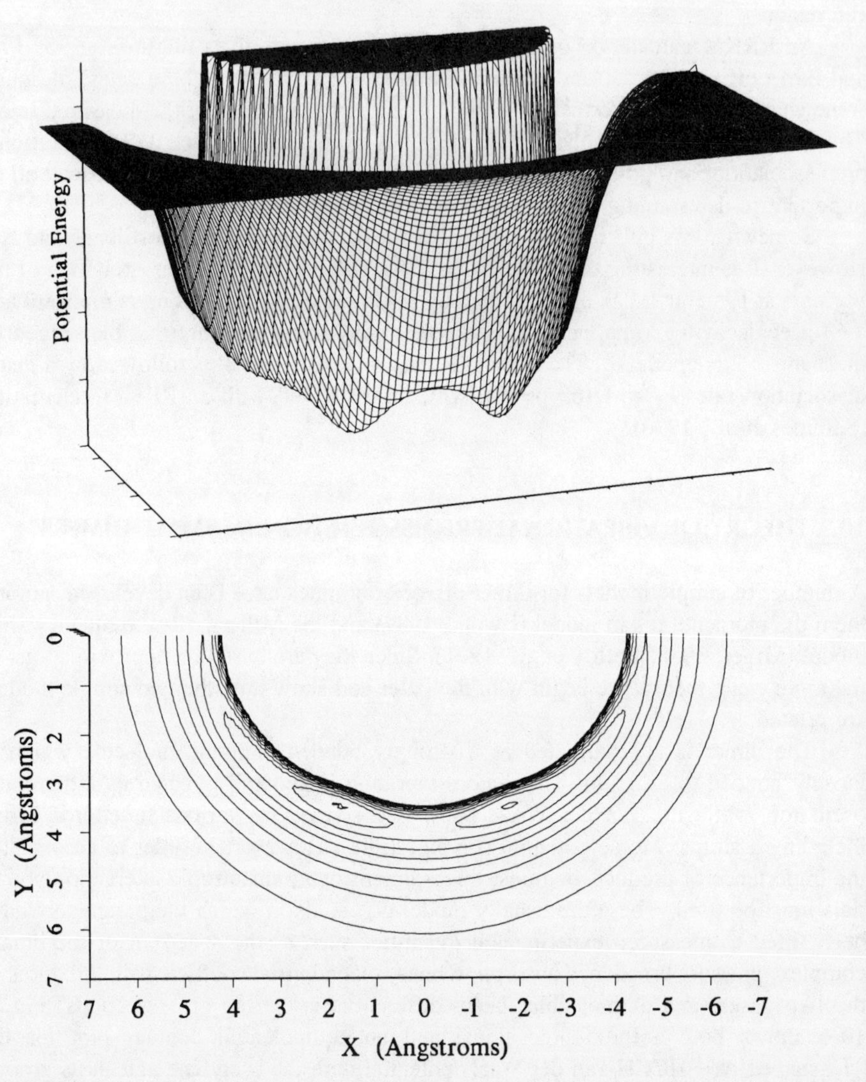

Figure 10.6 A 2-D potential energy surface and the corresponding contour plot of the "T" shaped conformation of Ar-HCCH in the acetylene ground vibrational energy level. This X-Y coordinate plot shows the wide amplitude motion of the Ar atom around the acetylene unit. Taken with permission from Bemish et al. (1993).

Waals bond coordinate) and r or q (the diatom bond coordinate). That is, the X coordinate in the Ar—HCCH potential of figure 10.6 is ignored. Figure 10.7 is a schematic illustration of the two potentials, $U_m(q)$ and $V_v(R)$ taken from Le Roy et al. (1991). Shown in this figure are also the various vibrational wave functions which enter into the problem. On the left, are the wave functions for the unperturbed diatomic in the $v'' = 0$ and $v' = 1$ states, while on the right are two van der Waals potential energy curves, the lower one for the ground state of the diatom ($v'' = 0$), and the upper curve

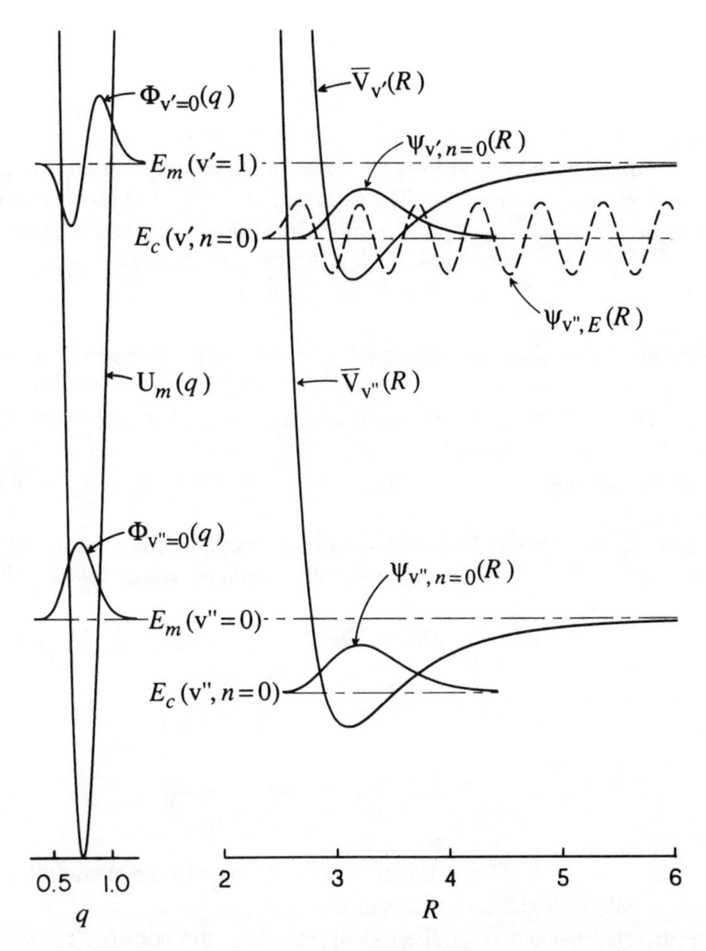

Figure 10.7 Schematic illustration of the potential energy curves and wave functions for an atom-diatom system governing the vibrational predissociation process. q and R are the coordinates for the diatom and the van der Waals bonds, respectively. The angle between the diatom and atom is held fixed. Taken with permission from LeRoy et al. (1991).

for the vibrationally excited diatom ($v' = 1$). Dissociation takes place when the vibrationally excited diatom transfers its energy to the van der Waals bond and converts the $\Psi_{v',n=0}(R)$ wave function into the $\Psi_{v'',E}(R)$ wave function, where n is the vibrational quantum number for the van der Waals mode. The final $\Psi_{v'',E}(R)$ wave function is a product of the vibrational ground state wavefunction and a quasi-free wave function of the departing fragments. In what follows, we summarize the theory as developed by Le Roy et al.(1991)

The Hamiltonian for this problem consists of three terms:

$$H(q, R) = H_m(q) + H_c(R) + \Delta H(q, R) , \tag{10.3}$$

where the first term is the Hamiltonian for the free diatom. Its eigenfunctions are $\Phi_v(q)$ and its eigenvalues are $E_m(v)$. The second term in Eq. (10.3) is for the van der Waals bond and is given by

$$H_c = -\frac{\hbar^2}{2\mu}\frac{d^2}{dR^2} + V_0(R) , \tag{10.4}$$

in which the interaction potential V_0 is for a frozen monomer potential which is independent of the monomer vibrational state. Its eigenfunctions are the zeroth-order van der Waals bond wavefunctions. The final term in the Hamiltonian couples the r and R motions. In its simplest form, it can be expressed as a product of two functions:

$$\Delta H(q, R) = f_1(q)V_1(R) . \tag{10.5}$$

As such it can be considered as the second term in an expansion of which the first term is the $V_0(R)$ in Eq. (10.4). It is this coupling term that drives the dissociation. Without it, the upper ($v' = 1$) van der Waals potential would be identical to the lower one ($v'' = 0$); the total wave function would be simply the product, $\Psi(q, R) = \Phi(q)\psi^{o}(R)$; and there would be no frequency shift in the $v'' \rightarrow v'$ absorption between the monomer and complexed diatom.

We can account for the dissociation with separable Schrödinger equations for the two coordinates if we construct a vibrationally averaged van der Waals potential:

$$\bar{V}_v(R) = V_0(R) + \langle \Phi_v(q)|f_1(q)V_1(R)|\Phi_v(q)\rangle \tag{10.6}$$

$$\approx V_0(R) + \langle f_1\rangle_{v,v}V_l(R) \tag{10.7}$$

in which,

$$\langle f_1\rangle_{v',v'} \equiv \langle \Phi_{v'}(q)|f_1(q)|\Phi_{v'}(q)\rangle \tag{10.8}$$

and a similar average is defined for the lower v'' state. In the averaging process, the integration is only over the q coordinate, and $\langle f_1\rangle_{v,v}$ is just a number. It is only in this average sense that the potential energy curves, $\bar{V}_v(R)$ are constructed in figure 10.7.

With this result for the averaged van der Waals potential energy curves, something can already be learned about the frequency shift in the diatom transition. The difference in the absorption frequency between the free and complexed diatom is equal to the difference in the binding energy of the van der Waals bond for $v'' = 0$ and $v' = 1$. This difference is given by

$$\Delta\bar{V}_{v',v''}(R) \equiv \bar{V}_{v'}(R) - \bar{V}_{v''}(R)$$

$$\approx [\langle f_1\rangle_{v',v'} - \langle f_1\rangle_{v'',v''}]V_1(R) . \tag{10.9}$$

The difference in the ground state energies between the $v' = 1$ and $v'' = 0$ van der Waals curves can now be derived from first order perturbation theory. If it is assumed that $\langle \Psi_{v',n}|V_1(R)|\Psi_{v',n}\rangle = \langle \Psi_{v'',n}|V_1(R)|\Psi_{v'',n}\rangle$, then the $v' \leftarrow v''$ shift between any two van der Waals vibrational levels, n, is given by:

$$\nu(v' \leftarrow v'') \propto [\langle f_1\rangle_{v',v'} - \langle f_1\rangle_{v'',v''}]\langle \Psi_{v',n}|V_1(R)|\Psi_{v',n}\rangle . \tag{10.10}$$

Thus, we conclude that upon forming the van der Waals dimer, the shift in the diatom absorption frequency is directly proportional to the coupling term, $V_1(R)$.

10.3.1 The Vibrational Predissociation Rate

Before deriving the vibrational predissociation rate, we need to specify more precisely the functions, $f(q)$ and $V_1(R)$. Although they depend very much on the system in question, a common feature is that they are small compared to $V_0(R)$. Consider the potentials shown in figure 10.8 for the case of Ar-HCl (Hutson, 1984). The van der Waals potential for $v'' = 0$ (solid lines) and $v' = 1$ (dashed lines) are very similar so that the coupling term, $V_1(R)$, is small. Not surprisingly, the perturbation depends upon the orientation of this diatom. The effect is much greater when the H atom forms the van der Waals bond because it is the dominant vibrator in the HCl molecule. Thus, accurate calculations need to take into account the anisotropy of the interaction potential.

Among some of the functional forms for $V_1(R)$ that have been considered, are $V_1(R) = -Ae^{-\beta R}$ and $V_1(R) = A dV_0/dR$ (LeRoy et al., 1991). In its simplest form, the function, f, can be defined to be zero for $v'' = 0$ and some constant value for $v'' = 1$.

A convenient method for deriving the dissociation rate is based on Fermi's Golden rule. Approximations inherent in this approach are similar to those required to separate (adiabatically via averaged interactions) the diatom and the van der Waals motions. The two states of interest are the excited complex in which the diatom has been excited to $v' = 1$, and the final dissociated state in which the diatom is back in the ground state ($v'' = 0$) with the energy released as translational (or rotational) energy. Within the approximation of separable motions, the system wave function in these two states are

$$\Psi_{v',n}(q, R) = \Phi_{v'}(q)\, \psi_{v',n}(R) \; , \tag{10.11}$$

$$\Psi_{v'',E}(q, R) = \Phi_{v''}(q)\psi_{v'',E}(R) \; , \tag{10.12}$$

where $\psi_{v'',E}$ is the continuum wave function for the departing fragments, and n is the van der Waals mode vibrational quantum number. The bound-state wave functions are obtained by solving the adiabatically separated Schrödinger equation in which the coupling between the two oscillators is taken into account in an average way. The continuum eigenfunction, $\psi_{v'',E}(R)$, can be approximated by a plane wave which is rapidly damped out at the repulsive wall of the ground state van der Waals potential. In case of weakly bound systems ($D_o \approx 300$ cm^{-1}) and relatively high excitation energies (ca. 3000 cm^{-1}), details of the van der Waals interaction potential will not affect greatly the final result. However, in cases where the binding energy is a significant fraction of the total available energy, the plane wave approximation will not be very good.

In case the molecular partner in the dimer complex has more than one internal degree of freedom, additional dissociation channels (product internal energies) must be taken into account. These can be accommodated by using different final vibrational state wavefunctions, $\Phi_{v''}(q)$ and correspondingly different translational energies for the departing fragments.

According to the Golden Rule, the predissociation rate for each final dissociation channel is then given by

$$k_{v''} = 2\pi c|\Psi_{v'',E}(q, R)|\langle f_1\rangle_{v,v} V_1(R)|\Psi_{v',n}(q, R)|^2 \; , \tag{10.13}$$

where, as before, the coupling term is the interaction averaged over the vibrational motion of the diatomic fragment.

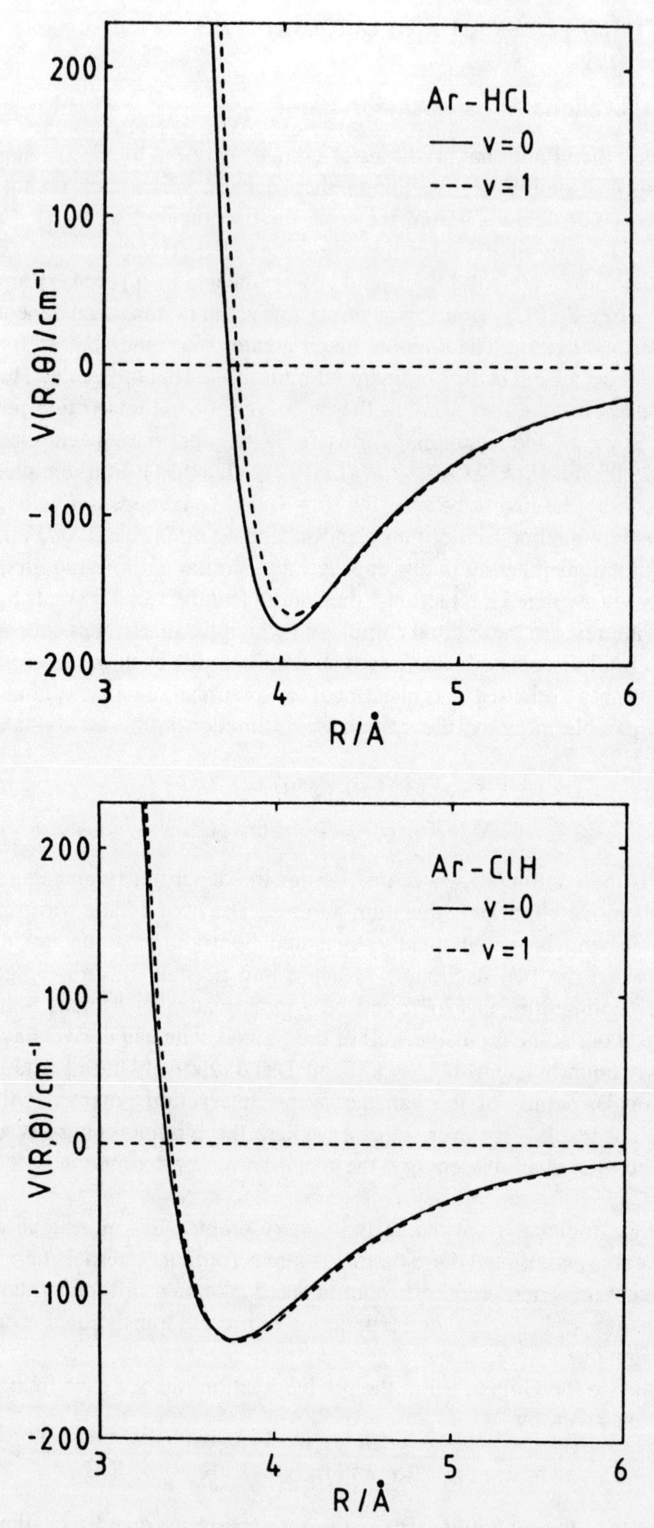

Figure 10.8 Van der Waals mode potential energy curves for the linear Ar-HCl(*v*) and Ar-ClH(*v*). Note the greater effect of HCl vibrational excitation in the upper geometry. Taken with permission from Hutson (1984).

10.3.2 The Miller-LeRoy Frequency Shift-Dissociation Rate Correlation

We can now state the relationship between the vibrational predissociation rate and the frequency shift of the monomer. Since the frequency shift is proportional to the interaction potential, and the dissociation rate is proportional to the square of the interaction potential, the VP rate constant, k_{vp}, is proportional to the square of the frequency shift:

$$k_{vp} \approx (\Delta\nu_{v',v''})^2 \tag{10.14}$$

Figure 10.9 compares Eq. (10.14) and the experimental data for a variety of van der Waals dimers involving HF, HCN, and HCCH units. This plot shows good agreement over three orders of magnitude in the dissociation rate constants. In particular it correlates the different dissociation rates of systems such as those in table 10.3 in which the different C—H stretch frequency shifts give very different rates. The scatter in the correlation is attributed to the approximations made in the derivation of Eq. (10.14) (LeRoy et al., 1991).

10.3.3 The Momentum Gap Law and the Ewing Model

Consider again the potential energy curves in figure 10.7 in which the vibrational predissociation is viewed in terms of an energy transfer from the diatom to the transla-

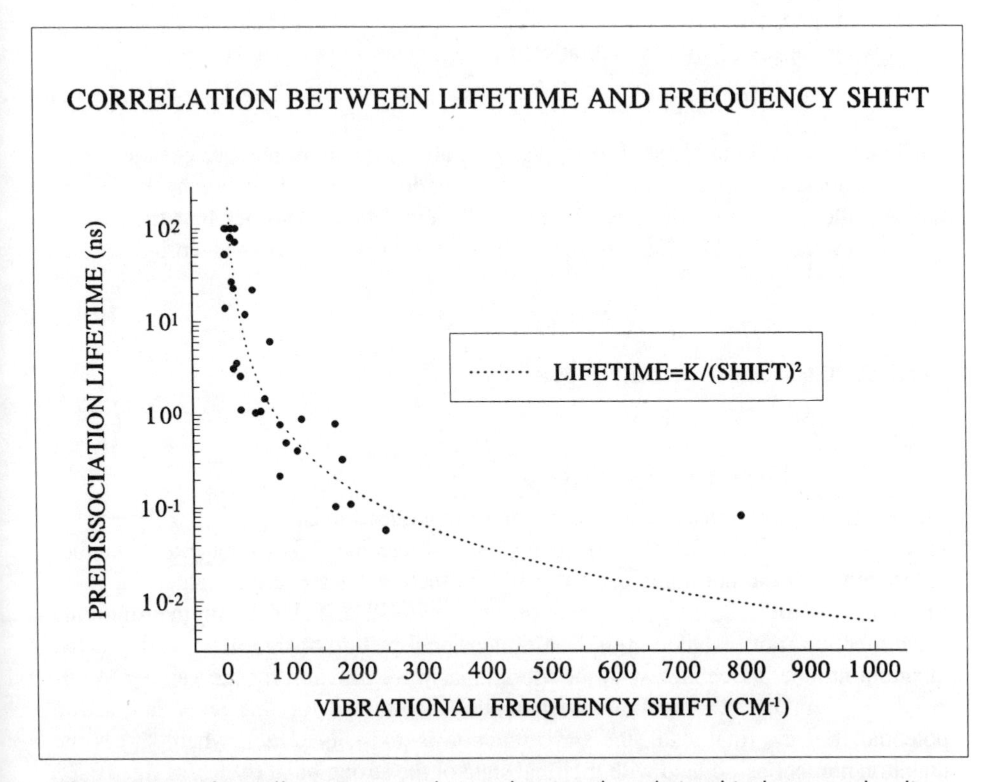

Figure 10.9 The Miller-LeRoy correlation showing the relationship between the dimer predissociation rate and the square of the shift in the absorption peak of the chromophore unit. Taken with permission from Miller (1994).

tional continuum of the van der Waals bond. Ewing (1987) has suggested a simple principle which governs the rate of energy flow from the initially excited state. This is that the rate of energy exchange is related in an exponential fashion to the change in the quantum numbers during the course of dissociation.

Consider the fate of energy flow in the dimer:

$$A \overset{q}{—} B \cdots \overset{R}{\cdots} C \quad (\theta)$$

where A, B, and C can be atoms or molecular units, q is again the coordinate for the strong bond, and R is the coordinate for the weak van der Waals bond. After excitation, the following events can ensue:

$$A - B^{*} \cdots C \rightarrow AB + C \qquad \Delta E_{\text{v-t}} , \tag{10.15}$$

$$A - B^{*} \cdots C \rightarrow AB^{\#} + C^{\#} \qquad \Delta E_{\text{v-r}} , \tag{10.16}$$

$$A - B^{*} \cdots C \rightarrow AB^{*} + C^{*} \qquad \Delta E_{\text{v-v}} . \tag{10.17}$$

The three reactions describe energy flow from vibrations to translation, rotations, and other vibrations. It is well known from collision studies that v-v energy transfer is much more efficient than v-r or v-t energy transfer (Gilbert and Smith, 1990). These propensities, which apply also to dissociation reactions, can be explained by the momentum gap law.

Ewing suggested that the vibrational predissociation rate is given by:

$$k = 10^{13} \exp[-\pi(\delta n_{\text{t}} + \delta n_{\text{r}} + \delta n_{\text{v}}) = 10^{13} \exp[-\pi \Delta n_{\text{T}}] , \tag{10.18}$$

where Δn_{T} is the sum of the absolute value of all the quantum number changes. This relationship is based on a number of approximations made in evaluating the Fermi Golden rule expression for energy transfer (Ewing, 1987). Consider first the case of V-T energy transfer. The Golden rule for $V \rightarrow T$ energy transfer is given by

$$\text{k} = \frac{4}{\hbar^2 u_{\text{f}}} \langle f|V_{\text{c}}|i\rangle^2 , \tag{10.19}$$

where the final relative velocity is given by

$$u_{\text{f}} = \sqrt{\frac{2E_t}{\mu_{\text{t}}}} . \tag{10.20}$$

The E_t term is the translational energy, and the reduced mass $\mu_{\text{t}} = m_{\text{AB}}m_{\text{C}}/(m_{\text{AB}} + m_{\text{C}})$. The initial wave function $|i\rangle = |\xi_i\rangle|\xi_{\text{r}}\rangle|\xi_{\text{t}}\rangle$ where the ξ_i is the quantum number associated with the harmonic oscillator wave function for the strong molecular bond ($\Phi_{\text{v}=1}$ in fig. 10.7), ξ_{r} is the rotational quantum number associated with the rotational wave function (not included in the one-dimensional picture of fig. 10.7), and ξ_{t} is the quantum number associated with the vibrational wave function for the van der Waals bond ($\Psi_{v',n}$ in fig. 10.7). A Morse potential is assumed for this latter interaction potential. For the final state, the wave function is $|f\rangle = |\xi_{\text{f}}\rangle|J\rangle|q_{\text{t}}\rangle$, where $|\xi_{\text{f}}\rangle$ is the quantum number associated with the final state of the strong bond ($\Phi_{v''=0}$ in fig. 10.7), the rotational quantum number, J, represents the rotational wave function (now a free rotor), and q_t is the translational quantum number for the final product translational wave function ($\Psi_{v'',E}$ in fig. 10.7). The latter is assumed to be a plane wave. The coupling term, $V_c(q,\theta,R)$ can be expressed as a difference in the Morse potential for the van der Waals interaction, $V(R - \alpha sq) - V(R) \approx - \alpha sq[dV/dR]$. It is a measure of the relative displacements of the atoms B and C during energy transfer. In common with the Schwartz, Slawsky, Herzfeld (1952) SSH theory for vibrational to translational energy transfer, the strong molecular and the weak van der Waals bonds are coupled only via the motions of B relative to C, while A is ignored. The factor $\alpha = m_A/(m_A + m_B)$, is a mass term that determines how much of the $A - B$ vibrational motion is due to the B atom, while the angular part of the interaction is approximated by $s = \exp(-\beta\cos m\theta)$, where β is a parameter related to the anisotropy of the angular potential and m is related to the symmetry of the complex. The derivative of the Morse potential is easily evaluated.

The desired matrix element finally reduces to

$$\langle \xi_f; J; q_t | V_c | \xi_i; \xi_r; \xi_t \rangle = \alpha \langle q \rangle \langle s \rangle \langle q_t | dV/dR | \xi_t \rangle \ . \tag{10.21}$$

When these terms are evaluated (Ewing, 1987), the rate of energy transfer for an "average" dimer is given by

$$k \approx 10^{12} \exp(-\pi q_t/2) \ , \tag{10.22}$$

where

$$q_t = \frac{(2\mu_t E_t)^{1/2}}{a\hbar} \ . \tag{10.23}$$

The a in the denominator is a range parameter from the Morse interaction potential. It is typically 2 Å^{-1}. The q_t term is a dimensionless quantity which characterizes the translational wave function and is related to the number of nodes in the wave function. It is through this function that we obtain the translational quantum number because the quantum number is also related to the number of nodes. In the vicinity of the Morse potential well, there will be approximately $q_t/2$ nodes so that the effective translational quantum number change is approximated by

$$\delta n_t \approx |q_t/2 - \xi_t| \ . \tag{10.24}$$

This relation expresses the difference in the translational quantum numbers and the number of stretching vibrational nodes in the van der Waals bond prior to dissociation.

In an analogous manner (albeit with more severe approximations), the change in the rotational quantum number can be "derived." It is given by

$$q_r = \frac{(2\mu_r E_r)^{1/2}}{\alpha a\hbar} \quad \text{and} \quad \delta n_r \approx |q_r/2 - \xi_r| \ , \tag{10.25}$$

where $\mu_r = m_A m_B/(m_A + m_B)$ and $J \approx q_r$.

Finally, the change in the vibrational quantum numbers is given by:

$$\delta n_v \approx \tau |v_f - v_i| \ . \tag{10.26}$$

The factor τ is related to the effectiveness for coupling the molecular bonds with the van der Waals bonds, and is generally assigned an average value of 1.

We see then that the final Ewing equation is a combination of simple models and approximations which are designed to give reasonable values for a number of systems. In figure 10.10 are plotted the measured dimer lifetimes versus the total quantum number changes for a number of systems. The correlation over many orders of magnitude in dimer lifetimes is impressive. However, so are the discrepancies. The Ewing model is useful precisely because physical insight can be derived from its successes as well as its failures. Consider first its successes.

The predissociation of Br_2—Ne has been studied by exciting the bromine molecule to the B(v = 27) state (Cline et al., 1987). One quantum of Bromine vibrational energy (64 cm^{-1}) is used to break the weak van der Waals bond (62 cm^{-1}) while the remaining 2 cm^{-1} of energy is imparted to the products as translational energy. With the a parameter set to 2 $Å^{-1}$ the q_t function is readily evaluated. Thus we find that δn_t is about 0.4. This small value for the quantum number is a reflection of the small energy transferred to the translational degree of freedom. Since the vibrational quantum number changed by -1, the total change in quantum numbers, $\Delta n_T = 1.4$. The calculated lifetime of 8 psec is very close to the measured value of 2 psec (point A in Figure 10.10).

Another interesting system is ClH···Ar. Although it has not been studied experimentally, a close coupling calculation has shown that this cluster dissociates with a rate of 4×10^{-6} sec^{-1} (Hutson, 1984). The HCl bond absorbs infrared radiation at 2886 cm^{-1}. Since the binding energy is 116 cm^{-1}, we have a total of 2770 cm^{-1} of energy to dispose. Suppose first that the decay converts all of this energy into translations. If

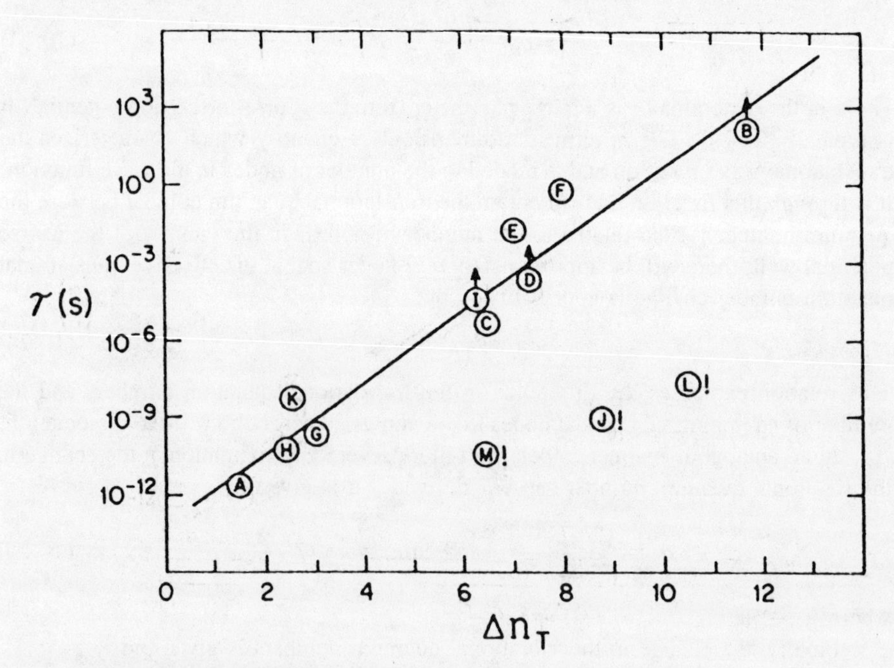

Figure 10.10 Plot of the dimer lifetime versus the total quantum number change Δn_T in the predissociation reaction. The three points that are well below the line are due to $(NO)_2$, $(HF)_2$, and *para*-difluorobenzene-Ar (p-DFB). The latter does not fit because of Fermi resonance which mixes the zero order bright state of p-DFB. Taken with permission from Ewing (1987).

we take the parameter a = 2.6 Å^{-1} (Ewing, 1987) and μ_t = 18.9 amu, we find that δn_t = 10.7 and since just one vibrational quantum number was lost, the total change in the quantum numbers is Δn_T = 11.7. This would give a lifetime of 3000 sec, which is orders of magnitude greater than the close coupling calculated lifetime of ca 1 μsec. However, because the rotational constant of HCl is rather large (10.6 cm^{-1}), it is clearly advantageous to convert as much of the available energy into rotations rather than translations. The maximum rotational energy that can be accommodated is when J = 15 so that $E_r = J(J + 1)B$ = 2544 cm^{-1}. If this is substituted into Eq. (10.25) (taking α = .97), we obtain q_r = 4.8 so that δn_r = 2.4. The remaining 226 cm^{-1} will then go into translation which gives a δn_t = 3.1. Thus the total Δn_T = 6.5 and the predicted lifetime becomes 7 × 10^{-5}, which is just a factor of 10 greater than the close coupling calculation (point C in Figure 10.10). Of greater interest is that the accurate calculations also predict that much of the energy goes into HCl rotations.

As shown in figure 10.8, the HCl—Ar complex can exist in two conformations, with the ClH—Ar more stable by 45 cm^{-1}. If the structure were HCl—Ar, the lifetime calculation would differ in only one respect. This is in the value for α, which in this case would be 0.028, rather than 0.97. The major effect is that the Cl atom is much less effective in imparting rotational energy to the departing fragments than is the H atom. Thus the value of q_r increases from 4.8 to 166. The net result is that the channel leading to rotational excitation of the products is shut off.

The Ewing model is based on a three-atom approximation of the dimer. As such it is best applied to dimers in which the initial excitation is in the bond that is part of the van der Waals bond. Hence, in the previously cited example of HCN—HCCH, the model is much more likely to succeed in the case of the linear isomer in which the acetylenic C—H stretch has been excited rather than the C—H stretch on the HCN molecule; and to the T-shaped isomer in which the C—H stretch in the HCN molecule has been excited. Suppose that the C—H stretch in HCN at ca. 3300 cm^{-1} is excited in the T-shaped isomer. If the 2045 mode in the HCN product is excited during the vibrational predissociation, then just 500 cm^{-1} remains in translation. Thus we obtain a Δn_T = 5.5 and predict a lifetime of about 1 nsec, which is precisely the measured lifetime of 1.3 nsec. However, it is evident that as the complexity of the molecules increases, there will be numerous combinations of product vibrations and rotations which could be excited. Which one is dominant is difficult to determine a priori.

The Ewing model has a serious problem with the HF dimer because so much energy is imparted to the rotations that the interaction potential (the Morse potential) changes dramatically during the course of dissociation (point J in Figure 10.10). The rapid rotation of one of the HF fragments averages out the dipole so that the interaction is converted into a dipole-induced dipole interaction. The concomitant reduction in the interaction potential decreases the δn_t so that the lifetime is much shorter than predicted.

It is interesting to compare the Ewing and Miller-Leroy methods for correlating the cluster dissociation data. One is based on the energy gap, the other on the absorption frequency shift. The energy gap law is expected to correlate the lifetime data if the energy gap is the primary factor that determines the lifetime. It only treats excitations to oscillators that are connected to the van der Waals bond. As a result, it cannot account for the different lifetimes of the $(HF)_2$ dimer when the different HF stretches are excited. Thus, numerous exceptions to the energy gap law are expected. On the other hand, the correlation as well as the exceptions provide considerable physical insight into the dissociation process. The model works best for small clusters in which the product energy spacings are large and enumerable.

The Miller-LeRoy correlation is much more closely related to theory so that few exceptions are expected. This is because ultimately, it is the coupling between the initial and final states that determines the rate of vibrational predissociation. However, the Miller-LeRoy relation offers less physical insight perhaps because few failures are expected. It simply correlates the dissociation rate with a more or less easily measured quantity, the frequency shift.

10.3.4 Vibrational Predissociation versus IVR

The Ewing model for vibrational predissociation treats the reaction in terms of the coupling between the AB chromophore and the $AB \leftrightarrow M$ translational continuum. The implied mechanism involves direct coupling to the translational continuum without redistribution of the internal energy. However, this model has been generalized by Kelley and Bernstein (1986) and applied to IVR as well. The details are discussed later in this chapter. The suggestion that the energy gap model might describe IVR as well as VP raises an interesting question. How do we know that the dissociation of the small dimers proceeds by VP and not by IVR followed by rapid statistical dissociation? It is very likely that examples of both will be found. However, it is not a simple matter to decide which mechanism is operative. A case which seems rather clear cut is the HF dimer dissociation. Bohac et al. (1992a) found that one of the HF products is formed rotationally excited, while the other HF unit retains the rotational energy it had from the initial IR excitation. Thus, the energy remains localized during the dissociation process. The reacting system has clearly avoided statistical redistribution of its internal energy.

On the other hand, recent experimental and theoretical results on the Ar—Cl_2 dimer indicate that this dimer can be made to dissociate via either VP or IVR followed by dissociation (Halberstadt et al., 1992). In this experiment Ar—Cl_2 dimer is excited by visible radiation to a long progression of Cl_2 vibrational levels in the $\tilde{B}$ state. At low vibrational levels, a $\Delta v = -1$ transition is sufficient to break the van der Waals bond so that the reaction is a classic case of VP and is well accommodated by the Ewing model. However, at higher vibrational energies, anharmonicity reduces the energy between the Cl_2 vibrational levels so that two quanta are required to break the van der Waals bond. Thus, the transfer of the first quantum of vibrational energy excites the van der Waals modes from which IVR can take place. The experimental evidence for the IVR process is contained in the absorption spectrum shown in figure 10.11. The large peak is a result of absorption to the $\tilde{B}$ ($v = 10$) level of the Cl_2 in the dimer. This zero-order bright state is mixed with the complete set of dimer, or van der Waals, modes. The weaker absorption peak at -176.69 cm^{-1} is due to excitation of the $v = 9$ Cl_2 level and the 36th van der Waals vibrational level. This is the dark state that was calculated to mix most strongly with the bright state. The much weaker peak at -177.95 cm^{-1} is due to the $v = 9$, $v_{vdw} = 35$ level. Similar IVR processes have been proposed for the ethylene dimer, but not for C_2H_4 rare gas dimers (Nesbitt et al., 1985).

Because of experimental limitations, most of the infrared studies have dealt with rate measurements at energies in the 3000 cm^{-1} region. However, as longer wavelength and intense infrared lasers become available (e.g., free electron lasers) it should

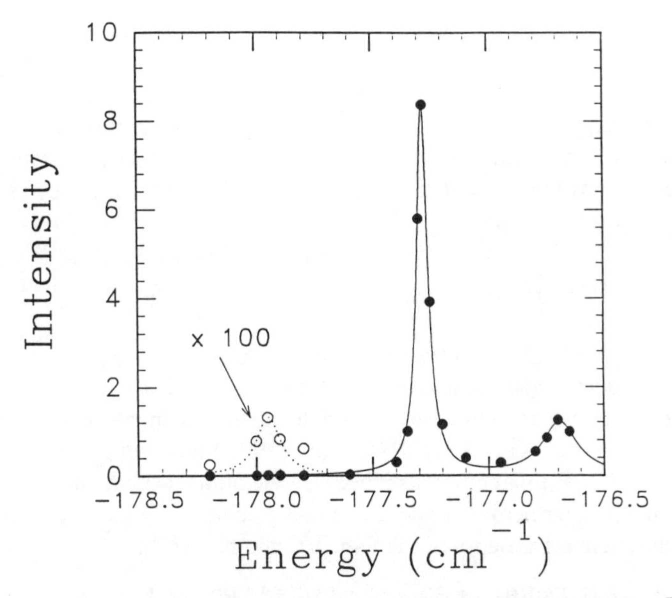

Figure 10.11 The absorption spectrum of Ar-Cl_2 to the $v = 10$ level of the $\tilde{B}$ state. The strong peak at -177.28 cm^{-1} (relative to the origin of the $\tilde{B}$ state) is the zero-order bright state at $v = 10$. The resonance at -176.69 cm^{-1} corresponds to a combination band consisting of $v(Cl_2) = 9$ and the 36th van der Waals vibrational level. This strong mixing of states is attributed to IVR. Taken with permission from Halberstadt et al. (1992).

be possible to photodissociate for instance the HCN—HCCH dimer via the 842 cm^{-1} transition as well. This infrared active transition exhibits a much larger shift (see table 10.2) upon dimer formation which suggests that it couples very well with the rest of the molecule so that IVR may rapidly equilibrate all of the energy. If this happens, then the dissociation step would be statistical with a calculated rate of about 10^9 sec^{-1}. If this low dissociation rate is less than the IVR rate, even the dissociation rate may be determined by RRKM.

10.3.5 IVR Followed by Statistical Dissociation in Larger Dimers

In large dimers, especially those having large binding energies, the dissociation mechanism is likely to be descried best by mechanism 2, which is

$$A^* \cdot\cdot\cdot B \underset{k_{\text{ivr}}}{\overset{\text{IVR}}{\longrightarrow}} A \cdot\cdot * \cdot\cdot B \underset{k_{\text{d}}}{\overset{\text{RRKM}}{\longrightarrow}} A + B\,. \tag{10.27}$$

The asterisk denotes excitation which may reside in the A chromophore or, after IVR, in the van der Waals modes. The reason for the expected change of mechanism in large dimers is that as the density of states increases, the IVR rate increases while the RRKM dissociation rate decreases. Thus, ultimately, large dimers will dissociate statistically.

The solutions to the differential equations that describe the sequential mechanism [Eq. (10.27)] are well known (Steinfeld et al., 1989; Berry et al., 1980). Several situations can arise depending upon the relative rates of the two reactions. If $k_{ivr} >> k_d$, the intermediate equilibrated dimer can survive for a considerable length of time. The over all rate of dissociation will then be given by the dissociation rate, k_d. If, on the other hand, $k_{ivr} << k_d$, then the equilibrated dimer, once formed, will dissociate rapidly and will be difficult to detect experimentally. In that case the rate of dissociation will be given by the rate of IVR. In general, the formation of dissociation products as a function of time is given by

$$P(t) = [D_0]\left(1 - \frac{k_d}{k_d - k_{ivr}}\, e^{-k_{ivr}t} + \frac{k_{ivr}}{k_d - k_{ivr}}\, e^{-k_d t}\right), \qquad (10.28)$$

where $[D_0]$ is the initial concentration of dimers.

Such a sequential dissociation mechanism is most likely to be found in polyatomic dimers. Examples which appear to dissociate in this fashion are aniline molecules complexed with either Ar or CH_4 (Bernstein, 1990; Nimlos et al., 1989). These two dimers offer an interesting comparison because the number of van der Waals modes in the case of the CH_4 partner is 6 while it is only 3 with the Ar as partner. Replacement of the Ar by the methane group will increase k_{ivr} and reduce k_d.

10.3.5.1 The Dissociation of Aniline—X (X=Ar or CH_4)

The dissociation of aniline—Ar (An—Ar) and aniline-methane (An—Me) has been investigated on the excited S_1 surface. Either dispersed fluorescence (Nimlos et al., 1989), or a time-resolved threshold photoelectron spectrum (TPES) obtained with a second laser (Smith et al., 1993) can distinguish among the following three species: An#*—X, An#—*—X, and An# + X, where the # denotes electronic excitation and the * denotes vibrational excitation. At low S_1 excitation energies and in the absence of IVR, the dimer fluoresces back to S_0 with only a few well defined (and assigned) transitions. Once the energy exceeds the dissociation limit, the free aniline S_1 state can also be observed at transition energies that are shifted from the complexed dimer energies. The energy at which the monomer fluorescence appears can be related to the dimer binding energy. Finally, in the case of An—Me dimer, dispersed fluorescence or time resolved TPES spectra can also be observed from the equilibrated dimer. These spectra are much broader showing contributions from the van der Waals modes. Figure 10.12 shows three dispersed fluorescence spectra from An—Me excited to three different energies (Nimlos et al., 1989). When the transition origin is excited, neither reaction nor IVR takes place and only emission from the initially prepared dimer is evident. At intermediate energies, just above the dissociation limit, broad emission from the equilibrated dimer is observed, while at very high energies, the dimer is nearly totally dissociated so that monomer aniline fluorescence dominates the spectrum. It is significant that the intermediate equilibrated dimer fluorescence is absent in the case of An—Ar dimer.

Two types of time resolved experiments can be carried out on such a system. The less ambiguous approach is that employed in the stilbene—He study (fig. 10.5) in which a second laser, time delayed from the pump laser, is used to excited the S_1 state. Smith and Knee (1993) have used picosecond time resolved TPES spectroscopy in which zero-energy electrons are collected as a function of the probe laser wavelength. [Laser based TPES has also been called zero kinetic energy (ZEKE) electron spectroscopy or pulsed field ionization (PFI) (Muller-Dethlefs and Schlag, 1991)]. Figure

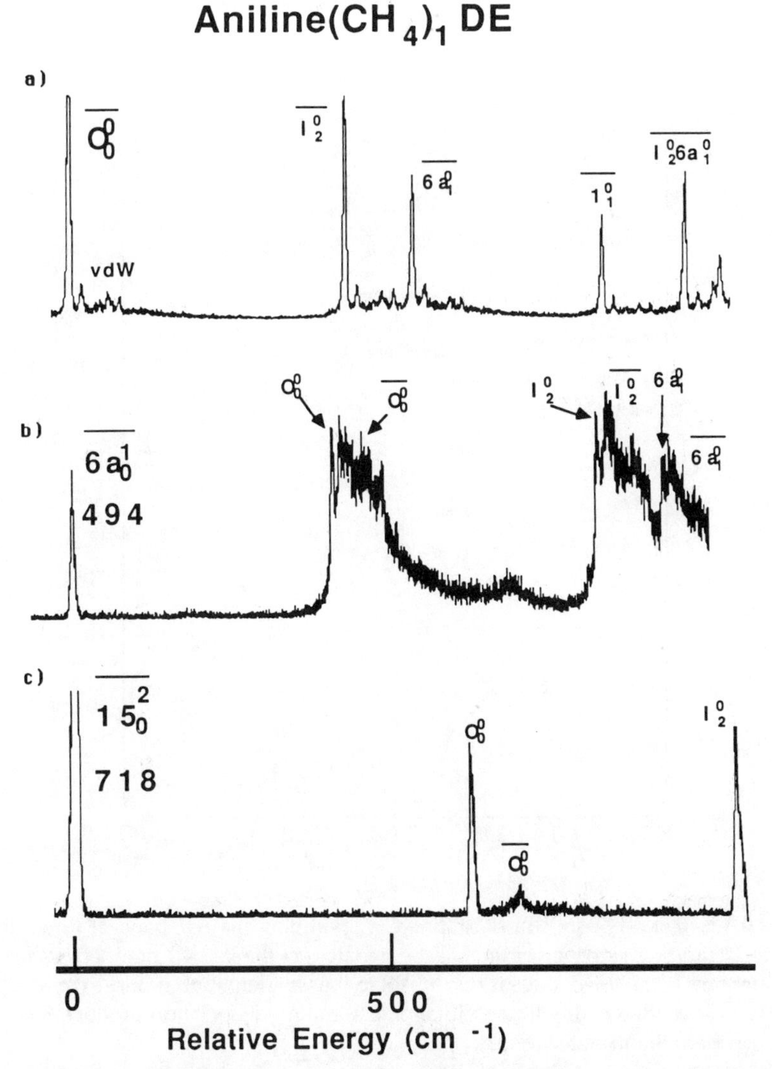

Figure 10.12 The dispersed emission spectra of aniline-CH_4 (A-M) system following excitation to the three indicated energies above the origin of the Ã state. (a) the emission spectrum is that of the aniline unit in the dimer. (The bar over the peak assignments indicates emission from the dimer) (b) Excitation to an energy just above the dimer dissociation limit shows some sharp monomer emission and much broader dimer emission. (c) Excitation at 718 cm^{-1} results in very rapid dissociation and thus mostly monomer emission. Taken with permission from Nimlos et al. (1989).

10.13 shows the three time resolved ZEKE-PFI spectra of An—Me with the pump laser fixed to excite the $6a_0^1$ level of the dimer (excitation energy for the middle spectrum of fig. 10.12). With no delay, the ZEKE-PFI spectrum has all the features associated with the An—Me dimer prior to IVR. At an intermediate time of 400 psec, the ZEKE-PFI spectrum is much broader indicating that energy has leaked into other

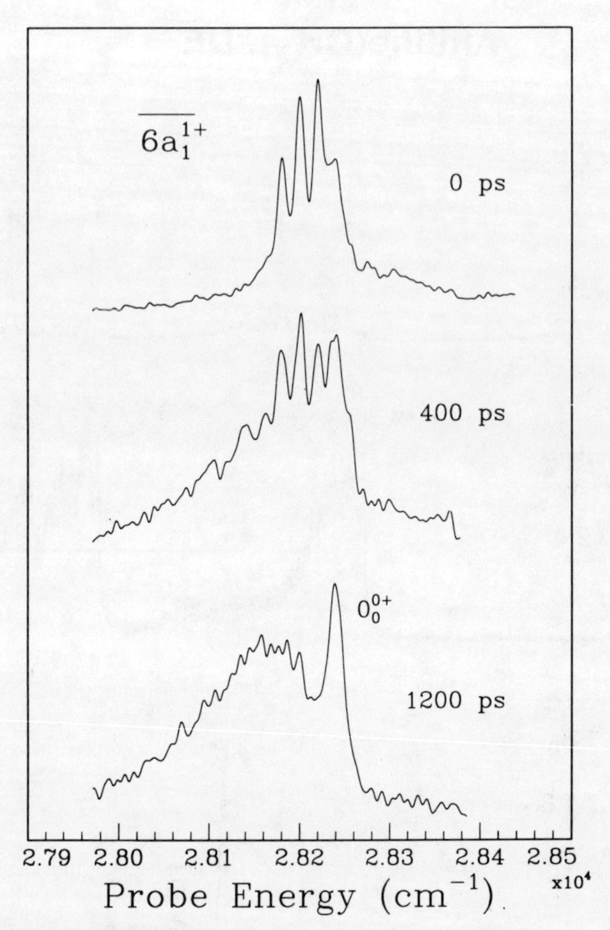

Figure 10.13 ZEKE-PFI spectra of aniline-CH_4 pumping the $\overline{6a}_0^1$ band at three different probe delay times. The top spectrum has the signature of the initially pumped complex, the broad structure in the lower trace is due to IVR to the van der Waals modes. The sharp peak at 1200 ps delay time is due to the vibrationless aniline dissociation product. Taken with permission from Smith et al. (1993).

modes. Finally, after 1200 psec, the single sharp peak associated with product aniline monomer begins to show. Although there is considerable overlap among the three species involved in this reaction, it is possible to partially resolve them. Thus, by monitoring the ZEKE-PFI signal at a fixed probe laser energy, the decay time of the initially produced dimer and the rise time of the relaxed dimer were measured and found to be 350 psec.

Another time resolved approach involves measuring the natural fluorescence of the aniline back to the ground state. Figure 10.14 shows the product aniline formed with a rise time of 240 psec at an excitation energy of 718 cm^{-1}. The slow decay is a result of the aniline fluorescence lifetime. However what governs the 240 psec rise time; is it IVR, or the RRKM dissociation rate? A clear answer to this question is not yet in hand in part because overlapping spectra for the dimer, the relaxed dimer, and

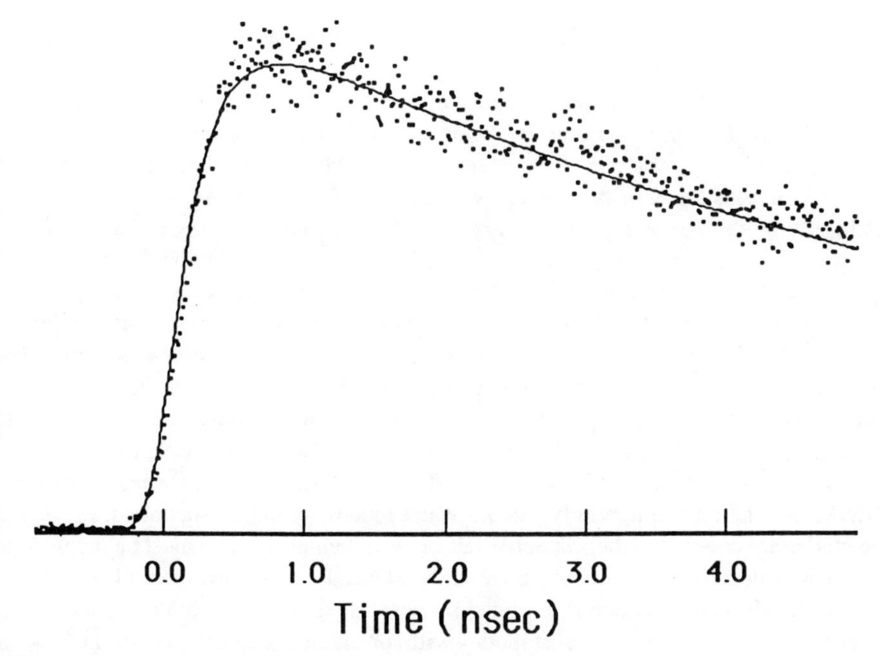

Figure 10.14 The emission kinetics of aniline-CH_4 dimer following excitation to the 15^2_0 level as shown in spectrum c) of Figure 10.12. The product aniline shows a rise time of 240 ps and a decay time of 7.6 ns (the aniline radiative lifetime). Taken with permission from Nimlos et al. (1989).

product aniline have prevented measuring the rates at a sufficient number of excitation energies. The time resolved ZEKE-PFI experiments indicate that the IVR rate does not vary rapidly with energy, where as the RRKM dissociation rates do (Smith et al., 1993) Hence, the dissociation step is rate-limiting at low energies, while IVR is rate-limiting at the higher excitation energies. More experimental work needs to be done in order to unravel the various decay rates involved in this reaction.

10.3.5.2 The Kelley-Bernstein Model for Sequential IVR-RRKM Dissociation

Kelley and Bernstein (1986) proposed a model to account for the dissociation of larger dimers by the sequential mechanism [Eq. (10.27)]. The model considers both the finite IVR and the RRKM dissociation rates. The IVR rate is treated by the previously mentioned energy (or momentum) gap law, however here slightly modified. The model distinguishes the van der Waals modes from the molecular modes by treating the former in terms of a density of states while treating the molecular modes individually. The vibrational excitation, which resides initially in one of the molecular modes, decays by an energy transfer to the van der Waals modes, leaving the molecule in a different final vibrational level. The IVR rate constant, k_{ivr}, associated with the transfer of energy ΔE from the molecular to the van der Waals modes is proportional to

$$k_{ivr} \propto A\, e^{-\Delta v}\, e^{-\Delta E/h\nu_{max}}\, \rho(\Delta E)\ , \tag{10.29}$$

where Δv is the change in the vibrational quantum number of the chromophore molecule, ν_{max} is the highest van der Waals vibrational mode, and $\rho(\Delta E)$ is the density of vibrational states in the van der Waals modes at the energy ΔE. The term $\Delta E/h\,\nu_{max}$ is a measure of the change in the vibrational quantum number in the van der Waals modes. This rather *ad hoc* expression borrows some of the features of the Ewing energy gap model, namely the exponential dependence on the change in the quantum numbers. However, it adds the density of states which accounts for the many ways that the low-frequency modes can accept the energy from the chromophore mode. The rotational and translational degrees of freedom are not involved in IVR. Although this equation appears reasonable, it is difficult to apply to real systems because of the many IVR pathways whose importance surely depend upon subtle and unknown coupling terms.

The RRKM rate for dimer dissociation must be treated with some care. For instance, it is not clear whether the van der Waals modes (six in the case of aniline—CH_4) are more strongly coupled to the dissociation continuum than they are to the aniline or the CH_4 modes. Kelley and Bernstein (1986) assume that the coupling between van der Waals modes and the rest of the dimer modes is weak so that the RRKM calculation includes only van der Waals modes. If on the other hand, the rest of the modes are coupled to the van der Waals modes, then all of them must be included in the RRKM density and sums of states calculations. In some cases it will make little difference which mechanism is assumed because the dissociation limit is often below an energy at which the molecular modes will contribute much density. In the case of aniline, only a few modes are below the dimer binding energy so that they contribute very little to the density of states.

If another X group is added onto the aniline molecule, the An—X_2 dissociation rate is more likely to be determined by the statistical dissociation step because with each additional nonlinear monomer the number of van der Waals modes increases by six. Because the van der Waals modes are extremely anharmonic and coupled to each other, a proper RRKM calculation should use anharmonic densities and sums. However, these are not yet generally available for the systems of interest. In all cases it is best to use the quantum density of states (i.e., RRKM) and not the classical approximation of it (RRK). With a binding energy of say 480 cm^{-1} and six oscillators, the average energy per van der Waals mode is 60 cm^{-1}. Since these frequencies typically vary between 20 and about 400 cm^{-1}, it is evident that the average number of quanta excited per mode is only about 1 or 2, which does not correspond to the classical limit.

It is tempting to assume that as the cluster size increases, the classical RRK approximation will become better. To show otherwise, we consider the dissociation of the $(H_2O)_6$ cluster, whose van der Waals vibrational frequencies have been calculated. (Li and Bernstein, 1991) These frequencies, arranged into groups of similar size are listed in Table 10.4.

The classical density of states is given by $E^{s-1}/[(s-1)!\,\Pi h\nu_i]$. At energies of 500 and 1000 cm^{-1}, these classical densities for the above set of oscillators are: 3.8 ×

Table 10.4. Van der Waals Frequencies (Degeneracy) for $(H_2O)_6$.

486(3)	376(3)	295(6)	197(4)	166(2)	81(1)
61(5)	43(1)	24(2)	16(1)	11(2)	

Adapted from the Li and Bernstein calculation (1991).

10^{-16} and 2×10^{-7} states/cm^{-1}. The true density of states at these same energies are 3.6×10^5 and 2.6×10^8 states/cm^{-1}. These errors of more than 15 orders of magnitude are considerably reduced when the zero-point energy is added to the energy. Thus, when we E is replaced by $E + ZPE$ (which is 2996 cm^{-1}), we obtain "classical" densities of 1.2×10^9 and 5.8×10^{10} states/cm^{-1}. These are now too large, but considerably closer to the true density of states.

It might be expected that the errors associated with the classical density of states calculations will cancel when evaluating the RRK rate of dissociation because a sum of states is divided by a density of states. Again a simple calculation suggests otherwise. In figure 10.15, RRK and RRKM rate constants are compared for the dissociation of the water cluster in which we have assumed that the binding energy is 400 cm^{-1} and that the critical coordinate is 166 cm^{-1}. This comparison shows that the error for the RRK rate with no account taken of the zero point energy, is over 15 orders of magnitude near the dissociation threshold, and that even at high energies, the errors are significant. If the RRK rate constant includes the ZPE $\{k(E) = \nu[(E - E_0 + \mathrm{ZPE}^{\#})/(E + \mathrm{ZPE})]^{s-1}\}$, where the $ZPE^{\#}$ is the zero-point energy in the transition state, the rates are now higher than, and considerably closer to the RRKM rates. The results of figure 10.15 show that although the vibrational frequencies in large clusters are often low, their numbers increase as well, so the fundamental problem of low occupation number per mode remains.

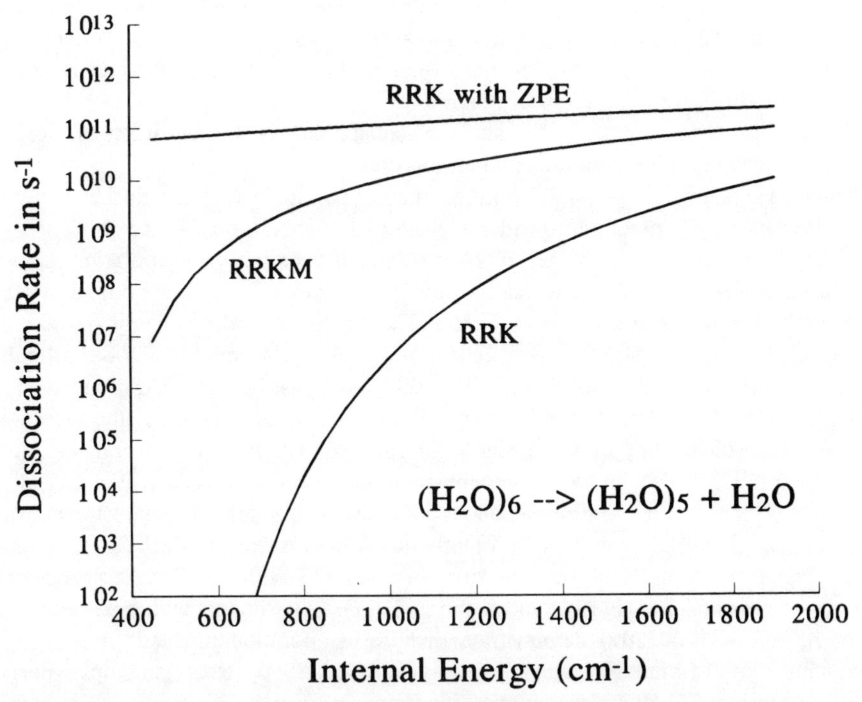

Figure 10.15 The RRK, RRKM, and RRK with ZPE decay rates for the dissociation of $(H_2O)_6$. The calculations used the van der Waals modes vibrational frequencies listed in Table 10.4.

10.4 THE DISSOCIATION OF LARGE CLUSTERS

Clusters provide the bridge between the gas phase in which molecules are isolated and independent of each other, and the condensed phase, where intermolecular forces provide each liquid its unique properties. It is thus of considerable interest to probe the physical properties of clusters from dimers to very large clusters as they approach the properties of the bulk liquid. Some of these attributes, such as the ionization energy or electronic energy levels, can be defined and measured for the molecule as well as the clusters. Other properties, such as melting temperature and surface tension, are not defined for the isolated molecules but become important for clusters of some size. (Berry, 1990; Wales and Berry, 1990)

In all cluster investigations, two parameters above all else must be controlled. These are the cluster size and the internal energy. Neutral clusters can be prepared cold; in fact their binding energy is often so low that they only exist if they are cold. Thus, when neutral clusters are excited by monoenergetic photons, their internal energy is known. However, as previously mentioned, the size of the neutral cluster is usually not known so that experiments are difficult to interpret. This is especially true when the cluster size exceeds about 10 monomer units.

The study of ionic clusters encounters the opposite problem. Ion studies have the distinct advantage that the cluster size can be selected by a mass filter. Furthermore, it can be hoped that as the cluster size increases the effect of the central ion core becomes insignificant, so that properties such as the monomer binding energy to a large cluster will not depend upon the presence or absence of a charge. Indeed Klots (1985) predicts that the evaporation rate of single monomers from a cluster of 100 water molecules is the same for ionic and neutral clusters, whereas for a cluster of 10 water molecules these rates differ by two orders of magnitude.

A major disadvantage in the study of cluster ions is that the internal energy is often not known. This is because when ion clusters are prepared by ionizing neutral clusters, the change in the equilibrium geometry upon ionization causes the cluster ion to relax, thereby generating considerable internal energy (heat). Often this energy is sufficient to dissociate the cluster, a process that has been termed evaporation. Consider the case of an Ar cluster in which the atoms in the neutral cluster are bound by about 80 cm^{-1} (Weerasinghe and Amar, 1991). [The Ar dimer binding energy is 100 cm^{-1} (Tang and Toennies, 1986)]. Upon ionization, an Ar_2^+ is formed which has a binding energy of 9700 cm^{-1} (Morioka et al., 1991). Because the Ar—Ar bond distance changes from 4.0 Å to 2.5 Å (Ma et al., 1993), adiabatic ionization to the vibrational ground state of the Ar_2^+ dimer is highly improbable. The peak in the Franck-Condon distribution lies more than half way to the dissociation energy (Pradeep et al., 1993), so that in a typical ionization process, about 5000 cm^{-1} are released to the van der Waals modes. This is sufficient to "boil" off numerous Ar atoms. Recent PEPICO photoionization experiments on Ar clusters confirm this. All Ar_n^+ with $n > 2$ were produced by dissociative ionization (Furuya and Kimura, 1992). Experiments with potassium clusters K_n^+ ($5 < n < 200$) demonstrate that the evaporation proceeds mostly via a sequential loss of monomer units (Brechignac et al., 1990). Reaction stops when the available energy is below the monomer binding energy (E_o). This leaves the remaining cluster ions with a distribution of internal energy ranging from 0 to E_o.

A similar evaporation mechanism is operative in the photodissociation of negative ion clusters. The absorption of a UV photon by $(CO_2)_n^-$ ions can lead to the loss of as many as 20 monomer units (Alexander et al., 1986). Because each evaporation step is associated with a variable amount of translational energy, the final internal energy of the stable ion is known only to be in the range from 0 to E_o. Nevertheless, with the application of unimolecular rate theory, it possible to extract useful information from such studies. Two approaches due to Engelking and Klots have been particularly fruitful.

10.4.1 Engelking Model for Relating the Binding Energy and $\langle \epsilon_t \rangle$

In many mass spectrometric experiments, the dissociation of meta-stable ions can be studied after mass analysis. The typical lifetime required is between 10 and 100 μsec. The primary data consist of a knowledge of the parent mass, the daughter mass, and either the average kinetic energy release $\langle E_t \rangle$ or the kinetic energy release distribution, KERD, and finally the mean lifetime of the parent ion. The major unknown quantity is the parent ion internal energy, which is generally in the form of a distribution. Engelking (1987) proposed a model which relates these observables and permits the determination of the binding energy. This model uses a highly approximate version of the phase space theory. Angular momentum is not conserved and the density of states are approximated by classical harmonic oscillator densities.

Consider the equilibrium between a cluster, AM and its dissociation products, $A + M$, in which A is a cluster of some size and M is the monomer which evaporates. Equivalently, the reaction can be considered as a recombination of A and M (Klots, 1971).

$$AM \leftrightharpoons A + M$$

By microscopic reversibility, the forward and backward rates are equal at equilibrium so that

$$k_u = k_b \frac{[A][B]}{[AB]} = \sigma v \frac{\rho_A(E,J)^*\rho_B(E,J)}{\rho_{AB}(E + E_0,J_0)}, \tag{10.30}$$

where we have expressed the equilibrium constant, $[A][B]/[AB]$ in terms of the density of states, and have replaced the bimolecular backward rate constant with the collision cross section times the relative velocity of A and B. Because the energy, E, is the total available energy to the products, the binding energy, E_o, must be added to the molecule. The translational density of states of the products is included in the numerator, which can thus be expressed more explicitly as

$$\rho_A(E, J)^*\rho_B(E, J) = \rho_t(E_t)\rho_{A,B}(E - E_t), \tag{10.31}$$

in which the $\rho_{A,B}$ density represents the total product density of vibrational states. If we now use the three dimensional translational density of states [Eq. (6.15)], the unimolecular rate constant can be rearranged to yield:

$$k(E,E_t) = \frac{\sigma g[8\pi/h^2]\mu E_t\, \rho_{A,M}(E - E_t)}{h\, \rho_{AM}(E + E_0)}, \tag{10.32}$$

where we have also introduced the reaction degeneracy factor (g), and for simplicity, have dropped reference to the rotational degrees of freedom. This equation bears some similarity to the PST equation in chapter 7. However, because rotations are ignored Eq. (10.32) is considerably easier to evaluate.

In order to maintain the benefits of analytical functions, classical expressions for the densities of states are substituted into Eq. (10.32). If it is assumed that the frequencies are the same for the product (A) and reactant (AM) we obtain

$$k(E, E_t) = 8\pi\sigma g\mu\nu^3(s-3)(s-2)(s-1)\frac{E_t(E-E_t)^{s-4}}{(E+E_0)^{s-1}} . \quad (10.33)$$

In obtaining this we assume that the AM cluster is made up of atoms with s normal modes so that the dissociated products, $A + M$, have three fewer vibrational modes. Note that the rotational degrees of freedom of the loose transition state are ignored. The average dissociation rate is then

$$k(E) = \int_0^E k(E, E_t)dE_t = 8\pi\sigma g\mu\nu^3(s-1)\frac{E^{s-2}}{(E+E_0)^{s-1}} . \quad (10.34)$$

Equation (10.33) can also be used to obtain the average kinetic energy release:

$$\langle E_t \rangle = \frac{\int_0^E E_t k(E, E_t)dE_t}{\int_0^E k(E, E_t)dE_t} = \frac{2}{s-1}E . \quad (10.35)$$

This result is very different from the modified prior distribution (chapter 9). It implies that the average kinetic energy release is given by $2k_BT^*$, that is, a four-dimensional translational energy distribution. When the expression for $E = [(s-1)/2]\langle E_t \rangle$ is substituted into Eq. (10.34), and the latter equation solved for E_o, we finally obtain

$$E_o = 0.5(s-1)[C^{1/(s-1)}\langle E_t \rangle^{(s-2/(s-1)} - \langle E_t \rangle] , \quad (10.36)$$

where $C = 16\pi\sigma g\mu\nu^3/k(E)$.

Equation (10.36) permits the calculation of the binding energy from the measured kinetic energy release and other known, measurable, or estimable quantities such as the recombination cross section, the reduced mass, the vibrational frequency and the rate of the reaction. Because the metastable ions are generally collected in about 10 μsec, the rate is fixed at 10^5 sec^{-1}. Engelking (1987) has estimated the other parameters for the case of $(CO_2)^+_{n+1} \rightarrow (CO_2)^+_n + CO_2$ reaction to be

$$\sigma = 100 \text{ Å}^2 \quad \nu = 20 \text{ cm}^{-1} \quad s = 5(n-1) \quad g = n-1 .$$

The values of s and g depend upon the structure of the complex. If it is assumed that the cluster is made up of a central core consisting of $(CO_2)^+_2$ [i.e., $(CO_2)^+_{n+1} = (CO_2)^+_2(CO_2)_{n-1}$] surrounded by loosely held CO_2 units, then any of the loosely held CO_2 units can evaporate. Thus it is assumed that the cluster symmetry number is $n-1$ while the transition state has a symmetry of 1, so that $g = n-1$. Because the intramolecular CO_2 modes are assumed too high to contribute to the density of states at the low energies involved in the metastable ion dissociation, the total number of vibrational modes is just $s = 5(n-1)$. (With the loss of each linear CO_2 unit, the cluster looses five van der Waals modes.) With these assumptions and the measured (Stace and Shukla, 1982) average kinetic energy releases, the binding energies for CO_2 monomers to the cluster ion were found to be about 150 meV (1200 cm^{-1}) and rather independent of the cluster size up to at least $n = 13$. A significant exception was $(CO_2)_4$ whose binding energy was found to be 200 meV.

Equation (10.36) is appealing precisely because of its simplicity. In addition, it appears to be reliable, a rather surprising result in the light of the drastic approximation involved in the use of the classical density of states and the neglect of rotations. Apparently, a happy cancellation of errors makes this equation insensitive to the density of states. This can be shown as follows. During the years before RRKM theory was routinely applied, the s parameter in the classical RRK theory was used as an adjustable parameter to fit experimental rate data. Invariably, the best value for s was considerably less than the actual number of oscillators in the molecule, a typical value being about 50% of the number of oscillators (Laidler, 1987). We can thus correct the Engelking equation for the error introduced with the use of the classical density of states by reducing the value of s. Consider the case of CO_2 evaporation from $(CO_2)_{12}^+$ for which $C = 1.66 \times 10^8$ eV. Suppose now that the measured kinetic energy release is $\langle E_t \rangle = 0.010$ eV. The derived binding energies (in eV) for various values of s are: 0.151 ($s = 50$); 0.161 ($s = 40$); 0.181 ($s = 30$); 0.20 ($s = 25$); 0.233 ($s = 20$).

The variation of the binding energy with assumed values of s is remarkably small. While the absolute rates obtained from the RRK theory vary by 10 orders of magnitude (depending on the values of E and E_o) when s is reduced from 50 to 25, the derived binding energies vary only by a factor of 1.3. The binding energy is similarly insensitive to other parameters. If for instance, the reaction degeneracy is reduced by a factor of 2 from $g = 10$ to $g = 5$, the derived binding energy (assuming $s = 50$) is reduced just 4% from 0.151 to 0.146 eV. This insensitivity is a result of the fact that binding energy and the kinetic energy release depend in a similar manner on the density of states calculation.

It is often the case that one or more of the parameters in Eq. (10.36) are not known. However, if the binding energy of at least one of the cluster ions is known from other experiments, then the constant C in the Engelking equation can be evaluated empirically. Once that is known, the binding energies of higher order cluster ions can be determined with much greater precision.

The evaporation of NH_3 units from protonated ammonia provides an excellent example of the utility of the Engelking equation. When ammonia clusters are ionized either by electron impact or by multiphotoionization in which a large amount of energy is deposited in the clusters. The following reactions take place:

$$(NH_3)_{n+1} + e^- \text{ or } h\nu \rightarrow [(NH_3)_nH^+]^{**} + NH_2 \,, \tag{10.37}$$

$$[(NH_3)_nH^+]^{**} \rightarrow [(NH_3)_{n-1}H^+]^* + NH_3 \,, \tag{10.38}$$

$$[(NH_3)_{n-1}H^+]^* \rightarrow (NH_3)_{n-2}H^+ + NH_3 \,, \tag{10.39}$$

etc.

With each evaporation, the internal energy of the cluster decreases. The last dissociation will be very slow taking place in the microsecond domain. The fragment ions from these slow dissociation are the metastable ions which can be detected in a mass spectrometer. Furthermore, the average kinetic energy released in this final dissociation can be measured with considerable accuracy (see Eq. 5.30).

Figure 10.16 shows a plot of the derived binding energy as a function of the number of ammonia monomers. A few literature values (Keese and Castleman, 1986) for the lower ammonia cluster binding energies obtained from equilibrium studies are also shown. In fact, the binding energies derived from the Engelking equation were normalized to the literature values at $n = 7$ by empirically fixing the C parameter in Eq. (10.36). In this manner, the binding energies up to $n = 17$ are determined. Of particular interest is the fact that certain cluster sizes have higher than normal binding energies. These are related to the magic numbers observed in mass spectra of cluster size distributions. They indicate particular stabilities associated with those ionic cluster sizes. Also shown in figure 10.16 are the derived binding energies from the Klots equations which will be discussed in the following section.

The Engelking model deals primarily with the average kinetic energy release. However, this approach can also be used to obtain the kinetic energy release distribution (KERD). The predicted KERD based on the models of Engelking, RRK, and classical PST have been compared to classical molecular dynamics simulation of the Ar_n ($n = 12 - 14$) evaporation (Weerasinghe and Amar, 1993). Figure 10.17 shows the results. The RRK theory in which the KERD is determined from the transition state structure closely resembles an exponential function. On the other hand, the Engelking model, in which three translational degrees of freedom were used for the translational phase space shows a clear dip at $E_t = 0$. Finally, the classical PST which treats the angular momentum conservation in detail is in between these limits and closely resembles the classical trajectory calculations. All of the models in figure 10.17 are classical so that the comparison is a valid one.

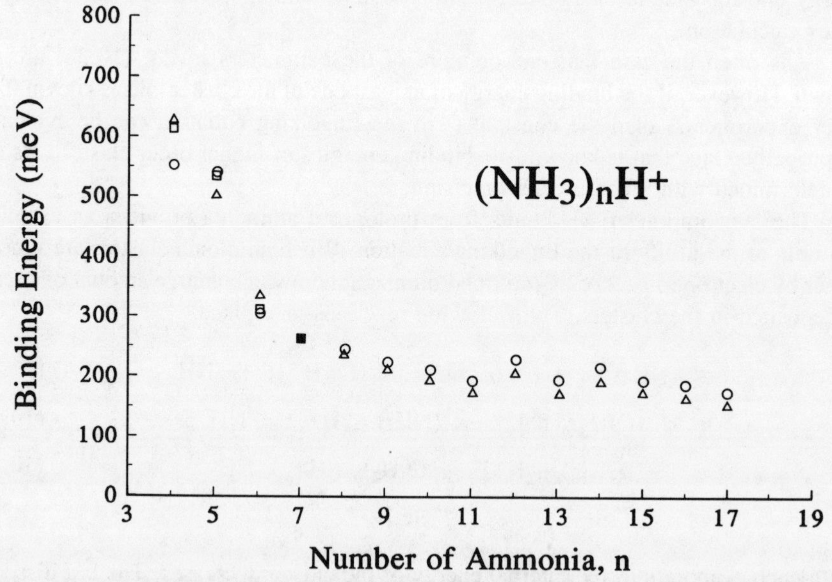

Figure 10.16 Plot of average binding energy as a function of cluster size for the $(NH_3)_nH^+$ system. The open squares are literature values, the circles are from the Engelking equation, while the triangles are from the Klots vaporization ensembles model. The data were normalized at $n = 7$. Adapted from Wei et al. (1990).

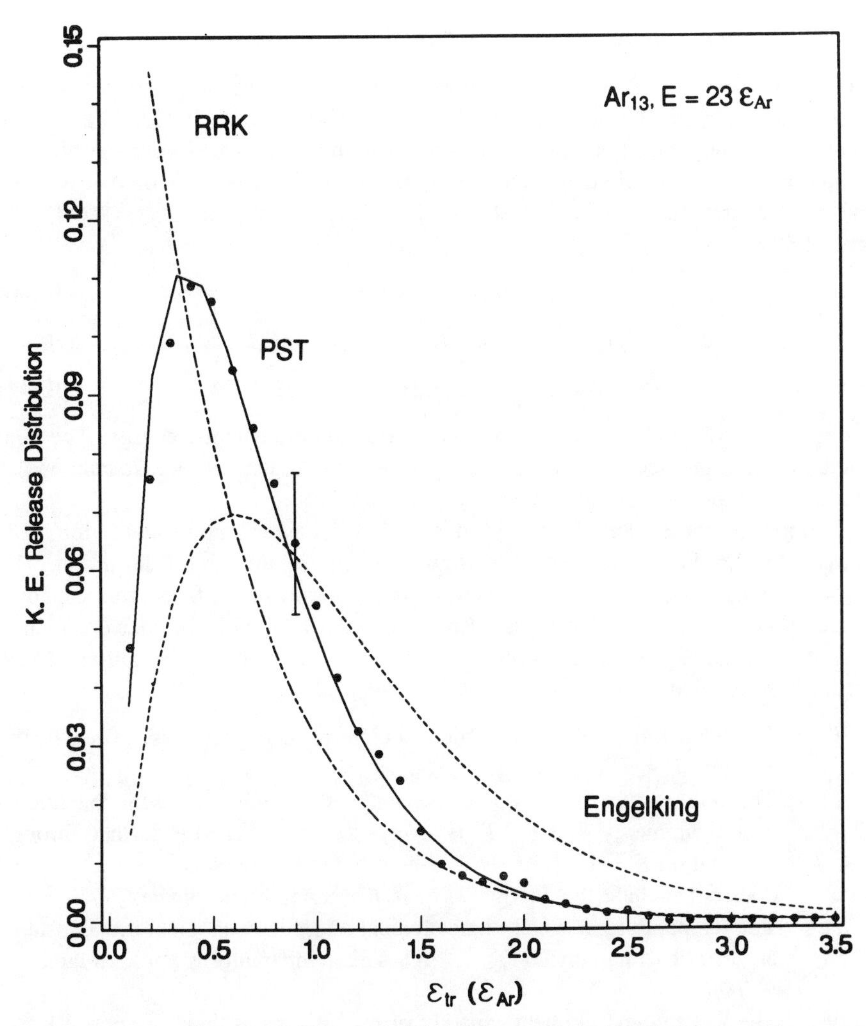

Figure 10.17 The kinetic energy release distribution (KERD) predicted for the evaporation of one Ar atom from Ar_{13}. All calculations are classical. The points are a result from a classical trajectory calculation. Taken with permission from Weerasinghe and Amar (1993).

10.4.2 A Thermodynamic Approach to Cluster Evaporation

Besides the kinetic energy release associated with cluster evaporation, it is also possible in a mass spectrometer (either the double focusing M/E type or in a reflectron TOF apparatus) to measure the ratio of the daughter to parent signal, that is, M/AM. A model that expresses this ratio as well as the kinetic energy release is one based on the Klots theory of cluster evaporation. Because this approach is very different from the microcanonical theory so far presented, some basic ideas of thermal kinetics must be discussed. Two excellent reviews of the basic theory (Klots, 1994) and their application to cluster evaporation (Lifshitz, 1993) provide most of the information needed to understand this field.

10.4.2.1 Thermal Kinetics in Small Systems

We begin this section by summarizing a series of relationships which connect microcanonical and canonical concepts as developed by Klots (1988, 1989, 1992). These ideas are intimately based on approximations implicit in the inversion of the partition function by the method of steepest descent (Hoare and Ruijgrok, 1970). We first list some standard statistical mechanics relations involving the partition function, Q (Berry et al., 1980):

$$d\ln Q/d\beta = -\langle E(T)\rangle \tag{10.40}$$

$$S = k_B \ln Q + k_B T d\ln Q/dT = k_b \ln Q + \langle E(T)\rangle/T \, , \tag{10.41}$$

$$d^2\ln Q/d\beta^2 = k_B c_v/(k_B\beta)^2 = C/(k_B\beta^2) \, , \tag{10.42}$$

where $\beta = 1/k_BT$, and $\langle E\rangle$ is the usual canonical thermodynamic energy. The heat capacity, c_v, is expressed in the usual energy/mol·K units, while C is that same molar heat capacity expressed in units of k_B (i.e., it is unitless).

In applying these thermodynamic ideas to a dissociating molecule or ion, it is necessary to introduce a number of quantities which require careful definition. The energy selected dissociating molecule is described in terms of the fixed, microcanonical energy, E, its 0 K dissociation energy, E_o, and a number of thermodynamic (canonical) variables such as the average energies $\langle E\rangle$ and their temperatures. These quantities are defined as follows:

- T — The temperature of the molecule at an internal energy E. It is defined by the relation: $E = \langle E(T)\rangle - k_BT$.
- $T^\ddagger$ — The temperature of the molecule at the transition state with the microcanonical energy $E - E_o$. $T^\ddagger$ is always less than T, and is defined through the relation $E - E_o = \langle E^\ddagger(T)\rangle$, where the transition-state energy, $\langle E^\ddagger(T)\rangle$, does not include the energy (k_BT) in the reaction coordinate.
- T_b — The temperature at which the canonical and microcanonical rate constants are identical, namely, $k(T_b) = k(E)$. This temperature is approximately $(T + T^\ddagger)/2$.
- c_v — The thermodynamic heat capacity of the clusters in units of energy/K.
- C — The heat capacity expressed in units of k_B (unitless).
- $C^\ddagger$ — The unitless thermodynamic heat capacity of the transition state.
- $\langle S\rangle$ — The entropy of the molecule.
- $\langle S^\ddagger\rangle$ — The entropy of the transition state.

The energy relations permit the calculation of the various temperatures. For instance, the temperature of the transition state is determined by the relation:

$$E - E_0 = \langle E^\ddagger(T^\ddagger)\rangle = \sum \frac{h\boldsymbol{\nu}_i^\ddagger}{\exp(h\boldsymbol{\nu}_i^\ddagger/kT^\ddagger) - 1} + \langle E_r(T^\ddagger)\rangle \, , \tag{10.43}$$

in which the sum is over all the vibrational modes of the transition state and in which the average rotational energy is included. The latter would be important for a very loose transition state in which some of the vibrations have already been converted into rotations. It might also include the K- rotor, but not the two overall rotational motions associated with J_o. The heat capacity in units of k_B is defined in terms of the usual relation:

$$C = \frac{dE}{d(k_B T)} = \frac{d\langle E(T)\rangle}{d(k_B T)} - 1 \; . \qquad (10.44)$$

To the list of well established thermodynamic relations (10.40)–(10.44), we now add two approximate relations involving the density and sums of states and the "temperature" of the system. These result from the approximations made in conversion of the partition function to the density of states when the inverse Laplace transform is solved by the method of steepest descent (Hoare and Ruijgrok, 1970):

$$N(E) = \frac{\rho(E)}{\beta} , \qquad (10.45)$$

$$\rho(E) = \sqrt{\frac{k_B}{2\pi c_v(T)}} \; \beta Q(T) \, e^{\beta E} , \qquad (10.46)$$

$$\ln\rho(E) = \ln Q(T) + \beta E - \ln \left\{ \frac{\sqrt{2\pi C}}{\beta} \right\} , \qquad (10.47)$$

where the heat capacity has been replaced by the unitless heat capacity, C, and:

$$\frac{d\ln\rho(E)}{dE} = \beta \; . \qquad (10.48)$$

The first three relationships [Eqs (10.45)–(10.47)], derived by Hoare and Ruijgrok (1970), are a direct result of the integration theorem for Laplace transforms and the steepest descent approximation. Equation (10.48) comes directly from the derivative of Eq. (10.47).

By combining Eqs. (10.41) and (10.47), an expression for the entropy of the energy selected molecule in terms of its density of states and the heat capacity is obtained:

$$S(T) = k_B[\ln(k_B T \rho(E)) + \ln \sqrt{(2\pi C)}] \; . \qquad (10.49)$$

This can be expressed in terms of the sum of states through use of relation (10.45):

$$S(T) = k_B[\ln N(E) + \ln\sqrt{(2\pi C)}] \; . \qquad (10.50)$$

All of the quantities in Eqs. (10.49) and (10.50) can be calculated with a knowledge of the molecule's vibrational frequencies. The entropy in these equations is a curious mixture of canonical and microcanonical quantities, which can be compared to the fully canonical form of Eq. (10.41).

A final important relation derived from Eq. (10.45) is:

$$\frac{d\ln N(E)}{dE} = \frac{1}{N(E)} \frac{dN(E)}{dE} = \frac{\rho(E)}{N(E)} = \beta \; . \qquad (10.51)$$

It states that the sum of states, or at least its logarithm [much like the density of states in Eq. (10.48)] increases with energy as $1/k_B T$.

10.4.2.2 The Daughter to Parent Intensity Ratio for Metastable Ions

We are now in a position to derive our first important result for evaporative ensembles of clusters. The logarithm of the RRKM molecular rate constant, $k(E)$ (without the reaction degeneracy) is given by

$$\ln k(E) = \ln N^{\ddagger}(E - E_o) - \ln \rho(E) - \ln h \; . \qquad (10.52)$$

Taking the derivative with respect to the energy yields

$$\frac{d\ln k(E)}{dE} = \frac{d\ln N^{\ddagger}(E)}{dE} - \frac{d\ln\rho(E)}{dE} . \tag{10.53}$$

By using Eqs. (10.48) and (10.51) this derivative becomes (Klots, 1988)

$$\frac{d\ln k(E)}{dE} = \beta^{\ddagger} - \beta = \frac{(T - T^{\ddagger})}{k_B T T^{\ddagger}} . \tag{10.54}$$

The temperature difference in the numerator of Eq. (10.54) can be replaced by a heat capacity through the general relation $\Delta T = \Delta E/(Ck_B)$. The heat capacity of the reacting molecule is not a constant over the rather large temperature change. The appropriate temperature is one intermediate between T and $T^{\ddagger}$, which is approximately T_b. As the cluster becomes large, the loss of a monomer unit will not greatly reduce the heat capacity so that any error in the choice of temperatures becomes negligible. The energy difference ΔE is the Arrhenius activation energy for evaporation of a single monomer unit. We can thus identify this ΔE with the ΔE_{vap} which yields:

$$\frac{d\ln k(E)}{dE} = \frac{\Delta E_{vap}}{C(k_B T^{\ddagger})(k_B T)} = \frac{\Delta E_{vap}}{C(k_B T_b)^2} , \tag{10.55}$$

where the approximate relation $T_b = (T^{\ddagger}T)^{1/2}$ has been used.

The evaporative ensemble model is one in which the cluster ions have been generated in a broad distribution of cluster sizes and internal energies. Hence, the rate constant, $k(E)$, is inversely related to the reaction time, t, so that $\ln(k) \approx -\ln(t)$ and $d\ln(k) = -d\ln(t)$. This permits us to convert the derivative in Eq. (10.55) from $d\ln(k)$ to $d\ln(t)$ (Lifshitz, 1993) as

$$dE = -\frac{C(k_B T_b)^2}{\Delta E_{vap}} \, d\ln t . \tag{10.56}$$

In the mass spectrometer, the metastable ions are those that are formed in a field free region between the characteristic times, t_i and t_f. The integral of Eq. (10.56) between these two limits gives δE, the ion internal energy range over which meta-stable dissociation takes place.

The clusters which are sufficiently long lived to reach the field free region must contain internal energy ranging from 0 to about ΔE_{vap}. Those with internal energy much above ΔE_{vap}, will have already dissociated in the ion source. Because the rate constant is a strong function of the internal energy, most of the sequential evaporation steps take place at very short times in the ion source. Only the final dissociation step which results in a stable cluster ion takes place in the field free region. Thus, the range of internal energies of those final ions is no greater than ΔE_{vap}. As a result, the ratio of daughter to total ions (daughter plus parent) can be approximated by the ratio of $D/(D + P) = \delta E/\Delta E_{vap}$. This ratio is thus given by

$$\frac{D}{D+P} = \frac{\delta E}{\Delta E_{vap}} = \frac{C_n k_B^2 T_b^2}{\Delta E_{vap}^2} \ln \frac{t_f}{t_i} . \tag{10.57}$$

Klots (1991) has suggested that the ratio $\Delta E_{vap}/k_b T_b = \gamma$ is approximately constant and has named it the Gspann parameter. [Gspann (1982) first proposed that the rate of cluster evaporation could be expressed in the form of an Arrhenius equation with $\Delta E_a = \Delta E_{vap}$.] Although this is not strictly true since the Arrhenius expression contains no reference to the cluster size, it turns out that the γ parameter is roughly constant at 23.5 $\pm$ 1.5 for rates which are approximately 10^5 sec^{-1}. The constancy of the Gspann parameter is closely related to Trouton's rule which states that the entropy of vaporization ($\Delta S_{vap} = \Delta H_v/T_b$, where T_b is the boiling point of the liquid) is approximately constant.

With the substitution of the Gspann parameter into Eq. (10.57), the fraction of dissociated clusters in the meta-stable time range is

$$\frac{D}{D + P} = \frac{C_n}{\gamma^2} \ln \frac{t_f}{t_i}. \tag{10.58}$$

It turns out that Eq. (10.58) is only a limiting form when $t_f - t_i$ goes to zero. A more nearly correct expression is (Klots, 1988)

$$\frac{D}{D + P} = \frac{C_n}{\gamma^2} \ln \left[\frac{t_f}{t_i + (t_f - t_i)\exp(-\gamma^2/C_n)}\right]. \tag{10.59}$$

Equations (10.58) and (10.59) cannot be plotted as a function of cluster size because the latter parameter does not appear in those equations. Rather, these equations provide a means for determining the cluster ion heat capacity from the ratio of daughter to total ions, with no adjustable parameters. Experiments have shown that the cluster heat capacities are related in rather simple and interesting ways to the cluster size. The heat capacities appear to be well represented by their classical equipartition limit. Thus, a rare gas cluster such as Xe_n^+, with its ion core of Xe_2^+ surrounded by $n - 1$ relatively weakly bound monomer units, would have a heat capacity given by: $C_n = 3(n - 1)$ in units of k_B. By substituting this functional form for C_n into Eq. (10.58), the ratio $D/(D + P)$ can be plotted as a function of n. Figure 10.18 is such a plot for the case of the mixed cluster: $[(NH_3)_nCH_3CN]H^+$ (Tzeng et al., 1991). The points are the experimental measurements of the daughter ion to total ion ratio as a function of cluster size. The solid lines are calculated using Eq. (10.59) with two functional forms for the heat capacity. Because the ammonia units are not linear, they each contribute six van der Waals modes. Of the two heat capacity functions that were tried, $C_n = 6(n - 1)$ and $C_n = 6(n - 1) + 8$, the latter gave a much better fit. The van der Waals modes associated with the acetonitrile unit contribute six of the additional modes. The origin of the final two modes is not clear, since CH_3CN has no low frequency modes. The functional form for the heat capacity is the only adjustable parameter. The theoretical lines in figure 10.18 are curved. Had the approximate Eq. (10.58) been plotted, they would be linear. It is evident that the data require the more correct form.

The heat capacity as a function of cluster size was analyzed in a similar manner for several other polyatomic clusters (Wei et al., 1990, 1991a) and the results show that $C_n[H^+(NH_3)_n] = 6(n - 1)$ (Tzeng et al., 1991), while $C_n[H^+(CH_3OCH_3)_n = 6(n - 1) + 2n$ (Wei et al., 1991b). If only the van der Waals modes contributed, the heat capacities would be $6(n - 1)$ for both clusters because there are $n - 1$ monomers attached to the protonated core ions. However, in the case of the protonated ether clusters, just as in the mixed ammonia-acetonitrile, there are evidently two low energy ether vibrations which contribute to raise the heat capacity of this cluster. On the other

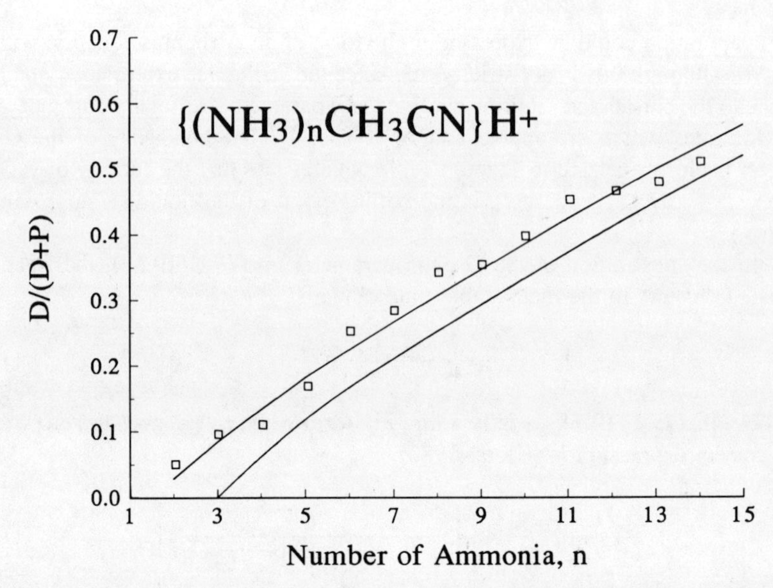

Figure 10.18 Plot of decay fraction $D/(D + P)$ [Daughter/(Daughter + Parent)] of $\{(NH_3)_nCH_3CN\}H^+$ as a function of n. Open squares are the experimental results, while the solid lines are calculated curves based on equation (10.59). The assumed functional forms for the heat capacity are $C_n = 6(n - 1) + 8$ (upper line) and $C_n = 6(n - 1)$ (lower line). Taken with permission from Tzeng et al. (1991).

hand, ammonia has no low-frequency modes so that only the van der Waals modes contribute.

10.4.2.3 Average Translational Energy Release as a Function of Cluster Size

According to the modified prior distribution, the average translational energy release in a dissociation is given by a two-dimensional Boltzmann distribution. Thus the average translational energy release [see Eq. (9.17)] is:

$$\langle E_t \rangle = k_B T^{\ddagger} . \tag{10.60}$$

Here the temperature is that of the transition state, while in the previous chapter the temperature referred to the products. In the case of a dissociation with no barrier and an orbiting transition state, these temperatures should be nearly the same. Because of dynamical constraints imposed by the centrifugal barrier, Eq. (10.60) is not general. In fact, Klots has shown that $k_B T^{\ddagger} \leq \langle E_t \rangle \leq 2k_B T^{\ddagger}$. However, for reactions with loose transition states which lose polarizable neutral monomers (i.e., not H atoms) the lower (two-dimensional) limit will be nearly correct.

With the definition of the cluster heat capacity as:

$$C_n = \Delta E_n / \{k_B(T - T^{\ddagger})\} \tag{10.61}$$

and the definition of $T_b = (T + T^{\ddagger})/2$, we can substitute $\langle E_t \rangle$ for $k_B T^{\ddagger}$ in Eq. (10.61), rearrange terms to finally express the average translational energy release as a function of ΔE_n and the cluster heat capacity:

$$\langle E_t \rangle = \frac{\Delta E_n}{\gamma}\left\{1 - \frac{\gamma}{2C_n}\right\}. \tag{10.62}$$

The heat of vaporization ΔE_{vap} has been replaced by ΔE_n to indicate that this quantity varies with cluster size. Interestingly, the Gspann parameter, $\gamma = \Delta E_{vap}/k_B T_b$, is nevertheless approximately constant because as the heat of vaporization changes with cluster size, so will T_b in such a manner that the two cancel. It is constant for the same reason that Trouton's rule holds for a large body of data.

Lifshitz and Louage (1990) have measured the average kinetic energy release in the dissociation of metastable ammonia cluster ions. The binding energies up to $n = 6$ are known so that the measured and predicted [Eq. (10.62)] kinetic energy release could be compared. Figure 10.19 is a plot of $\langle E_t \rangle \gamma / \Delta E_n$ vs. C_n. The points are obtained from the experimentally measured $\langle E_t \rangle$, the known ΔE_n, and the assumed functional form for the heat capacity, which for ammonia is $C_n = 6(n - 1)$. The solid line is derived from Eq. (10.62) in which the known ΔE_n values are used. The agreement between experiment and the theory (with no adjustable parameters) is impressive.

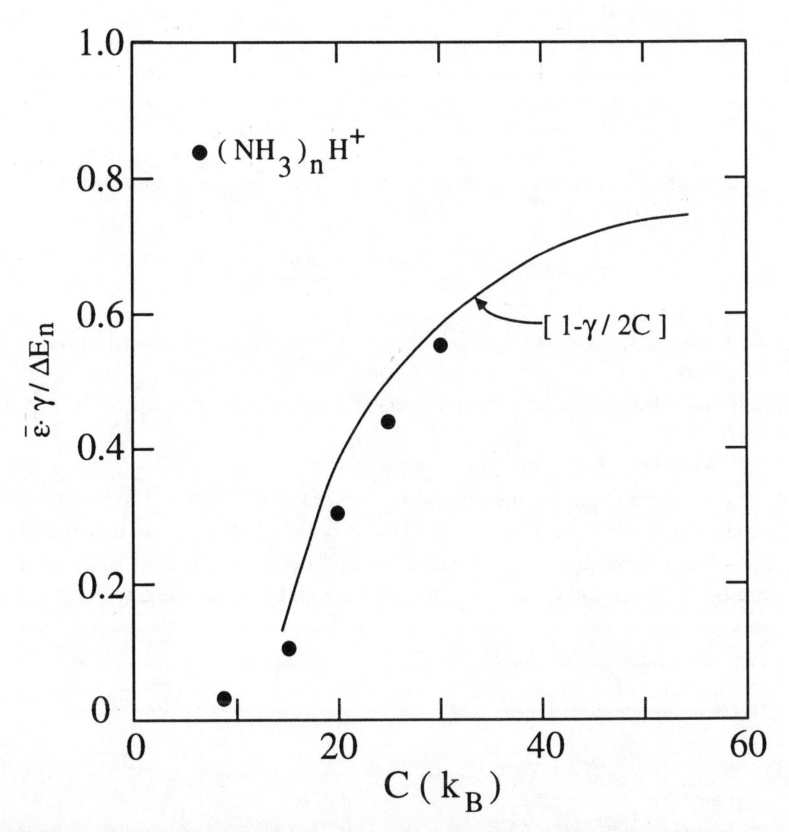

Figure 10.19 A plot of $\langle E_t \rangle \gamma/\Delta E_n$ versus C_n, the heat capacity in units of k_B. The points are experimental values using measured kinetic energy release and binding energies and the following heat capacity function: $C_n = 5n$. The solid line is based on equation (10.62). Taken with permission from Lifshitz and Louage (1990).

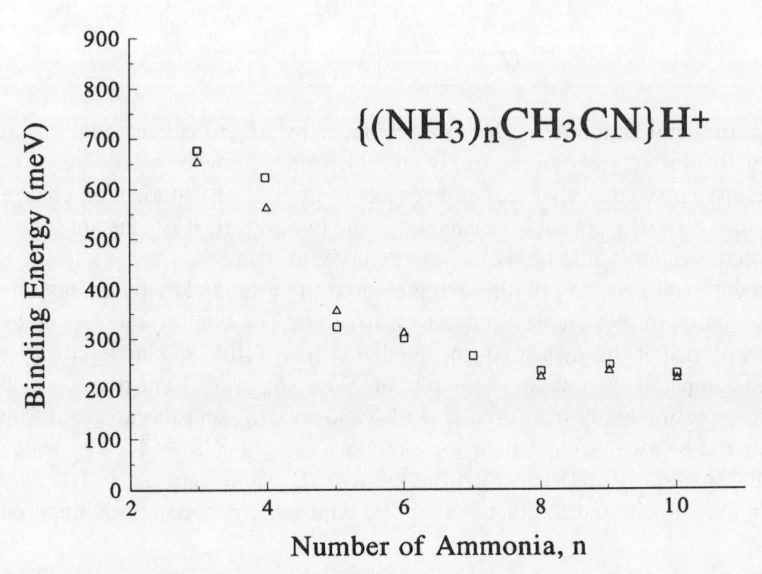

Figure 10.20 Plot of the calculated binding energies of ammonia to the cluster ions $\{(NH_3)_nCH_3CN\}H^+$ as a function of n. The triangles are deduced from the Klots equation (10.62), while the open squares are obtained from the Engelking equation (10.36). Taken with permission from Tzeng et al. (1991).

As in the previous derivation of $D/(D + P)$, Eq. (10.62) is approximate. A more nearly correct expression is

$$\langle E_t \rangle = \frac{\Delta E_n}{\gamma} \left\{ 1 - \frac{\gamma}{2C_n} + \frac{\gamma^2}{12C_n^2} + \cdots \right\}. \tag{10.63}$$

It is evident that as n increases, the limiting form for the kinetic energy release is just $\Delta E_n/\gamma$.

Equations (10.62) can be expressed as $\Delta E_n = \gamma\langle E_t \rangle/[1 - \gamma/2C_n]$. This is plotted in figure 10.20 for the case of $(NH_3)_nCH_3CN)H^+$, where $C_n = 6(n - 1) + 8$ (Tzeng et al., 1991). Also shown are the predictions of the Engelking model. As pointed out before, once the constant in the Engelking model is evaluated for one of the data points, this equation works quite well over the range of data so far investigated. The advantage of the Klots evaporative ensemble approach over that of Engelking is that the cluster heat capacity and the binding energies can be determined with no adjustable parameters.

REFERENCES

Alexander, M.L., Johnson, M.A., Levinger, N.E., and Lineberger, W.C. (1986). *Phys. Rev. Lett.* **57,** 976.

Bemish, R.J., Block, P.A., Pedersen, L.G., Yang, W., and Miller, R.E. (1993). *J. Chem. Phys.* **99,** 8585.

Bernstein, E.R. (1990). In *Dynamics of Polyatomic van der Waals Complexes*, Halberstadt, N., and Janda, K.C., Eds. Plenum Press, New York, pp. 295–320.

Berry, M.T., Brustein, M.R., and Lester, M.I. (1990). *J. Chem. Phys.* **92,** 6469.
Berry, R.S., Rice, S.A., and Ross, J. (1980) *Physical Chemistry*. Wiley, New York.
Berry, R.S. (1990). *J. Chem. Soc. Faraday Trans.* **86,** 2343.
Block, P.A., Jucks, K.W., Pedersen, L.G., and Miller, R.E. (1989). *Chem. Phys.* **139,** 15.
Bohac, E.J., Marshall, M.D., and Miller, R.E. (1992a). *J. Chem. Phys.* **96,** 6681.
Bohac, E.J., Marshall, M.D., and Miller, R.E. (1992b). *J. Chem. Phys.* **97,** 4890.
Booze, J.A., and Baer, T. (1992). *J. Chem. Phys.* **96,** 5541.
Brechignac, C., Cahuzac, Ph., Carlier, F., de Frutos, M., and Leygnier, J. (1990). *J. Chem. Phys.* **93,** 7449.
Brechignac, C., Cahuzac, Ph., Carlier, F., de Frutos, M., Leygnier, J., and Roux, J.Ph. (1995). *J. Chem. Phys.* **102,** 763.
Brutschy, B. (1990). *J. Phys. Chem.* **94,** 8637.
Buck, U. (1988). *J. Phys. Chem.* **92,** 1023.
Buck, U., Gu, X.J., Lauenstein, Ch., and Rudolph, A. (1990). *J. Chem. Phys.* **92,** 6017.
Butz, K.W., Catlett, D.L., Ewing, G.E., Krajnovich, D., and Parmenter, C.S. (1986). *J. Phys. Chem.* **90,** 3533.
Casassa, M.P., Woodward, A.M., Stephenson, J.C., and King, D.S. (1986). *J. Chem. Phys.* **85,** 6235.
Clary, D.C. (1992). *J. Chem. Phys.* **96,** 90.
Cline, J.I., Evard, D.D., Reid, B.P., Sivakumar, N., Thommen, F., and Janda, K.C. (1987). In *Structure and Dynamics of Weakly Bound Molecular Complexes*, Weber, A., Ed. Reidel, Dordrecht.
Dayton, D.C., Pedersen, L.G., and Miller, R.E. (1990). *J. Chem. Phys.* **93,** 4560.
DeHaan, D.O., Holton, A.L., and Zwier, T.S. (1989). *J. Chem. Phys.* **90,** 3952.
Disselkamp, R., and Bernstein, E.R. (1993). *J. Chem. Phys.* **98,** 4339.
Draves, J.A., Zaida, L.S., Liu, W.L., and Lisy, J.M. (1990). *J. Chem. Phys.* **93,** 4589.
Drobits, J.C., and Lester, M.I. (1988). *J. Chem. Phys.* **89,** 4716.
Engelking, P.C. (1987). *J. Chem. Phys.* **87,** 936.
Ewing, G.E. (1987). *J. Phys. Chem.* **91,** 4662.
Farrell, J.T., Sneh, O., McIlroy, A., Knight, A.E., and Nesbitt, D.J. (1992). *J. Chem. Phys.* **97,** 7967.
Felker, P.M., Lambert, W.R., and Zewail, A.H. (1985). *J. Chem. Phys.* **82,** 3003.
Felker, P.M., and Zewail, A.H. (1988). *Adv. Chem. Phys.* **70,** 265.
Fraser, G.T., Nelson, D.D., Peterson, K.I., and Klemperer, W. (1986). *J. Chem. Phys.* **84,** 2472.
Furuya, K., and Kimura, K. (1992). *J. Chem. Phys.* **97,** 1022.
Gilbert, R.G., and Smith, S.C. (1990). *Theory of Unimolecular and Recombination Reactions*. Blackwell Scientific, Oxford.
Grover, J.R., Walters, E.A., and Hui, E.T. (1987). *J. Phys. Chem.* **91,** 3233.
Grover, J.R., Walters, E.A., and Baumgartel, H. (1989). *J. Phys. Chem.* **93,** 7534.
Gspann, J. (1982). In *Physics of Electronic and Atomic Collisions*, Datz, S., Ed. North-Holland, Amsterdam, pp. 79–96.
Halberstadt, N., Beswick, J.A., and Janda, K.C. (1987). *J. Chem. Phys.* **87,** 3966.
Halberstadt, N., Beswick, J.A., Roncero, O., and Janda, K.C. (1992). *J. Chem. Phys.* **96,** 2404.
Herzberg, G. (1967). *Molecular Structure and Molecular Spectroscopy III. Electronic Spectra and Electronic Structure of Polyatomic Molecules*. Van-Nostrand, Princeton, NJ.
Hetzler, J.R., Casassa, M.P., and King, D.S. (1991). *J. Phys. Chem.* **95,** 8086.
Hoare, M.R., and Ruijgrok, Th.W. (1970). *J. Chem. Phys.* **52,** 113.
Huber, K.P., and Herzberg, G. (1979). *Molecular Spectra and Molecular Structure IV. Constants of Diatomic Molecules*. Van Nostrand Reihhold, New York.
Hutson, J.M. (1984). *J. Chem. Phys.* **81,** 2357.

Hye-Keun, O., Parmenter, C.S., and Su, M.C. (1988). *Ber. Bunsenges. Phys. Chem.* **92,** 253.
Jucks, K.W., and Miller, R.E. (1988). *J. Chem. Phys.* **88,** 6059.
Jungwirth, P., and Bally, T. (1993). *J. Am. Chem. Soc.* **115,** 5783.
Keese, R.G., and Castleman, A.W. (1986). *J. Phys. Chem. Ref. Data* **15,** 1011.
Kelley, D.F., and Bernstein, E.R. (1986). *J. Phys. Chem.* **90,** 5164.
Khundkar, L.R., Marcus, R.A., and Zewail, A.H. (1983). *J. Phys. Chem.* **87,** 2473.
Klots, C.E. (1971). *J. Phys. Chem.* **75,** 1526.
Klots, C.E. (1985). *J. Chem. Phys.* **83,** 5854.
Klots, C.E. (1988). *J. Phys. Chem.* **92,** 5864.
Klots, C.E. (1989). *J. Chem. Phys.* **90,** 4470.
Klots, C.E. (1991). *Z. Phys. D. -Atoms, Molecules and Clusters* **20,** 105.
Klots, C.E. (1992). *J. Phys. Chem.* **96,** 1733.
Klots, C.E. (1994). In *The Wiley Series in Ion Chemistry and Physics: Cluster Ions*, Baer, T., Ng, C.Y., and Powis, I., Eds. Wiley, New York.
Klots, T.D., Ruoff, R.S., and Gutowsky, H.S. (1989). *J. Chem. Phys.* **90,** 4217.
Krause, H., Ernstberger, B., and Neusser, H.J. (1991). *Chem. Phys. Lett.* **184,** 411.
Krause, H., Ernstberger, B., and Neusser, H.J. (1992). *Ber. Bunsenges. Phys. Chem.* **96,** 1183.
Laidler, K.J. (1987). *Chemical Kinetics, 3rd ed.* Harper & Row, New York.
Leisner, T., Echt, O., Kandler, O., Yan, X.J., and Recknagel, E. (1988). *Chem. Phys. Lett.* **148,** 386.
LeRoy, R.J., Davies, M.R., and Lam, M.E. (1991). *J. Phys. Chem.* **95,** 2167.
Li, S., and Bernstein, E.R. (1991). *J. Chem. Phys.* **95,** 1577.
Lifshitz, C. (1993). In *Cluster Ions*, Baer, T., Ng, C.Y., and Powis, I., Eds. Wiley, Chichester, pp. 121–164.
Lifshitz, C., and Louage, F. (1990). *Int. J. Mass Spectrom. Ion. Proc.* **101,** 101.
Liu, W.L., and Lisy, J.M. (1988). *J. Chem. Phys.* **89,** 605.
Lovejoy, C.M., Schuder, M.D., and Nesbitt, D.J. (1986). *J. Chem. Phys.* **85,** 4890.
Lovejoy, C.M., and Nesbitt, D.J. (1990). *J. Chem. Phys.* **93,** 5387.
Ma, N.L., Li, W.K., and Ng, C.Y. (1993). *J. Chem. Phys.* **99,** 3617.
Majors, T.J., Even, U., and Jortner, J. (1984). *J. Chem. Phys.* **95,** 295.
Miller, R.E. (1988). *Science* **240,** 447.
Miller, R.E. (1990). *Acc. Chem. Res.* **23,** 10.
Miller, R.E. (1994). Unpublished.
Mons, M., LeClave, J., Piuzzi, F., and Dimicoli, I. (1990). *J. Chem. Phys.* **92,** 2155.
Morioka, Y., Ogawa, M., Matsumoto, T., Ito, K., Tanaka, K., and Hayaishi, T. (1991). *J. Phys. B:Atom. Mol. Phys.* **24,** 791.
Muller-Dethlefs, K., and Schlag, E.W. (1991). *Ann. Rev. Phys. Chem.* **42,** 109.
Nelson, D.D., Fraser, G.T., and Klemperer, W. (1985). *J. Chem. Phys.* **83,** 6201.
Nesbitt, D.J., Petek, H., Foltz, H.F., Filseth, S.V., Bamford, D.J., and Moore, C.B. (1985). *J. Chem. Phys.* **83,** 223.
Nesbitt, D.J., and Lovejoy, C.M. (1990). *J. Chem. Phys.* **93,** 7716.
Nimlos, M.R., Young, M.A., Bernstein, E.R., and Kelley, D.F. (1989). *J. Chem. Phys.* **91,** 5268.
Norwood, K., Guo, J.H., and Ng, C.Y. (1989). *J. Chem. Phys.* **90,** 2995.
Perry, J.W., Scherer, N.F., and Zewail, A.H. (1983). *Chem. Phys. Lett.* **103,** 1.
Polik, W.F., Moore, C.B., and Miller, W.H. (1988). *J. Chem. Phys.* **89,** 3584.
Pradeep, T., Niu, B., and Shirley, D.A. (1993). *J. Chem. Phys.* **98,** 5269.
Quack, M., and Suhm, M.A. (1991). *J. Chem. Phys.* **95,** 28.
Riehn, Ch., Lahmann, Ch., Brutschy, B., and Baumgartel, H. (1992a). *Ber. Bunsenges. Phys. Chem.* **96,** 1164.

Riehn, Ch., Lahmann, Ch., Wassermann, B., and Brutschy, B. (1992b). *Chem. Phys. Lett.* **197,** 443.

Roncero, O., Beswick, J.A., Halberstadt, N., Villarreal, P., and Delgado-Barrio, G. (1990). *J. Chem. Phys.* **92,** 3348.

Ruoff, R.S., Klots, T.D., Emilsson, T., and Gutowsky, H.S. (1990). *J. Chem. Phys.* **93,** 3142.

Schwartz, R.N., Slawsky, Z.I., and Herzfeld, K.F. (1952). *J. Chem. Phys.* **20,** 1591.

Semmes, D.H., Baskin, J.S., and Zewail, A.H. (1987). *J. Am. Chem. Soc.* **109,** 4104.

Semmes, D.H., Baskin, J.S., and Zewail, A.H. (1990). *J. Chem. Phys.* **92,** 3359.

Smith, J.M., Zhang, X., and Knee, J.L. (1993). *J. Chem. Phys.* **99,** 2550.

Stace, A.J., and Shukla, A.K. (1982). *Chem. Phys. Lett.* **85,** 157.

Steinfeld, J.I., Francisco, J.S., and Hase, W.L. (1989). *Chemical Kinetics and Dynamics.* Prentice-Hall, Englewood Cliffs, NJ.

Syage, J.A., Lambert, Wm.R., Felker, P.M., Zewail, A.H., and Hochstrasser, R.M. (1982). *Chem. Phys. Lett.* **88,** 266.

Tang, K.T., and Toennies, J.P. (1986). *Z. Phys. D-Atoms,Molecules and Clusters* **1,** 91.

Tiller, A.R., Peet, A.C., and Clary, D.C. (1989). *Chem. Phys.* **129,** 125.

Troe, J. (1985). *Chem. Phys. Lett.* **114,** 241.

Truhlar, D.G. (1990). In *Dynamics of Polyatomic van der Waals Complexes*, Halberstadt, N., and Janda, K.C., Eds. Plenum Press, New York, pp. 159–185.

Tsuzuki, S., and Tanabe, K. (1992). *J. Phys. Chem.* **96,** 10804.

Tzeng, W.B., Wei, S., and Castleman, A.W. (1991). *J. Phys. Chem.* **95,** 5757.

Uzer, T., and Miller, W.H. (1991). *Phys. Rep.* **199,** 73.

van der Avoird, A. (1993). *J. Chem. Phys.* **98,** 5327.

Wales, D.J., and Berry, R.S. (1990). *J. Chem. Phys.* **92,** 4283.

Weerasinghe, S., and Amar, F.G. (1991). *Z. Phys. D. -Atoms,Molecules and Clusters* **20,** 167.

Weerasinghe, S., and Amar, F.G. (1993). *J. Chem. Phys.* **98,** 4967.

Wei, S., Tzeng, W.B., and Castleman, A.W. (1990). *J. Chem. Phys.* **93,** 2506.

Wei, S., Tzeng, W.B., and Castleman, A.W. (1991a). *J. Phys. Chem.* **95,** 8306.

Wei, S., Tzeng, W.B., and Castleman, A.W. (1991b). *J. Phys. Chem.* **95,** 5080.

Zwier, T.S., Carrasquillo, E., and Levy, D.H. (1983). *J. Chem. Phys.* **78,** 5493.

Appendix

RRKM PRGRAM IN BASIC USING STEEPEST DESCENT

This program calculates densities and sums of ro-vibrational states. It is a translation into BASIC of a program written in APL [Forst, W. (1973). *Theory of Unimolecular Reactions*. Academic Press, pp. 395–403]. The density or sum is calculated simply by changing the variable R as defined below. In order to increase speed, frequencies can be combined by replacing a set of n frequencies by their geometric mean. This is especially useful for the high frequencies which do not contribute much to the density of states. It is best not to combine those frequencies below about 500 cm^{-1}. All energies are expressed in cm^{-1}.

DEFINITION OF VARIABLES (variables terminating with "#" are double precision)

Nfreq	Number of unique vibrational frequencies
Nrot	Number of free rotors (assumed to be one-dimensional)
DEN(E)	The density of ro-vibrational states and sums of states vectors
E(I)	Ion Internal Energy
H	Planck's constant in units of cm^{-1}.
R	Nrot/2 for density of states and $R = (N\text{rot}/2) + 1$ for sum of states
QR	is a reduced rotational partition function, which Forst (p. 99) calls Q_r'. It is the rotational parition function (without the rotational symmetry) and divided by $(kT)^{N_{rot}/2}$. The constant, 0.18634, converts the energy into cm^{-1}.

The intermediate variables, **F0#, F1#, QV, SU#, TH# ≡ θ, T1#,** and **EP#** are functions whose meanings are defined by Forst p. 396–7.

INPUT

Number of unique Frequencies (**Nfreq**);
Number of free Rotors (**Nrot**);
Initial Energy (**Einit**);
Activation Energy (**Eo**);
Energy Increment (**Edel**);
Number of Energies to be Calculated (**NUM**)

Molecular Frequencies (**Freq**) in cm^{-1}, and their degeneracies (**Deg**)
Moments of inertia in amu-A^2 (**MOM**)

THE PROGRAM

```
    DIM Freq(50), Deg(50), DEN(100), E(100), MOM(5), SUM(100)
    T1# = .9995: EP# = 5E-08: QR = 1: H = 3.33644E-11
    FOR I = 1 TO Nrot: QR = QR * SQR(MOM(I) * .18634): NEXT I
    E(1) = Einit
    FOR K = 1 TO NUM 'scan over the number of energies
    N = 1
20  F0# = 0: SU# = 0: R = Nrot / 2: QV = 1
    FOR I = 1 TO Nfreq
    F0# = F0# + Deg(I) * Freq(I) * T1# ^ Freq(I) / (1 - T1# ^ Freq(I))
    SU# = SU# + Deg(I) * Freq(I) ^ 2 * T1# ^ (Freq(I) - 1) / (1 - T1# ^ Freq(I)) ^ 2
    NEXT I
    F0# = F0# + R / LOG(1 / T1#) - E(K)
    F1# = SU# + R / (T1# * (LOG(1 / T1#)) ^ 2)
    T2# = T1# - F0# / F1#
    IF ABS(T2# - T1#) < = EP# OR N > 10 THEN GOTO 30 'Convergence Test
    T2# = ABS(T2#)
    IF T2# < 1 THEN T1# = T2#
    IF T2# > 1 THEN T1# = .9999 'This insures that calculation won't diverge
    N = N + 1: GOTO 20
30  TH# = T1#
    FOR I = 1 TO Nfreq: QV = QV * (1 - TH# ^ Freq(I)) ^ (-Deg(I)): NEXT I
    DN = TH# ^ E(K) * (LOG(1 / TH#)) ^ R * SQR(6.2832 * F1# * TH#)
    DEN(K) = QV * QR / DN
    E(K + 1) = E(K) + Edel: NEXT K 'increment energy and repeat
```

The above calculation is for the density (DEN) because R was set to $N_{rot}/2$. At this point, the calculation is repeated with $R = (N_{rot}/2) + 1$ using the set of transition state frequencies (which includes one less than that used for the density of states calculation). The energy is redefined as $E = E - E_o$. This results in SUM(K). The RRKM dissociation rate constant is then evaluated by:

$$\text{RATE}(K) = \text{SUM}(K)/(H^*\text{DEN}(K)) ,$$

where $H = 3.3364\text{E-11}\ cm^{-1}$ sec.

Author Index

Subject Index